De Gruyter Graduate

Paul E. Bland

Rings and Their Modules

De Gruyter

Mathematics Subject Classification 2010: Primary: 16-01; Secondary: 16D10, 16D40, 16D50, 16D60, 16D70, 16E05, 16E10, 16E30.

ISBN 978-3-11-025022-0
e-ISBN 978-3-11-025023-7

Library of Congress Cataloging-in-Publication Data

Bland, Paul E.
　　Rings and their modules / by Paul E. Bland.
　　　p. cm. – (De Gruyter textbook)
　　Includes bibliographical references and index.
　　ISBN 978-3-11-025022-0 (alk. paper)
　　1. Rings (Algebra)　2. Modules (Algebra)　I. Title.
　　QA247.B545　2011
　　512'.4–dc22

2010034731

Bibliographic information published by the Deutsche Nationalbibliothek

The Deutsche Nationalbibliothek lists this publication in the Deutsche Nationalbibliografie; detailed bibliographic data are available in the Internet at http://dnb.d-nb.de.

© 2011 Walter de Gruyter GmbH & Co. KG, Berlin/New York

Typesetting: Da-TeX Gerd Blumenstein, Leipzig, www.da-tex.de
Printing and binding: Hubert & Co. GmbH & Co. KG, Göttingen
∞ Printed on acid-free paper

Printed in Germany

www.degruyter.com

Preface

The goal of this text is to provide an introduction to the theory of rings and modules that goes beyond what one normally obtains in a beginning graduate course in abstract algebra. The author believes that a text directed to a study of rings and modules would be deficient without at least an introduction to homological algebra. Such an introduction has been included and topics are intermingled throughout the text that support this introduction. An effort has been made to write a text that can, for the most part, be read without consulting references. No attempt has been made to present a survey of rings and/or modules, so many worthy topics have been omitted in order to hold the text to a reasonable length.

The theme of the text is the interplay between rings and modules. At times we will investigate a ring by considering a given set of conditions on the modules it admits and at other times we will consider a ring of a certain type to see what structure is forced on its modules.

About the Text

It is assumed that the reader is familiar with concepts such as Zorn's lemma, commutative diagrams and ordinal and cardinal numbers. A brief review of these and other basic ideas that will hold throughout the text is given in the chapter on preliminaries to the text and in Appendix A. We also assume that the reader has a basic knowledge of rings and their homomorphisms. No such assumption has been made with regard to modules.

In the first three sections of Chapter 1, the basics of ring theory have been provided for the sake of completeness and in order to give readers quick access to topics in ring theory that they might require to refresh their memory. An introduction to the fundamental properties of (unitary) modules, submodules and module homomorphisms is also provided in this chapter. Chapters 1 through 6 present what the author considers to be a "standard" development of topics in ring and module theory, culminating with the Wedderburn–Artin structure theorems in Chapter 6. These theorems, some of the most beautiful in all of abstract algebra, present the theory of semisimple rings. Over such a ring, modules exhibit properties similar to those of vector spaces: submodules are direct summands and every module decomposes as a direct sum of simple submodules. Topics are interspersed throughout the first six chapters that support the development of semisimple rings and the accompanying Wedderburn-Artin theory. For example, concepts such as direct products, direct sums, free modules and tensor products appear in Chapter 2.

Another goal of the text is to give a brief introduction to category theory. For the most part, only those topics necessary to discuss module categories are developed. However, enough attention is devoted to categories so that the reader will have at least a passing knowledge of this subject. Topics in category theory form the substance of Chapter 3.

Central to any study of semisimple rings are ascending and descending chain conditions on rings and modules, and injective and projective modules. These topics as well as the concept of a flat module are covered in Chapters 4 and 5.

Our investigation of semisimple rings begins in Chapter 6 with the development of the Jacobson radical of a ring and the analogous concept for a module. Simple artinian rings and primitive rings are also studied here and it is shown that a ring is semisimple if and only if it is a finite ring direct product of $n \times n$ matrix rings each with entries from a division ring.

The remainder of the text is a presentation of various topics in ring and module theory, including an introduction to homological algebra. These topics are often related to concepts developed in Chapters 1 through 6.

Chapter 7 introduces injective envelopes and projective covers. Here it is shown that every module has a "best approximation" by an injective module and that, for a particular type of ring, every module has a "best approximation" by a projective module. Quasi-injective and quasi-projective covers are also developed and it is established that every module has a projective cover if and only if every module has a quasi-projective cover.

In Chapter 8 a localization procedure is developed that will produce a ring of fractions (or quotients) of a suitable ring. This construction, which is a generalization of the method used to construct the field of fractions of an integral domain, plays a role in the study of commutative algebra and, in particular, in algebraic geometry.

Chapter 9 is devoted to an introduction to graded rings and modules. Graded rings and modules are important in the study of commutative algebra and in algebraic geometry where they are used to gain information about projective varieties. Many of the topics introduced in Chapters 1 through 8 are reformulated and studied in this "new" setting.

Chapter 10 deals with reflexive modules. The fact that a vector space is reflexive if and only it is finite dimensional, leads naturally to the question, "What are the rings over which every finitely generated module is reflexive?" To this end, quasi-Frobenius rings are defined and it is proved that if a ring is quasi-Frobenius, then every finitely generated module is reflexive. However, the converse fails. Consideration of a converse is taken up in Chapter 12 after our introduction to homological algebra has been completed and techniques from homological algebra are available.

The substance of Chapter 11 is homological algebra. Projective and injective resolutions of modules are established and these concepts are used to investigate the left and right derived functors of an additive (covariant/contravariant) functor $\mathcal{F}$. These results are then applied to obtain Ext_R^n, the nth right derived functors of Hom, and Tor_n^R, the nth left derived functors of $\otimes_R$.

The text concludes with Chapter 12 where an injective, a projective and a flat dimension of a module are defined. A right global homological dimension of a ring is also developed as well as a global flat (or weak) dimension of a ring. It is shown that these dimensions can be used to gain information about some of the rings and modules studied in earlier chapters. In particular, using homological methods it is shown that if a ring is left and right noetherian and if every finitely generated module is reflexive, then the ring is quasi-Frobenius.

There are several excellent texts given in the bibliography that can be consulted by the reader who wishes additional information on rings and modules and homological algebra.

Problem Sets

Problem sets follow each section of the text and new topics are sometimes introduced in the problem sets. These new topics are related to the material given in the section and they are often an extension of that material. *For some of the exercises, a hint as to how one might begin to write a solution for the exercise is given in brackets at the end of the exercise. Often such a hint will point to a result in the text that can be used to solve the exercise, or the hint will point to a result in the text whose proof will suggest a technique for writing a solution. It is left to the reader to decide which is the case.* Finally, the exercises presented in the problem sets for Sections 1.1, 1.2 and 1.3 are intended as a review. The reader may select exercises from these problem sets according to their interests or according to what they feel might be necessary to refresh their memories on the arithmetic of rings and their homomorphisms.

Cross Referencing

The chapters and sections of the text have been numbered consecutively while propositions and corollaries have been numbered consecutively within each section. Examples have also been numbered consecutively within each section and referenced by example number and section number of each chapter unless it is an example in the current section. For instance, Example 3 means the third example in the current section while Example 3 in Section 4.1 has the obvious meaning. Similar remarks hold for the exercises in the problem sets. Also, some equations have been numbered on their right. In each case, the equations are numbered consecutively within each chapter without regard for the chapter sections.

Backmatter

The backmatter for the text is composed of a bibliography, a section on ordinal and cardinal numbers, a list of symbols used in the text, and a subject matter index. The page number attached to each item in the index refers to the page in the text where the subject was first introduced.

Acknowledgements

This text was prepared using Scientific WorkPlace and the diagrams were rendered using Paul Taylor's software package for diagrams in a category. The software package along with a user's manual can be found at ctan.org.

Danville, KY, November 2010   Paul E. Bland

Contents

Chapter 0

Preliminaries

Before beginning Chapter 1, we state the definitions and assumptions that will hold throughout the text.

0.1 Classes, Sets and Functions

Throughout the text we will on occasion need to distinguish between *classes* and *proper classes*. Assuming that all collections are sets quickly leads, of course, to the well-known *Russell–Whitehead Paradox*: If $\mathcal{S}$ is the collection of all sets X such that $X \notin X$, then assuming that $\mathcal{S}$ is a set gives $\mathcal{S} \in \mathcal{S}$ and $\mathcal{S} \notin \mathcal{S}$. Consequently, $\mathcal{S}$ cannot be a set. However, the term "collection" will subsequently mean that the collection of objects in question is a set. When a collection in question may or may not be a set, we will call it a *class*. A collection that is not a set will be referred to as a *proper class*. Throughout the text, we will also deal with indexed family of sets, rings and modules, etc. It will always be assumed that the indexing set is nonempty unless stated otherwise.

The notations $\subseteq$ and $\subsetneqq$ will have their usual meanings while $\cup$ and $\bigcup$, and $\cap$ and $\bigcap$ will be used to designate *union* and *intersection*, respectively. A similar observation holds for $\supseteq$ and $\supsetneqq$. The collection of all subsets of a set X, the *power set* of X, will be denoted by $\wp(X)$ and if X and Y are sets, then $X - Y$ is the *complement* of Y relative to X. If $\{X_\alpha\}_{\alpha \in \Delta}$ is an indexed family of sets, then $\{X_\alpha\}_{\alpha \in \Delta}$, $\bigcup_{\alpha \in \Delta} X_\alpha$ and $\bigcap_{\alpha \in \Delta} X_\alpha$ will often be shortened to $\{X_\alpha\}_\Delta$, $\bigcup_\Delta X_\alpha$ and $\bigcap_\Delta X_\alpha$, respectively. Likewise, $\prod_\Delta X_\alpha$ will denote the *Cartesian product* of a family of sets $\{X_\alpha\}_\Delta$. For such a family, an element (x_α) of $\prod_\Delta X_\alpha$ is referred to as a Δ-*tuple* and if $\Delta = \{1, 2, \ldots, n\}$, then $(x_\alpha) \in \prod_{\alpha=1}^{n} X_\alpha = X_1 \times X_2 \times \cdots \times X_n$ is an *n-tuple*. For a set X, X^Δ will denote the product $\prod_\Delta X_\alpha$, with $X_\alpha = X$ for each $\alpha \in \Delta$.

If $f : X \to Y$ is a function, then $f(x) = y$ will be indicated from time to time by $x \mapsto y$ and the *image of f* will often be denoted by Im f or by $f(X)$. The notation $f|_S$ will indicate that f has been restricted to $S \subseteq X$. We will also write gf for the composition of two functions $f : X \to Y$ and $g : Y \to Z$. The identity function $X \to X$ defined by $x \mapsto x$ will be denoted by id_X. A function that is *one-to-one* will be called *injective* and a function that is *onto* is a *surjective function*. A function that is injective and surjective is said to be a *bijective function*. At times, a bijective function will be referred to as a *one-to-one correspondence*.

Finally, we will always assume that an indexing set is nonempty and we will consider the empty set $\emptyset$ to be unique and a subset of every set.

Partial Orders and Equivalence Relations

There are two relations among elements of a class that are ubiquitous in mathematics, namely partial orders and equivalence relations. Both are used extensively in abstract algebra.

If X is a class, then a *partial order* on X is an order relation $\leq$ defined on X such that

1. $\leq$ is *reflexive*: $x \leq x$ for all $x \in X$,

2. $\leq$ is *anti-symmetric*: $x \leq y$ and $y \leq x$ imply $x = y$ for $x, y \in X$, and

3. $\leq$ is *transitive*: $x \leq y$ and $y \leq z$ imply $x \leq z$ for $x, y, z \in X$.

If we say that X is a *partially ordered class*, then it is to be understood that there is an order relation $\leq$ defined on X that is a *partial order* on X.

A relation $\sim$ defined on a class X is said to be an *equivalence relation* on X if

1. $\sim$ is *reflexive*: $x \sim x$ for all $x \in X$,

2. $\sim$ is *symmetric*: $x \sim y$ implies that $y \sim x$ for $x, y \in X$, and

3. $\sim$ is *transitive*: $x \sim y$ and $y \sim z$ imply that $x \sim z$ for $x, y, z \in X$.

An equivalence relation $\sim$ partitions X into disjoint *equivalence classes* $[x] = \{y \in X \mid y \sim x\}$. The element x displayed in $[x]$ is said to be the *representative of the equivalence class*. Since $[x] = [y]$ if and only if $x \sim y$, any element of an equivalence class can be used as its representative.

Throughout the text we let

$$\mathbb{N} = \{1, 2, 3, \ldots\},$$
$$\mathbb{N}_0 = \{0, 1, 2, \ldots\},$$
$$\mathbb{Z} = \{\ldots, -2, -1, 0, 1, 2, \ldots\}, \quad \text{and}$$
$$\mathbb{Q}, \ \mathbb{R} \text{ and } \mathbb{C}$$

will be used exclusively for the sets of rational, real and complex numbers, respectively. An integer $p \in \mathbb{N}$, $p \neq 1$, will be referred to as a *prime number* if the only divisors in $\mathbb{N}$ of p are 1 and p.

A nonempty subclass C of a partially ordered class X is said to be a *chain* in X if whenever $x, y \in C$, either $x \leq y$ or $y \leq x$. If X itself is a chain, then X is said to be *linearly ordered* or *totally ordered* by $\leq$. If S is a nonempty subclass of a partially ordered class X, then an element $b \in X$ is an *upper* (a *lower*) *bound* for S if $x \leq b$ ($b \leq x$) for all $x \in S$. If there is a necessarily unique upper (lower) bound b^* of S

such that $b^* \le b$ ($b \le b^*$) for every upper (lower) bound b of S, then b^* is said to be the *least upper (greatest lower) bound* for S. The notation

$$\sup S \quad \text{and} \quad \inf S$$

will be used for the least upper bound and the greatest lower bound, respectively, whenever they can be shown to exist. If X is a partially ordered class, then an element $m \in X$ is said to be a *maximal (minimal) element* of X, if whenever $x \in X$ and $m \le x$ ($x \le m$), then $m = x$. If X is a partially ordered class, then a nonempty subclass S of X is said to have a *first (last) element*, if there is an element $f \in S$ ($l \in S$) such that $f \le x$ ($x \le l$) for all $x \in S$. A first element f or a last element l of S may also be referred to from time to time as a *smallest* or *largest element* of S, respectively. Both f and l are clearly unique when they exist.

Zorn's Lemma and the Well-Ordering Principle

We can now state Zorn's lemma, a concept that is almost indispensable in mathematics as it is currently practiced. Zorn published his "maximum principle" in a short paper entitled *A remark on method in transfinite algebra* in the Bulletin of the American Mathematical Society in 1935. Today we know that the Axiom of Choice (See Exercise 3), the Well-Ordering Principle, and Zorn's Lemma are equivalent.

Zorn's Lemma. *If X is a nonempty partially ordered set and if every chain in X has an upper bound in X, then X has at least one maximal element.*

If a nonempty partially ordered set X has the property that every chain in X has an upper bound in X, then X is said to be *inductive*. With this in mind, Zorn's lemma is often stated as "Every inductive partially ordered set has at least one maximal element."

If a partially ordered class X is such that every nonempty subclass of X has a first element, then X is said to be *well ordered*. It is easy to show that if X is well ordered by $\le$, then $\le$ is a linear ordering of X. We assume the following as an axiom.

Well-Ordering Principle. *If X is a set, then there is at least one partial order $\le$ on X that is a well ordering of X or, more briefly, any set can be well ordered.*

Note that the empty set $\varnothing$ can be well ordered. Actually, any relation $\le$ on $\varnothing$ is a partial ordering of $\varnothing$. For example, $x \le x$ for all $x \in \varnothing$, since if not, there would be an $x \in \varnothing$ such that $x \not\le x$, a clear absurdity. Similar arguments show $\le$ is anti-symmetric and transitive and it follows that $\le$ is a well ordering of $\varnothing$.

0.2   Ordinal and Cardinal Numbers

It is assumed that the reader is familiar with the proper class **Ord** of ordinal numbers and the proper class **Card** of cardinal numbers. A brief discussion of these classes of numbers can be found in Appendix A.

0.3   Commutative Diagrams

Commutative diagrams will be used throughout the text wherever they are appropriate. For example, if $f_1 : A \to B$, $f_2 : C \to D$, $g_1 : A \to C$ and $g_2 : B \to D$ are functions, then to say that the diagram

$$
\begin{array}{ccc}
A & \xrightarrow{\;f_1\;} & B \\
\Big\downarrow{\scriptstyle g_1} & & \Big\downarrow{\scriptstyle g_2} \\
C & \xrightarrow{\;f_2\;} & D
\end{array}
$$

is commutative means that $g_2 f_1 = f_2 g_1$. Showing that $g_2 f_1 = f_2 g_1$ is often referred to as *chasing the diagram*. Similarly, for a diagram of the form

$$
\begin{array}{ccc}
A & \xrightarrow{\;f\;} & B \\
 & \underset{g_1}{\searrow} \quad \underset{g_2}{\nearrow} & \\
 & C &
\end{array}
$$

In a diagram such as

$$
\begin{array}{ccc}
A & \xrightarrow{\;f\;} & B \\
 & \underset{g_1}{\searrow} \quad \overset{}{\nwarrow}{\scriptstyle g_2} & \\
 & C &
\end{array}
$$

the dotted arrow indicates that we can find a mapping $g_2 : B \to C$ such that $g_2 f = g_1$. If $g_2 : B \to C$ is such that $g_2 f = g_1$, then we say that g_2 *completes the diagram commutatively*. Commutative diagrams of differing complexities will be formed using various algebraic structures developed in the text. A diagram will be considered to be commutative if all triangles and/or rectangles that appear in the diagram are commutative.

0.4   Notation and Terminology

At this point a word about notation and terminology is in order. When we say that a condition holds *for almost all* $\alpha \in \Delta$, then we mean that the condition holds for all $\alpha \in \Delta$ with at most a finite number of exceptions. Throughout the text we will encounter sums of elements of an abelian group such as $\sum_{\alpha \in \Delta} x_\alpha$, where $x_\alpha = 0$ for almost all $\alpha \in \Delta$. This simply means that there are at most a finite number of nonzero x_α in the sum $\sum_{\alpha \in \Delta} x_\alpha$. If there is at least one $x_\alpha \neq 0$, then $\sum_{\alpha \in \Delta} x_\alpha$ is to be viewed as the finite sum of the nonzero terms and if $x_\alpha = 0$ for all $\alpha \in \Delta$, then we set $\sum_{\alpha \in \Delta} x_\alpha = 0$. To simplify notation, $\sum_{\alpha \in \Delta} x_\alpha$ will be written as $\sum_\Delta x_\alpha$ and sometimes it will be convenient to express $\sum_\Delta x_\alpha$ as $\sum_{i=1}^n x_i$.

From this point forward, all such sums $\sum_\Delta x_\alpha$ are to be viewed as finite sums and the expression "$x_\alpha = 0$ for almost all $\alpha \in \Delta$" will be omitted unless required for clarity.

Problem Set

1. Let $f : X \rightarrow Y$ be a function and suppose that X_1 is a nonempty subset of X and Y_1 is a nonempty subset of Y. If $f(X_1) = \{f(x) \mid x \in X_1\}$ and $f^{-1}(Y_1) = \{x \in X \mid f(x) \in Y_1\}$, show that each of the following hold.
   (a) $X_1 \subseteq f^{-1}(f(X_1))$
   (b) $Y_1 \supseteq f(f^{-1}(Y_1))$
   (c) $f(X_1) = f(f^{-1}(f(X_1)))$
   (d) $f^{-1}(Y_1) = f^{-1}(f(f^{-1}(Y_1)))$

2. (a) Let $f : X \rightarrow Y$ and $g : Y \rightarrow Z$ be functions. If gf is injective, show that f is injective and if gf is surjective, show that g is surjective.
   (b) Prove that a function $f : X \rightarrow Y$ is injective if and only if $f^{-1}(f(X_1)) = X_1$ for each nonempty subset X_1 of X and that f is surjective if and only if $f(f^{-1}(Y_1)) = Y_1$ for each nonempty subset Y_1 of Y.
   (c) A function $f^{-1} : Y \rightarrow X$ is said to be an *inverse function* for a function $f : X \rightarrow Y$ if $ff^{-1} = \mathrm{id}_Y$ and $f^{-1}f = \mathrm{id}_X$. Show that a function $f : X \rightarrow Y$ has an inverse function if and only if f is a bijection. Show also that an inverse function for a function is unique whenever it exists.

3. Let S be a nonempty set and suppose that $\wp(X)^*$ is the set of nonempty subsets of S. A *choice function* for S is a function $c : \wp(X)^* \rightarrow S$ such that $c(A) \in A$ for each $A \in \wp(X)^*$. The *Axiom of Choice* states that every nonempty set S has at least one choice function. Prove that assuming the Axiom of Choice is equivalent to assuming that if $\{X_\alpha\}_\Delta$ is an indexed family of nonempty sets, then $\prod_\Delta X_\alpha \neq \varnothing$.

4. Let X be a nonempty partially ordered set. Zorn's lemma states that if every chain in X has an upper bound in X, then X has at least one maximal element. This is sometimes referred to as *Zorn's lemma going up. Zorn's lemma going down* states that if every chain in X has a lower bound in X, then X has at least one minimal element. Prove that Zorn's lemma going up implies Zorn's lemma

going down and conversely. Conclude that the two forms of Zorn's lemma are equivalent.

5. Suppose that X is a nonempty set. Let

$$f : \wp(X) \times \wp(X) \to \wp(X)$$

be defined by $f(A, B) = A \cup B$, for all $A, B \in \wp(X)$. Also let

$$g : \wp(X) \times \wp(X) \to \wp(X)$$

be given by $g(A, B) = (X - A) \cap (X - B)$, $A, B \in \wp(X)$, and suppose that

$$h : \wp(X) \to \wp(X)$$

is such that $h(A) = X - A$ for all $A \in \wp(X)$. Is the diagram

commutative?

6. (a) Consider the cube

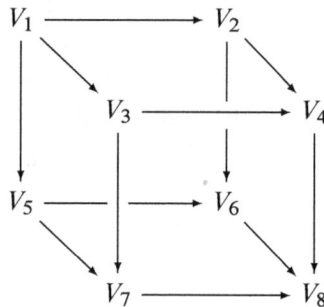

where the arrows indicate functions $f_{ij} : V_i \to V_j$. Assume that the top square and all squares that form the sides of the cube are commutative diagrams. If the function $f_{15} : V_1 \to V_5$ is a surjective mapping, prove that the bottom square is a commutative diagram. [Hint: Chase the commutative faces of the cube to show that $f_{78} f_{57} f_{15} = f_{68} f_{56} f_{15}$ and then use the fact that f_{15} is surjective.]

(b) If all the faces of the cube are commutative except possibly the top and f_{48} is an injection, then is the top face commutative?

Chapter 1
Basic Properties of Rings and Modules

The first three sections of this chapter along with their problem sets contain a brief review of the basic properties of rings and their homomorphisms as well as the definitions and terminology that will hold throughout the text. These sections are presented in order to provide for a smooth transition to the concept of a module. As such these sections can be read quickly. However, the reader should, at the very least, familiarize him or herself with the notation and terminology contained in these sections. In addition, these sections may contain concepts and examples that the reader may not have previously encountered. It is assumed that the reader is familiar with the arithmetic of groups and their homomorphisms.

Commutative ring theory began with algebraic geometry and algebraic number theory. Central to the development of these subjects were the rings of integers in algebraic number fields and the rings of polynomials in two or more variables. Noncommutative ring theory began with attempts to extend the field of complex numbers to various hypercomplex number systems. These hypercomplex number systems were identified with matrix rings by Joseph Wedderburn [71] and later generalized by Michael Artin [47], [48]. The theory of commutative and noncommutative rings dates from the early nineteenth century to the present. The various areas of ring theory continue to be an active area of research.

1.1 Rings

Definition 1.1.1. A *ring* R is a nonempty set together with two binary operations $+$ and $\cdot$, called *addition* and *multiplication*, respectively, such that the following conditions hold:

R1. *R together with addition forms an additive abelian group.*

R2. *Multiplication is associative: $a(bc) = (ab)c$ for all $a, b, c \in R$.*

R3. *Multiplication is distributive over addition from the left and the right: $a(b+c) = ab + ac$ and $(b + c)a = ba + ca$ for all $a, b, c \in R$.*

If $ab = ba$ for all $a, b \in R$, then R is said to be a *commutative ring* and if there is a necessarily unique element $1 \in R$ such that $a1 = 1a = a$ for all $a \in R$, then R is a *ring with identity*. The element 1 is the *multiplicative identity* of R, denoted by 1_R if there is a need to emphasize the ring. If R is a ring with identity and a is a nonzero element of R, then an element $b \in R$ (should it exist) is said to be a *right (left) inverse* for a if $ab = 1$ $(ba = 1)$. An element of R that is a left and a right inverse for a is said to be a *multiplicative inverse of a*. If a has a multiplicative inverse, then it will be denoted by a^{-1}. If $a \in R$ has a multiplicative inverse in R, then a is said to be an *invertible element* of R or a *unit* in R.

One trivial example of a ring with identity is the zero ring $R = \{0\}$, where 0 is both the additive identity and the multiplicative identity of R. In order to eliminate this ring from our considerations, we assume from this point forward that all rings have an identity $1 \neq 0$. Because of this assumption, every ring considered will have at least two elements and the expression "for all rings" will mean "for all rings with an identity."

Definition 1.1.2. A nonzero element $a \in R$ is said to be a *left (right) zero divisor* if there is a nonzero element $b \in R$ such that $ab = 0$ $(ba = 0)$. A nonzero element of $a \in R$ will be referred to as a *zero divisor* if there is a nonzero element $b \in R$ such that $ab = ba = 0$. A ring R in which every nonzero element has a multiplicative inverse is a *division ring*. A commutative division ring is a *field*. A commutative ring that has no zero divisors is an *integral domain*. If S is a nonempty subset of a ring R, then S is said to be a *subring* of R if S is a ring under the operations of addition and multiplication on R. Due to our assumption that all rings have an identity, if S is to be a subring of R, then S must have an identity and we also require that $1_S = 1_R$ before we will say that S is a subring of R.

It is easy to see that every division ring is free of left and right zero divisors, so every field is an integral domain. The integral domain $\mathbb{Z}$ of integers shows that the converse is false. However, every finite ring without zero divisors is a division ring.

Examples

1. (a) If $\{R_\alpha\}_\Delta$ is an indexed family of rings, then $\prod_\Delta R_\alpha$ is a ring under *componentwise addition*,

    $$(a_\alpha) + (b_\alpha) = (a_\alpha + b_\alpha),$$

    and *componentwise multiplication*,

    $$(a_\alpha)(b_\alpha) = (a_\alpha b_\alpha).$$

    The ring $\prod_\Delta R_\alpha$ is called the *ring direct product* of $\{R_\alpha\}_\Delta$. Since each R_α has an identity, $\prod_\Delta R_\alpha$ has an identity (1_α), where each 1_α is the identity of R_α. $\prod_\Delta R_\alpha$ is commutative if and only if each R_α is commutative, but $\prod_\Delta R_\alpha$ is never an integral domain even if each R_α is. For example, $\mathbb{Z}$ is an integral domain, but $(a, 0)(0, b) = (0, 0)$ in $\mathbb{Z} \times \mathbb{Z}$.

    (b) $\mathbb{Z} \times 0$ is a ring with identity $(1, 0)$ under coordinatewise addition and coordinatewise multiplication and $\mathbb{Z} \times 0 \subseteq \mathbb{Z} \times \mathbb{Z}$. However, we do not consider $\mathbb{Z} \times 0$ to be a subring of $\mathbb{Z} \times \mathbb{Z}$ since $\mathrm{id}_{\mathbb{Z} \times 0} = (1, 0) \neq (1, 1) = \mathrm{id}_{\mathbb{Z} \times \mathbb{Z}}$.

2. **Matrix Rings.** The set $\mathbb{M}_n(R)$ of all $n \times n$ matrices whose entries are from R is a noncommutative ring under addition and multiplication of matrices. Moreover, $\mathbb{M}_n(R)$ has zero divisors, so it is not an integral domain even if R is such a domain. $\mathbb{M}_n(R)$ is referred to as the $n \times n$ *matrix ring over* R. Elements of $\mathbb{M}_n(R)$ will be denoted by (a_{ij}), where a_{ij} represents the entry in the ith row and the jth column. When considering matrix rings, it will always be the case, unless stated otherwise, that $n \geq 2$.

3. **Triangular Matrix Rings.** Consider the matrix ring $\mathbb{M}_n(R)$ of Example 2. A matrix $(a_{ij}) \in \mathbb{M}_n(R)$ is said to be an *upper triangular matrix* if $a_{ij} = 0$ when $i > j$. If $\mathbb{T}_n(R)$ denotes the set of upper triangular matrices, then $\mathbb{T}_n(R)$ is a subring of $\mathbb{M}_n(R)$. If an upper triangular matrix has zeros for diagonal entries then the matrix is said to be a *strictly upper triangular matrix*.

4. **Rings of Integers Modulo n.** Let $\mathbb{Z}_n$ denote the set of equivalence classes $[a]$, $a \in \mathbb{Z}$, determined by the equivalence relation defined on $\mathbb{Z}$ by $a \equiv b \bmod n$. Then $\mathbb{Z}_n = \{[0], [1], \ldots, [n-1]\}$ is a ring with identity under the operations

$$[a] + [b] = [a + b]$$
$$[a][b] = [ab].$$

$\mathbb{Z}_n$, called the *ring of integers modulo n*, will have zero divisors if n is a composite integer and $\mathbb{Z}_n$ is a field if and only if n is a prime number. It will always be assumed that $n \geq 2$ when considering the ring $\mathbb{Z}_n$.

5. **Left Zero Divisors Need Not Be Right Zero Divisors.** Consider the matrix ring

$$\begin{pmatrix} \mathbb{Z} & \mathbb{Z}_2 \\ 0 & \mathbb{Z} \end{pmatrix} = \left\{ \begin{pmatrix} a & [b] \\ 0 & c \end{pmatrix} \,\middle|\, a, c \in \mathbb{Z} \text{ and } [b] \in \mathbb{Z}_2 \right\}.$$

Then

$$\begin{pmatrix} 2 & [0] \\ 0 & 1 \end{pmatrix} \begin{pmatrix} 0 & [1] \\ 0 & 0 \end{pmatrix} = \begin{pmatrix} 0 & [0] \\ 0 & 0 \end{pmatrix} \quad \text{and yet}$$

$$\begin{pmatrix} 0 & [1] \\ 0 & 0 \end{pmatrix} \begin{pmatrix} 2 & [0] \\ 0 & 1 \end{pmatrix} \neq \begin{pmatrix} 0 & [0] \\ 0 & 0 \end{pmatrix}.$$

Hence, $\begin{pmatrix} 2 & [0] \\ 0 & 1 \end{pmatrix}$ is a left zero divisor, but not a right zero divisor.

6. **The Opposite Ring.** If R is a ring, then we can construct a new ring called the *opposite ring* of R, denoted by R^{op}. As sets, $R = R^{op}$ and the additive structures on both rings are the same. Multiplication $\circ$ is defined on R^{op} by $a \circ b = ba$, where ba is multiplication in R. Clearly, if R is commutative, then R and R^{op} are the same ring.

7. **Endomorphism Rings.** If G is an additive abelian group and $\operatorname{End}_{\mathbb{Z}}(G)$ denotes the set of all group homomorphisms $f : G \to G$, then $\operatorname{End}_{\mathbb{Z}}(G)$ is a ring with identity under function addition and function composition. The additive identity of $\operatorname{End}_{\mathbb{Z}}(G)$ is the zero homomorphism and the multiplicative identity is the identity homomorphism $\operatorname{id}_G : G \to G$. $\operatorname{End}_{\mathbb{Z}}(G)$ is called the *endomorphism ring* of G.

8. **Polynomial Rings.** If R is a ring and $R[X]$ is the set of polynomials $a_0 + X a_1 + \cdots + X^n a_n$ with their coefficients in R, then $R[X]$ is a ring under the usual operations of addition and multiplication of polynomials. More formally, if $a_0 + X a_1 + \cdots + X^m a_m$ and $b_0 + X b_1 + \cdots + X^n b_n$ are polynomials in $R[X]$, then addition is accomplished by adding coefficients of like terms of the polynomials and

$$(a_0 + X a_1 + \cdots + X^m a_m)(b_0 + X b_1 + \cdots + X^n b_n)$$
$$= c_0 + X c_1 + \cdots + X^{m+n} c_{m+n},$$

where $c_k = a_k b_0 + a_{k-1} b_1 + \cdots + a_0 b_k$ for $k = 0, 1, \ldots, m+n$. If R is commutative, then $R[X]$ is commutative as well. $R[X]$ is the *ring of polynomials over* R. We assume that $aX = Xa$ for all $a \in R$. In this case, X is said to be a *commuting indeterminate*. The coefficient a_n is said to be the *leading coefficient* of $a_0 + X a_1 + \cdots + X^n a_n$ and any polynomial with 1 as its leading coefficient is said to be a *monic polynomial*. More generally, we have a polynomial ring $R[X_1, X_2, \ldots, X_n]$, where $X_1, X_2, \ldots, X_n$ are commuting indeterminates. The set $R[[X]]$ of all formal power series $a_0 + X a_1 + \cdots + X^n a_n + \cdots$ can be made into a ring in a similar fashion. $R[[X]]$ is called the *ring of formal power series over* R. A similar observation holds for $R[[X_1, X_2, \ldots, X_n]]$.

Remark. The usual practice is to write a polynomial as $a_0 + a_1 X + \cdots + a_n X^n$ with the coefficients from R written on the left of the powers of X. Since we will work primarily with right R-modules, to be introduced later in this chapter, we will write the coefficients on the right. When X is a commutating indeterminate, this is immaterial since $a_0 + X a_1 + \cdots + X^n a_n = a_0 + a_1 X + \cdots + a_n X^n$. Throughout the remainder of the text, when an indeterminate is considered it will always be assumed to be a commutating indeterminate unless indicated otherwise.

9. **Differential Polynomial Rings.** If R is a ring, let $R[X]$ denote the set of all right polynomials $p(X) = \sum_{k=0}^{n} X^k a_k$, where we do not assume that $aX = Xa$ for each $a \in R$, that is, X is a *noncommuting indeterminate*. If $R[X]$ is to be made into a ring, then it is necessary to commute a past X in expressions such as $XaXb$ that arise in polynomial multiplication. Assuming that the associative and distributive properties hold, let $\delta : R \to R$ be a function and set $aX =$

$Xa + \delta(a)$ for all $a \in R$. Then $(a + b)X = X(a + b) + \delta(a + b)$ and $aX + bX = X(a + b) + \delta(a) + \delta(b)$, so

$$\delta(a + b) = \delta(a) + \delta(b) \tag{1.1}$$

for all $a, b \in R$. Likewise,

$$(ab)X = X(ab) + \delta(ab) \quad \text{and}$$
$$(ab)X = a(bX) = a(Xb + \delta(b))$$
$$= (aX)b + a\delta(b)$$
$$= (Xa + \delta(a))b + a\delta(b)$$
$$= X(ab) + \delta(a)b + a\delta(b).$$

Thus,

$$\delta(ab) = \delta(a)b + a\delta(b) \tag{1.2}$$

for all $a, b \in R$. A function $\delta : R \to R$ satisfying conditions (1.1) and (1.2) is said to be a *derivation* on R. Given a derivation $\delta : R \to R$, the set of all right polynomials can be made into a ring $R[X, \delta]$ by setting $aX = Xa + \delta(a)$ for all $a \in R$. $R[X, \delta]$ is referred to as a *differential polynomial ring* over R. Differential polynomial rings can be defined symmetrically for *left polynomials* over a ring R in a noncommuting indeterminate.

10. **Quadratic Fields.** Let n be a *square free integer* (n has no factors other than 1 that are perfect squares), set

$$Q(n) = \{a + b\sqrt{n} \mid a, b \in \mathbb{Q}\}$$

and define addition and multiplication on $Q(n)$ by

$$(a + b\sqrt{n}) + (a' + b'\sqrt{n}) = (a + a') + (b + b')\sqrt{n} \quad \text{and}$$
$$(a + b\sqrt{n})(a' + b'\sqrt{n}) = (aa' + bb'n) + (ab' + a'b)\sqrt{n}$$

for all $a + b\sqrt{n}, a' + b'\sqrt{n} \in Q(n)$. Then $Q(n)$ is a commutative ring with identity and, in fact, $Q(n)$ is a field called a *quadratic field*. Each nonzero element $a + b\sqrt{n}$ in $Q(n)$ has

$$(a + b\sqrt{n})^{-1} = \frac{a}{a^2 - b^2 n} - \frac{b}{a^2 - b^2 n}\sqrt{n}$$

as a multiplicative inverse. Note that $Q(n)$ is a subfield of $\mathbb{C}$ and if n is positive, then $Q(n)$ is a subfield of $\mathbb{R}$.

Problem Set 1.1

In each of the following exercises R denotes a ring.

1. (a) If a has a left inverse b' and a right inverse b, show that $b = b'$. Conclude that an element of R with a left and a right inverse is a unit in R.

   (b) Suppose that $U(R)$ denotes the set of units of R. Prove that $U(R)$ is a group under the multiplication defined on R. $U(R)$ is called the *group of units* of R.

   (c) If R is such that every nonzero element of R has a right (left) inverse, prove that R is a division ring.

   (d) Prove that if R is a finite ring without zero divisors, then R is a division ring. Conclude that a finite integral domain is a field.

2. The *characteristic* of R, denoted by char(R), is the smallest positive integer n such that $na = 0$ for all $a \in R$. If no such positive integer n exists, then char(R) = 0.

   (a) Prove that char(R) $= n$ if and only if n is the smallest positive integer n such that $n1 = 0$.

   (b) If R is an integral domain, prove that the characteristic of R is either 0 or a prime number.

3. (a) If $a \in R$ and $\delta_a : R \to R$ is defined by $\delta_a(b) = ab - ba$ for all $b \in R$, show that δ_a is a derivation on R. Such a derivation on R is said to be an *inner derivation*. A derivation that is not an inner derivation is an *outer derivation* on R.

   (b) Let $\delta : R \to R$ be a derivation on R. If

   $$R_\delta = \{c \in R \mid \delta(c) = 0\},$$

   prove that R_δ is a subring of R. Show also that if $c \in R_\delta$ has a multiplicative inverse in R, then $c^{-1} \in R_\delta$. Elements of R_δ are called δ-*constants* of R. Conclude that if R is a field, then R_δ is a subfield of R.

   (c) Let $\delta : R \to R$ be a derivation on a commutative ring R and let $R[X]$ be the polynomial ring over R in a commuting indeterminate X. For each $X^n a$ in $R[X]$, set $\hat{\delta}(X^n a) = X^n \delta(a) + nX^{n-1} a\delta(X)$, where $\delta(X)$ is a fixed, but arbitrarily chosen polynomial in $R[X]$. Extend $\hat{\delta}$ to $R[X]$ by setting $\hat{\delta}(p(X)) = \hat{\delta}(a_0) + \hat{\delta}(Xa_1) + \cdots + \hat{\delta}(X^n a_n)$ for each $p(X) \in R[X]$. Show that $\hat{\delta} : R[X] \to R[X]$ is a derivation on $R[X]$. Show also that if $p(X) \in R[X]$, then $\hat{\delta}(p(X)) = p^\delta(X) + p'(X)\delta(X)$, where $p^\delta(X)$ represents $p(X)$ with δ applies to its coefficients and $p'(X)$ is the usual formal derivative of $p(X)$. Conclude that if $p(X) \in R_\delta[X]$ and $\delta(X)$ is chosen to be the constant polynomial 1, then $\hat{\delta}(p(X)) = p'(X)$. Conclude that $\hat{\delta}$ generalizes the usual formal derivative of a polynomial in $R[X]$.

4. (a) An element e of R is said to be *idempotent* if $e^2 = e$. Note that R always has at least two idempotents, namely 0 and 1. Let e be an idempotent element of R. Show that the set eRe of all finite sums $\sum_{i=1}^{n} ea_i e$, where $a_i \in R$ for $i = 1, 2, \ldots, n$, is a ring with identity e. The integer n is not fixed, that is, any two finite sums such as $\sum_{i=1}^{m} ea_i e$ and $\sum_{i=1}^{n} eb_i e$, with $m \neq n$, are in eRe. Show also that $eRe = \{a \in R \mid ea = a = ae\}$.

(b) If e and f are idempotents of R, then e and f are said to be *orthogonal idempotents* if $ef = fe = 0$. If $\{e_1, e_2, \ldots, e_n\}$ is a set of pairwise orthogonal idempotents of R, prove that $e = e_1 + e_2 + \cdots + e_n$ is an idempotent of R.

(c) If every element of R is idempotent, then R is said to be a *boolean ring*. Prove that a Boolean ring R is commutative. [Hint: Consider $(a + a)^2$ and $(a + b)^2$, where $a, b \in R$.]

5. An element $a \in R$ is said to be *nilpotent* if there is a positive integer n such that $a^n = 0$. The smallest positive integer n such that $a^n = 0$ is called the *index of nilpotency* of a.

(a) If a is a nilpotent element of R, prove that $1 - a$ has a multiplicative inverse in R. [Hint: Factor $1 - a^n$, where n is the index of nilpotency of a.]

(b) If $\mathbb{M}_n(R)$ is the matrix ring of Example 2, let $(a_{ij}) \in \mathbb{M}_n(R)$ be such that $a_{ij} = 0$ if $i \geq j$. Then (a_{ij}) is a strictly upper triangular matrix. Show that (a_{ij}) is nilpotent. In fact, show that $(a_{ij})^n = 0$. Conclude that, in general, $\mathbb{M}_n(R)$ has an abundance of nonzero nilpotent elements.

6. Prove that if $\mathcal{C}$ is the set of 2×2 matrices of the form $\left(\begin{smallmatrix} a & b \\ -b & a \end{smallmatrix} \right)$, $a, b \in \mathbb{R}$, then $\mathcal{C}$ is a subring of the matrix ring $\mathbb{M}_2(\mathbb{R})$.

7. (a) Suppose that $q(\mathbb{R})$ is the set of all 2×2 matrices of the form

$$\begin{pmatrix} a + bi & c + di \\ -c + di & a - bi \end{pmatrix},$$

where $a, b, c, d \in \mathbb{R}$ and $i = \sqrt{-1}$. Prove that $q(\mathbb{R})$ is a noncommutative division ring that is a subring of $\mathbb{M}_2(\mathbb{C})$. $q(\mathbb{R})$ is the *ring of real quaternions*. If a, b, c and d are rational numbers, then $q(\mathbb{Q})$ is also a division ring called the *ring of rational quaternions* and if a, b, c and d are integers, then $q(\mathbb{Z})$ is the *ring of integral quaternions*. The ring $q(\mathbb{Z})$ has no zero divisors, but $q(\mathbb{Z})$ is not a division ring.

(b) Let $\mathbf{1} = \left(\begin{smallmatrix} 1 & 0 \\ 0 & 1 \end{smallmatrix} \right)$, $\mathbf{i} = \left(\begin{smallmatrix} i & 0 \\ 0 & -i \end{smallmatrix} \right)$, $\mathbf{j} = \left(\begin{smallmatrix} 0 & 1 \\ -1 & 0 \end{smallmatrix} \right)$ and $\mathbf{k} = \left(\begin{smallmatrix} 0 & i \\ i & 0 \end{smallmatrix} \right)$. Prove that every element of $q(\mathbb{R})$ has a unique expression of the form $\mathbf{1}a_1 + \mathbf{i}a_2 + \mathbf{j}a_3 + \mathbf{k}a_4$, where $a_1, a_2, a_3, a_4 \in \mathbb{R}$.

(c) Show $\mathbf{i}^2 = \mathbf{j}^2 = \mathbf{k}^2 = -\mathbf{1}$ and that

$$\mathbf{ij} = -\mathbf{ji} = \mathbf{k},$$
$$\mathbf{jk} = -\mathbf{kj} = \mathbf{i}, \quad \text{and}$$
$$\mathbf{ki} = -\mathbf{ik} = \mathbf{j}.$$

Observe that the ring $\mathcal{C}$ of Exercise 6 is a subring of $q(\mathbb{R})$.

8. (a) Let n be an integer such that $n = m^2 n'$ for some integers m and n', where n' is square free. The integer n' is said to be the *square free part* of n. If the requirement that n is a square free integer is dropped from Example 10, show that $Q(n)$ and $Q(n')$ are the same quadratic fields. Conclude that one need only use the square free part of n to compute $Q(n)$.

   (b) If $Q[n] = \{a + b\sqrt{n} \mid a, b \in \mathbb{Z}\}$, show that $Q[n]$ is a subring of $Q(n)$. Conclude that $Q[n]$ is an integral domain.

9. (a) If $\{R_\alpha\}_\Delta$ is a family of subrings of R, prove that $\bigcap_\Delta R_\alpha$ is a subring of R.

   (b) Suppose that the subrings of a ring R are ordered by $\subseteq$. If S is a subset of R and $\{R_\alpha\}_\Delta$ is the family of subrings of R that contain S, show that $\bigcap_\Delta R_\alpha$ is the smallest subring of R containing S. The subring $\bigcap_\Delta R_\alpha$ is called the *subring of R generated by S*.

10. Let R be a ring and let $\text{cent}(R) = \{b \in R \mid ba = ab \text{ for all } a \in R\}$.

    (a) If $a \in R$, show that $C(a) = \{b \in R \mid ba = ab\}$ is a subring of R.

    (b) Verify that $\text{cent}(R) = \bigcap_R C(a)$. The subring $\text{cent}(R)$ is called the *center of R*.

    (c) If $a \in R$ is a unit in R, prove that if $a \in \text{cent}(R)$, then $a^{-1} \in \text{cent}(R)$. Conclude that if R is a division ring, then $\text{cent}(R)$ is a field.

1.2   Left and Right Ideals

We now turn our attention to subgroups of a ring that are closed under multiplication by ring elements.

Definition 1.2.1. An additive subgroup A of the additive group of R is said to be a *right (left) ideal* of R if $ab \in A$ $(ba \in A)$ for all $a \in A$ and all $b \in R$. If I is a left and a right ideal of R, then I is an *ideal* of R. A right ideal (A left ideal, An ideal) A of R is said to be *proper* if $A \subsetneq R$. A proper right ideal (proper left ideal, proper ideal) $\mathfrak{m}$ of R is a *maximal right ideal (maximal left ideal, maximal ideal)* of R if whenever A is a right ideal (left ideal, an ideal) of R such that $\mathfrak{m} \subseteq A \subseteq R$, then $\mathfrak{m} = A$ or $A = R$. A right (left) ideal A of R is said to be *minimal right (left) ideal* if $\{0\}$ and A are the only right (left) ideals of R contained in A. The symbol 0 will be used to denote both the *zero ideal* $\{0\}$ and the additive identity of R. The context of the discussion will indicate which is being considered. The right ideal $aR = \{ab \mid b \in R\}$ of R is the *principal right ideal* of R generated by a. Similarly, $Ra = \{ba \mid b \in R\}$ is the *principal left ideal* generated by a. A principal ideal aR in a commutative ring R will often be denoted simply by (a). A proper ideal $\mathfrak{p}$ of commutative ring R is said to be a *prime ideal* if whenever $a, b \in R$ are such that $ab \in \mathfrak{p}$, then either $a \in \mathfrak{p}$ or $b \in \mathfrak{p}$. Finally, a commutative ring R is said to be a *local ring* if it has a unique maximal ideal $\mathfrak{m}$.

Examples

1. **Simple Rings.** Every ring R has at least two ideals, namely the zero ideal and the ring R. If these are the only ideals of R, then R is said to be a *simple ring*. Every division ring is a simple ring.

2. **Left and Right Ideals in a Matrix Ring.** If $\mathbb{M}_n(R)$ is the ring of $n \times n$ matrices over R, then for each integer k, $1 \le k \le n$, let $c_k(R)$ be the set of kth *column matrices* (a_{ij}) defined by $a_{ij} = 0$ if $j \neq k$. Then $c_k(R)$ is just the set of all matrices with arbitrary entries from R in the kth column and zeroes elsewhere. The set $c_k(R)$ is a left ideal but not a right ideal of $\mathbb{M}_n(R)$. Likewise, for each k, $1 \le k \le n$, the set $r_k(R)$ of kth *row matrices* (a_{ij}) with $a_{ij} = 0$ if $i \neq k$ is a right ideal but not a left ideal of $\mathbb{M}_n(R)$. If D is a division ring, then for each k, $c_k(D)$ is a minimal left ideal of $\mathbb{M}_n(D)$ and $r_k(D)$ is a minimal right ideal of $\mathbb{M}_n(D)$. Furthermore, one can show that if $\bar{I}$ is an ideal of $\mathbb{M}_n(R)$, then there is a uniquely determined ideal I of R such that $\bar{I} = \mathbb{M}_n(I)$. It follows that if R a simple ring, then $\mathbb{M}_n(R)$ is a simple ring. Hence, if D is a division ring, then $\mathbb{M}_n(D)$ is a simple ring. However, $\mathbb{M}_n(D)$ is not a division ring since $\mathbb{M}_n(D)$ has zero divisors.

3. **Principal Ideal Rings.** Every ideal of the ring $\mathbb{Z}$ is a principal ideal and an ideal (p) of $\mathbb{Z}$ is prime if and only if it is a maximal ideal if and only if p is a prime number. A commutative ring in which every ideal is principal is said to be a *principal ideal ring* and an integral domain with this property is a *principal ideal domain*.

4. The ring $\mathbb{Z}_{p^n}$, where p is a prime number and n is a positive integer is a local ring. The ideals of $\mathbb{Z}_{p^n}$ are linearly ordered and $([p])$ is the unique maximal ideal of $\mathbb{Z}_{p^n}$. A field is also a local ring with maximal ideal 0.

5. **Sums of Right Ideals.** If $\{A_\alpha\}_\Delta$ is a family of right ideals of R, then

$$\sum_\Delta A_\alpha = \left\{ \sum_\Delta a_\alpha \mid a_\alpha \in A_\alpha \text{ for all } \alpha \in \Delta \right\}$$

is a right ideal of R. A similar observation holds for left ideals and ideals of R.

6. If S is a nonempty subset of R, then $\mathrm{ann}_r(S) = \{a \in R \mid Sa = 0\}$ is a right ideal of R called the *right annihilator* of S. The *left annihilator* $\mathrm{ann}_\ell(S) = \{a \in R \mid aS = 0\}$ of S is a left ideal of R. It also follows that $\mathrm{ann}_r(A)$ $(\mathrm{ann}_\ell(A))$ is an ideal of R whenever A is a right (left) ideal of R.

The proof of the following proposition is left as an exercise.

Proposition 1.2.2. *The following hold in any ring R.*

(1) *Let A and B be nonempty subsets of R. If B is a right ideal (If A is a left ideal) of R, then*

$$AB = \left\{ \sum_{i=1}^{n} a_i b_i \mid a_i \in A, \ b_i \in B \ for \ i = 1, 2, \ldots, n, \ n \geq 1 \right\}$$

is a right ideal (a left ideal) of R.

(2) *For any $r \in R$,*

$$RrR = \left\{ \sum_{i=1}^{n} a_i r b_i \mid a_i, b_i \in R \ for \ i = 1, 2, \ldots, n, \ n \geq 1 \right\}$$

is an ideal of R.

Notation. At this point it is important to point out notational differences that will be used throughout the text. If A is a right ideal of R and $n \geq 2$ is an integer, then $A^n = AA \cdots A$ will denote the set of all finite sums of products $a_1 a_2 \cdots a_n$ of n elements from A. The notation $A^{(n)}$ will be used for $A \times A \times \cdots \times A$, with n factors of A. As pointed out in the exercises, A^n is a right ideal of R while $A^{(n)}$ is an R-module, a concept defined in Section 1.4.

The following proposition is often referred to as Krull's lemma [66]. The proof involves our first application of Zorn's lemma.

Proposition 1.2.3 (Krull). *Every proper right ideal (left ideal, ideal) A of a ring R is contained in a maximal right ideal (maximal left ideal, maximal ideal) of R.*

Proof. Let A be a proper right ideal of R and suppose that $\mathcal{S}$ is the collection of proper right ideals B of R that contain A. Then $\mathcal{S} \neq \emptyset$ since $A \in \mathcal{S}$. If $\mathcal{C}$ is a chain in $\mathcal{S}$, then $\bar{A} = \bigcup_\mathcal{C} B$ is a right ideal of R that contains A. Since R has an identity, $\bar{A} \neq R$ and so $\bar{A}$ is an upper bound in $\mathcal{S}$ for $\mathcal{C}$. Thus, $\mathcal{S}$ is inductive and Zorn's lemma indicates that $\mathcal{S}$ has a maximal element, say $\mathfrak{m}$, which, by the definition of $\mathcal{S}$, contains A. If $\mathfrak{m}$ is not a maximal right ideal of R, then there is a right ideal B of R such that $\mathfrak{m} \subsetneq B \subsetneq R$. But B is then a proper right ideal of R that contains A and this contradicts the maximality of $\mathfrak{m}$ in $\mathcal{S}$. Therefore, $\mathfrak{m}$ is a maximal right ideal of R. A similar proof holds if A is a left ideal or an ideal of R. $\qquad \square$

Corollary 1.2.4. *Every ring R has at least one maximal right ideal (maximal left ideal, maximal ideal).*

Factor Rings

Definition 1.2.5. If I is an ideal of R, then R/I, the set of cosets of I in R, is a ring under coset addition and coset multiplication defined by

$$(a + I) + (b + I) = (a + b) + I \quad \text{and} \quad (a + I)(b + I) = ab + I.$$

R/I is said to be the *factor ring* (or *quotient ring*) of R formed by factoring out I. The additive identity of R/I will usually be denoted by 0 rather than $0 + I$ and the multiplicative identity of R/I is $1 + I$.

Remark. Since the zero ring has been eliminated from our discussion by assuming that all rings have an identity $1 \neq 0$, we do not permit $I = R$ when forming the factor ring R/I, unless this should arise naturally in our discussion.

The following well-known proposition demonstrates the connection between prime (maximal) ideals in a commutative ring and integral domains (fields). The proofs of the proposition and its corollaries are left to the reader.

Proposition 1.2.6. *If R is a commutative ring, then:*

(1) *$R/\mathfrak{p}$ is an integral domain if and only if $\mathfrak{p}$ is a prime ideal of R.*

(2) *$R/\mathfrak{m}$ is a field if and only if $\mathfrak{m}$ is a maximal ideal of R.*

Corollary 1.2.7. *A commutative ring R is a field if and only if the zero ideal is a maximal ideal of R.*

Corollary 1.2.8. *If R is a commutative ring, then every maximal ideal of R is prime.*

Problem Set 1.2

1. Prove that the following are equivalent for a nonempty subset A of a ring R.

   (a) A is a right (left) ideal of R.

   (b) A is closed under subtraction and under multiplication on the right (left) by ring elements.

   (c) A is closed under addition and under multiplication on the right (left) by ring elements.

2. Prove Proposition 1.2.2.

3. Prove Proposition 1.2.6 and its corollaries.

4. Prove that the following are equivalent.

   (a) The ideal (p) of $\mathbb{Z}$ is maximal.

   (b) The ideal (p) of $\mathbb{Z}$ is prime.

   (c) p is a prime number.

5. Prove that R is a division ring if and only if 0 is a maximal right (left) ideal of R.

6. Let R be a commutative ring.

   (a) If $\{\mathfrak{p}_\alpha\}_\Delta$ is a chain of prime ideals of R, prove that $\bigcup_\Delta \mathfrak{p}_\alpha$ and $\bigcap_\Delta \mathfrak{p}_\alpha$ are prime ideals of R.

   (b) If $\{\mathfrak{p}_\alpha\}_\Delta$ is a family of prime ideals of R that is not necessarily a chain, then must $\bigcap_\Delta \mathfrak{p}_\alpha$ be prime?

   (c) Use Zorn's lemma going down to prove that R has at least one minimal prime ideal.

   (d) Prove that every prime ideal of R contains a minimal prime ideal.

   (e) If $\{A_i\}_{i=1}^n$ is a family of ideals of R and $\mathfrak{p}$ is a prime ideal of R such that $A_1 A_2 \cdots A_n \subseteq \mathfrak{p}$, prove that $\mathfrak{p}$ contains at least one of the A_i.

   (f) Use Zorn's lemma to show that every prime ideal of R is contained in a maximal prime ideal of R. Can the same result be achieved by using Proposition 1.2.3 and Corollary 1.2.8? Conclude that if $\mathfrak{p}$ is a prime ideal of R and $\bar{\mathfrak{p}}$ is a maximal prime ideal of R containing $\mathfrak{p}$, then $\bar{\mathfrak{p}}$ is a maximal ideal of R.

7. Suppose that R is a commutative ring. Show that the principal ideal generated by X in the polynomial ring $R[X]$ is a prime (maximal) ideal of $R[X]$ if and only if R is an integral domain (a field).

8. If A is a right ideal of R, prove that A^n, $n \geq 1$, is a right ideal of R.

9. If R is a commutative ring and I is an ideal of R, then $\sqrt{I} = \{a \in R \mid a^n \in I$ for some integer $n \geq 1\}$ is called the *radical of I*.

   (a) Prove that $\sqrt{I}$ is an ideal of R such that $\sqrt{I} \supseteq I$.

   (b) An ideal I is said to be a *radical ideal* if $\sqrt{I} = I$. Prove that every prime ideal of R is a radical ideal.

10. If R is a commutative ring, does the set of all nilpotent elements of R form an ideal of R?

11. Let R be a commutative ring.

   (a) Prove that an ideal $\mathfrak{p}$ of R is prime if and only if whenever I_1 and I_2 are ideals of R such that $I_1 \cap I_2 \subseteq \mathfrak{p}$, then either $I_1 \subseteq \mathfrak{p}$ or $I_2 \subseteq \mathfrak{p}$.

   (b) If R is a local ring with unique maximal ideal $\mathfrak{m}$, show that $\mathfrak{m}$ is the set of nonunits of R.

12. Let R be a ring and consider the set $\{E_{ij}\}$, $1 \le i, j \le n$, where E_{ij} is the $n \times n$ matrix with 1 as its (i, j)th entry and zeroes elsewhere. The E_{ij} are known as the *matrix units* of $\mathbb{M}_n(R)$.

(a) Show that $E_{ij} E_{kl} = E_{il}$ if $j = k$ and $E_{ij} E_{kl} = 0$ when $j \ne k$. Conclude that E_{ij} is nilpotent when $i \ne j$. Note also that $a E_{ij} = E_{ij} a$ for all $a \in R$.

(b) Deduce that E_{ii} is an idempotent for $i = 1, 2, \ldots, n$, that $\sum_{i=1}^{n} E_{ii}$ is the identity matrix of $\mathbb{M}_n(R)$ and that $(a_{ij}) = \sum_{i,j=1}^{n} a_{ij} E_{ij} = \sum_{i,j=1}^{n} E_{ij} a_{ij}$ for each $(a_{ij}) \in \mathbb{M}_n(R)$.

(d) Show that if $\bar{I}$ is an ideal of $\mathbb{M}_n(R)$, then there is a uniquely determined ideal I of R such that $\bar{I} = \mathbb{M}_n(I)$. [Hint: If $\bar{I}$ is an ideal of $\mathbb{M}_n(R)$, let I be the set of all elements $a \in R$ that appear as an entry in the first row and first column of a matrix in $\bar{I}$. I is clearly an ideal of R and if (a_{ij}) is a matrix in $\bar{I}$, then, $a_{st} E_{11} = E_{1s}(a_{ij}) E_{t1} \in \bar{I}$, so $a_{st} \in I$ for all s and t such that $1 \le s, t \le n$. Thus, (a_{ij}) is a matrix formed from elements of I, so $\bar{I} \subseteq \mathbb{M}_n(I)$. Now show that $\mathbb{M}_n(I) \subseteq \bar{I}$ and that I is unique.]

(e) Prove that $I \mapsto \bar{I}$ is a one-to-one correspondence between the ideals of R and the ideals of $\mathbb{M}_n(R)$ such that if $I_1 \subseteq I_2$, then $\bar{I}_1 \subseteq \bar{I}_2$.

(f) Prove that if R is a simple ring, then so is $\mathbb{M}_n(R)$. Conclude that if D is a division ring, then $\mathbb{M}_n(D)$ is a simple ring.

13. Let $\mathbb{M}_n(R)$ be the ring of $n \times n$ matrices over R.

(a) Show that for each k, $1 \le k \le n$, the set $c_k(R)$ of kth column matrices is a left ideal but not a right ideal of $\mathbb{M}_n(R)$. Show also that if D is a division ring, then $c_k(D)$ is a minimal left ideal of $\mathbb{M}_n(D)$.

(b) Prove that for each k, $1 \le k \le n$, the set $r_k(R)$ of kth row matrices is a right ideal but not a left ideal of $\mathbb{M}_n(R)$. If D is a division ring, show for each k that $r_k(D)$ is a minimal right ideal of $\mathbb{M}_n(D)$.

14. (a) Show that a commutative simple ring is a field.

(b) Prove that the center of a simple ring is a field.

(c) If R is a simple ring, must $R[X]$ be a simple ring?

(d) Show by example that a subring of a simple ring need not be simple.

15. If p is a prime number and n is a positive integer, show that the ideals of $\mathbb{Z}_{p^n}$ are linearly ordered by $\subseteq$ and that $\mathbb{Z}_{p^n}$ is a local ring with unique maximal ideal $([p])$.

16. Let K be a field. Prove that $K[X]$ is a principal ideal domain. [Hint: If I is an ideal of $K[X]$, consider a monic polynomial $p(X)$ in I of minimal degree.]

17. Let $\{A_\alpha\}_\Delta$ be a family of right ideals of a ring R.

(a) Prove that $\bigcap_\Delta A_\alpha$ is a right ideal of R.

(b) Let S be a subset of R. If $\{A_\alpha\}_\Delta$ is the family of right ideals of R each of which contains S, prove that $\bigcap_\Delta A_\alpha$ is the smallest right ideal of R that contains S.

(c) Prove that

$$\sum_\Delta A_\alpha = \left\{ \sum_\Delta x_\alpha \mid x_\alpha \in A_\alpha \text{ for all } \alpha \in \Delta \right\}$$

is a right ideal of R.

(d) Suppose that $A_\beta \cap \sum_{\alpha \neq \beta} A_\alpha = 0$ for all $\beta \in \Delta$. If $\sum_\Delta x_\alpha \in \sum_\Delta A_\alpha$ is such that $\sum_\Delta x_\alpha = 0$, prove that $x_\alpha = 0$ for all $\alpha \in \Delta$.

(e) Suppose that $A_\beta \cap \sum_{\alpha \neq \beta} A_\alpha = 0$ for all $\beta \in \Delta$ and that $x \in \sum_\Delta A_\alpha$. If x can be written in $\sum_\Delta A_\alpha$ as $x = \sum_\Delta x_\alpha$ and as $x = \sum_\Delta y_\alpha$, prove that $x_\alpha = y_\alpha$ for all $\alpha \in \Delta$.

18. (a) If S is a nonempty subset of R, show that

$$\operatorname{ann}_r(S) = \{a \in R \mid Sa = 0\}$$

is a right ideal of R and that

$$\operatorname{ann}_\ell(S) = \{a \in R \mid aS = 0\}$$

of S is a left ideal of R.

(b) If A is a right ideal of R, show that $\operatorname{ann}_r(A)$ is an ideal of R. Conclude also that if A is a left ideal of R, then $\operatorname{ann}_\ell(A)$ is an ideal of R.

(c) If S and T are nonempty subsets of R such that $S \subseteq T$, show that $\operatorname{ann}_r(T) \subseteq \operatorname{ann}_r(S)$ and that $\operatorname{ann}_\ell(T) \subseteq \operatorname{ann}_\ell(S)$. Show also that $S \subseteq \operatorname{ann}_\ell(\operatorname{ann}_r(S))$ and $S \subseteq \operatorname{ann}_r(\operatorname{ann}_\ell(S))$.

(d) If S is a nonempty subset of R, prove that $\operatorname{ann}_r(\operatorname{ann}_\ell(\operatorname{ann}_r(S))) = \operatorname{ann}_r(S)$ and that $\operatorname{ann}_\ell(\operatorname{ann}_r(\operatorname{ann}_\ell(S))) = \operatorname{ann}_\ell(S)$.

(e) If $\{A_i\}_{i=1}^n$ is a set of right ideals of R, show that $\operatorname{ann}_r(\sum_{i=1}^n A_i) = \bigcap_{i=1}^n \operatorname{ann}_r(A_i)$ and $\operatorname{ann}_\ell(\sum_{i=1}^n A_i) = \bigcap_{i=1}^n \operatorname{ann}_\ell(A_i)$.

(f) If $\{A_i\}_{i=1}^n$ is a set of right ideals of R, show that $\sum_{i=1}^n \operatorname{ann}_r(A_i) \subseteq \operatorname{ann}_r(\bigcap_{i=1}^n A_i)$ and $\sum_{i=1}^n \operatorname{ann}_\ell(A_i) \subseteq \operatorname{ann}_\ell(\bigcap_{i=1}^n A_i)$.

19. If A, B and C are right ideals of ring R, prove each of the following.

(a) $(A : a) = \{b \in R \mid ab \in A\}$ is a right ideal of R for each $a \in R$.

(b) $((A : a) : b) = (A : ab)$ for all $a, b \in R$.

(c) $(B : A) = \{a \in R \mid Aa \subseteq B\}$ is an ideal of R.

(d) $(C : AB) = ((C : A) : B)$.

(e) If $\{A_\alpha\}_\Delta$ is a family of right ideals of R, then

$$\left(A : \sum_\Delta A_\alpha \right) = \bigcap_\Delta (A : A_\alpha).$$

20. Let R be a commutative ring. A nonempty subset S of a ring R is said to be *multiplicatively closed* if $ab \in S$ whenever $a, b \in S$. If S is a multiplicatively closed subset of R and if $1 \in S$ and $0 \notin S$, then S is a *multiplicative system* in R.

(a) If $\mathfrak{p}$ is a proper ideal of R, prove that $\mathfrak{p}$ is prime if and only if $S = R - \mathfrak{p}$ is a multiplicative system.

(b) Let S be a multiplicative system in R. If I is an ideal of R such that $I \cap S = \emptyset$, use Zorn's lemma to prove that there is a prime ideal $\mathfrak{p}$ of R such that $\mathfrak{p} \supseteq I$ and $\mathfrak{p} \cap S = \emptyset$.

1.3   Ring Homomorphisms

A fundamental concept in the study of rings is that of a ring homomorphism. Its importance lies in the fact that a ring homomorphism $f : R \to S$ provides for the transfer of algebraic information between the rings R and S.

Definition 1.3.1. If R and S are rings, not necessarily with identities, then a function $f : R \to S$ is a *ring homomorphism* if $f(a + b) = f(a) + f(b)$ and $f(ab) = f(a)f(b)$ for all $a, b \in R$. The identity map $\mathrm{id}_R : R \to R$ is a ring homomorphism called the *identity homomorphism*. A ring homomorphism that is injective and surjective is a *ring isomorphism*. If $f : R \to S$ is a surjective ring homomorphism, then S is said to be a *homomorphic image* of R. If f is an isomorphism, then we say that R and S are *isomorphic rings* and write $R \cong S$. If R and S have identities and $f(1_R) = 1_S$, then f is said to be an *identity preserving ring homomorphism*.

> We now assume that all ring homomorphisms are identity preserving. Unless stated otherwise, this assumption will hold throughout the remainder of the text.

The proof of the following proposition is standard.

Proposition 1.3.2. *Let $f : R \to S$ be a ring homomorphism. Then:*

(1) $f(0_R) = 0_S$ *and* $f(-a) = -f(a)$ *for each $a \in R$.*

(2) *If $a \in R$ has a multiplicative inverse in R, then $f(a)$ has a multiplicative inverse in S and $f(a^{-1}) = f(a)^{-1}$.*

The proof of the following proposition is an exercise.

Proposition 1.3.3. *If $f : R \to S$ is a ring homomorphism, then:*

(1) *If R' is a subring of R, then $f(R')$ is a subring of S.*

(2) *If S' is a subring of S, then $f^{-1}(S')$ is a subring of R.*

(3) *If f is a surjection and A is a right ideal (a left ideal, an ideal) of R, then f(A) is a right ideal (a left ideal, an ideal) of S.*

(4) *If B is a right ideal (a left ideal, an ideal) of S, then f^{-1}(B) is a right ideal (a left ideal, an ideal) of R.*

Definition 1.3.4. If I is an ideal of R, then the mapping $\eta : R \to R/I$ defined by $\eta(a) = a + I$ is a surjective ring homomorphism called the *canonical surjection* or the *natural mapping*. If $f : R \to S$ is a ring homomorphism, then the set Ker $f = \{a \in R \mid f(a) = 0\}$ is the *kernel* of f.

The next proposition is one of the cornerstones of ring theory. Part (3) of the proposition shows that every homomorphic image of a ring R is, up to isomorphism, a factor ring of R.

Proposition 1.3.5 (First Isomorphism Theorem for Rings). *If $f : R \to S$ is a ring homomorphism, then*

(1) *Ker f is an ideal of R,*

(2) *f is an injection if and only if Ker $f = 0$, and*

(3) *$R/$ Ker $f \cong f(R)$.*

Proof. The proofs of (1) and (2) are straightforward.

(3) If $\varphi : R/$ Ker $f \to S$ is defined by $\varphi(a+$ Ker $f) = f(a)$, then it is easy to show that φ is a ring isomorphism.   □

Corollary 1.3.6. *If $f : R \to S$ is a surjective ring homomorphism, then $R/$ Ker $f \cong S$.*

If $f : R \to S$ is an injective ring homomorphism, then $f(R)$ is a subring of S that is isomorphic to R. When this is the case we say that R *embeds in* S and that S *contains a copy of* R. Note also that if f is an isomorphism, then it follows that the inverse function $f^{-1} : S \to R$ is also a ring isomorphism.

Proposition 1.3.7 (Second Isomorphism Theorem for Rings). *If I and K are ideals of a ring R such that $I \subseteq K$, then K/I is an ideal of R/I and $(R/I)/(K/I) \cong R/K$.*

Proof. It is easy to verify that K/I is an ideal of R/I and that the mapping $f : R/I \to R/K$ given by $a + I \mapsto a + K$ is a well-defined surjective ring homomorphism with kernel K/I. The result follows from Corollary 1.3.6.   □

Proposition 1.3.8 (Third Isomorphism Theorem for Rings). *If I and K are ideals of a ring R, then $I/(I \cap K) \cong (I + K)/K$.*

Proof. The mapping $f : I \to (I + K)/K$ defined by $f(a) = a + K$ is a well-defined surjective ring homomorphism with kernel $I \cap K$. Corollary 1.3.6 shows that $I/(I \cap K) \cong (I + K)/K$.   □

Examples

1. **Embedding Maps.** (a) The mapping $f : R \to R[X]$ given by $f(a) = a$, where a is viewed as a constant polynomial in $R[X]$, is an injective ring homomorphism.

   (b) The mapping $f : R \to \mathbb{M}_n(R)$ defined by $f(a) = (a_{ij})$, where $a_{ij} = 0$ if $i \neq j$ and $a_{ii} = a$ for $i = 1, 2, \ldots, n$, is an injective ring homomorphism. Thus, the $n \times n$ matrix ring $\mathbb{M}_n(R)$ contains a copy of the ring R.

   (c) If $\mathcal{C} = \{\left(\begin{smallmatrix} a & b \\ -b & a \end{smallmatrix}\right) \mid a, b \in \mathbb{R}\}$, then $\mathcal{C}$ is a subring of $M_2(\mathbb{R})$. The mapping $a + bi \mapsto \left(\begin{smallmatrix} a & b \\ -b & a \end{smallmatrix}\right)$ from $\mathbb{C}$ to $\mathcal{C}$ is a ring isomorphism. Thus, the matrix ring $M_2(\mathbb{R})$ contains a copy of the field of complex numbers.

2. **Ring Homomorphisms and Skew Polynomial Rings.** If R is a ring, let $R[X]$ denote the set of all right polynomials $p(X) = \sum_{k=0}^{n} X^k a_k$, where X is a non-commuting indeterminate. If $R[X]$ is to be made into a ring, then when forming the product of two right polynomials in $R[X]$, products such as $X^j a_j X^k b_k$ will be encountered. In order to write $X^j a_j X^k b_k$ as $X^{j+k} c$ for some $c \in R$, it is necessary to commute a past X in aX. Let $\sigma : R \to R$ be a ring homomorphism and set $aX = X\sigma(a)$ for all $a \in R$. Then $aX^k = X^k \sigma^k(a)$, for each integer $k \geq 1$, and we see that $X^j a_j X^k b_k = X^{j+k} \sigma^k(a_j) b_k$, where σ^k denotes the composition of σ with itself k times. If addition of right polynomials is defined in the usual fashion by adding coefficients of "like terms" and if the product of two right polynomials is given by

$$p(X)q(X) = \left(\sum_{j=0}^{m} X^j a_j \right)\left(\sum_{k=0}^{n} X^k b_k \right) = \sum_{j=0}^{m} \sum_{k=0}^{n} X^{j+k} \sigma^k(a_j) b_k,$$

   then the set of right polynomials $R[X]$ is a ring under these binary operations. This ring, denoted by $R[X, \sigma]$, is called the *ring of skew polynomials* or the *ring of twisted polynomials*. Given a ring homomorphism $\sigma : R \to R$, a similar ring can be constructed from the set of left polynomials over R in a noncommuting indeterminate X. The *skew power series ring* $R[[X, \sigma]]$ of all right power series with coefficients in R has a similar definition.

We conclude our brief review of rings with the following proposition whose proof is left as an exercise.

Proposition 1.3.9 (Correspondence Property of Rings). *If $f : R \to S$ is a surjective ring homomorphism, then there is a one-to-one correspondence between the following sets.*

(1) *The set of right (left) ideals in R that contain* Ker f *and the set of right (left) ideals in S.*

(2) *The set of maximal right (left) ideals of R that contain* Ker *f and the set of maximal right (left) ideals of S.*

(3) *The set of ideals in R that contain* Ker *f and the set of ideals in S.*

(4) *The set of maximal ideals of R that contain* Ker *f and the set of maximal ideals of S.*

(5) *The set of subrings of R that contain* Ker *f and the set of subrings of S.*

Problem Set 1.3

1. Verify (a) through (c) of Example 1.

2. Consider the ring $\mathbb{T}_2(\mathbb{Z})$ of 2×2 upper triangular matrices and the ring $\mathbb{Z} \times \mathbb{Z}$ under componentwise addition and multiplication. Show that the map $\left(\begin{smallmatrix} a & b \\ 0 & d \end{smallmatrix} \right) \mapsto (a, d)$ from $\mathbb{T}_2(\mathbb{Z})$ to $\mathbb{Z} \times \mathbb{Z}$ is a surjective ring homomorphism and compute its kernel.

3. Let R, R_1 and R_2 be rings and suppose that $R \cong R_1 \times R_2$ is the ring direct product of R_1 and R_2. Prove that there are ideals I_1 and I_2 of R such that $R/I_1 \cong R_2$ and $R/I_2 \cong R_1$.

4. Show that an integral domain R either contains a copy of the ring of integers or a copy of a field $\mathbb{Z}_p$, for some prime number p. [Hint: Consider the map $n \mapsto n1_R$ from $\mathbb{Z}$ to R.]

5. If $\{E_{ij}\}$, $1 \le i, j \le n$, is the set of matrix units of $M_n(R)$, verify that $E_{ii} M_n(R) E_{ii}$ and R are isomorphic rings, for $i = 1, 2, \ldots, n$.

6. Prove Propositions 1.3.2 and 1.3.3.

7. Complete the proof of Proposition 1.3.5.

8. Prove Proposition 1.3.9.

9. Let $f : R \to S$ be a surjective ring homomorphism. If R is commutative, show that S is commutative as well and then prove that there is a one-to-one correspondence between the set of prime ideals of R that contain Ker f and the set of prime ideals of S.

10. Let R and S be rings, not necessarily having identities, and suppose that $f : R \to S$ is a surjective ring homomorphism.

    (a) If R has an identity, prove that S also has an identity.

    (b) If R is a division ring, deduce that S must be a division ring.

11. Let R and S be rings and suppose that $f : R \to S$ is a nonzero ring homomorphism. If $f(1_R) \ne 1_S$, show that S has zero divisors. Conclude that if S is an integral domain, then f must be identity preserving.

12. (a) Suppose that R is a ring that does not have an identity. Show that there is a ring S with identity that contains a subring isomorphic to R. [Hint: Let $S = R \times \mathbb{Z}$ and define addition and multiplication on S by $(a, m) + (b, n) = (a + b, m + n)$ and $(a, m)(b, n) = (ab + an + bm, mn)$ for all $(a, m), (b, n) \in R \times \mathbb{Z}$. Show that S is a ring with identity under these operations and then consider $R' = \{(a, 0) \mid a \in R\}$.] Conclude that a ring not having an identity can be embedded in one that does.

    (b) If the ring R has an identity to begin with, then does the ring S constructed in (a) have zero divisors?

13. Let $f : R \to S$ and $g : S \to T$ be ring homomorphisms. Prove each of the following.

    (a) If gf is an injective ring homomorphism, then so is f.

    (b) If f and g are injective ring homomorphisms, then gf is an injective ring homomorphism.

    (c) If gf is a surjective ring homomorphism, then so is g.

    (d) If f and g are surjective ring homomorphisms, then gf is also a surjective ring homomorphism.

    (e) If f is a ring isomorphism, then the inverse function $f^{-1} : S \to R$ is a ring isomorphism.

14. Let $f : R \to S$ be a nonzero ring homomorphism and suppose that I is an ideal of R such that $I \subseteq \operatorname{Ker} f$. Show that there is a ring homomorphism $\bar{f} : R/I \to S$. The map $\bar{f}$ is said to be the *ring homomorphism induced by f*.

15. Prove that S is a simple ring if and only if every nonzero ring homomorphism $f : S \to R$ is injective for every ring R.

16. Prove that the matrix ring $\mathbb{M}_n(R/I)$ is isomorphic to $\mathbb{M}_n(R)/\mathbb{M}_n(I)$. [Hint: Consider the mapping $f : \mathbb{M}_n(R) \to \mathbb{M}_n(R/I)$ given by $(a_{ij}) \mapsto (a_{ij} + I)$ for all $(a_{ij}) \in \mathbb{M}_n(R)$.]

1.4 Modules

In a vector space, the scalars taken from a field act on the vectors by scalar multiplication, subject to certain rules. In a module, the scalars need only belong to a ring, so the concept of a module is a significant generalization. Much of the theory of modules is concerned with extending the properties of vector spaces to modules. However, module theory can be much more complicated than that of vector spaces. For example, every vector space has a basis and the cardinalities of any two bases of the vector space are equal. However, a module need not have a basis and even if it does, then it may be the case that the module has two or more bases with differing cardinalities.

Modules are central to the study of commutative algebra and homological algebra. Moreover, they are used widely in algebraic geometry and algebraic topology.

Definition 1.4.1. If M is an additive abelian group, then M is said to be a (unitary) *right R-module* if there is a binary operation $M \times R \to M$ such that if $(x, a) \mapsto xa$, then the following conditions hold for all $x, y \in M$ and $a, b \in R$.

(1) $x(a + b) = xa + xb$

(2) $(x + y)a = xa + ya$

(3) $x(ab) = (xa)b$

(4) $x1 = x$

Left R-modules are defined analogously but with the ring elements operating on the left of elements of M. If M is a right R-module, then a nonempty subset N of M is said to be a *submodule* of M if N is a subgroup of the additive group of M and $xa \in N$ whenever $x \in N$ and $a \in R$. If N is a submodule of M and $N \neq M$, then N is a *proper submodule* of M. Finally, if R and S are rings and M is at once a left R-module and a right S-module such that $a(xb) = (ax)b$ for all $a \in R, b \in S$ and $x \in M$, then M is said to be an (R, S)-*bimodule*.

If R is a noncommutative ring, then a right R-module M cannot be made into a left R-module by setting $a \cdot x = xa$. The left-hand versions of properties (1), (2) and (4) of Definition 1.4.1 carry over, but the offending condition is property (3). If setting $a \cdot x = xa$, for all $x \in M$ and $a \in R$, were to make M into a left R-module, then we would have

$$(ab) \cdot x = x(ab) = (xa)b = b \cdot (xa) = b \cdot (a \cdot x) = (ba) \cdot x$$

for all $x \in M$ and $a, b \in R$. Examples of left R-modules over a noncommutative ring abound, where $(ab) \cdot x \neq (ba) \cdot x$, so we have a contradiction. Of course, if the ring is commutative, this difficulty disappears and M can be made into a left R-module in exactly this fashion. Even though a right R-module cannot be made into a left R-module using the method just described, a right R-module M can be made into a left R^{op}-module by setting $a \cdot x = xa$ for all $x \in M$ and $a \in R$. If multiplication in R^{op} is denoted by $\circ$, $x \in M$ and $a, b \in R^{\mathrm{op}}$, then

$$(a \circ b) \cdot x = (ba) \cdot x = x(ba) = (xb)a = a \cdot (xb) = a \cdot (b \cdot x),$$

so the left-hand version of property (3) holds and it is easy to check that the left-hand versions of properties (1), (2) and (4) hold as well.

If the ring R is replaced by a division ring D, then Definition 1.4.1 yields the definition of a *right vector space* V over D and when R is a field K, V is a *vector space* over K. If V is a right vector space over a division ring D, then a submodule of V will often be referred to as a *subspace* of V.

Terminology. To simplify terminology, the expression "R-module" or "module" will mean right R-module and "vector space over D", D a division ring, will mean right vector space over D. When M is an R-module we will also, on occasion, refer to the multiplication $M \times R \to M$ given by $(x, a) \mapsto xa$ as the R-*action* on M.

Examples

1. **Finite Sums of Submodules.** If $M_1, M_2, \ldots, M_n$ are submodules of an R-module M, then $M_1 + M_2 + \cdots + M_n = \{x_1 + x_2 + \cdots + x_n \mid x_i \in M_i$ for $i = 1, 2, \ldots, n\}$ is a submodule of M for each integer $n \geq 1$.

2. **The R-module $\mathbb{M}_n(R)$.** If $\mathbb{M}_n(R)$ is the set of $n \times n$ matrices over R, then $\mathbb{M}_n(R)$ is an additive abelian group under matrix addition. If $(a_{ij}) \in \mathbb{M}_n(R)$ and $a \in R$, then the operation $(a_{ij})a = (a_{ij}a)$ makes $\mathbb{M}_n(R)$ into an R-module. $\mathbb{M}_n(R)$ is also a left R-module under the operation $a(a_{ij}) = (aa_{ij})$.

3. **Left and Right Ideals as Submodules.** The ring R is a left and a right R-module under the operation of multiplication defined on R. Note that A is a right (left) ideal of R if and only if A is a submodule of R when R is viewed as a right (left) R-module.

4. **The Annihilator of a Module.** If M is an R-module and

$$A = \mathrm{ann}_r(M) = \{a \in R \mid xa = 0 \text{ for all } x \in M\},$$

then A is an ideal of R, referred to as the *annihilator in R of M*. If there is a need to emphasize the ring, then we will write $\mathrm{ann}_r^R(M)$ for $\mathrm{ann}_r(M)$. For example, if S is a subring of R and if M is an R-module, then $\mathrm{ann}_r^S(M)$ will indicate that the annihilator is taken in S. If A is a left ideal of R, then

$$\mathrm{ann}_\ell^M(A) = \{x \in M \mid xa = 0 \text{ for all } a \in A\}$$

is a submodule of M, called the *annihilator in M of A*.

If I is an ideal of R such that $I \subseteq \mathrm{ann}_r(M)$, then M is also an R/I-module under the addition already present on M and the R/I-action on M defined by $x(a + I) = xa$ for all $x \in M$ and $a + I \in R/I$.

5. **Modules over Endomorphism Rings.** If G is any additive abelian group, then $\mathrm{End}_{\mathbb{Z}}(G)$, the set of all group homomorphisms $f : G \to G$, is a ring under addition and composition of group homomorphisms. The group G is a left $\mathrm{End}_{\mathbb{Z}}(G)$-module if we set $fx = f(x)$ for all $x \in G$ and all $f \in \mathrm{End}_{\mathbb{Z}}(G)$. G is also a right $\mathrm{End}_{\mathbb{Z}}(G)$-module under the operation $xf = (x)f$, where we agree to write each group homomorphism on the right of its argument and in expressions such as $(x)fg$ we first apply f and then g. Since G is a $\mathbb{Z}$-module, it follows that G is an $(\mathrm{End}_{\mathbb{Z}}(G), \mathbb{Z})$-bimodule and a $(\mathbb{Z}, \mathrm{End}_{\mathbb{Z}}(G))$-bimodule.

6. **Submodules formed from Column and Row Matrices.** Let $c_k(R)$ and $r_k(R)$, $1 \leq k \leq n$, be the sets of kth column matrices and kth row matrices, respectively, as defined in Example 2 in Section 1.2. Then for each k, $c_k(R)$ and $r_k(R)$ are submodules of the R-module $\mathbb{M}_n(R)$.

7. **Cyclic Modules.** If M is an R-module and $x \in M$, then $xR = \{xa \mid a \in R\}$ is a submodule of M called the *cyclic submodule of M generated by x*. If $M = xR$ for some $x \in R$, then M is said to be a *cyclic module*.

8. **The Module $R[X]$.** If $R[X]$ is the set of all polynomials in X with their coefficients in R, then $R[X]$ is an additive abelian group under polynomial addition. $R[X]$ is an R-module via the R-action on $R[X]$ defined by $(a_0 + Xa_1 + \cdots + X^n a_n)a = (a_0 a) + X(a_1 a) + \cdots + X^n(a_n a)$.

9. **Change of Rings.** If $f : R \to S$ is a ring homomorphism and M is an S-module and we let $xa = xf(a)$ for all $x \in M$ and $a \in R$, then M is an R-module. We say that M has been made into an R-module by *pullback along f*. Clearly if M is an R-module and S is a subring of R, then M is an S-module by pullback along the canonical injection $i : S \to R$.

10. **Modular Law.** There is one property of modules that is often useful. It is known as the *modular law* or as the *modularity property of modules*. If M, N and X are submodules of an R-module and $X \supseteq M$, then $X \cap (M + N) = M + (X \cap N)$. (See Exercise 9.)

There are several elementary properties of modules that hold regardless of whether they are left or right modules. Stated for an R-module M,

$$x0_R = 0_M,$$

$$x(-1) = -x,$$

$$x(a - b) = xa - xb,$$

$$x(-a)b = (xa)(-b) = -x(ab) \quad \text{and}$$

$$(x - y)a = xa - ya$$

for all $x, y \in M$ and $a, b \in R$.

Proposition 1.4.2. *A nonempty subset N of an R-module M is a submodule of M if and only if $x + y \in N$ and $xa \in N$ for all $x, y \in N$ and $a \in R$.*

Proof. If N is a submodule of M, there is nothing to prove, so suppose that $x + y \in N$ and $xa \in N$ for all $x, y \in N$ and $a \in R$. Then $x - y = x + y(-1) \in N$, so N is a subgroup of the additive group of M. Conditions (1) through (4) of Definition 1.4.1 must hold for all $x, y \in N$ and $a, b \in R$, since these conditions hold for all $x, y \in M$ and $a, b \in R$.   □

Definition 1.4.3. Let N be a submodule of an R-module M and suppose that S is a subset of N. If every element of $x \in N$ can be expressed as $x = \sum_{i=1}^n x_i a_i$, where $x_i \in S$ and $a_i \in R$ for each $i = 1, 2, \ldots, n$, then we say that S is a set of *generators* of N or that N is generated by S. If N is generated by S, then we write $N = \sum_S xR$ and when S is a finite set, we say that N is *finitely generated*.

Every R-module M has a least one set of generators, namely the set M.

Proposition 1.4.4. *The following hold for any R-module M.*

(1) *If $\{M_\alpha\}_\Delta$ is a family of submodules of M, then $\bigcap_\Delta M_\alpha$ is a submodule of M.*

(2) *If $\{M_\alpha\}_\Delta$ is a family of submodules of M, then*

$$\sum_\Delta M_\alpha = \left\{ \sum_\Delta x_\alpha \;\middle|\; x_\alpha \in M_\alpha \text{ for all } \alpha \in \Delta \right\}$$

is a submodule of M.

(3) *Let S be a subset of M and suppose that $\{M_\alpha\}_\Delta$ is the family of submodules of M each of which contains S. Then S is a set of generators for the submodule $\bigcap_\Delta M_\alpha$.*

Proof. We prove (3) and leave the proofs of (1) and (2) as exercises.

(3) Let $\{M_\alpha\}_\Delta$ be the family of submodules of M each of which contains S. If $x \in S$, then $x \in \bigcap_\Delta M_\alpha$, so $xR \subseteq \bigcap_\Delta M_\alpha$. Hence, $\sum_S xR \subseteq \bigcap_\Delta M_\alpha$. For the reverse containment, note that $\sum_S xR$ is a submodule of M that contains S. Thus, $\bigcap_\Delta M_\alpha \subseteq \sum_S xR$, so we have $\bigcap_\Delta M_\alpha = \sum_S xR$. □

Corollary 1.4.5. *The empty set is a set of generators for the zero module.*

Proof. Part (3) of the proposition shows that if $S = \varnothing$ and if $\{M_\alpha\}_\Delta$ is the family of all submodules of M, then $\varnothing \subseteq M_\alpha$ for each $\alpha \in \Delta$. Hence, $\bigcap_\Delta M_\alpha = 0$, so $\sum_\varnothing xR = 0$. □

Factor Modules

If N is a submodule of an R-module M, then N is a subgroup of the additive group of M. So if N is viewed as a subgroup of the abelian group M, then we know from group theory that we can form the set of cosets $\{x + N\}_{x \in M}$ of N in M, where $x + N = \{x + n \mid n \in N\}$. If M/N denotes this set of cosets, then we also know that M/N can be made into an additive abelian group if coset addition is defined by

$$(x + N) + (y + N) = (x + y) + N.$$

With this in mind M/N can now be made into an R-module by defining an R-action on M/N by $(x + N)a = xa + N$ for $x + N \in M/N$ and $a \in R$. This leads to the following definition.

Definition 1.4.6. If N is a submodule of an R-module M, then M/N together with the operations

$$(x + N) + (y + N) = (x + y) + N \quad \text{and} \quad (x + N)a = xa + N$$

for $x + N, y + N$ and $a \in R$ is called the *factor module* (or the *quotient module*) of M formed by factoring out N.

Examples

11. Consider the R-module $R[X]$. If P is the set of polynomials in $R[X]$ with zero constant term, then P is a submodule of $R[X]$. Every element of the factor module $R[X]/P$ can be expressed as $a_0 + Xa_1 + \cdots + X^na_n + P$. But since $Xa_1 + \cdots + X^na_n \in P$, we see that

$$a_0 + Xa_1 + \cdots + X^na_n + P = a_0 + P.$$

Consequently, the operations on $R[X]/P$ are given by

$$(a + P) + (b + P) = (a + b) + P \quad \text{and}$$
$$(a + P)c = ac + P,$$

where $a, b, c \in R$.

12. Consider the R-module $\mathbb{M}_3(R)$. The set of all matrices of the form $\begin{pmatrix} a_{11} & 0 & a_{13} \\ a_{21} & 0 & a_{23} \\ a_{31} & 0 & a_{33} \end{pmatrix}$ is a submodule N of $\mathbb{M}_3(R)$. It follows that elements of the factor module $\mathbb{M}_3(R)/N$ can be expressed in the form

$$\begin{pmatrix} 0 & a_{12} & 0 \\ 0 & a_{22} & 0 \\ 0 & a_{32} & 0 \end{pmatrix} + N.$$

Definition 1.4.7. Every nonzero right R-module has at least two submodules, namely M and the *zero submodule* $\{0\}$, denoted simply by 0. A nonzero right R-module S that has only 0 and S for its submodules is said to be a *simple module*. The set of all submodules of a right R-module M is partially ordered by $\subseteq$, that is, by inclusion. Under this ordering a *minimal submodule* of M is just a simple submodule of M. A proper submodule N of M is said to be a *maximal submodule* of M if whenever N' is a submodule of M such that $N \subseteq N' \subseteq M$, either $N = N'$ or $N' = M$. Clearly A is a minimal right ideal of R if and only if A_R is a simple R-module.

Example

13. If N is a maximal submodule of M, then it follows that M/N is a simple R-module. In particular, a nonempty subset A of a ring R is a right ideal of R if and only if A is a submodule of R when R is viewed as an R-module. Hence, we can form the factor module R/A. So if $\mathfrak{m}$ is a maximal right ideal of R, then $R/\mathfrak{m}$ is a simple R-module.

Problem Set 1.4

1. If M is an R-module, prove that each of the following hold for all $x, y \in M$ and $a, b \in R$.

(a) $x0_R = 0_M$

(b) $x(-1) = -x$

(c) $x(a - b) = xa - xb$

(d) $x(-a)b = (xa)(-b) = -x(ab)$

(e) $(x - y)a = xa - ya$

2. (a) Let A be a right ideal of R. If addition and an R-action are defined on $A^{(n)}$ by

$$(a_1, a_2, \ldots, a_n) + (b_1, b_2, \ldots, b_n) = (a_1 + b_1, a_2 + b_2, \ldots, a_n + b_n) \quad \text{and}$$

$$(a_1, a_2, \ldots, a_n)r = (a_1 r, a_2 r, \ldots, a_n r),$$

prove that $A^{(n)}$ is an R-module.

(b) If M is an R-module and A is a right ideal of R, prove that

$$MA = \left\{ \sum_{i=1}^{n} x_i a_i \mid x_i \in M, \ a_i \in A, \ n \geq 1 \right\}$$

is a submodule of M.

(c) If N is a submodule of an R-module M and A is a right ideal of R, show that $(M/N)A = (N + MA)/N$.

3. (a) Prove (1) and (2) of Proposition 1.4.4.

(b) If $\{M_\alpha\}_\Delta$ is a family of submodules of M, prove that $\sum_\Delta M_\alpha$ is the intersection of all the submodules of M that contain the set $\bigcup_\Delta M_\alpha$. Conclude that $\sum_\Delta M_\alpha$ is the "smallest" submodule of M generated by $\bigcup_\Delta M_\alpha$.

4. Let M be an R-module.

(a) Prove that $\mathrm{ann}_r(M)$ is an ideal of R.

(b) If I is an ideal of R such that $I \subseteq \mathrm{ann}_r(M)$, prove that M is also an R/I-module via the operation given by $x(a + I) = xa$, for all $x \in M$ and $a + I \in R/I$, and the addition already present on M.

5. Let M be an R-module.

(a) If A is a left ideal of R, show that $\mathrm{ann}_\ell^M(A)$ is a submodule of M.

(b) If N is a submodule of M, deduce that $N \subseteq \mathrm{ann}_\ell^M(\mathrm{ann}_r^R(N))$. Does $\mathrm{ann}_r^R(N) = \mathrm{ann}_r^R(\mathrm{ann}_\ell^M(\mathrm{ann}_r^R(N))$?

(c) If A is a left ideal of R, show that $A \subseteq \mathrm{ann}_r^R(\mathrm{ann}_\ell^M(A))$. Does $\mathrm{ann}_\ell^M(A) = \mathrm{ann}_\ell^M(\mathrm{ann}_r^R(\mathrm{ann}_\ell^M(A))$?

6. Let $\mathbb{M}_{m \times n}(R)$ be the set of all $m \times n$ matrices with entries from R.

(a) Show that $R^{(m)}\mathbb{M}_{m \times n}(R)$ is a left R-submodule of $R^{(n)}$.

(b) If $R^{(k)T}$ denotes the k-tuples of $R^{(k)}$ written in column form, show that $\mathbb{M}_{m \times n}(R) R^{(n)T}$ is a submodule of the R-module $R^{(m)T}$.

(c) If $(a_{ij}) \in \mathbb{M}_{m \times n}(R)$, show that $\text{ann}_\ell^{R^{(m)}}(a_{ij})$ is an R-submodule of $R^{(m)}$ that is the solution set for a system of homogeneous linear equations with their coefficients in R. Make a similar observation for $\text{ann}_r^{R^{(n)T}}(a_{ij})$.

7. Suppose that M is an (R, R)-bimodule and consider the set $S = R \times M$. Show that if addition and multiplication are defined on S by

$$(a, x) + (b, y) = (a + b, x + y) \quad \text{and}$$
$$(a, x)(b, y) = (ab, ay + xb)$$

for all $(a, x), (b, y) \in S$, then S is a ring with identity.

8. (a) A family $\{M_\alpha\}_\Delta$ of submodules of M is a *chain of submodules* of M if $M_\alpha \subseteq M_\beta$ or $M_\beta \subseteq M_\alpha$ for all $\alpha, \beta \in \Delta$. Prove that if $\{M_\alpha\}_\Delta$ is a chain of submodules of M, then $\bigcup_\Delta M_\alpha$ is a submodule of M.

(b) Prove or find a counterexample to: If $\{M_\alpha\}_\Delta$ is a family of submodules of M such that $\bigcup_\Delta M_\alpha$ is a submodule of M, then $\{M_\alpha\}_\Delta$ is a chain of submodules of M.

9. Prove the modular law for modules of Example 10.

10. (a) If N is a submodule of an R-module M, prove that M/N is a simple R-module if and only if N is a maximal submodule of M.

(b) Prove that R is a division ring if and only if R_R is a simple R-module.

11. Let N be a proper submodule of a nonzero R-module M. If $x \in M - N$, use Zorn's lemma to show that there is a submodule N' of M containing N that is maximal with respect to $x \notin N'$.

12. (a) Find an example of a finitely generated R-module that has a submodule that is not finitely generated.

(b) If M is a finitely generated R-module and N is a submodule of M, show that M/N is finitely generated.

1.5 Module Homomorphisms

One important facet of ring homomorphisms is that they provide for the transfer of algebraic information between rings. In order to transfer algebraic information between modules, we need the concept of a module homomorphism.

Definition 1.5.1. If M and N are R-modules, then a mapping $f : M \to N$ is said to be an *R-module homomorphism* or an *R-linear mapping* if

(1) $f(x + y) = f(x) + f(y)$ and
(2) $f(xa) = f(x)a$

for all $x, y \in M$ and $a \in R$. A mapping $f : M \to N$ that satisfies (1) is said to be an *additive function*. The identity mapping $M \to M$ defined by $x \mapsto x$ is an R-module homomorphism that will be denoted by id_M. An R-module homomorphism that is an injective function will be referred to simply as a *monomorphism* and an R-module homomorphism that is a surjective function will be called an *epimorphism*. If $f : M \to N$ is an epimorphism, then N is said to be a homomorphic image of M. If $f : M \to M$ is an R-linear mapping, then f is an *endomorphism* of M. An *iso-morphism* is an R-linear mapping that is injective and surjective and an isomorphism $f : M \to M$ is said to be an *automorphism* of M. If $f : M \to N$ is an isomorphism, then M and N are said to be *isomorphic R-modules*, denoted by $M \cong N$. The set of all R-linear mappings from M to N will be denoted by $\mathrm{Hom}_R(M, N)$ and when $M = N$, $\mathrm{Hom}_R(M, M)$ will be written as $\mathrm{End}_R(M)$. If U and V are vector spaces over a division ring D, then a D-linear mapping $f : U \to V$ is referred to as a *linear transformation*.

Examples

1. **$\mathrm{Hom}_R(M, N)$ as a Left R-module.** If M and N are R-modules, then $\mathrm{Hom}_R(M, N)$ is an additive abelian group under function addition, that is, $\mathrm{Hom}_R(M, N)$ is a $\mathbb{Z}$-module. In general, $\mathrm{Hom}_R(M, N)$ is not an R-module if fa is defined as $(fa)(x) = f(xa)$ since

$$(f(ab))(x) = f(x(ab)) = f((xa)b) = (fb)(xa) = ((fb)a)(x),$$

   so $f(ab) = (fb)a$. However, what is required is that $f(ab) = (fa)b$, so condition (3) of Definition 1.4.1 fails to hold. If R is commutative, then we immediately see that $\mathrm{Hom}_R(M, N)$ can be made into an R-module in precisely this manner. Even though $\mathrm{Hom}_R(M, N)$ cannot be made into an R-module by setting $(fa)(x) = f(xa)$, $\mathrm{Hom}_R(M, N)$ can be made into a left R-module using this technique. If $a \in R$ and $x \in M$, let $(af)(x) = f(xa)$. Then for $a, b \in R$ and $x \in M$ we see that

$$((ab)f)(x) = f(x(ab)) = f((xa)b) = (bf)(xa) = a(bf)(x).$$

   Hence, $(ab)f = a(bf)$, so the left-hand version of condition (3) of Definition 1.4.1 does indeed hold. It is easy to check that the left-hand version of conditions (1), (2) and (4) also hold, so when M and N are R-modules, $\mathrm{Hom}_R(M, N)$ is a left R-module. Similarly, when M and N are left R-modules, then $\mathrm{Hom}_R(M, N)$ can be made into an R-module by setting $(fa)(x) = f(ax)$ for all $a \in R$ and $x \in M$.

2. **The Endomorphism Ring of a Module.** If M is an R-module, then $\mathrm{End}_{\mathbb{Z}}(M)$ and $\mathrm{End}_R(M)$ are rings under function addition and function composition called

the $\mathbb{Z}$-*endomorphism ring* of M and the R-*endomorphism ring* of M, respectively. If we defined $fx = f(x)$ for each $f \in \mathrm{End}_{\mathbb{Z}}(M)$ and all $x \in M$, then this makes M into a left $\mathrm{End}_{\mathbb{Z}}(M)$-module. Since $\mathrm{End}_R(M)$ is a subring of $\mathrm{End}_{\mathbb{Z}}(M)$, M is also a left $\mathrm{End}_R(M)$-module under the same operation.

Remark. Bimodule structures yield various module structures on Hom.

(1) If M is an (R, S)-bimodule and N is a left R-module, then $\mathrm{Hom}_R({}_RM_S,{}_R N)$ is a left S-module. If sf is defined by $(sf)(x) = f(xs)$ for $f \in \mathrm{Hom}_R({}_RM_S, {}_RN)$, $s \in S$ and $x \in M$, then $sf \in \mathrm{Hom}_R({}_RM_S, {}_RN)$. In fact, if $a \in R$, then

$$(sf)(ax) = f((ax)s) = f(a(xs)) = af(xs) = a(sf)(x).$$

(2) If M is an (R, S)-bimodule and N is a right S-module, then $\mathrm{Hom}_S({}_RM_S, N_S)$ is a right R-module. If $(fa)(x) = f(ax)$ for $f \in \mathrm{Hom}_S({}_RM_S, N_S)$, $a \in R$ and $x \in M$, then it follows that $fa \in \mathrm{Hom}_S({}_RM_S, N_S)$ because for $s \in S$ we have

$$(fa)(xs) = f(a(xs)) = f((ax)s) = f(ax)s = (fa)(x)s.$$

(3) If N is an (R, S)-bimodule and M is a left R-module, then $\mathrm{Hom}_R({}_RM, {}_RN_S)$ is a right S-module. To see this, let $(fs)(x) = f(x)s$ for $f \in \mathrm{Hom}_R({}_RM, {}_RN_S)$, $s \in S$ and $x \in M$. Then the map fs is in $\mathrm{Hom}_R({}_RM, {}_RN_S)$ due to the fact that if $a \in R$, then

$$(fs)(ax) = f(ax)s = af(x)s = a(fs)(x).$$

(4) If N is an (R, S)-bimodule and M is a right S-module, then $\mathrm{Hom}_S(M_S, {}_RN_S)$ is a left R-module. If af is defined by $(af)(x) = af(x)$ for $f \in \mathrm{Hom}_S(M_S, {}_RN_S)$, $a \in R$ and $x \in M$, then af is in $\mathrm{Hom}_S(M_S, {}_RN_S)$ since if $s \in S$, then

$$(af)(xs) = af(xs) = af(x)s = (af)(x)s.$$

The proofs of the following two propositions are straightforward. The proof of each is an exercise.

Proposition 1.5.2. *If $f : M \to N$ is an R-linear mapping, then $f(0_M) = 0_N$ and $f(-x) = -f(x)$ for each $x \in M$.*

The next proposition shows that submodules and the inverse image of submodules are preserved under module homomorphisms.

Proposition 1.5.3. *Let $f : M \to N$ be an R-module homomorphism.*

(1) *If M' is a submodule of M, then $f(M')$ is a submodule of N.*

(2) *If N' is a submodule of N, then $f^{-1}(N')$ is a submodule of M.*

Part 2 in the preceding Remark shows that if M is an R-module, then $\operatorname{Hom}_R(R, M)$ is an R-module. This fact is responsible for the "as R-modules" in the following proposition.

Proposition 1.5.4. *If M is an R-module, then $\operatorname{Hom}_R(R, M) \cong M$ as R-modules.*

Proof. Let $\varphi : \operatorname{Hom}_R(R, M) \to M$ be such that $\varphi(f) = f(1)$. Then $\varphi(f + g) = (f + g)(1) = f(1) + g(1) = \varphi(f) + \varphi(g)$ and $\varphi(fa) = (fa)(1) = f(a) = f(1)a = \varphi(f)a$, so φ is R-linear. If $f \in \operatorname{Ker}\varphi$, then $0 = \varphi(f) = f(1)$ clearly implies that $f = 0$, so φ is injective. If $x \in M$, then $f_x : R \to M$ defined by $f_x(a) = xa$ is R-linear and $\varphi(f_x) = x$. Thus, φ is also surjective.   □

Definition 1.5.5. If M is an R-module and N is a submodule of M, then the mapping $\eta : M \to M/N$ defined by $\eta(x) = x + N$ is an epimorphism called the *canonical surjection* or the *natural mapping*. If $f : M \to N$ is an R-module homomorphism, then

$$\operatorname{Ker} f = \{x \in M \mid f(x) = 0\}$$

is the *kernel* of f and the R-module $N/\operatorname{Im} f$ is called the *cokernel* of f. The cokernel of f will be denoted by $\operatorname{Coker} f$.

The following three propositions are of fundamental importance. The first of these propositions is one of the most useful results in the theory of modules.

Proposition 1.5.6 (First Isomorphism Theorem for Modules). *If $f : M \to N$ is an R-linear mapping, then*

(1) $\operatorname{Ker} f$ *is a submodule of M,*

(2) f *is a monomorphism if and only if* $\operatorname{Ker} f = 0$*, and*

(3) $M/\operatorname{Ker} f \cong f(M)$.

Proof. (1) If $x, y \in \operatorname{Ker} f$ and $a \in R$, then $f(x + y) = f(x) + f(y) = 0$ and $f(xa) = f(x)a = 0a = 0$, so $x + y$ and xa are in $\operatorname{Ker} f$. Proposition 1.4.2 gives the result.

(2) If f is a monomorphism and $x \in \operatorname{Ker} f$, then $f(x) = 0 = f(0)$, so $x = 0$ since f is an injection. Thus, $\operatorname{Ker} f = 0$. Conversely, suppose that $\operatorname{Ker} f = 0$ and $x, y \in M$ are such that $f(x) = f(y)$. Then $f(x - y) = 0$, so $x - y$ is in $\operatorname{Ker} f = 0$. Hence, $x = y$, so f is an injection.

(3) Define $\varphi : M/\operatorname{Ker} f \to f(M)$ by $\varphi(x + \operatorname{Ker} f) = f(x)$. If $x + \operatorname{Ker} f = y + \operatorname{Ker} f$, then $x - y \in \operatorname{Ker} f$, so $f(x) = f(y)$, so φ is well defined. It follows easily that φ is an epimorphism and if $\varphi(x + \operatorname{Ker} f) = 0$, then $f(x) = 0$, so $x \in \operatorname{Ker} f$. Thus, $x + \operatorname{Ker} f = 0$, so φ is an injection and we have that φ is an isomorphism.   □

Corollary 1.5.7. *If $f : M \to N$ is an epimorphism, then $M/\operatorname{Ker} f \cong N$.*

Part (2) of Proposition 1.5.6 shows that an R-linear mapping $f : M \to N$ is a monomorphism if and only if $\operatorname{Ker} f = 0$. It also follows that f is an epimorphism if and only if $\operatorname{Coker} f = 0$.

If $f : M \to N$ is a monomorphism, then $f(M)$ is a submodule of N that is isomorphic to M. When this is the case we say that M *embeds in* N and that N *contains a copy of* M.

Proposition 1.5.8 (Second Isomorphism Theorem for Modules). *If M_1 and M_2 are submodules of an R-module M such that $M_1 \subseteq M_2$, then M_2/M_1 is a submodule of M/M_1 and $(M/M_1)/(M_2/M_1) \cong M/M_2$.*

Proof. It is straightforward to show that M_2/M_1 is a submodule of M/M_1. Moreover, the mapping $f : M/M_1 \to M/M_2$ given by $f(x + M_1) = x + M_2$ is well defined. Indeed, if $x + M_1 = y + M_1$, then $x - y \in M_1 \subseteq M_2$, so $f(x + M_1) = x + M_2 = y + M_2 = f(y + M_2)$. Moreover, it is easy to show that f is an epimorphism with kernel M_2/M_1, so the result follows from Corollary 1.5.7. □

Proposition 1.5.9 (Third Isomorphism Theorem for Modules). *If M_1 and M_2 are submodules of an R-module M, then $M_1/(M_1 \cap M_2) \cong (M_1 + M_2)/M_2$.*

Proof. The epimorphism $f : M_1 \to (M_1 + M_2)/M_2$ defined by $f(x) = x + M_2$ has kernel $M_1 \cap M_2$. Corollary 1.5.7 gives the desired result. □

Proposition 1.5.10 (Correspondence Property of Modules). *If $f : M \to N$ is an epimorphism, then there is a one-to-one correspondence between the submodules of M that contain $\operatorname{Ker} f$ and the submodules of N.*

Proof. If $\bar{N}$ is a submodule of $M/\operatorname{Ker} f$ and $\eta : M \to M/\operatorname{Ker} f$ is the natural mapping, then there is a unique submodule $N = \eta^{-1}(\bar{N})$ of M such that $N \supseteq \operatorname{Ker} f$ and $N/\operatorname{Ker} f = \bar{N}$. The result now follows from the observation that $M/\operatorname{Ker} f \cong N$. □

Problem Set 1.5

1. Prove Propositions 1.5.2 and 1.5.3.

2. If M_1 and M_2 are submodules of an R-module M such that $M_1 \subseteq M_2$, show that there is an R-linear mapping $M/M_1 \to M/M_2$.

3. If $f : M \to N$ is a bijective function and M is an R-module, show that N can be made into an R-module in a way that turns f into an isomorphism.

4. (a) If $f : M \to N$ and $g : N \to M$ are R-linear mappings such that $gf = \mathrm{id}_M$, prove that f is injective and that g is surjective.

(b) If $f : M \to N$ and $g : N \to M$ are monomorphisms, show that gf is a monomorphism.

(c) If $f : M \to N$ and $g : N \to M$ are epimorphisms, show that gf is an epimorphism.

5. (a) If $f : M \to N$ is a monomorphism, then f is an injective function and as such has a left inverse $g : N \to M$. Show by example that g need not be an R-linear mapping.

(b) If $f : M \to N$ is an epimorphism, then f is a surjective function and so has a right inverse $g : N \to M$. Show by example that g may not be an R-linear mapping.

(c) If $f : M \to N$ is an isomorphism, prove that the inverse function $f^{-1} : N \to M$ for f is an isomorphism.

Note that even though the observations of (a) and (b) hold, the inverse function for an R-linear bijection is an R-linear bijection.

6. (a) Let M be a $\mathbb{Z}$-module and suppose that $x \in M$. Deduce that the mapping $\mathbb{Z}_n \to M$ defined by $[a] \mapsto xa$ is a well-defined $\mathbb{Z}$-linear mapping if and only if $nx = 0$.

(b) Show that $N = \{x \in M \mid nx = 0\}$ is a submodule of M and that $\mathrm{Hom}_{\mathbb{Z}}(\mathbb{Z}_n, M) \cong N$.

(c) If (m, n) denotes the greatest common divisor of two positive integers m and n, prove that $\mathrm{Hom}_{\mathbb{Z}}(\mathbb{Z}_m, \mathbb{Z}_n) \cong \mathbb{Z}_{(m,n)}$.

7. (a) If S and S' are simple R-modules and $f : S \to S'$ is an R-linear mapping, show that if $f \neq 0$, then f is an isomorphism.

(b) Prove that an R-module S is simple if and only if there is a maximal right ideal $\mathfrak{m}$ of R such that $R/\mathfrak{m} \cong S$.

8. If $\mathcal{M}_R$ is the class of all R-modules, define the relation $\sim$ on $\mathcal{M}_R$ by $M \sim N$ if and only there is an isomorphism $f : M \to N$. Show that $\sim$ is an equivalence relation on $\mathcal{M}_R$.

9. Verify each of the module structures induced on Hom in the Remark of this section.

10. Let M be an R-module and suppose that I is an ideal of R such that $MI = 0$. Then M is an R/I-module via the operation of $x(a + I) = xa$.

(a) Show there is a one-to-one correspondence between the submodules of M as an R-module and the submodules of M as an R/I-module.

(b) Prove that if N is also an R-module such that $NI = 0$, then $f : M \to N$ is an R-linear mapping if and only if it is R/I-linear.

11. If R and S are rings, then an additive mapping $f : R \to S$ is said to be an *anti-ring homomorphism* if $f(ab) = f(b)f(a)$ for all $a, b \in R$. If f is a bijection, then R and S are said to be *anti-isomorphic*. Clearly, if $f : R \to S$ and $g : S \to T$ are anti-ring homomorphisms, then $gf : R \to T$ is a ring homomorphism.

(a) Consider the ring R to be a right R-module and let $\operatorname{End}_R(R_R)$ denote the ring of R-linear mappings $f : R_R \to R_R$. If $\varphi : R \to \operatorname{End}_R(R_R)$ is defined by $\varphi(x) = f_x$ for each $x \in R$, where $f_x : R_R \to R_R$ is such that $f_x(a) = xa$ for all $a \in R$, show that φ is a ring isomorphism. Conclude that R and $\operatorname{End}_R(R_R)$ are isomorphic rings.

(b) Consider R to be a left R-module and let $\operatorname{End}_R(_R R)$ denote the ring of all R-linear mappings $f : _R R \to _R R$. Prove that $\varphi : R \to \operatorname{End}_R(_R R)$ given by $\varphi(x) = f_x$ for each $x \in R$, where $f_x : R \to R$ is such that $f_x(a) = ax$ for all $a \in R$, is an anti-ring isomorphism. Conclude that R and $\operatorname{End}_R(_R R)$ are anti-isomorphic rings.

(c) Prove that in (b) if we agree to write f_x as $(a)f_x = ax$ for each $a \in R$, then φ is a ring isomorphism.

12. (a) If M is an R-module, show that M is an $(\operatorname{End}_R(M), R)$-bimodule.

(b) If M is an R-module and $H = \operatorname{End}_R(M)$, then from (a) we see that M is a left H-module. If $B = \operatorname{End}_H(M)$, show that M can be viewed as an (H, B)-bimodule. B is the *biendomorphism ring* of M.

Chapter 2
Fundamental Constructions

We now introduce four concepts that are ubiquitous in abstract algebra: direct products, direct sums, free modules and tensor products. Each concept can be used to produce an object that possesses a property known as the *universal mapping property*. At this point, we are not in a position to give a definition of this property, so we will have to be satisfied with pointing out when it holds. The concept will be made more definitive in the following chapter where categories are introduced. Additional constructions will be shown to have this property at subsequent points in the text.

2.1 Direct Products and Direct Sums

Direct Products

If $\{M_\alpha\}_\Delta$ is a family of R-modules, then the Cartesian product $\prod_\Delta M_\alpha$ can be made into an R-module by defining

$$(x_\alpha) + (y_\alpha) = (x_\alpha + y_\alpha) \quad \text{and} \quad (x_\alpha)a = (x_\alpha a)$$

for all $(x_\alpha), (y_\alpha) \in \prod_\Delta M_\alpha$ and $a \in R$, where we set $\prod_\Delta M_\alpha = 0$ if $\Delta = \emptyset$. The addition and the R-action on $\prod_\Delta M_\alpha$ are said to be *componentwise operations*. The mapping $\pi_\beta : \prod_\Delta M_\alpha \to M_\beta$ such that $\pi_\beta((x_\alpha)) = x_\beta$ for all $(x_\alpha) \in \prod_\Delta M_\alpha$ is an epimorphism called the *βth canonical projection* and the mapping $i_\beta : M_\beta \to \prod_\Delta M_\alpha$ defined by $i_\beta(x) = (x_\alpha)$, where $x_\alpha = 0$ if $\alpha \neq \beta$ and $x_\beta = x$, is an R-linear injection called the *βth canonical injection*. The R-module $\prod_\Delta M_\alpha$ together with the family of *canonical projections* $\{\pi_\alpha : \prod_\Delta M_\alpha \to M_\alpha\}_\Delta$ is said to be a *direct product* of the family $\{M_\alpha\}_\Delta$. Such a product will be denoted by $(\prod_\Delta M_\alpha, \pi_\alpha)$ or more simply by $\prod_\Delta M_\alpha$ with the family of mappings $\{\pi_\alpha\}_\Delta$ understood. We will often refer to $\prod_\Delta M_\alpha$ as a direct product. The mappings $\{\pi_\alpha\}_\Delta$ are called the *structure maps* of $\prod_\Delta M_\alpha$. Notice that the canonical injections $i_\beta : M_\beta \to \prod_\Delta M_\alpha$ have not been mentioned. This is not an oversight, we will soon see that these mappings are actually determined by the π_β.

Example

1. A direct product of a family $\{M_\alpha\}_\Delta$ of R-modules is particularly simple when Δ is finite. For example, let $\Delta = \{1, 2, 3\}$ and suppose that M_1, M_2 and M_3

are R-modules. Then

$$\prod_\Delta M_i = M_1 \times M_2 \times M_3$$

and the operations on $\prod_\Delta M_i$ are given by

$$(x_1, x_2, x_3) + (y_1, y_2, y_3) = (x_1 + y_1, x_2 + y_2, x_3 + y_3) \quad \text{and}$$
$$(x_1, x_2, x_3)a = (x_1 a, x_2 a, x_3 a)$$

for all $(x_1, x_2, x_3), (y_1, y_2, y_3) \in \prod_\Delta M_i$ and $a \in R$. In this case,

$$\pi_1((x_1, x_2, x_3)) = x_1, \quad i_1(x) = (x, 0, 0),$$
$$\pi_2((x_1, x_2, x_3)) = x_2, \quad i_2(x) = (0, x, 0),$$
$$\pi_3((x_1, x_2, x_3)) = x_3, \quad \text{and} \quad i_3(x) = (0, 0, x).$$

We now need the following concept: Let $\{M_\alpha\}_\Delta$ be a family of R-modules and suppose that for each $\alpha \in \Delta$ there is an R-linear mapping $f_\alpha : N \to M_\alpha$, N a fixed R-module. Then

$$f : N \to \prod_\Delta M_\alpha \quad \text{defined by } f(x) = (f_\alpha(x))$$

is a well-defined R-linear mapping, called the *product of the family of mappings* $\{f_\alpha\}_\Delta$.

The following proposition gives a fundamental property of a direct product.

Proposition 2.1.1. *If* $\{M_\alpha\}_\Delta$ *is a family of R-modules, then a direct product* $(\prod_\Delta M_\alpha, \pi_\alpha)$ *has the property that for every R-module N and every family* $\{f_\alpha : N \to M_\alpha\}_\Delta$ *of R-linear mappings there is a unique R-linear mapping* $f : N \to \prod_\Delta M_\alpha$ *such that for each $\alpha \in \Delta$ the diagram*

$$
\begin{array}{ccc}
& N & \\
& \big\downarrow^{\textstyle f} \;\;\searrow^{\textstyle f_\alpha} & \\
& & M_\alpha \\
& \nearrow_{\textstyle \pi_\alpha} & \\
\prod_\Delta M_\alpha & &
\end{array}
$$

is commutative.

Proof. Let N be an R-module and suppose that, for each $\alpha \in \Delta$, $f_\alpha : N \to M_\alpha$ is an R-linear mapping. If $f : N \to \prod_\Delta M_\alpha$ is the product of the family of mappings $\{f_\alpha\}_\Delta$, then it is easy to check that f is such that $\pi_\alpha f = f_\alpha$ for each $\alpha \in \Delta$. Now suppose that $g : N \to \prod_\Delta M_\alpha$ is also an R-linear mapping such that $\pi_\alpha g = f_\alpha$ for each $\alpha \in \Delta$. If $g(x) = (x_\alpha)$, then $f_\alpha(x) = \pi_\alpha g(x) = \pi_\alpha((x_\alpha)) = x_\alpha$, so $(x_\alpha) = (f_\alpha(x)) = f(x)$. Hence, $f = g$. $\square$

Proposition 2.1.1 and the preceding discussion provide the motivation for the formal definition of a direct product of a family of R-modules.

Definition 2.1.2. An R-module P together with a family of R-linear mappings $\{p_\alpha : P \to M_\alpha\}_\Delta$ is said to be a *direct product* of the family $\{M_\alpha\}_\Delta$ of R-modules if for every R-module N and every family of R-linear mappings $\{f_\alpha : N \to M_\alpha\}_\Delta$, there is a unique R-linear mapping $f : N \to P$ such that, for each $\alpha \in \Delta$, the diagram

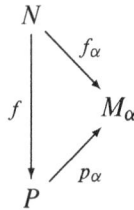

$$
\begin{array}{ccc}
N & & \\
 & \searrow^{f_\alpha} & \\
f\downarrow & & M_\alpha \\
 & \nearrow_{p_\alpha} & \\
P & &
\end{array}
$$

is commutative. The mappings of the family $\{p_\alpha\}_\Delta$ are the *structure maps* for the direct product. A direct product $(P, p_\alpha)_\Delta$ is universal in the sense that given any family $\{f_\alpha : N \to M_\alpha\}_\Delta$ of R-linear mappings there is always a unique R-linear mapping $f : N \to P$ such that $p_\alpha f = f_\alpha$ for each $\alpha \in \Delta$. It is in this sense that we say that $(P, p_\alpha)_\Delta$ has the *universal mapping property*.

Proposition 2.1.1 shows that every family $\{M_\alpha\}_\Delta$ of R-modules has a direct product. An important result of the universal mapping property is that for a direct product $(P, p_\alpha)_\Delta$ the R-module P is unique up to isomorphism .

Proposition 2.1.3. *If* $(P, p_\alpha)_\Delta$ *and* $(P', p'_\alpha)_\Delta$ *are direct products of the family* $\{M_\alpha\}_\Delta$ *of R-modules, then there is unique isomorphism* $\varphi : P' \to P$ *such that* $p_\alpha\varphi = p'_\alpha$ *for each* $\alpha \in \Delta$.

Proof. Consider the diagram

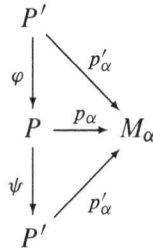

$$
\begin{array}{ccc}
P' & & \\
\varphi\downarrow & \searrow^{p'_\alpha} & \\
P & \xrightarrow{p_\alpha} & M_\alpha \\
\psi\downarrow & \nearrow_{p'_\alpha} & \\
P' & &
\end{array}
$$

where φ and ψ are the unique maps given by the definition of a direct product. Since $p_\alpha\varphi = p'_\alpha$ and $p'_\alpha\psi = p_\alpha$, we see that $p'_\alpha\psi\varphi = p'_\alpha$ for each $\alpha \in \Delta$. Hence, we have a commutative diagram

$$
\begin{array}{c}
P' \\
\downarrow{\scriptstyle \psi\varphi} \qquad \searrow{\scriptstyle p'_\alpha} \\
\qquad\qquad M_\alpha \\
\nearrow{\scriptstyle p'_\alpha} \\
P'
\end{array}
$$

But $\mathrm{id}_{P'}$ is such that $p'_\alpha \mathrm{id}_{P'} = p'_\alpha$, so it follows from the uniqueness of the map $\psi\varphi$ that $\psi\varphi = \mathrm{id}_{P'}$. Similarly, it follows that $\varphi\psi = \mathrm{id}_P$. Thus, φ is the required isomorphism.   □

In view of Proposition 2.1.3, we see that $(\prod_\Delta M_\alpha, \pi_\alpha)$ is actually a model for every direct product of a family $\{M_\alpha\}_\Delta$ of R-modules. If $(P, p_\alpha)_\Delta$ is also a direct product of $\{M_\alpha\}_\Delta$, then there is a unique isomorphism $\varphi : P \to \prod_\Delta M_\alpha$ such that $\pi_\alpha\varphi = p_\alpha$ for each $\alpha \in \Delta$. Since π_α is an epimorphism, it follows that p_α is also an epimorphism. Thus, if $(P, p_\alpha)_\Delta$ is a direct product of a family $\{M_\alpha\}_\Delta$ of R-modules, then each of its structure maps is an epimorphism. A direct product $(P, p_\alpha)_\Delta$ also determines a family $\{u_\alpha : M_\alpha \to P\}_\Delta$ of R-linear injections into P.

Proposition 2.1.4. *If $(P, p_\alpha)_\Delta$ is a direct product of a family $\{M_\alpha\}_\Delta$ of R-modules, then there is a unique family $\{u_\alpha : M_\alpha \to P\}_\Delta$ of injective R-linear mappings such that $p_\alpha u_\alpha = \mathrm{id}_{M_\alpha}$ for each $\alpha \in \Delta$ and $p_\beta u_\alpha = 0$ when $\alpha \neq \beta$.*

Proof. Suppose that $(P, p_\alpha)_\Delta$ is a direct product of the family $\{M_\alpha\}_\Delta$ of R-modules and let α be a fixed element of Δ. Let $N = M_\alpha$ and for each $\beta \in \Delta$ define $f_\beta : N \to M_\beta$ by $f_\beta = 0$ if $\beta \neq \alpha$ and $f_\alpha = \mathrm{id}_{M_\alpha}$. Then we have a family of R-linear mapping $\{f_\beta : N \to M_\beta\}_\Delta$, so it follows from the definition of a direct product that there is a unique R-linear mapping $u_\alpha : N \to P$ such that $p_\beta u_\alpha = f_\beta$ for each $\beta \in \Delta$. Hence, $p_\beta u_\alpha = 0$, if $\alpha \neq \beta$ and $p_\alpha u_\alpha = \mathrm{id}_{M_\alpha}$. Since $p_\alpha u_\alpha = \mathrm{id}_{M_\alpha}$, we have that u_α is an injective mapping and that p_α is surjective, a fact observed earlier.   □

The family $\{u_\alpha\}_\Delta$ of injective R-linear mappings given in Proposition 2.1.4 is said to be the *family of canonical injections* (*uniquely*) *determined* by $(P, p_\alpha)_\Delta$. It follows that the family $\{i_\beta : M_\beta \to \prod_\Delta M_\alpha\}_\Delta$ of canonical injections is uniquely determined by $(\prod_\Delta M_\alpha, \pi_\alpha)$.

If $\{R_\alpha\}_\Delta$ is an indexed family of rings, then as indicated in Chapter 1 the Cartesian product $\prod_\Delta R_\alpha$ is a ring with identity (1_α) under *componentwise addition*

$$(a_\alpha) + (b_\alpha) = (a_\alpha + b_\alpha)$$

and *componentwise multiplication*

$$(a_\alpha)(b_\alpha) = (a_\alpha b_\alpha).$$

The *ring direct product* $\prod_\Delta R_\alpha$ is such that the canonical injection $i_\beta : R_\beta \to \prod_\Delta R_\alpha$ is a non-identity preserving ring homomorphism. However, each canonical projection $\pi_\beta : \prod_\Delta R_\alpha \to R_\beta$ does preserve identities.

External Direct Sums

If $\{M_\alpha\}_\Delta$ is a family of R-modules, then $\prod_\Delta M_\alpha$ has a submodule called the *external direct sum* of the family $\{M_\alpha\}_\Delta$. It is not difficult to show that

$$\bigoplus_\Delta M_\alpha = \left\{ (x_\alpha) \in \prod_\Delta M_\alpha \mid x_\alpha = 0 \text{ for almost all } \alpha \in \Delta \right\}$$

is a submodule of $\prod_\Delta M_\alpha$. Furthermore, for each $\beta \in \Delta$, there is an R-linear injection $i_\beta : M_\beta \to \bigoplus_\Delta M_\alpha$ given by $i_\beta(x) = (x_\alpha)$, where $x_\alpha = 0$ if $\alpha \neq \beta$ and $x_\beta = x$. The map i_β is the βth *canonical injection* into $\bigoplus_\Delta M_\alpha$.

The R-module $\bigoplus_\Delta M_\alpha$ together with the family $\{i_\alpha\}_\Delta$ of canonical injections is an example of an *external direct sum* of $\{M_\alpha\}_\Delta$. Such a sum will be denoted by $(\bigoplus_\Delta M_\alpha, i_\alpha)$ and the mappings $\{i_\alpha\}_\Delta$ are the *structure maps* for $\bigoplus_\Delta M_\alpha$.

Earlier we defined the product of a family of mappings. We now need the concept of a sum of a family of mappings: If $f_\alpha : M_\alpha \to N$ is an R-linear mapping for each $\alpha \in \Delta$, where N is a fixed R-module, then $f : \bigoplus_\Delta M_\alpha \to N$ defined by $f((x_\alpha)) = \sum_\Delta f_\alpha(x)$ is a well-defined R-linear mapping called the *sum of the family* $\{f_\alpha\}_\Delta$.

Proposition 2.1.5. *If $\{M_\alpha\}_\Delta$ is a family of R-modules, then $(\bigoplus_\Delta M_\alpha, i_\alpha)$ has the property that for every R-module N and every family $\{f_\alpha : M_\alpha \to N\}_\Delta$ of R-linear mappings there is a unique R-linear mapping $f : \bigoplus_\Delta M_\alpha \to N$ such that for each $\alpha \in \Delta$ the diagram*

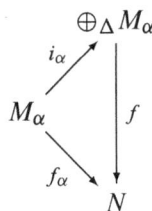

is commutative.

Proof. Let N be an R-module and suppose that, for each $\alpha \in \Delta$, $f_\alpha : M_\alpha \to N$ is an R-linear mapping. If $f : \bigoplus_\Delta M_\alpha \to N$ is the sum of the family of mappings $\{f_\alpha\}_\Delta$, then it is easy to check such that $f i_\alpha = f_\alpha$ for each $\alpha \in \Delta$. Now suppose that $g : \bigoplus_\Delta M_\alpha \to N$ is also such that $g i_\alpha = f_\alpha$ for each $\alpha \in \Delta$. If $(x_\alpha) \in \bigoplus_\Delta M_\alpha$, then $f((x_\alpha)) = \sum_\Delta f_\alpha(x_\alpha) = \sum_\Delta g i_\alpha(x_\alpha) = g \sum_\Delta i_\alpha(x_\alpha) = g((x_\alpha))$, so $f = g$. $\square$

Now for the formal definition of an external direct sum.

Definition 2.1.6. An R-module S together with a family of R-linear mappings $\{u_\alpha : M_\alpha \to S\}_\Delta$ is said to be an *external direct sum* of the family $\{M_\alpha\}_\Delta$ of R-modules if for every R-module N and for every family of R-linear mappings $\{f_\alpha : M_\alpha \to N\}_\Delta$, there is a unique R-linear mapping $f : S \to N$ such that for each $\alpha \in \Delta$ the diagram

$$
\begin{array}{ccc}
 & & S \\
 & \overset{u_\alpha}{\nearrow} & \big| \\
M_\alpha & & \big| f \\
 & \underset{f_\alpha}{\searrow} & \big\downarrow \\
 & & N
\end{array}
$$

is commutative. An external direct sum will be denoted by $(S, u_\alpha)_\Delta$. The mappings of the family $\{u_\alpha\}_\Delta$ are called the *structure maps* for the external direct sum. Because the unique map $f : S \to N$ always exists, an external direct sum $(S, u_\alpha)_\Delta$ is said to have the *universal mapping property*.

Note that the proof of Proposition 2.1.5 is just the proof of Proposition 2.1.1 with the arrows reversed and the necessary adjustments made. In a similar manner the proofs of the following propositions can be modeled after the proofs of "corresponding" propositions given for direct products.

Proposition 2.1.7. *If $(S, u_\alpha)_\Delta$ and $(S', u'_\alpha)_\Delta$ are direct sums of the family $\{M_\alpha\}_\Delta$ of R-modules, then there is unique isomorphism $\varphi : S \to S'$ such that $\varphi u_\alpha = u'_\alpha$ for each $\alpha \in \Delta$.*

Because of Proposition 2.1.7 we see that $(\bigoplus_\Delta M_\alpha, i_\alpha)$ is a model for an external direct sum of a family $\{M_\alpha\}_\Delta$ of R-modules. If $(S, u_\alpha)_\Delta$ is also an external direct sum of $\{M_\alpha\}_\Delta$, then there is a unique isomorphism $\varphi : S \to \bigoplus_\Delta M_\alpha$ such that $\varphi u_\alpha = i_\alpha$ for each $\alpha \in \Delta$. Since i_α is an injection, it follows that u_α is also an injection. Thus, if $(S, u_\alpha)_\Delta$ is an external direct sum of a family $\{M_\alpha\}_\Delta$ of R-modules, then each of its structure maps is an injection. An external direct sum $(S, u_\alpha)_\Delta$ also determines a family of projections $\{p_\alpha : S \to M_\alpha\}_\Delta$.

Proposition 2.1.8. *If $(S, u_\alpha)_\Delta$ is an external direct sum of a family $\{M_\alpha\}_\Delta$ of R-modules, then there is a family $\{p_\alpha : S \to M_\alpha\}_\Delta$ of surjective R-linear mappings such that $p_\alpha u_\alpha = \mathrm{id}_{M_\alpha}$ for each $\alpha \in \Delta$ and $p_\beta u_\alpha = 0$ when $\alpha \neq \beta$.*

The mappings of the family $\{p_\alpha\}_\Delta$ of surjective R-linear mappings given by Proposition 2.1.8 are called the *canonical projections (uniquely) determined* by $(S, u_\alpha)_\Delta$. From this we see that the canonical projections $\pi_\beta : \bigoplus_\Delta M_\alpha \to M_\beta$ are uniquely determined by $(\bigoplus_\Delta M_\alpha, i_\alpha)$.

Remark. Earlier we defined a product and a sum of a family $\{f_\alpha\}_\Delta$ of R-linear mappings. We also have the concepts of the direct product and direct sum of a family of mappings: Let $\{M_\alpha\}_\Delta$ and $\{N_\alpha\}_\Delta$ be families of R-modules and suppose that for each $\alpha \in \Delta$ there is an R-linear mapping $f_\alpha : M_\alpha \to N_\alpha$. Then

$$\prod_\Delta f_\alpha : \prod_\Delta M_\alpha \to \prod_\Delta N_\alpha \quad \text{given by} \quad \left(\prod_\Delta f_\alpha\right)((x_\alpha)) = (f_\alpha(x_\alpha))$$

is a well-defined R-linear mapping called the *direct product of the family of mappings* $\{f_\alpha\}_\Delta$. Likewise,

$$\bigoplus_\Delta f_\alpha : \bigoplus_\Delta M_\alpha \to \bigoplus_\Delta N_\alpha \quad \text{given by} \quad \left(\bigoplus_\Delta f_\alpha\right)((x_\alpha)) = (f_\alpha(x_\alpha))$$

is a well-defined R-linear mapping called the *direct sum of the family*.

Internal Direct Sums

Part (2) of Proposition 1.4.4 shows that the sum $\sum_\Delta M_\alpha$ of a family $\{M_\alpha\}_\Delta$ of submodules of M is a submodule of M. We now define the internal direct sum of a family $\{M_\alpha\}_\Delta$ of submodules of M. These sums will prove to be important in the study of the internal structure of modules. A direct product and a direct sum of a family $\{M_\alpha\}_\Delta$ of R-modules will now be denoted more simply by $\prod_\Delta M_\alpha$ and $\bigoplus_\Delta M_\alpha$, respectively, with the canonical maps associated with each structure understood.

Before defining the internal direct sum of a family $\{M_\alpha\}_\Delta$ of submodules of M, we consider an example that motivates the definition. Let N_1, N_2 and N_3 be R-modules and consider the R-module $M = N_1 \times N_2 \times N_3$. Then

$$M_1 = N_1 \times 0 \times 0 = \{(x_1, 0, 0) \mid x_1 \in N_1\},$$
$$M_2 = 0 \times N_2 \times 0 = \{(0, x_2, 0) \mid x_2 \in N_2\} \quad \text{and}$$
$$M_3 = 0 \times 0 \times M_3 = \{(0, 0, x_3) \mid x_3 \in N_3\}$$

are submodules of M such that

$$M_1 \cap (M_2 + M_3) = 0,$$
$$M_2 \cap (M_1 + M_3) = 0 \quad \text{and}$$
$$M_3 \cap (M_1 + M_2) = 0.$$

It follows that $(x_1, x_2, x_3) = (x_1, 0, 0) + (0, x_2, 0) + (0, 0, x_3)$ and each summand in this expression for (x_1, x_2, x_3) is unique. To indicate that each element of M can be written uniquely as a sum of elements from M_1, M_2 and M_3 we write $M = M_1 \oplus M_2 \oplus M_2$ and we say that M is the *internal direct sum* of M_1, M_2 and M_3. Since $M_i \cap M_j = 0$, whenever $i \neq j$, we also write $M_i \oplus M_j$ and say that this sum is direct.

Definition 2.1.9. Suppose that $\{M_\alpha\}_\Delta$ is a family of submodules of an R-module M such that

$$M_\beta \cap \sum_{\alpha \neq \beta} M_\alpha = 0 \quad \text{for each } \beta \in \Delta.$$

Then the sum $\sum_\Delta M_\alpha$ is said to be the *internal direct sum* of the family $\{M_\alpha\}_\Delta$. The notation $\bigoplus_\Delta M_\alpha$ will indicate that the sum $\sum_\Delta M_\alpha$ is direct. If $M = \bigoplus_\Delta M_\alpha$, then M is said to be the internal direct sum of the family $\{M_\alpha\}_\Delta$ and $\bigoplus_\Delta M_\alpha$ is a *direct sum decomposition* of M. If N is a submodule of M and if there is a submodule N' of M such that $N \cap N' = 0$ and $M = N + N'$, then $M = N \oplus N'$ and N is referred to as a *direct summand* of M.

Remark. The external direct sum $\bigoplus_\Delta M_\alpha$ of a family $\{M_\alpha\}_\Delta$ of R-modules was given in Definition 2.1.6 while the internal direct sum $\bigoplus_\Delta M_\alpha$ of a family of sub-modules of M was given in the preceding definition. We use the same notation for both and refer to each simply as a direct sum, leaving it to the reader to determine if the direct sum is internal or external from the context of the discussion.

Examples

2. **Matrices.** Consider the R-module $\mathbb{M}_n(R)$ of $n \times n$ matrices over R. Then $\mathbb{M}_n(R) = \bigoplus_{i=1}^n r_i(R)$, and $\mathbb{M}_n(R) = \bigoplus_{i=1}^n c_i(R)$. For example, if $(a_{ij}) \in \mathbb{M}_2(R)$, then

$$\begin{pmatrix} a_{11} & a_{12} \\ a_{21} & a_{22} \end{pmatrix} = \begin{pmatrix} a_{11} & a_{12} \\ 0 & 0 \end{pmatrix} + \begin{pmatrix} 0 & 0 \\ a_{21} & a_{22} \end{pmatrix} \quad \text{and}$$

$$\begin{pmatrix} a_{11} & a_{12} \\ a_{21} & a_{22} \end{pmatrix} = \begin{pmatrix} a_{11} & 0 \\ a_{21} & 0 \end{pmatrix} + \begin{pmatrix} 0 & a_{12} \\ 0 & a_{22} \end{pmatrix}.$$

Clearly $\mathbb{M}_2(R) = r_1(R) \oplus r_2(R) = c_1(R) \oplus c_2(R)$.

3. **$\mathbb{Z}$-modules.** The $\mathbb{Z}$-module $\mathbb{Z}_6$ is such that $\mathbb{Z}_6 \cong \mathbb{Z}_2 \oplus \mathbb{Z}_3$. In general, if s and t are relatively prime integral divisors of n such that $st = n$, then $\mathbb{Z}_n \cong \mathbb{Z}_s \oplus \mathbb{Z}_t$. Moreover, if $n = p_1^{k_1} p_2^{k_2} \cdots p_t^{k_t}$ is a factorization of n as a product of distinct prime numbers, then $\mathbb{Z}_n$ and $\mathbb{Z}_{p_1^{k_1}} \oplus \mathbb{Z}_{p_2^{k_2}} \oplus \cdots \oplus \mathbb{Z}_{p_t^{k_t}}$ are isomorphic as $\mathbb{Z}$-modules as well as rings. These observations are the result of the Chinese Remainder Theorem given in Exercise 4.

4. **Polynomials.** If $R[X]$ is the R-module of polynomials in X with their coefficients in R and $R_k = \{X^k a_k \mid a_k \in R\}$, then R_k is a submodule of $R[X]$ for $k = 0, 1, 2, \ldots$, where $X^0 = 1$. It follows easily that $R[X] = \bigoplus_{k=0}^\infty R_k$. The submodules R_k are called the *homogeneous submodules* of $R[X]$.

5. **Finite Direct Products.** If $\{M_i\}_{i=1}^n$ is a finite set of R-modules, then for each k, $1 \le k \le n$,

$$N_k = \left\{ (x_i) \in \prod_{i=1}^n M_i \mid x_i = 0 \text{ if } i \ne k \right\}$$

is a submodule of $\prod_{i=1}^n M_i$ and

$$\prod_{i=1}^n M_i = \bigoplus_{i=1}^n N_i.$$

6. **Direct Sums and Factor Modules.** Let M_1 and M_2 be submodules of an R-module M such that $M = M_1 \oplus M_2$. Then $M/M_1 \cong M_2$ and $M/M_2 \cong M_1$. In general, if $\{M_\alpha\}_\Delta$ is a family of submodules of M such that $M = \bigoplus_\Delta M_\alpha$, then $M/M_\beta \cong \bigoplus_{\alpha \ne \beta} M_\alpha$ and $M/\bigoplus_{\alpha \ne \beta} M_\alpha \cong M_\beta$.

Proposition 2.1.10. *Let $\{M_\alpha\}_\Delta$ be a family of submodules of an R-module M such that $M = \sum_\Delta M_\alpha$. Then the sum $\sum_\Delta M_\alpha$ is direct if and only if each $x \in M$ can be written in one and only one way as a sum $x = \sum_\Delta x_\alpha$, where $x_\alpha \in M_\alpha$ for all $\alpha \in \Delta$.*

Proof. Let the sum $\sum_\Delta M_\alpha$ be direct and suppose that $x \in M$ can be written as $x = \sum_\Delta x_\alpha$ and as $x = \sum_\Delta y_\alpha$. Then $\sum_\Delta x_\alpha = \sum_\Delta y_\alpha$, so for each $\beta \in \Delta$ we have

$$x_\beta - y_\beta = \sum_{\alpha \ne \beta} (y_\alpha - x_\alpha) \in M_\beta \cap \sum_{\alpha \ne \beta} M_\alpha = 0.$$

Thus, $x_\beta = y_\beta$ for each $\beta \in \Delta$, so the expression $x = \sum_\Delta x_\alpha$ is unique.

Conversely, suppose that each element x of M can be written in one and only one way as a finite sum of elements from the M_α. If $x \in M_\beta \cap \sum_{\alpha \ne \beta} M_\alpha$, then $x = x_\beta$ for some $x_\beta \in M_\beta$ and $x = \sum_{\alpha \ne \beta} x_\alpha$, where $x_\alpha \in M_\alpha$ for each $\alpha \in \Delta$, $\alpha \ne \beta$. Thus, if $\sum_\Delta y_\alpha$ is such that $y_\beta = -x_\beta$ and $y_\alpha = x_\alpha$ when $\alpha \ne \beta$, then $\sum_\Delta y_\alpha = 0$. But 0 can be written as $0 = \sum_\Delta 0_\alpha$, where $0 = 0_\alpha \in M_\alpha$ for each $\alpha \in \Delta$. Hence, $\sum_\Delta y_\alpha = \sum_\Delta 0_\alpha$ gives $x_\alpha = 0$ for each $\alpha \in \Delta$, so $x = 0$. Consequently, the sum $\sum_\Delta M_\alpha$ is direct. $\qquad\square$

Remark. It is also the case that if $\{M_\alpha\}_\Delta$ is a family of submodules of an R-module M, then the sum $\sum_\Delta M_\alpha$ is direct if and only if for each finite sum $\sum_\Delta x_\alpha$ such that $\sum_\Delta x_\alpha = 0$ we have $x_\alpha = 0$ for each $\alpha \in \Delta$. Note that if the sum $\sum_\Delta M_\alpha$ is direct and $\sum_\Delta x_\alpha = 0$, then $\sum_\Delta x_\alpha = \sum_\Delta 0_\alpha$ and the proposition above gives $x_\alpha = 0$ for each $\alpha \in \Delta$. Conversely, suppose that $\sum_\Delta x_\alpha = 0$ implies that $x_\alpha = 0$ for each $\alpha \in \Delta$. If $x \in M_\alpha \cap \sum_{\beta \ne \alpha} M_\beta$, then $x = x_\alpha$ and $x = \sum_{\beta \ne \alpha} x_\beta$, so $x_\alpha + \sum_{\beta \ne \alpha} (-x_\beta) = 0$ gives $x = x_\alpha = 0$, so the sum $\sum_\Delta M_\alpha$ is direct.

There is a fundamental connection between external direct sums and internal direct sums. Since internal direct sums and external direct sums are both involved in the following proposition, for the purposes of this proposition and its proof, $\oplus^i$ and $\oplus^e$ will denote an internal and an external direct sum, respectively.

Proposition 2.1.11. *The following hold for any R-module M.*

(1) *If $\{M_\alpha\}_\Delta$ is a family of submodules of M such that $M = \oplus^i_\Delta M_\alpha$, then $M \cong \oplus^e_\Delta M_\alpha$.*

(2) *If there is a family $\{N_\alpha\}_\Delta$ of R-modules such that $M \cong \oplus^e_\Delta N_\alpha$, then there is a family $\{M_\alpha\}_\Delta$ of submodules of M such that $M = \oplus^i_\Delta M_\alpha$ and $N_\alpha \cong M_\alpha$ for each $\alpha \in \Delta$.*

Proof. (1) If $M = \oplus^i_\Delta M_\alpha$, then each element of M can be written as a finite sum $\sum_\Delta x_\alpha$, where the $x_\alpha \in M_\alpha$ are unique. It follows that $\varphi : \oplus^i_\Delta M_\alpha \to \oplus^e_\Delta M_\alpha$ defined by $\varphi(\sum_\Delta x_\alpha) = (x_\alpha)$ is an isomorphism.

(2) If $\varphi : \oplus^e_\Delta N_\alpha \to M$ is an isomorphism, let $M_\alpha = \varphi i_\alpha(N_\alpha)$ for each $\alpha \in \Delta$, where $i_\alpha : N_\alpha \to \oplus^e_\Delta N_\alpha$ is the canonical injection. It is easy to check that $M = \oplus^i_\Delta M_\alpha$. The fact that $N_\alpha \cong M_\alpha$ for each $\alpha \in \Delta$ is clear. $\square$

The following isomorphisms of abelian groups will be useful.

Proposition 2.1.12. *Let $\{M_\alpha\}_\Delta$ be a family of R-modules. Then*

(1) $$\mathrm{Hom}_R\left(\bigoplus_\Delta M_\alpha, N\right) \cong \prod_\Delta \mathrm{Hom}_R(M_\alpha, N) \quad and$$

(2) $$\mathrm{Hom}_R\left(N, \prod_\Delta M_\alpha\right) \cong \prod_\Delta \mathrm{Hom}_R(N, M_\alpha)$$

for any R-module N.

Proof. We prove (1) and omit the proof of (2) since it is similar. Let

$$\varphi : \mathrm{Hom}_R\left(\bigoplus_\Delta M_\alpha, N\right) \to \prod_\Delta \mathrm{Hom}_R(M_\alpha, N)$$

be such that $\varphi(f) = (f i_\alpha)$, where $i_\alpha : M_\alpha \to \oplus_\Delta M_\alpha$ is the canonical injection for each $\alpha \in \Delta$. Then φ is a group homomorphism and if $f \in \mathrm{Ker}\,\varphi$, then $f i_\alpha = 0$ for each $\alpha \in \Delta$. So if $(x_\alpha) \in \oplus_\Delta M_\alpha$, then $(x_\alpha) = \sum_\Delta i_\alpha(x_\alpha)$ gives $f((x_\alpha)) = f(\sum_\Delta i_\alpha(x_\alpha)) = \sum_\Delta f i_\alpha(x_\alpha) = 0$. Thus, $f = 0$, so φ is an injection. If $(g_\alpha) \in \prod_\Delta \mathrm{Hom}_R(M_\alpha, N)$, let $\sum_\Delta g_\alpha \in \mathrm{Hom}_R(\oplus_\Delta M_\alpha, N)$ be such that $(\sum_\Delta g_\alpha)((x_\alpha)) = \sum_\Delta g_\alpha(x_\alpha)$ for each $(x_\alpha) \in \oplus_\Delta M_\alpha$. Then, $\varphi(\sum_\Delta g_\alpha) = ((\sum_\Delta g_\alpha)i_\alpha) \in \prod_\Delta \mathrm{Hom}_R(M_\alpha, N)$. But $(\sum_\Delta g_\alpha)i_\alpha = g_\alpha$ for each $\alpha \in \Delta$, so $\varphi(\sum_\Delta g_\alpha) = (g_\alpha)$. Hence, φ is also a surjection. $\square$

Problem Set 2.1

1. Let $\{M_\alpha\}_\Delta$ be a family of R-modules.

   (a) Verify that $\bigoplus_\Delta M_\alpha$ is a submodule of $\prod_\Delta M_\alpha$.

   (b) Show that $\prod_\Delta M_\alpha = \bigoplus_\Delta M_\alpha$ if and only if Δ is a finite set.

2. Let $\{R_\alpha\}_\Delta$ be a family of rings. Consider the ring direct product $\prod_\Delta R_\alpha$ and let A_α be a right ideal of R_α for each $\alpha \in \Delta$.

   (a) Show that $\prod_\Delta A_\alpha$ is a right ideal of $\prod_\Delta R_\alpha$.

   (b) Is $\bigoplus_\Delta A_\alpha$ a right ideal of $\prod_\Delta R_\alpha$?

   (c) Is $\bigoplus_\Delta R_\alpha$ an ideal of $\prod_\Delta R_\alpha$?

3. (a) If $R_1 \times R_2 \times \cdots \times R_n$ is a ring direct product, show that every right ideal of the ring $R_1 \times R_2 \times \cdots \times R_n$ is of the form $A_1 \times A_2 \times \cdots \times A_n$, where each A_i is a right ideal of R_i. [Hint: Let $n = 2$ and suppose that A is a right ideal of $R_1 \times R_2$. If $A_1 = \pi_1(A)$, $A_2 = \pi_2(A)$ and $a \in A$, then $\pi_1(a) = a_1$ and $\pi_2(a) = a_2$ for some $a_1 \in A_1$ and $a_2 \in A_2$. Hence, $a = (i_1\pi_1 + i_2\pi_2)(a) = (a_1, 0) + (0, a_2) = (a_1, a_2)$ and so $A \subseteq A_1 \times A_2$. Conversely, if $a_1 \in A_1$ and $a_2 \in A_2$, then there are $a, b \in A$ such that $\pi_1(a) = a_1$ and $\pi_1(b) = a_2$. Hence, $a(1, 0) + b(0, 1) = (i_1\pi_1 + i_2\pi_2)(a(1, 0) + b(0, 1)) = (a_1, a_2)$, so $A_1 \times A_2 \subseteq A$. Show this procedure holds for an arbitrary n.]

   (b) Show that (a) is false for modules. That is, show that if $\{M_i\}_{i=1}^n$ is a family of R-modules and N is a submodule of the R-module $M_1 \times M_2 \times \cdots \times M_n$, then there may not be submodules N_i of M_i for $i = 1, 2, \ldots, n$ such that $N = N_1 \times N_2 \times \cdots \times N_n$. [Hint: Consider the diagonal $\{(x, x) \mid x \in M\}$ of $M \times M$.]

4. **Chinese Remainder Theorem.** Two ideals I_1 and I_2 of a ring R are said to be *comaximal* if $R = I_1 + I_2$.

   (a) If I_1 and I_2 are comaximal ideals of R, show that $I_1 I_2 = I_1 \cap I_2$ and that

   $$R/I_1 I_2 = R/(I_1 \cap I_2) \cong R/I_1 \oplus R/I_2.$$

   [Hint: Consider the map $f : R \to R/I_1 \oplus R/I_2$ given by $a \mapsto (a+I_1, a+I_2)$.]

   (b) Let $\{I_i\}_{i=1}^n$ be a family of pairwise comaximal ideals of R. Prove that

   $$I_1 I_2 \cdots I_n = I_1 \cap I_2 \cap \cdots \cap I_n \quad \text{and that}$$

   $$R/I_1 I_2 \cdots I_n = R/\bigcap_{i=1}^n I_i \cong \bigoplus_{i=1}^n R/I_i.$$

   (c) Verify Example 3.

5. Consider the additive abelian group $G = \prod_{\mathbb{N}} \mathbb{Z}_i$, where $\mathbb{Z}_i = \mathbb{Z}$ for each $i \in \mathbb{N}$ and addition is defined componentwise. Suppose also that $R = \text{End}_{\mathbb{Z}}(G)$.

(a) Define $f : G \to G$ by $f((a_1, a_2, a_3, \ldots)) = (a_2, a_3, a_4, \ldots)$ and $g : G \to G$ by $g((a_1, a_2, a_3, \ldots)) = (0, a_1, a_2, \ldots)$. Show that f and g are in R.

(b) Show that $fg = \text{id}_G$, but that $gf \neq \text{id}_G$. Conclude that g has a left inverse f, but that f is not a right inverse for g.

(c) Find an element $h \in R$, $h \neq 0$, such that $fh = 0$. Conclude that f is a left zero divisor in R.

(d) Prove that it is not possible to find a nonzero $h \in R$ such that $hf = 0$. In view of (c), conclude that f is a left zero divisor but not a right zero divisor.

6. Let $\{M_\alpha\}_\Delta$ be a family of R-modules and suppose that N is a submodule of $\prod_\Delta M_\alpha$. Prove that $\text{ann}_r(N) = \bigcap_\Delta \text{ann}_r(\pi_\alpha(N))$, where π_α is the canonical projection from $\prod_\Delta M_\alpha$ to M_α for each $\alpha \in \Delta$.

7. If $\{M_\alpha\}_\Delta$ and $\{N_\alpha\}_\Delta$ are families of R-modules and $f_\alpha : M_\alpha \to N_\alpha$ is an R-linear mapping for each $\alpha \in \Delta$, prove that the mapping $\prod_\Delta f_\alpha : \prod_\Delta M_\alpha \to \prod_\Delta N_\alpha$ defined by $(\prod_\Delta f_\alpha)((x_\alpha)) = (f_\alpha(x_\alpha))$ for each $(x_\alpha) \in \prod_\Delta M_\alpha$ is an R-linear mapping. Show also that $\prod_\Delta f_\alpha$ is a monomorphism (an epimorphism, an isomorphism) if and only if each f_α is a monomorphism (an epimorphism, an isomorphism).

8. Verify Examples 4 and 5.

9. Prove Proposition 2.1.7.

10. Prove Proposition 2.1.8.

11. Let $(P, p_\alpha)_\Delta$ be a direct product of a family $\{M_\alpha\}_\Delta$ of R-modules and suppose that $\{u_\alpha : M_\alpha \to P\}_\Delta$ is the family of associated injections. Prove that there is a submodule S of P such that $(S, u_\alpha)_\Delta$ is a direct sum of $\{M_\alpha\}_\Delta$.

12. Verify Example 6.

13. (a) If $\{M_\alpha\}_\Delta$ of a family of R-modules and $\Gamma \subseteq \Delta$, prove that there is a monomorphism $f : \prod_\Gamma M_\alpha \to \prod_\Delta M_\alpha$.

(b) Let $\{M_\alpha\}_\Delta$ be a family of R-modules and suppose that $\Delta = \Gamma \cup \Lambda$. If $\Gamma \cap \Lambda = \varnothing$, verify that $\prod_\Delta M_\alpha \cong \prod_\Gamma M_\alpha \oplus \prod_\Lambda M_\alpha$.

14. Let $\{M_\alpha\}_\Delta$ be a family of R-modules and suppose that for each $\alpha \in \Delta$, N_α is a submodule of M_α. Prove that $(\bigoplus_\Delta M_\alpha)/(\bigoplus_\Delta N_\alpha) \cong \bigoplus_\Delta M_\alpha/N_\alpha$.

15. (a) Let M be a nonzero R-module and suppose that $\{M_i\}_\mathbb{N}$ is the family of R-modules such that $M_i = M$ for each $i \geq 1$. Show that $\bigoplus_{i \geq 1} M_i \cong \bigoplus_{i \geq n} M_i$ for each $n \geq 1$.

(b) Find a nonzero R-module M such that $M \cong M \oplus M$. [Hint: Consider (a).]

16. Prove the second isomorphism of Proposition 2.1.12.

17. Let $\{M_\alpha\}_\Delta$ be a family of R-modules and suppose that $(B, u_\alpha, p_\alpha)_\Delta$ is such that $(B, u_\alpha)_\Delta$ is a direct sum and $(B, p_\alpha)_\Delta$ is a direct product of $\{M_\alpha\}_\Delta$. Then $(B, u_\alpha, p_\alpha)_\Delta$ is said to be a *biproduct* of $\{M_\alpha\}_\Delta$. Prove that if $\{M_\alpha\}_\Delta$ is a family of nonzero R-modules, then $(B, u_\alpha, p_\alpha)_\Delta$ is a biproduct of $\{M_\alpha\}_\Delta$ if and only if Δ is a finite set.

2.2   Free Modules

Recall that if N is a submodule of an R-module M, then a subset X of N such that $N = \sum_X xR$ is said to be a *set of generators* for N and we say that N *is generated by* X. If X is a finite set, then N is said to be *finitely generated*. Recall also that if N is generated by X and if $\{N_\alpha\}_\Delta$ is the family of submodules of M that contain X, then $N = \bigcap_\Delta N_\alpha = \sum_X xR$.

Every submodule N of an R-module M has at least one set of generators, namely the set N. Moreover, a submodule can have more than one set of generators. For example, the submodule $N = \{[0], [2], [4], [6]\}$ of the $\mathbb{Z}$-module $\mathbb{Z}_8$ is generated by $\{[2]\}$, by $\{[6]\}$ and by $\{[2], [4]\}$. Hence, a set of generators of a submodule need not be unique and it is not necessary for generating sets for a submodule to have the same cardinality. If M is an R-module and X is a set of generators of M, then we say that X is a *minimal set of generators* for M if M is not generated by a proper subset of X. The sets $\{[2]\}$ and $\{[6]\}$ are both minimal sets of generators of $\{[0], [2], [4], [6]\}$.

Notation. If $\{M_\alpha\}_\Delta$ is a family of R-modules such that $M_\alpha = M$ for each $\alpha \in \Delta$, then $\prod_\Delta M_\alpha$ and $\bigoplus_\Delta M_\alpha$ will be denoted by M^Δ and $M^{(\Delta)}$, respectively. Recall that $M^{(n)}$ represents the direct product $M \times M \times \cdots \times M$ of n factors of M.

Definition 2.2.1. If M is an R-module, then a set $\{x_\alpha\}_\Delta$ of elements of M is said to be *linearly independent* if the only way that a finite sum $\sum_\Delta x_\alpha a_\alpha$ of elements of $\{x_\alpha\}_\Delta$ can be such that $\sum_\Delta x_\alpha a_\alpha = 0$ is for $a_\alpha = 0$ for all $\alpha \in \Delta$. If there is a finite sum $\sum_\Delta x_\alpha a_\alpha$ of elements of $\{x_\alpha\}_\Delta$ such that $\sum_\Delta x_\alpha a_\alpha = 0$ with at least one $a_\alpha \neq 0$, then the set $\{x_\alpha\}_\Delta$ is *linearly dependent*. If an R-module F has a linearly independent set $\{x_\alpha\}_\Delta$ of generators, then $\{x_\alpha\}_\Delta$ is said to be a *basis* for F and F is said to be a *free R-module* with basis $\{x_\alpha\}_\Delta$. A set X of linearly independent elements of an R-module M is said to be a *maximal linearly independent set* of elements of M if no set of linearly independent elements of M properly contains X.

Examples

1. Note that the empty set $\varnothing$ is a linearly independent set in every R-module M. To assume otherwise would mean that there are elements $x_1, x_2, \ldots, x_n \in \varnothing$

and $a_1, a_2, \ldots, a_n \in R$ such that

$$x_1 a_1 + x_2 a_2 + \cdots + x_n a_n = 0$$

with at least one $a_i \neq 0$, a clear absurdity. We have seen earlier that the zero module is generated by $\emptyset$, so we will consider the zero module to be a free R-module with basis $\emptyset$.

2. If the ring R is viewed as an R-module, then R is a free R-module with basis $\{1\}$.

3. The R-module $R^{(n)}$ is a free R-module with basis $\{e_i\}_{i=1}^n$, where

$$e_1 = (1, 0, 0, \ldots, 0),$$
$$e_2 = (0, 1, 0, \ldots, 0),$$
$$e_3 = (0, 0, 1, \ldots, 0), \quad \text{and}$$
$$\vdots$$
$$e_n = (0, 0, 0, \ldots, 1).$$

In general, if Δ is a nonempty set, then $R^{(\Delta)}$ is a free R-module with basis $\{e_\alpha\}_\Delta$, where $e_\alpha = (a_\beta)$ with $a_\beta = 1$ if $\beta = \alpha$ and $a_\beta = 0$ when $\beta \neq \alpha$. The basis $\{e_\alpha\}_\Delta$ will be referred to as the *canonical basis* for $R^{(\Delta)}$. Note that $R^{(\Delta)}$ is also a free left R-module with the same basis $\{e_\alpha\}_\Delta$.

4. The matrix ring $\mathbb{M}_n(R)$ is a free R-module. One basis of $\mathbb{M}_n(R)$ is the set of matrix units $\{E_{ij}\}_{i,j=1}^n$ with n^2 elements. For example, if $(a_{ij}) \in \mathbb{M}_2(R)$, then $(a_{ij}) = E_{11}a_{11} + E_{12}a_{12} + E_{21}a_{21} + E_{22}a_{22}$.

To avoid discussing trivialities, we assume that if F is a free R-module, then $F \neq 0$.

Proposition 2.2.2. *The following are equivalent for a subset $\{x_\alpha\}_\Delta$ of an R-module M.*

(1) *$\{x_\alpha\}_\Delta$ is a basis for M.*

(2) *$\{x_\alpha\}_\Delta$ is (a) a maximal linearly independent subset of M and (b) a minimal set of generators of M.*

Proof. (1) $\Rightarrow$ (2). Suppose that $\{x_\alpha\}_\Delta$ is a basis for M and that $\{x_\alpha\}_\Delta$ is not a maximal set of linearly independent elements of M. Let $\{y_\beta\}_\Gamma$ be a set of linearly independent elements of M such that $\{x_\alpha\}_\Delta \subsetneqq \{y_\beta\}_\Gamma$. If $y \in \{y_\beta\}_\Gamma - \{x_\alpha\}_\Delta$, then since $\{x_\alpha\}_\Delta$ generates M we see that $y = \sum_\Delta x_\alpha a_\alpha$, where $a_\alpha = 0$ for almost all $\alpha \in \Delta$. But then $y + \sum_\Delta x_\alpha(-a_\alpha) = 0$ is a linear combination of elements in $\{y_\beta\}_\Gamma$ and not all of the a_α can be zero, since $y \neq 0$. Thus, $\{y_\beta\}_\Gamma$ is linearly dependent, a contradiction, so we have (a). Suppose next that $\{x_\alpha\}_\Delta$ is not a minimal set of generators of M. Then there is a set $\{y_\beta\}_\Gamma \subsetneq \{x_\alpha\}_\Delta$ such that $\{y_\beta\}_\Gamma$ generates M. If $x \in \{x_\alpha\}_\Delta - \{y_\beta\}_\Gamma$, then we can write $x = \sum_\Gamma y_\alpha a_\alpha$. But then $x + \sum_\Gamma y_\alpha(-a_\alpha) = 0$ is a linear combination of elements in $\{x_\alpha\}_\Delta$ and not all of the a_α can be zero, contradicting the fact that $\{x_\alpha\}_\Delta$ is linearly independent. Hence, we have (b).

(2) $\Rightarrow$ (1). Let $\{x_\alpha\}_\Delta$ be a maximal linearly independent subset of M which is at the same time a minimal set of generators of M. Then it is immediate that $\{x_\alpha\}_\Delta$ is a basis for M. $\square$

Proposition 2.2.3. *The following are equivalent for a subset $\{x_\alpha\}_\Delta$ of an R-module M.*

(1) *$\{x_\alpha\}_\Delta$ is a basis for M.*

(2) *Each element $x \in M$ can be written as $x = \sum_\Delta x_\alpha a_\alpha$, where each a_α is unique and $a_\alpha = 0$ for almost all $\alpha \in \Delta$.*

(3) *$M = \bigoplus_\Delta x_\alpha R$.*

Proof. (1) $\Rightarrow$ (2). Since $\{x_\alpha\}_\Delta$ is a set of generators of M, it is certainly the case that each $x \in M$ can be expressed as $x = \sum_\Delta x_\alpha a_\alpha$, where $a_\alpha = 0$ for almost all $\alpha \in \Delta$. If $x = \sum_\Delta x_\alpha b_\alpha$ is another such expression for x, then $\sum_\Delta x_\alpha a_\alpha = \sum_\Delta x_\alpha b_\alpha$ implies that $\sum_\Delta x_\alpha(a_\alpha - b_\alpha) = 0$. Consequently, $a_\alpha - b_\alpha = 0$ for each $\alpha \in \Delta$, since the set $\{x_\alpha\}_\Delta$ is linearly independent. Hence, $a_\alpha = b_\alpha$ for each $\alpha \in \Delta$.

(2) $\Rightarrow$ (1). Since each element $x \in M$ can be written as $x = \sum_\Delta x_\alpha a_\alpha$, $\{x_\alpha\}_\Delta$ is a set of generators for M. Now suppose that $\sum_\Delta x_\alpha a_\alpha = 0$. Since 0 can be written as $\sum_\Delta x_\alpha 0_\alpha$, where $0_\alpha = 0$ for each $\alpha \in \Delta$, the uniqueness of coefficients shows that $a_\alpha = 0$ for each $\alpha \in \Delta$. Hence, $\{x_\alpha\}_\Delta$ is a basis for M.

The proof of the equivalence of (1) and (3) is equally straightforward and so is omitted. $\square$

Corollary 2.2.4. *An R-module F is free if and only if there is a set Δ such that $F \cong R^{(\Delta)}$.*

Proof. Suppose that F is a free R-module with basis $\{x_\alpha\}_\Delta$. Then $F = \bigoplus_\Delta x_\alpha R$, so let $f : F \to R^{(\Delta)}$ be such that $f(\sum_\Delta x_\alpha a_\alpha) = (a_\alpha)$. We claim that f is an isomorphism. If $\sum_\Delta x_\alpha a_\alpha, \sum_\Delta x_\alpha b_\alpha \in \bigoplus_\Delta x_\alpha R$ and $a \in R$, then

$$f\left(\sum_\Delta x_\alpha a_\alpha + \sum_\Delta x_\alpha b_\alpha\right) = f\left(\sum_\Delta x_\alpha(a_\alpha + b_\alpha)\right)$$

$$= (a_\alpha + b_\alpha) = (a_\alpha) + (b_\alpha)$$

$$= f\left(\sum_\Delta x_\alpha a_\alpha\right) + f\left(\sum_\Delta x_\alpha b_\alpha\right) \quad \text{and}$$

$$f\left(\left(\sum_\Delta x_\alpha a_\alpha\right)a\right) = f\left(\sum_\Delta x_\alpha(a_\alpha a)\right)$$

$$= (a_\alpha a) = (a_\alpha)a$$

$$= \left(f\left(\sum_\Delta x_\alpha a_\alpha\right)\right)a,$$

so f is R-linear. If $\sum_\Delta x_\alpha a_\alpha \in \text{Ker } f$, then $(a_\alpha) = 0$ indicates that $a_\alpha = 0$ for each $\alpha \in \Delta$. Thus, $\sum_\Delta x_\alpha a_\alpha = 0$, so (2) of Proposition 1.5.6 shows that f is an injection. If $(a_\alpha) \in R^{(\Delta)}$, then $\sum_\Delta x_\alpha a_\alpha \in F$ and $f(\sum_\Delta x_\alpha a_\alpha) = (a_\alpha)$, so f is also a surjection.

Conversely, suppose that there is a set Δ such that $F \cong R^{(\Delta)}$. First, note that $R^{(\Delta)}$ is a free R-module with basis $\{e_\alpha\}_\Delta$, where $e_\alpha = (a_\beta)$ with $a_\beta = 1$ if $\beta = \alpha$ and $a_\beta = 0$ when $\beta \neq \alpha$. If $f : R^{(\Delta)} \to F$ is an isomorphism and $f(e_\alpha) = x_\alpha$, then it follows that $\{x_\alpha\}_\Delta$ is a basis for F.   □

The basis elements of a free R-module F uniquely determine the R-linear mappings with domain F.

Proposition 2.2.5. *Let F be a free R-module with basis $\{x_\alpha\}_\Delta$.*

(1) *If $f, g : F \to M$ are R-linear mappings such that $f(x_\alpha) = g(x_\alpha)$ for all $\alpha \in \Delta$, then $f = g$.*

(2) *If $f : \{x_\alpha\}_\Delta \to M$ is a function, then there exists a unique R-linear mapping $g : F \to M$ such that $g(x_\alpha) = f(x_\alpha)$ for all $\alpha \in \Delta$.*

Proof. (1) If $x \in F$ and $x = \sum_\Delta x_\alpha a_\alpha$, suppose that $f, g : F \to M$ are R-linear mappings such that $f(x_\alpha) = g(x_\alpha)$ for all $\alpha \in \Delta$. Then $f(x) = f(\sum_\Delta x_\alpha a_\alpha) = \sum_\Delta f(x_\alpha)a_\alpha = \sum_\Delta g(x_\alpha)a_\alpha = g(\sum_\Delta x_\alpha a_\alpha) = g(x)$.

(2) If $g : F \to M$ is defined by $g(x) = \sum_\Delta f(x_\alpha)a_\alpha$ for each $x = \sum_\Delta x_\alpha a_\alpha \in F$, then g is the desired R-linear mapping. The uniqueness of g follows from (1).   □

The mapping $g : F \to M$ constructed in the proof of (2) of Proposition 2.2.5 is said to be obtained by *extending f linearly to F*.

The following proposition will be used frequently throughout the text.

Proposition 2.2.6. *Every R-module M is the homomorphic image of a free R-module. Furthermore, if M is finitely generated, then the free module can be chosen to be finitely generated.*

Proof. Let M be an R-module and suppose that $\{x_\alpha\}_\Delta$ is a set of generators of M. If $f : R^{(\Delta)} \to M$ is defined by $f((a_\alpha)) = \sum_\Delta x_\alpha a_\alpha$, then

$$f((a_\alpha) + (b_\alpha)) = f((a_\alpha + b_\alpha))$$
$$= \sum_\Delta x_\alpha (a_\alpha + b_\alpha) = \sum_\Delta x_\alpha a_\alpha + \sum_\Delta x_\alpha b_\alpha$$
$$= f((a_\alpha)) + f((b_\alpha)) \quad \text{and}$$
$$f((a_\alpha)a) = f((a_\alpha a))$$
$$= \sum_\Delta x_\alpha (a_\alpha a) = \left(\sum_\Delta x_\alpha a_\alpha \right) a$$
$$= f((a_\alpha))a$$

for all $(a_\alpha), (b_\alpha) \in R^{(\Delta)}$ and $a \in R$. Hence, f is R-linear. If $x \in M$, then x can be written as $x = \sum_\Delta x_\alpha a_\alpha$, where $a_\alpha = 0$ for almost all $\alpha \in \Delta$. It follows that $(a_\alpha) \in R^{(\Delta)}$, so $f((a_\alpha)) = \sum_\Delta x_\alpha a_\alpha$ and f is thus an epimorphism.

If M is finitely generated, then Δ is a finite set. If $\Delta = \{1, 2, \ldots, n\}$, then M is a homomorphic image of the finitely generated free R-module $R^{(n)}$. $\qquad\square$

Remark. Because of Corollary 2.2.4 we see that any set X determines a free R-module $R^{(X)}$. If $X = \varnothing$, then we have seen that $R^{(X)} = 0$ is a free R-module with basis $\varnothing$. If $X \neq \varnothing$, then X can also be considered to be a basis for $R^{(X)}$. Indeed, let $\delta : X \times X \to R$ be such that $\delta(x, y) = \delta_{xy}$, where $\delta_{xx} = 1$ and $\delta_{xy} = 0$ when $x \neq y$. The function δ is called the *Kronecker delta function* defined on $X \times X$. If $e_x = (\delta_{xy})$ for each $x \in X$, then $\{e_x\}_X$ is a basis for $R^{(X)}$. The function $f : X \to \{e_x\}_{x \in X}$ such that $f(x) = e_x$ is a bijection, so if we identify each $x \in X$ with e_x, then every element of $R^{(X)}$ can be expressed uniquely as $\sum_X x a_x$. Under this identification, X is a basis for $R^{(X)}$ and $R^{(X)}$ together with the function f is called the *free module on X*. In this setting, the free module $(R^{(X)}, f)$ has the universal mapping property in the sense that if M is any R-module and if $g : X \to M$ is any function to the underlying set of M, then there is a unique R-linear mapping $\varphi : R^{(X)} \to M$ such that $\varphi f = g$. Consequently, a free module $R^{(X)}$ on a set X is unique up to isomorphism.

Since an R-module F is free if and only if there is a set Δ such that $F \cong R^{(\Delta)}$, every nonzero free R-module must have cardinality at least as large as that of R. For this reason, the $\mathbb{Z}$-module $\mathbb{Z}_n$ is not a free $\mathbb{Z}$-module for any integer $n \geq 2$, so even though every R-module has a set of generators, not every R-module is free.

In general, the cardinality of a basis of a free module is not unique. The following example illustrates this fact.

Example

5. If $M = \bigoplus_{\mathbb{N}} \mathbb{Z}$, then it follows that $M \cong M \oplus M$. So if $R = \text{End}_{\mathbb{Z}}(M)$, then

$$R = \text{Hom}_{\mathbb{Z}}(M, M) \cong \text{Hom}_{\mathbb{Z}}(M, M \oplus M)$$
$$\cong \text{Hom}_{\mathbb{Z}}(M, M) \oplus \text{Hom}_{\mathbb{Z}}(M, M) \cong R \oplus R.$$

Thus, R has a basis with one element and a basis with two elements.

An R-module M is said to be *directly finite* if M is not isomorphic to a direct summand of itself. If M is not directly finite, then it is called *directly infinite*.

Rings with Invariant Basis Number

Because of Example 5, it is possible for a free R-module to have bases with distinct cardinalities. Actually, as we will see in the exercises, it is possible for a free module to have a basis with n elements for each positive integer n. This brings up the question, is it possible for a ring R to have the property that every basis of a free R-module has the same cardinality?

Definition 2.2.7. Let R be a ring such that for every free R-module F, any two bases of F have the same cardinality. Then R is said to have the *invariant basis number* property and R will be referred to as an *IBN-ring*. For such a ring the cardinality of the basis of a free R-module F is called the *rank* of F and denoted by $\text{rank}(F)$. If $\text{rank}(F) = n$ for some nonnegative integer n, then F is said to have *finite rank*. Otherwise, F is said to have *infinite rank*.

Remark. It follows that if R is an IBN-ring, then a free R-module with finite rank is directly finite and a free R-module with infinite rank is directly infinite. Directly finite and directly infinite modules have been investigated by many authors. The reader who wishes additional information on these module can consult [16].

The following proposition shows, somewhat unexpectedly, that if F is a free R-module that has a basis $\mathcal{B}$ with an infinite cardinal number, then every basis of F will have the same cardinality.

Proposition 2.2.8. *Let F be a free R-module with an infinite basis $\mathcal{B}$. Then every basis of F will have the same cardinality as that of $\mathcal{B}$.*

Proof. Let $\mathcal{B}$ be a basis of F and suppose that $\text{card}(\mathcal{B})$ is an infinite cardinal. Let $\mathcal{B}'$ be another basis of F and suppose that $\text{card}(\mathcal{B}')$ is a finite cardinal. Since $\mathcal{B}$ generates F, every element of $\mathcal{B}'$ can be written as a linear combination of a finite number of the elements of $\mathcal{B}$. But $\mathcal{B}'$ is finite, so there is a finite set $\{x_1, x_2, \ldots, x_n\}$ in $\mathcal{B}$ that generates $\mathcal{B}'$. But this implies that $\{x_1, x_2, \ldots, x_n\}$ will also generate F

which contradicts the fact that $\mathcal{B}$ is a minimal set of generators of F. Hence, card$(\mathcal{B}')$ cannot be a finite cardinal.

Next, we claim that card$(\mathcal{B})$ = card$(\mathcal{B}')$. We begin by considering the set $F(\mathcal{B}')$ of all finite subsets of $\mathcal{B}'$ and by defining $\varphi : \mathcal{B} \to F(\mathcal{B}')$ by $x \mapsto \{y_1, y_2, \ldots, y_n\}$, where $x = y_1 a_1 + y_2 a_2 + \cdots + y_n a_n$ with each y_i in $\mathcal{B}'$ and $a_i \in R$ is nonzero for $i = 1, 2, \ldots, n$. The map φ is clearly well defined since $\mathcal{B}'$ is a basis for F. Since Im φ generates $\mathcal{B}$ and $\mathcal{B}$ generates F, we see that Im φ generates F. It follows that Im φ cannot be a finite set for if so, then F would have a finite set of generators contained in $\mathcal{B}'$ which would contradict the fact that the basis $\mathcal{B}'$ is a minimal set of generators for F. Thus, card(Im φ) is an infinite cardinal number, so it must be the case that card(Im φ) $\geq \aleph_0$.

Now let $S \in \text{Im } \varphi$. We claim that $\varphi^{-1}(S)$ is a finite subset of $\mathcal{B}$. Note first that $\varphi^{-1}(S)$ is contained in the submodule $\sum_S yR$ of F generated by S, since in view of the way φ is defined, each element of $\varphi^{-1}(S)$ is a finite linear combination of elements of S. Now each $y \in S$ is a finite linear combination of elements of $\mathcal{B}$, so since S is finite, there is a finite subset T of $\mathcal{B}$ such that $\varphi^{-1}(S) \subseteq \sum_S yR \subseteq \sum_T xR$. Thus, if $x' \in \varphi^{-1}(S)$, then $x' \in \sum_T xR$, so x' is a finite linear combination of elements of $\mathcal{B}$. But this contradicts the fact that $\mathcal{B}$ is linearly independent unless $x' \in T$. Therefore, $\varphi^{-1}(S) \subseteq T$ and we have that $\varphi^{-1}(S)$ is a finite set.

Finally, for each $S \in \text{Im } \varphi$ order the elements of $\varphi^{-1}(S)$ by $\{x_1, x_2, \ldots, x_n\}$ and define $f_S : \varphi^{-1}(S) \to \mathbb{N} \times \text{Im } \varphi$ by $f_S(x_k) = (k, S)$. Now the sets $\varphi^{-1}(S)$, $S \in \text{Im } \varphi$, form a partition of $\mathcal{B}$, so define $\psi : \mathcal{B} \to \mathbb{N} \times \text{Im } \varphi$ by $\psi(x) = f_S(x)$, where $x \in \varphi^{-1}(S)$. It follows that ψ is an injective function, so we have card$(\mathcal{B}) \leq$ card$(\mathbb{N} \times \text{Im } \varphi)$. Using property (2) of the proposition given in the discussion of cardinal numbers in Appendix A, we see that

$$\text{card}(\mathcal{B}) \leq \text{card}(\mathbb{N} \times \text{Im } \varphi) = \aleph_0 \, \text{card}(\text{Im } \varphi) = \text{card}(\text{Im } \varphi)$$
$$\leq \text{card}(F(\mathcal{B}')) \leq \text{card}(\mathcal{B}').$$

Interchanging $\mathcal{B}$ and $\mathcal{B}'$ in the argument shows that card$(\mathcal{B}') \leq$ card$(\mathcal{B})$, so card$(\mathcal{B})$ = card$(\mathcal{B}')$. $\qquad \square$

Proposition 2.2.8 and Example 5 illustrate the fact that a free R-module F can have bases with different cardinalities only if F has a finite basis. We will now show that division rings and commutative rings are IBN-rings. Additional information on IBN-rings can be found in [2], [13] and [40].

We first show that every division ring is an IBN-ring. This result will be used to show that every commutative ring is an IBN-ring.

Proposition 2.2.9. *The following hold for every division ring D.*

(1) *If V is a vector space over D and if X is a linearly independent set of vectors of V and $y \in V$ is such that $y \notin \sum_X xR$, then $X \cup \{y\}$ is linearly independent.*

(2) *If V is a vector space over D and if X is a set of linearly independent vectors of V, then there is a basis $\mathcal{B}$ of V such that $X \subseteq \mathcal{B}$.*

(3) *Every vector space over D has a basis.*

Proof. (1) Suppose to the contrary that $X \cup \{y\}$ is linearly dependent. Then there must exist vectors $x_1, x_2, \ldots, x_n \in X$ and scalars $a_1, a_2, \ldots, a_{n+1} \in D$ such that

$$x_1 a_1 + x_2 a_2 + \cdots + x_n a_n + y a_{n+1} = 0$$

and at least one of the a_i is nonzero. If $a_{n+1} = 0$, then

$$x_1 a_1 + x_2 a_2 + \cdots + x_n a_n = 0$$

which implies that $a_1 = a_2 = \cdots = a_n = 0$ since X is linearly independent, a contradiction. Hence, it must be the case that $a_{n+1} \neq 0$. This gives

$$y = -x_1 a_1 a_{n+1}^{-1} - x_2 a_2 a_{n+1}^{-1} - \cdots - x_n a_n a_{n+1}^{-1}$$

which implies that $y \in \sum_X xR$, again a contradiction. Thus, $X \cup \{y\}$ must be linearly independent.

(2) Let X be a set of linearly independent elements of V and suppose that $\mathcal{S}$ is the collection of all linearly independent subsets of V that contain X. Partially order $\mathcal{S}$ by inclusion and note that $\mathcal{S} \neq \varnothing$ since $X \in \mathcal{S}$. If $\mathcal{C}$ is a chain in $\mathcal{S}$, then we claim that $X^* = \bigcup_{\mathcal{C}} X'$ is also in $\mathcal{S}$. Clearly $X \subseteq X^*$, so let $x_1, x_2, \ldots, x_n \in X^*$ and $a_1, a_2, \ldots, a_n \in D$ be such that

$$x_1 a_1 + x_2 a_2 + \cdots + x_n a_n = 0.$$

Since $\mathcal{C}$ is a chain, there is an $X' \in \mathcal{S}$ such that $x_1, x_2, \ldots, x_n \in X'$. But X' is linearly independent, so

$$x_1 a_1 + x_2 a_2 + \cdots + x_n a_n = 0$$

implies that $a_1 = a_2 = \cdots = a_n = 0$. Hence, X^* is linearly independent and X^* is an upper bound for $\mathcal{C}$. Thus, $\mathcal{S}$ is inductive, so Zorn's lemma indicates that $\mathcal{S}$ has a maximal element, say $\mathcal{B}$. We claim that $\mathcal{B}$ is a basis for V. If $\mathcal{B}$ is not a basis for V, then $\sum_{\mathcal{B}} xR \neq V$, so if $y \in V$ is such that $y \notin \sum_{\mathcal{B}} xR$, then $y \notin \mathcal{B}$. By (1), $\mathcal{B} \cup \{y\}$ is linearly independent and this contradicts the maximality of $\mathcal{B}$. Thus, it must be the case that $\sum_{\mathcal{B}} xR = V$ and we have that $\mathcal{B}$ is a basis for V.

(3) If $x \in V$, $x \neq 0$, then $\{x\}$ is a linearly independent subset of V. Hence, by (2), there is a basis $\mathcal{B}$ of V that contains $\{x\}$.   □

Proposition 2.2.10. *Every division ring is an IBN-ring.*

Proof. Let V be a vector space over a division ring D. Proposition 2.2.8 treats the case when V has an infinite basis and shows, in addition, that if V has a finite basis, then every basis of V must be finite. So let $\mathcal{B} = \{x_1, x_2, \ldots, x_m\}$ and $\mathcal{B}' = \{y_1, y_2, \ldots, y_n\}$ be bases of V and suppose that $n < m$. Since $\mathcal{B}$ is a basis of V, there are $a_i \in R$ such that

$$y_n = x_1 a_1 + x_2 a_2 + \cdots + x_m a_m.$$

If a_k is the first a_i that is nonzero, then

$$x_k = y_n a_k^{-1} - x_1 a_1 a_k^{-1} - \cdots - x_{k-1} a_{k-1} a_k^{-1} - x_{k+1} a_{k+1} a_k^{-1} - \cdots - x_m a_m a_k^{-1}.$$

It follows that the set

$$\{x_1, x_2, \ldots, x_{k-1}, x_{k+1}, \ldots, x_m, y_n\}$$

is a basis of V. Thus, we have replaced x_k in $\mathcal{B}$ by y_n and obtained a new basis of V. If this procedure is repeated for y_{n-1} with

$$\{x_1, x_2, \ldots, x_{k-1}, x_{k+1}, \ldots, x_m, y_n\}$$

as the basis, then an x_j can be replaced by y_{n-1} and the result is again a basis

$$\{x_1, x_2, \ldots, x_{j-1}, x_{j+1}, \ldots, x_{k-1}, x_{k+1}, \ldots x_m, y_n, y_{n-1}\}$$

of V. Repeating this procedure n times gives a set

$$\{\text{the } x_i \text{ not eliminated}\} \cup \{y_n, y_{n-1}, \ldots, y_1\}$$

that is a basis of V. But $\mathcal{B}'$ is a maximal set of linearly independent vectors of V, so it cannot be the case that $n < m$. Hence, $n \geq m$. Interchanging $\mathcal{B}$ and $\mathcal{B}'$ in the argument gives $m \geq n$ and this completes the proof. $\qquad\square$

Remark. If V is a vector space over a division ring D, it is standard practice to refer to the rank of V as the *dimension* of V and to denote the rank of V by $\dim(V)$ or by $\dim_D(V)$ if there is a need to emphasize D. If $\dim(V) = n$ for some nonnegative integer n, then V is said to be a *finite dimensional vector space*; otherwise V is an *infinite dimensional vector space*. (If $\dim(V) = 0$, then V is the zero vector space over D with basis $\varnothing$.)

We conclude this section by showing that every commutative ring is an IBN-ring.

Proposition 2.2.11. *Every commutative ring is an IBN-ring.*

Proof. Let R be a commutative ring and suppose that F is a free R-module with basis $\{x_\alpha\}_\Delta$. If $\mathfrak{m}$ is a maximal ideal of R, then $R/\mathfrak{m}$ is a field and $F/F\mathfrak{m}$ is an $R/\mathfrak{m}$-vector space. We claim that $\{x_\alpha + F\mathfrak{m}\}_\Delta$ is a basis for the vector space $F/F\mathfrak{m}$. If $x + F\mathfrak{m} \in F/F\mathfrak{m}$, then $x = \sum_\Delta x_\alpha a_\alpha$, where $a_\alpha = 0$ for almost all $\alpha \in \Delta$, so

$$x + F\mathfrak{m} = \left(\sum_\Delta x_\alpha a_\alpha \right) + F\mathfrak{m} = \sum_\Delta ((x_\alpha + F\mathfrak{m})(a_\alpha + \mathfrak{m})).$$

Thus, $\{x_\alpha + F\mathfrak{m}\}_\Delta$ generates $F/F\mathfrak{m}$. If $\sum_\Delta (x_\alpha + F\mathfrak{m})(a_\alpha + \mathfrak{m}) = 0$, then $(\sum_\Delta x_\alpha a_\alpha) + F\mathfrak{m} = 0$, so $\sum_\Delta x_\alpha a_\alpha \in F\mathfrak{m}$. But $F\mathfrak{m} = \bigoplus_\Delta x_\alpha \mathfrak{m}$, so there are $b_\alpha \in \mathfrak{m}$ such that $\sum_\Delta x_\alpha a_\alpha = \sum_\Delta x_\alpha b_\alpha$. Thus, $a_\alpha = b_\alpha$ for all $\alpha \in \Delta$ and we have $a_\alpha + \mathfrak{m} = 0$ for all $\alpha \in \Delta$. Therefore, $\{x_\alpha + F\mathfrak{m}\}_\Delta$ is a basis for the $R/\mathfrak{m}$-vector space $F/F\mathfrak{m}$ and $\dim_{R/\mathfrak{m}}(F/F\mathfrak{m}) = \mathrm{card}(\Delta)$. If $\{\bar{x}_\beta\}_\Gamma$ is another basis for the free R-module F, then exactly the same argument shows that $\{\bar{x}_\beta + F\mathfrak{m}\}_\Gamma$ is a basis for $F/F\mathfrak{m}$ and that $\dim_{R/\mathfrak{m}}(F/F\mathfrak{m}) = \mathrm{card}(\Gamma)$. Since Proposition 2.2.10 shows that vector space dimension is unique, we see that $\mathrm{card}(\Delta) = \mathrm{card}(\Gamma)$ which proves the proposition. $\qquad \square$

It is now easy to prove that if R is an IBN-ring, then two free R-modules are isomorphic if and only if they have the same rank. We leave this as an exercise.

Problem Set 2.2

1. (a) If M and N are finitely generated R-modules, prove that $M \oplus N$ is finitely generated. Extend this to a finite number of R-modules.

   (b) Find a family $\{M_\alpha\}_\Delta$ of R-modules each of which is finitely generated but $\bigoplus_\Delta M_\alpha$ is not finitely generated.

   (c) Prove that $\bigoplus_\Delta M_\alpha$ is finitely generated if and only if each M_α is finitely generated and $M_\alpha = 0$ for almost all $\alpha \in \Delta$.

2. (a) Let $M_1 \subset M_2 \subset M_3 \subset \cdots$ be a strictly increasing chain of finitely generated submodules of an R-module M. Show by example that the submodule $M = \bigcup_\mathbb{N} M_i$ need not be finitely generated.

   (b) Prove that if M is an R-module which is not finitely generated, then M has a submodule N that has a set of generators X such that $\mathrm{card}(X) = \aleph_0$.

3. (a) If F is a free R-module and $f : F \to M$ is an isomorphism, prove that M is a free R-module.

   (b) If $\mathcal{B}$ is a basis for F, show that M has a basis with the same cardinality as that of $\mathcal{B}$.

4. If R is an IBN-ring, prove that if F_1 and F_2 are free R-modules, then $F_1 \cong F_2$ if and only if $\mathrm{rank}(F_1) = \mathrm{rank}(F_2)$.

5. Let V be a vector space over a division ring D.

(a) If $f : V \to V$ is a linear transformation that is a monomorphism, show that f is an isomorphism.

(b) If $f : V \to V$ is a linear transformation that is an epimorphism, prove that f is an isomorphism.

6. (a) Prove that if U_1 is a subspace of a vector space V over a division ring D, then there is a subspace U_2 of V such that $V = U_1 \oplus U_2$. Conclude that every subspace of a vector space is a direct summand.

(b) Show by example that submodules of free R-modules need not be free.

(c) Show by example that (a) does not hold for free R-modules.

7. Let F be a free R-module and suppose we have a diagram

$$
\begin{array}{ccc}
 & & F \\
 & {\scriptstyle h}\nearrow & \downarrow {\scriptstyle f} \\
M_1 & \xrightarrow{\ g\ } & M_2
\end{array}
$$

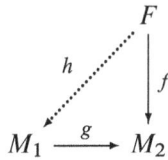

of R-modules and R-linear mappings, where g is an epimorphism. Prove that there is an R-linear mapping $h : F \to M_1$ which makes the diagram commutative.

8. (a) If F_1 and F_2 are free R-modules, prove that $F_1 \times F_2$ is free.

(b) If $\{F_\alpha\}_\Delta$ is a family of free R-modules, prove that the R-module $\bigoplus_\Delta F_\alpha$ is free.

9. The procedure followed in the proof Proposition 2.2.10 is often referred to as a *replacement procedure*. Show that each of the sets

$$\{x_1, x_2, \ldots, x_{k-1}, x_{k+1}, \ldots, x_m, y_n\}$$
$$\{x_1, x_2, \ldots, x_{j-1}, x_{j+1}, \ldots, x_{k-1}, x_{k+1}, \ldots x_m, y_n, y_{n-1}\}$$

$$\vdots$$

$$\{\text{the } x_i \text{ not eliminated}\} \cup \{y_n, y_{n-1}, \ldots, y_1\}$$

constructed in the proof of Proposition 2.2.10 is a basis of V.

10. Prove that every finitely generated free R-module has a finite basis.

11. (a) Verify Example 5.

(b) If R is the ring of Example 5, show that $R \cong R^{(n)}$ for each positive integer n. Conclude that R, as an R-module, has a basis with n elements for each positive integer n.

12. (a) Consider the first Remark of this section. Show that a free R-module on a set X has the universal mapping property and that such a module is unique up to isomorphism.

(b) Suppose that the definition of a free R-module on a set X is restated as follows: Let X be an arbitrary set. Call an R-module F, together with an injective mapping $f : X \to F$, a free R-module on X if for every R-module M and every function $g : X \to M$ into the underlying set of M, there is a unique R-linear mapping $\varphi : F \to M$ such that $\varphi f = g$. If (F, f) is a free R-module on X, prove that $F \cong R^{(X)}$.

13. Prove the equivalence of (1) and (3) of Proposition 2.2.3.

14. Show that $\mathbb{M}_n(R)$ is a free R-module that has a basis with n^2 elements.

15. Let R be an integral domain. An element x of an R-module M is said to be a *torsion element* of M if there is an nonzero element $a \in R$ such that $xa = 0$.

(a) If $t(M)$ is the set of all torsion elements of M, prove that $t(M)$ is a submodule of M. $t(M)$ is called the *torsion submodule* of M.

(b) An R-module M is said to be a *torsion module* if $t(M) = M$ and a *torsion free module* if $t(M) = 0$. Show that $M/t(M)$ is a torsion free R-module.

(c) Prove that every free R-module is torsion free.

16. Let F be a free R-module. Prove that there is a set Δ such that for any R-module M,

$$\mathrm{Hom}_R(F, M) \cong M^{\Delta}.$$

17. (a) Prove that R is an IBN-ring if and only if $R^{(m)} \cong R^{(n)}$ implies that $m = n$ for every pair of positive integers m and n. (Note, that if Δ and Λ are infinite sets, then the case for $R^{(\Delta)} \cong R^{(\Lambda)}$ implies that $\mathrm{card}(\Delta) = \mathrm{card}(\Lambda)$ follows from Proposition 2.2.8.)

(b) Let R be a ring, A a proper ideal of R, F a free R-module with basis $\mathcal{B}$ and $\eta : F \to F/FA$ the canonical epimorphism. Show that F/FA is a free R/A-module with basis $\eta(\mathcal{B})$ and $\mathrm{card}(\eta(\mathcal{B})) = \mathrm{card}(\mathcal{B})$. [Hint: If $\mathcal{B} = \{x_\alpha\}_\Delta$, then the proof that $\eta(\mathcal{B})$ is a basis for F/FA follows along the same lines as that of Proposition 2.2.11. Now show that $\mathrm{card}(\eta(\mathcal{B})) = \mathrm{card}(\mathcal{B})$.]

(c) If $f : R \to S$ is a nonzero surjective ring homomorphism and S is an IBN-ring, show that R is also an IBN-ring. [Hint: Let $\mathcal{B}_1$ and $\mathcal{B}_2$ be bases of a free R-module F and use (c).]

18. (a) Let W be a finite dimensional vector space over a division ring D. If U_1 and U_2 are subspaces of W, prove that

$$\dim(U_1 + U_2) = \dim(U_1) + \dim(U_2) - \dim(U_1 \cap U_2).$$

(Note that all dimensions involved are finite.) [Hint: If $U_1 \cap U_2 = 0$, the result is clear, so suppose that $U_1 \cap U_2 \neq 0$. Since $U_1 \cap U_2$ is a subspace of U_1, if V is a subspace of U_1 such that $(U_1 \cap U_2) \oplus V = U_1$, then $\dim(U_1 \cap U_2) + \dim(V) = \dim(U_1)$. Show that $U_1 + U_2 = V + U_2$ and that $V \cap U_2 = 0$, so that $\dim(U_1 + U_2) = \dim(V) + \dim(U_2)$.]

(b) Let R, S and T be division rings and suppose that R is a subring of S and S is a subring of T. Prove that $\dim_R(T) = \dim_R(S) \dim_S(T)$.

(c) Use (b) to show that a field K cannot exist such that $\mathbb{R} \subsetneq K \subsetneq \mathbb{C}$.

2.3 Tensor Products of Modules

Next, we develop the concept of a tensor product of two modules and, as we will see, this construction yields an additive abelian group that is unique up to isomorphism. The assumption that the expression "R-module" means right R-module continues and the notations M_R and $_R N$ will have the obvious meaning.

Let M be an R-module, N a left R-module and G an additive abelian group. Then a mapping $\rho : M \times N \to G$ is said to be R-*balanced* if

$$\rho(x_1 + x_2, y) = \rho(x_1, y) + \rho(x_2, y),$$
$$\rho(x, y_1 + y_2) = \rho(x, y_1) + \rho(x, y_2) \quad \text{and}$$
$$\rho(xa, y) = \rho(x, ay)$$

for all $x, x_1, x_2 \in M$, $y, y_1, y_2 \in N$ and $a \in R$. (R-balanced mappings are also called *bilinear mappings*.)

R-balanced mappings play a central role in the development of the tensor product of modules.

Definition 2.3.1. If M is an R-module and N is a left R-module, then an additive abelian group T together with an R-balanced mapping $\rho : M \times N \to T$ is said to be a *tensor product* of M and N, if whenever G is an additive abelian group and $\rho' : M \times N \to G$ is an R-balanced mapping, there is a unique group homomorphism $f : T \to G$ that completes the diagram

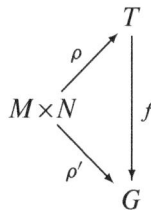

$$
\begin{array}{ccc}
 & & T \\
 & \overset{\rho}{\nearrow} & \downarrow f \\
M \times N & & \\
 & \underset{\rho'}{\searrow} & \downarrow \\
 & & G
\end{array}
$$

commutatively. The map ρ is called the *canonical R-balanced map* from $M \times N$ to T. A tensor product of M_R and $_R N$ will be denoted by (T, ρ).

Proposition 2.3.2. *If a tensor product (T, ρ) of M_R and $_R N$ exists, then T is unique up to group isomorphism.*

Proof. Let (T, ρ) and (T', ρ') be tensor products of M and N. Then there are group homomorphisms $f : T \to T'$ and $f' : T' \to T$ such that the diagram

$$
\begin{array}{c}
T \\
\rho \nearrow \quad \downarrow f \\
M \times N \xrightarrow{\rho'} T' \\
\rho \searrow \quad \downarrow f' \\
T
\end{array}
$$

is commutative. But $\mathrm{id}_T : T \to T$ makes the outer triangle commute, so the uniqueness of id_T gives $f' f = \mathrm{id}_T$. Similarly, $f f' = \mathrm{id}_{T'}$ and so f is a group isomorphism. $\qquad\square$

Our next task is to show that tensor products always exist.

Proposition 2.3.3. *Every pair of modules M_R and $_R N$ has a tensor product.*

Proof. Consider the free $\mathbb{Z}$-module $F = \mathbb{Z}^{(M \times N)}$ on $M \times N$. In view of the observations in the Remark immediately following the proof of Proposition 2.2.6, we can write

$$
F = \Big\{ \sum n_{(x,y)}(x, y) \mid n_{(x,y)} \in \mathbb{Z}, (x, y) \in M \times N
$$
$$
\text{and almost all } n_{(x,y)} = 0 \Big\}.
$$

Now let H be the subgroup of F generated by elements of the form

(1) $(x_1 + x_2, y) - (x_1, y) - (x_2, y),$
(2) $(x, y_1 + y_2) - (x, y_1) - (x, y_2)$ and
(3) $(xa, y) - (x, ay),$

where $x, x_1, x_2 \in M, y, y_1, y_2 \in N$ and $a \in R$. If $\rho : M \times N \to F/H$ is such that $\rho((x, y)) = (x, y) + H$, then ρ is R-balanced and we claim that $(F/H, \rho)$ is a tensor product of M and N. Suppose that $\rho' : M \times N \to G$ is an R-balanced mapping. Since $M \times N$ is a basis for F, Proposition 2.2.3 shows that ρ' can be extended uniquely to a group homomorphism $g : F \to G$. But ρ' is R-balanced, so $g(H) = 0$. Hence, there is an induced group homomorphism $f : F/H \to G$ such that the diagram

$$
\begin{array}{c}
F/H \\
\rho \nearrow \quad \Big\downarrow f \\
M \times N \\
\rho' \searrow \\
G
\end{array}
$$

is commutative. Since g is unique, f is unique, so $(F/H, \rho)$ is a tensor product of M and N, as asserted.   □

The additive abelian group F/H, constructed in the preceding proof, will now be denoted by $M \otimes_R N$ and we will call $M \otimes_R N$ the tensor product of M and N.
   If the cosets $(x, y) + H$ in $M \otimes_R N$ are denoted by $x \otimes y$, then

$$(x_1 + x_2) \otimes y = x_1 \otimes y + x_2 \otimes y,$$

$$x \otimes (y_1 + y_2) = x \otimes y_1 + x \otimes y_2 \quad \text{and}$$

$$xa \otimes y = x \otimes ay$$

for all $x, x_1, x_2 \in M, y, y_1, y_2 \in N$ and $a \in R$. Under this notation the canonical R-balanced mapping $\rho : M \times N \to M \otimes_R N$ is now given by $\rho((x, y)) = x \otimes y$.
   In general, the additive abelian group $M \otimes_R N$ is neither a left nor a right R-module. It is simply a $\mathbb{Z}$-module. Recall that if R and S are rings, then an additive abelian group M that is a left R-module and an S-module is said to be an (R, S)-bimodule, denoted by $_R M_S$ if $(ax)b = a(xb)$ for all $a \in R, x \in M$ and $b \in S$. If we are given an (R, S)-bimodule $_R M_S$ and a left S-module $_S N$, then $M \otimes_R N$ is a left R-module under the operation $a(x \otimes y) = ax \otimes y$ for all $a \in R$ and $x \otimes y \in M \otimes_R N$. Likewise, given an (R, S)-bimodule $_R N_S$ and an R-module M, then $M \otimes_R N$ is an S-module via $(x \otimes y)b = x \otimes yb$. We also point out that not every element of $M \otimes_R N$ can be written as $x \otimes y$. The set $\{x \otimes y\}_{(x,y) \in M \times N}$ is a set of generators of $M \otimes_R N$, so a general element of $M \otimes_R N$ is written as $\sum_{i=1}^{m} n_i (x_i \otimes y_i)$, where $n_i \in \mathbb{Z}$ for $i = 1, 2, \ldots, m$. If $0_M, 0_N$ and 0_R denote the additive identities of M, N and R respectively, then for $x \in M$ and $y \in N$

$$(x \otimes y) + (0_M \otimes 0_N) = (x \otimes y) + (0_M \otimes 0_R y) = (x \otimes y) + (0_M 0_R \otimes y)$$

$$= (x \otimes y) + (0_M \otimes y) = (x + 0_M) \otimes y = x \otimes y.$$

Simplifying notation, it follows that

$$(0 \otimes 0) + (x \otimes y) = (x \otimes y) + (0 \otimes 0) = x \otimes y,$$

so $0 \otimes 0$ is the additive identity of $M \otimes_R N$. Similarly,

$$x \otimes 0 = 0 \otimes y = 0 \otimes 0$$

for any $x \in M$ and $y \in N$.

Remark. Care must be taken when attempting to define a $\mathbb{Z}$-linear mapping with domain $M \otimes_R N$ by specifying the image of each element of $M \otimes_R N$. For example, if $f : M \to M'$ is an R-linear mapping, then one may be tempted to define a group homomorphism $g : M \otimes_R N \to M' \otimes_R N$ by setting $g(\sum_{i=1}^{m} n_i (x_i \otimes y_i)) = $

$\sum_{i=1}^{m} n_i (f(x_i) \otimes y_i)$. As we will see, this map actually works, but when g is specified in this manner it is difficult to show that it is well defined. This difficulty can be avoided by working with an R-balanced mapping and going through an intermediate step. To see this, consider the commutative diagram

$$
\begin{array}{ccc}
M \times N & \xrightarrow{\ \rho\ } & M \otimes_R N \\[2pt]
{\scriptstyle f \times \mathrm{id}_N}\downarrow & \searrow{\scriptstyle h} & \downarrow{\scriptstyle g} \\[2pt]
M' \times N & \xrightarrow{\ \rho'\ } & M' \otimes_R N
\end{array}
$$

where ρ and ρ' are the canonical R-balanced mappings. Since the map $h = \rho'(f \times \mathrm{id}_N)$ is an R-balanced map, the existence of the group homomorphism g displayed in the diagram is guaranteed by the definition of the tensor product $M \otimes_R N$. Thus,

$$g(x \otimes y) = \rho'(f \times \mathrm{id}_N)((x,y)) = \rho'(f(x),y) = f(x) \otimes y, \quad \text{so}$$

$$g\left(\sum_{i=1}^{m} n_i(x_i \otimes y_i)\right) = \sum_{i=1}^{m} n_i(f(x_i) \otimes y_i) \quad \text{as expected.}$$

Proposition 2.3.4. *If M is an R-module, then $M \otimes_R R \cong M$ as R-modules.*

Proof. If $\rho' : M \times R \to M$ is defined by $\rho'(x,a) = xa$, then ρ' is an R-balanced mapping. Thus, there is a unique group homomorphism $f : M \otimes_R R \to M$ such that $f\rho = \rho'$, where $\rho : M \times R \to M \otimes_R R$ is the canonical R-balanced map. Hence, we see that $f(x \otimes a) = xa$ for every generator $x \otimes a$ of $M \otimes_R R$. Note that if $b \in R$, then

$$f((x \otimes a)b) = f(x \otimes ab) = x(ab) = (xa)b = f(x \otimes a)b,$$

so f is R-linear. Now define $f' : M \to M \otimes_R R$ by $f'(x) = x \otimes 1$. Then f' is clearly well defined and additive. Furthermore,

$$f'(xa) = xa \otimes 1 = x \otimes a = (x \otimes 1)a = f'(x)a,$$

so f' is R-linear. Since $f'f(x \otimes a) = f'(xa) = xa \otimes 1 = x \otimes a$ for each generator $x \otimes a$ of $M \otimes_R R$, we see that $f'f = \mathrm{id}_{M \otimes_R R}$. Similarly, $ff'(x) = f(x \otimes 1) = x1 = x$, so $ff' = \mathrm{id}_M$. Hence, f is an isomorphism and we have $M \otimes_R R \cong M$. $\square$

The proof of Proposition 2.3.4 is clearly symmetrical, so we have $R \otimes_R N \cong N$ for every left R-module N.

Proposition 2.3.5. *If $f : M_R \to M'_R$ and $g :_R N \to_R N'$ are R-linear mappings, then there is a unique group homomorphism*

$$f \otimes g : M \otimes_R N \to M' \otimes_R N'$$

such that $(f \otimes g)(x \otimes y) = f(x) \otimes g(y)$ for all $x \otimes y \in M \otimes_R N$.

Proof. Consider the commutative diagram,

$$
\begin{array}{ccc}
M \times N & \xrightarrow{\ \rho\ } & M \otimes_R N \\
{\scriptstyle f \times g}\big\downarrow & \searrow^{h} & \big\downarrow{\scriptstyle f \otimes g} \\
M' \times N' & \xrightarrow{\ \rho'\ } & M' \otimes_R N'
\end{array}
$$

where ρ and ρ' are the canonical R-balanced maps. Moreover, $h = \rho'(f \times g)$ is R-balanced and the unique group homomorphism $f \otimes g$ is given by the tensor product $M \otimes_R N$. From this we see that $(f \otimes g)\rho = \rho'(f \times g)$, so if $(x, y) \in M \times N$, then

$$
\begin{aligned}
(f \otimes g)(x \otimes y) &= (f \otimes g)\rho((x, y)) \\
&= \rho'(f \times g)((x, y)) = \rho'((f(x), g(x))) \\
&= f(x) \otimes g(x).
\end{aligned}
$$
□

The following proposition shows that tensor products and direct sums enjoy a special relationship, that is, they commute.

Proposition 2.3.6. *If M is an R-module and $\{N_\alpha\}_\Delta$ is a family of left R-modules, then*

$$
M \otimes_R \left(\bigoplus_\Delta N_\alpha \right) \cong \bigoplus_\Delta (M \otimes_R N_\alpha).
$$

Furthermore, the group isomorphism is unique and given by $x \otimes (y_\alpha) \mapsto (x \otimes y_\alpha)$.

Proof. The mapping $\rho' : M \times (\bigoplus_\Delta N_\alpha) \to \bigoplus_\Delta (M \otimes_R N_\alpha)$ defined by $\rho'(x, (y_\alpha)) = (x \otimes y_\alpha)$ is R-balanced, so the definition of a tensor product produces a unique group homomorphism

$$
f : M \otimes_R \left(\bigoplus_\Delta N_\alpha \right) \to \bigoplus_\Delta (M \otimes_R N_\alpha)
$$

such that $f\rho = \rho'$, where $\rho : M \times (\bigoplus_\Delta N_\alpha) \to M \otimes_R (\bigoplus_\Delta N_\alpha)$ is the canonical R-balanced map given by $\rho(x, (y_\alpha)) = x \otimes (y_\alpha)$. It follows that $f(x \otimes (y_\alpha)) = (x \otimes y_\alpha)$ for each generator $x \otimes (y_\alpha) \in M \otimes_R (\bigoplus_\Delta N_\alpha)$.
    The proof will be complete if we can find a $\mathbb{Z}$-linear mapping

$$
f' : \bigoplus_\Delta (M \otimes_R N_\alpha) \to M \otimes_R \left(\bigoplus_\Delta N_\alpha \right)
$$

that serves as an inverse for f. For this, let $\bar{\imath}_\beta : M_\beta \to \bigoplus_\Delta M_\alpha$ be the canonical injection for each $\beta \in \Delta$. Then for each $\beta \in \Delta$ we have a $\mathbb{Z}$-linear mapping $f_\beta = \mathrm{id}_M \otimes \bar{\imath}_\beta : M \otimes_R M_\beta \to M \otimes_R (\bigoplus_\alpha M_\alpha)$ defined by $x \otimes x_\beta \mapsto x \otimes (y_\alpha)$, where

$y_\alpha = x_\beta$ when $\alpha = \beta$ and $y_\alpha = 0$ if $\alpha \neq \beta$. If $i_\beta : M \otimes_R N_\beta \to \bigoplus_\Delta (M \otimes_R N_\alpha)$ is the canonical injection for each $\beta \in \Delta$, then the definition of a direct sum produces a unique $\mathbb{Z}$-linear map $f' : \bigoplus_\Delta (M \otimes_R N_\alpha) \to M \otimes_R (\bigoplus_\Delta N_\alpha)$ such that $f' i_\beta = f_\beta$ for each $\beta \in \Delta$. It is not difficult to verify that

$$ff'((x \otimes y_\alpha)) = f(x \otimes (y_\alpha)) = (x \otimes y_\alpha) \quad \text{and}$$
$$f'f(x \otimes (y_\alpha)) = f'((x \otimes y_\alpha)) = x \otimes (y_\alpha),$$

so $ff' = \mathrm{id}_{\bigoplus_\Delta (M \otimes_R N_\alpha)}$ and $f'f = \mathrm{id}_{M \otimes_R (\bigoplus_\Delta N_\alpha)}$. Hence, f is an isomorphism.
$\square$

By symmetry, we see that if $\{M_\alpha\}_\Delta$ is a family of R-modules, then $(\bigoplus_\Delta M_\alpha) \otimes_R N \cong \bigoplus_\Delta (M_\alpha \otimes_R N)$ for any left R-module N.

Corollary 2.3.7. *If F is a free R-module and M is a left R-module, then there is a set Δ such that $F \otimes_R M \cong M^{(\Delta)}$.*

Proof. Since F is a free R-module, there is a set Δ such that $F \cong R^{(\Delta)}$. Hence, we have $F \otimes_R M \cong R^{(\Delta)} \otimes_R M \cong (R \otimes_R M)^{(\Delta)} \cong M^{(\Delta)}$.
$\square$

Corollary 2.3.8. *If $\{M_\alpha\}_\Delta$ is a family of R-modules and $\{N_\beta\}_\Gamma$ is a family of left R-modules, then $(\bigoplus_\Delta M_\alpha) \otimes_R (\bigoplus_\Gamma N_\beta) \cong (M_\alpha \otimes_R N_\beta)^{(\Delta \times \Gamma)}$.*

Proof.
$$\left(\bigoplus_\Delta M_\alpha\right) \otimes_R N \cong \bigoplus_\Delta (M_\alpha \otimes_R N) \cong \bigoplus_\Delta \left(M_\alpha \otimes_R \left(\bigoplus_\Gamma N_\beta\right)\right)$$
$$\cong \bigoplus_\Delta \bigoplus_\Gamma (M_\alpha \otimes_R N_\beta) \cong (M_\alpha \otimes_R N_\beta)^{(\Delta \times \Gamma)}.$$
$\square$

Examples

1. **Monomorphisms under Tensor Products.** If $i : \mathbb{Z} \to \mathbb{Q}$ is the canonical injection and $\mathrm{id}_{\mathbb{Z}_6} : \mathbb{Z}_6 \to \mathbb{Z}_6$ is the identity map, then $i \otimes \mathrm{id}_{\mathbb{Z}_6} : \mathbb{Z} \otimes_{\mathbb{Z}} \mathbb{Z}_6 \to \mathbb{Q} \otimes_{\mathbb{Z}} \mathbb{Z}_6$ is not an injection. Thus, injective maps are not, in general, preserved under tensor products.

2. **Epimorphisms under Tensor Products.** If $f : M_R \to M'_R$ and $g : {}_R N \to {}_R N'$ are epimorphisms, then

   $$f \otimes g : M \otimes_R N \to M' \otimes_R N'$$

   is a group epimorphism.

3. **Composition of Maps under Tensor Products.** If

   $$f : M_R \to M'_R, \ f' : M'_R \to M''_R, \ g : {}_R N \to {}_R N' \text{ and } g' : {}_R N' \to {}_R N''$$

   are R-linear mappings, then $(f' \otimes g')(f \otimes g) = (f'f) \otimes (g'g)$.

4. **Tensor Products Preserve Isomorphisms.** If $M_R \cong M'_R$ and $_R N \cong_R N'$, then $M \otimes_R N \cong M' \otimes_R N'$. This follows easily from Example 3, since if $f : M_R \to M'_R$ and $g :\ _R N \to\ _R N'$ are R-isomorphisms, then $f \otimes g : M \otimes_R N \to M' \otimes_R N'$ is a group homomorphism with inverse $f^{-1} \otimes g^{-1} :$ $M' \otimes_R N' \to M \otimes_R N$.

5. **Change of Rings.** If R is a subring of S and if M is an R-module, then $M \otimes_R S$ is an S-module. This procedure is referred to as a *change of rings* or as an *extension of scalars* from R to S. For example, if V is a vector space over $\mathbb{R}$, then $V \otimes_\mathbb{R} \mathbb{C}$ is a vector space over $\mathbb{C}$.

Problem Set 2.3

1. Show that each of the following hold.

   (a) If d is the greatest common divisor of two positive integers m and n, then $\mathbb{Z}_m \otimes_\mathbb{Z} \mathbb{Z}_n = \mathbb{Z}_d$. [Hint: Show that the mapping $\rho' : \mathbb{Z}_m \times \mathbb{Z}_n \to \mathbb{Z}_d$ defined by $\rho'(([a], [b])) = [ab]$ is a well defined $\mathbb{Z}$-balanced mapping and then consider the mapping $f : \mathbb{Z}_m \otimes_\mathbb{Z} \mathbb{Z}_n \to \mathbb{Z}_d$ given by the tensor product $\mathbb{Z}_m \otimes_\mathbb{Z} \mathbb{Z}_n$.] Conclude that if m and n are relatively prime, then $\mathbb{Z}_m \otimes_\mathbb{Z} \mathbb{Z}_n = 0$.

   (b) $\mathbb{Q}/\mathbb{Z} \otimes_\mathbb{Z} \mathbb{Q}/\mathbb{Z} = 0$

   (c) $\mathbb{Q} \cong \mathbb{Q} \otimes_\mathbb{Z} \mathbb{Q}$

   (d) If G is a torsion $\mathbb{Z}$-module, show that $G \otimes_\mathbb{Z} \mathbb{Q} = 0$.

2. If I is a left ideal of R and M is an R-module, prove that $M \otimes_R (R/I) \cong M/MI$. [Hint: Show that the mapping $h : M \to M \otimes_R (R/I)$ defined by $h(x) = x \otimes (1 + I)$ is an epimorphism. Next, show that $MI \subseteq \operatorname{Ker} h$ so that we have an induced epimorphism $g : M/MI \to M \otimes_R (R/I)$. Now the map $\rho' : M \times R/I \to M/MI$ given by $\rho'((x, a + I)) = xa + MI$ is R-balanced, so if $f : M \otimes_R (R/I) \to M/MI$ is the map given by the tensor product $M \otimes_R (R/I)$, show that $fg = \operatorname{id}_{M/MI}$ and $gf = \operatorname{id}_{M \otimes_R (R/I)}$.]

3. Let R and S be rings and consider the modules L_R, $_R M_S$ and $_S N$. Prove that $(L \otimes_R M) \otimes_S N \cong L \otimes_R (M \otimes_S N)$.

4. If R is a commutative ring and M and N are R-modules, then are $M \otimes_R N$ and $N \otimes_R M$ isomorphic?

5. Verify the assertions of Examples 1 through 5.

6. If I_1 and I_2 are ideals of R, prove that $R/I_1 \otimes_R R/I_2 \cong R/(I_1 + I_2)$. [Hint: The balanced map $\rho' : R/I_1 \times R/I_2 \to R/(I_1 + I_2)$ defined by $\rho'((a + I_1, b + I_2)) = ab + I_1 + I_2$ gives a group homomorphism $f : R/I_1 \otimes_R R/I_2 \to R/(I_1 + I_2)$ such that $f(a + I_1 \otimes b + I_2) = ab + I_1 + I_2$. So show that $g : R/(I_1 + I_1) \to R/I_1 \otimes_R R/I_2$ given by $f(a + I_1 + I_2) = 1 + I_1 \otimes a + I_2$ is a well defined group homomorphism such that $gf = \operatorname{id}_{R/I_1 \otimes_R R/I_2}$ and $fg = \operatorname{id}_{R/(I_1 + I_2)}$.]

7. Let R and S be rings and suppose that M is an (R, S)-bimodule. Let $R \times S$ be the ring direct product and make M into an $(R \times S, R \times S)$-bimodule by setting $(a, b)x = ax$ and $x(a, b) = xb$ for all $x \in M$ and $(a, b) \in R \times S$. Prove that $M \otimes_{R \times S} M = 0$.

8. Let I be an ideal of R and suppose that M and N are right and left R/I-modules, respectively. Then M is an R-module by pullback along the canonical surjection $\eta : R \to R/I$. Similarly, N is a left R-module. Prove that $M \otimes_R N \cong M \otimes_{R/I} N$.

9. (a) If F_1 and F_2 are free R-modules with bases $\{x_\alpha\}_\Gamma$ and $\{y_\beta\}_\Delta$, respectively, prove that $F_1 \otimes_R F_2$ is free R-module with basis $\{x_\alpha \otimes y_\beta\}_{(\alpha,\beta) \in \Gamma \times \Delta}$.

(b) If $f : R \to S$ is a ring homomorphism and F is a free R-module with basis $\{x_\alpha\}_\Delta$, prove that $F \otimes_R S$ is a free S-module with basis $\{x_\alpha \otimes 1\}_\Delta$.

(c) Determine a basis of the $\mathbb{R}$-vector space $\mathbb{C} \otimes_\mathbb{R} \mathbb{C}$.

(d) Show that $\mathbb{C} \otimes_\mathbb{R} \mathbb{C}$ and $\mathbb{C} \otimes_\mathbb{C} \mathbb{C}$ are $\mathbb{R}$-vector spaces, but that they cannot be isomorphic as $\mathbb{R}$-modules.

Chapter 3
Categories

One important use of category theory is to delineate areas of mathematics and to make important connections among these areas. Categories could just as well have been introduced in the first chapter but since our primary interest is in the study of rings and modules, we have chosen to introduce categories following a discussion of modules. We will not make extensive use of category theory, so only basic concepts will be developed. A more extensive development of category theory can be found in [3], [7], [32] and [41].

3.1 Categories

Definition 3.1.1. A *category* $\mathcal{C}$ consists of a class $\mathcal{O}$ of *objects* and a class $\mathcal{M}$ of *morphisms* such that the following conditions hold.

C1 For all $A, B \in \mathcal{O}$ there is a (possibly empty) set $\mathrm{Mor}(A, B)$, called the *set of morphisms* $f : A \to B$ from A to B, such that

$$\mathrm{Mor}(A, B) \cap \mathrm{Mor}(A', B') = \varnothing \quad \text{if } (A, B) \neq (A', B').$$

C2 If $A, B, C \in \mathcal{O}$, then there is a *rule of composition*

$$\mathrm{Mor}(A, B) \times \mathrm{Mor}(B, C) \to \mathrm{Mor}(A, C)$$

such that if $(f, g) \mapsto gf$, then:

(a) *Associativity*: If $f : A \to B$, $g : B \to C$ and $h : C \to D$ are morphisms in $\mathcal{C}$, then $(hg)f = h(gf)$.

(b) *Existence of Identities*: For each $A \in \mathcal{O}$ there is an *identity morphism* $\mathrm{id}_A : A \to A$ such that $f \mathrm{id}_A = f$ and $\mathrm{id}_A g = g$ for any morphisms $f : A \to B$ and $g : C \to A$ of $\mathcal{C}$.

A morphism $f : A \to B$ in $\mathcal{C}$ is said to be an *isomorphism* if there is a morphism $g : B \to A$ in $\mathcal{C}$ such that $fg = \mathrm{id}_B$ and $gf = \mathrm{id}_A$.

A category $\mathcal{C}$ is a *subcategory* of a category $\mathcal{D}$ if the following conditions are satisfied.

(1) Every object of $\mathcal{C}$ is an object of $\mathcal{D}$.

(2) If A and B are objects of $\mathcal{C}$, then $\mathrm{Mor}_{\mathcal{C}}(A, B) \subseteq \mathrm{Mor}_{\mathcal{D}}(A, B)$.

(3) The composition of morphisms in $\mathcal{C}$ is the same as the composition of morphisms in $\mathcal{D}$.

(4) For every object A of $\mathcal{C}$, the identity morphism $\mathrm{id}_A : A \to A$ in $\mathcal{C}$ is the same as the identity morphism $\mathrm{id}_A : A \to A$ in $\mathcal{D}$.

If $\mathcal{C}$ is a subcategory of $\mathcal{D}$ and $\mathrm{Mor}_{\mathcal{C}}(A, B) = \mathrm{Mor}_{\mathcal{D}}(A, B)$ for all objects A and B of $\mathcal{C}$, then $\mathcal{C}$ is said to be a *full subcategory* of $\mathcal{D}$.

A category $\mathcal{C}$ is *additive* if $\mathrm{Mor}(A, B)$ has the structure of an additive abelian group for each pair A, B of objects of $\mathcal{C}$ and if $g(f_1 + f_2) = gf_1 + gf_2$ and $(g_1 + g_2)f = g_1 f + g_2 f$ for all $f, f_1, f_2 \in \mathrm{Mor}(A, B)$ and all $g, g_1, g_2 \in \mathrm{Mor}(B, C)$. An object 0 of a category $\mathcal{C}$ is said to be a *zero object* of $\mathcal{C}$ if $\mathrm{Mor}(0, A)$ and $\mathrm{Mor}(A, 0)$ each contain a single morphism for each object A of $\mathcal{C}$.

Remark.

(1) Note that if $\mathcal{C}$ is an additive category, then $\mathrm{Mor}(A, B) \neq \varnothing$ for each pair A, B of objects of $\mathcal{C}$ since the *zero morphism* $0_{AB} : A \to B$ is in $\mathrm{Mor}(A, B)$.

(2) Morphisms in a category are usually denoted by $f : A \to B$ or by $A \xrightarrow{f} B$. This suggests that morphisms are functions of some type. Although this is true in many categories, the first example below shows that this need not be the case.

Examples

Categories are ubiquitous in mathematics. The reader should determine which of the following categories are additive and which subcategories are full.

1. **The Category P of a Partially Ordered Set.** Suppose that X is a set, partially ordered by $\leq$, and let $\mathcal{O} = X$. If $x, y \in X$, write $f : x \to y$ to indicate that $x \leq y$. The set $\mathrm{Mor}(x, y)$ is the single morphism $f : x \to y$ when $x \leq y$ and $\mathrm{Mor}(x, y) = \varnothing$ if $x \not\leq y$. The rule of composition is given by the transitive property of the partial order: if $f : x \to y$ and $g : y \to z$, then $gf : x \to z$ since $x \leq y$ and $y \leq z$ imply $x \leq z$.

2. **The Category Set of Sets.** For this category $\mathcal{O}$ is the class of all sets. If $A, B \in \mathcal{O}$, then $\mathrm{Mor}(A, B)$ is the set of all functions from A to B. The rule of composition is function composition. A set X is said to be *pointed* if there is a fixed point x^* of X called a *distinguished element* of X. A function $f : X \to Y$ is said to be a *pointed function* if $f(x^*) = y^*$ whenever X and Y are pointed sets with distinguished elements x^* and y^*, respectively. The category **Set*** of pointed sets and pointed functions is a subcategory of **Set**.

3. **The Category Grp of Groups.** If $\mathcal{O}$ is the class of all groups and $G, H \in \mathcal{O}$, then $\mathrm{Mor}(G, H)$ is the set of all group homomorphisms from G to H. The rule of composition is the composition of group homomorphisms. The category **Ab** $(= \mathbf{Mod}_{\mathbb{Z}})$ of abelian groups is a subcategory of **Grp**.

4. **The Category DivAb of Divisible Abelian Groups.** An abelian group G is said to be *divisible*, if for every $y \in G$ and for every positive integer n, there is an $x \in G$ such that $nx = y$. We can form a category **DivAb** whose objects are divisible abelian groups and whose morphisms are group homomorphisms $f : G \to H$. The rule of composition for **DivAb** is composition of group homomorphisms.

5. **The Category Mod$_R$ of Right R-modules.** For this category, $\mathcal{O}$ is the class of all (unitary) right R-modules and if $M, N \in \mathcal{O}$, then $\text{Mor}(M, N) = \text{Hom}_R(M, N)$. The rule of composition is the composition of R-linear mappings. Similarly, $_R\textbf{Mod}$ denotes the category of (unitary) left R-modules.

6. **The Category Rng of Rings.** The class $\mathcal{O}$ of objects consists of rings not necessarily having an identity and if $R, S \in \mathcal{O}$, then $\text{Mor}(R, S)$ is the set of all ring homomorphisms from R to S. The rule of composition is the composition of ring homomorphisms. If this category is denoted by **Rng** and if **Ring** denotes the category of all rings with an identity and identity preserving ring homomorphisms, then **Ring** is a subcategory of **Rng**.

7. **The Categories $\mathcal{C}^M$ and $\mathcal{C}_M$.** Let M be a fixed R-module and form the category $\mathcal{C}^M$ whose objects are R-linear mappings $f : M \to N$. If $f : M \to N_1$ and $g : M \to N_2$ are objects in $\mathcal{C}^M$, then a morphism in $\mathcal{C}^M$ is an R-linear mapping $h : N_1 \to N_2$ such that the diagram

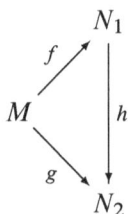

$$
\begin{array}{ccc}
 & & N_1 \\
 & \nearrow^{f} & \big\downarrow h \\
M & & \\
 & \searrow_{g} & \downarrow \\
 & & N_2
\end{array}
$$

is commutative. This category is sometimes referred to as *the category from M*. We can also form the category $\mathcal{C}_M$ whose objects are R-linear mappings $f : N \to M$ and whose morphisms are R-linear mappings $h : N_2 \to N_1$ such that the diagram

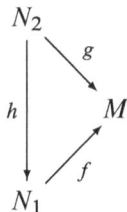

$$
\begin{array}{ccc}
N_2 & & \\
\big\downarrow h & \searrow^{g} & \\
 & & M \\
\downarrow & \nearrow_{f} & \\
N_1 & &
\end{array}
$$

is commutative. $\mathcal{C}_M$ is *the category to M*. More generally, if $\mathcal{C}$ is a category, then a category $\mathcal{C}(\to)$ can be formed from $\mathcal{C}$ as follows. An object of $\mathcal{C}(\to)$ is a morphism $f : A \rightarrow B$ in $\mathcal{C}$. If $f : A \rightarrow B$ and $g : C \rightarrow D$ are objects in $\mathcal{C}(\rightarrow)$, then the morphism set

$$\mathrm{Mor}\,(f : A \rightarrow B, g : C \rightarrow D)$$

in $\mathcal{C}(\rightarrow)$ is the set of all pairs $(\alpha : A \rightarrow C, \beta : B \rightarrow D)$ of morphisms from $\mathcal{C}$ such that the diagram

$$
\begin{array}{ccc}
A & \xrightarrow{\ f\ } & B \\[4pt]
\Big\downarrow{\scriptstyle \alpha} & & \Big\downarrow{\scriptstyle \beta} \\[4pt]
C & \xrightarrow[\ g\ ]{} & D
\end{array}
$$

is commutative.

The following definition fulfills a promise made in the previous chapter to make the concept of a universal mapping property more precise.

Definition 3.1.2. If $\mathcal{C}$ is a category, then an object A of $\mathcal{C}$ is said to be an *initial object* (or a *universal repelling object*) of $\mathcal{C}$ if $\mathrm{Mor}(A, B)$ has exactly one morphism for each object B of $\mathcal{C}$. Likewise, an object B of $\mathcal{C}$ is said to be a *final object* (or a *universal attracting object*) of $\mathcal{C}$ if $\mathrm{Mor}(A, B)$ has exactly one morphism for each object A of $\mathcal{C}$. (An object that is at the same time universal attracting and universal repelling is often referred to as a *universal object*.) If an object in a category is an initial or a final object in the category, then we say that the object has the *universal mapping property*. Conversely, if we say that a mathematical object has the universal mapping property, then we mean that a category can be formed in which the object is an initial or a final object.

The proof of the following proposition is left as an exercise.

Proposition 3.1.3.

(1) *If A is an initial or a final object in a category $\mathcal{C}$, then A is unique up to iso-morphism.*

(2) *An object 0 of a category $\mathcal{C}$ is a zero object of $\mathcal{C}$ if and only if 0 is an initial and a final object of $\mathcal{C}$.*

Examples

8. Let $\{M_\alpha\}_\Delta$ be a family of R-modules. Suppose that $\mathcal{C}_{\{M_\alpha\}_\Delta}$ is the category whose objects are pairs $(M, f_\alpha)_\Delta$, where M is an R-module and $\{f_\alpha\}_\Delta$ is a family of R-linear mappings such that $f_\alpha : M \rightarrow M_\alpha$ for each $\alpha \in \Delta$. A morphism

set $\text{Mor}((M, f_\alpha)_\Delta, (N, g_\alpha)_\Delta)$ is composed of R-linear mappings $f : M \to N$ such that $g_\alpha f = f_\alpha$ for each $\alpha \in \Delta$. Consequently, a direct product $(P, p_\alpha)_\Delta$ of the family $\{M_\alpha\}_\Delta$ is a final object in this category, so $(P, p_\alpha)_\Delta$ has the universal mapping property. Thus, the R-module P is unique up to isomorphism in $\mathbf{Mod}_R$. Similarly, one can form a category $\mathcal{C}^{\{M_\alpha\}_\Delta}$ with objects $(M, f_\alpha)_\Delta$, where M is an R-module and $f_\alpha : M_\alpha \to M$ is an R-linear mapping for each $\alpha \in \Delta$. The morphism set $\text{Mor}((N, g_\alpha)_\Delta, (M, f_\alpha)_\Delta)$ is the set of all R-linear mappings $f : N \to M$ such that $fg_\alpha = f_\alpha$ for each $\alpha \in \Delta$. Hence, we see that a direct sum $(S, u_\alpha)_\Delta$ of the family $\{M_\alpha\}_\Delta$ is an initial object in $\mathcal{C}^{\{M_\alpha\}_\Delta}$, so the R-module S is unique up to isomorphism in $\mathbf{Mod}_R$.

9. Let M and N be an R-module and a left R-module, respectively. Let $\mathcal{C}^{M \times N}$ be the category defined as follows: The objects are pairs (G, ρ), where G is an additive abelian group and $\rho : M \times N \to G$ is an R-balanced mapping. The morphism sets are defined by

$$h \in \text{Mor}((G_1, \rho_1), (G_2, \rho_2))$$

if and only if $h : G_1 \to G_2$ is a group homomorphism such that $h\rho_1 = \rho_2$. By Definition 2.3.1, (T, ρ) is a tensor product of M and N if and only if (T, ρ) is an initial object in $\mathcal{C}^{M \times N}$. Consequently, (T, ρ) has the universal mapping property in $\mathcal{C}^{M \times N}$ and it follows that the abelian group T is unique up to isomorphism in $\mathbf{Ab}$.

Functors

It was indicated earlier that an important use of category theory is to delineate areas of mathematics and to make connections among these areas. In order to make these connections, we need to have a method of passing information from one category to another that is, in some sense, structure preserving. The concept of a functor meets this requirement.

Definition 3.1.4. If $\mathcal{C}$ and $\mathcal{D}$ are categories, then a *covariant functor* $\mathcal{F} : \mathcal{C} \to \mathcal{D}$ is a rule that assigns to each object A in $\mathcal{C}$ exactly one object $\mathcal{F}(A)$ of $\mathcal{D}$ and to each morphism $f : A \to B$ of $\mathcal{C}$ exactly one morphism $\mathcal{F}(f) : \mathcal{F}(A) \to \mathcal{F}(B)$ in $\mathcal{D}$ such that the following conditions are satisfied:

(1) $\mathcal{F}(\text{id}_A) = \text{id}_{\mathcal{F}(A)}$ for each object A of $\mathcal{C}$.

(2) If $f : A \to B$ and $g : B \to C$ are morphisms in $\mathcal{C}$, then $\mathcal{F}(gf) = \mathcal{F}(g)\mathcal{F}(f)$ in $\mathcal{D}$.

If $\mathcal{C}$ is a category, then the *opposite category* $\mathcal{C}^{\text{op}}$ of $\mathcal{C}$ is defined as follows. The objects of $\mathcal{C}^{\text{op}}$ are the objects of $\mathcal{C}$ and for objects A and B in $\mathcal{C}^{\text{op}}$, $\text{Mor}^{\text{op}}(B, A)$ is the set of morphisms $\text{Mor}(A, B)$ in $\mathcal{C}$. Thus, if $f^{\text{op}} : B \to A$ and $g^{\text{op}} : C \to B$

are morphisms in $\mathcal{C}^{op}$, then $f : A \to B$ and $g : B \to C$ are morphisms in $\mathcal{C}$. The morphism f^{op} is called the *opposite of f*. The morphisms in $\mathcal{C}^{op}$ are said to be obtained from those of $\mathcal{C}$ by *reversing the arrows*. The rule of composition in $\mathcal{C}$ gives $gf : A \to C$ while the rule of composition in $\mathcal{C}^{op}$ gives $f^{op}g^{op} : C \to A$. Hence, $(gf)^{op} = f^{op}g^{op}$. Clearly, $(\mathcal{C}^{op})^{op} = \mathcal{C}$ and a statement about a category $\mathcal{C}$ involving arrows can often be dualized to a corresponding dual statement in $\mathcal{C}^{op}$ by reversing the arrows and making the necessary changes in terminology.

A *contravariant functor* $\mathcal{F} : \mathcal{C} \to \mathcal{D}$ has exactly the same properties as a covariant functor except that $\mathcal{F}$ assigns to each morphism $f : A \to B$ in $\mathcal{C}$ the morphism $\mathcal{F}(f) : \mathcal{F}(B) \to \mathcal{F}(A)$ in $\mathcal{D}$. If $\mathcal{F} : \mathcal{C} \to \mathcal{D}$ is contravariant and $f : A \to B$ and $g : B \to C$ are morphisms in $\mathcal{C}$, then $\mathcal{F}(g) : \mathcal{F}(C) \to \mathcal{F}(B), \mathcal{F}(f) : \mathcal{F}(B) \to \mathcal{F}(A)$ and $\mathcal{F}(gf) : \mathcal{F}(C) \to \mathcal{F}(A)$ in $\mathcal{D}$. Hence, $\mathcal{F}(gf) = \mathcal{F}(f)\mathcal{F}(g)$.

If $\mathcal{C}$ and $\mathcal{D}$ are additive categories, then a (covariant or contravariant) functor $\mathcal{F} : \mathcal{C} \to \mathcal{D}$ is said to be *additive* if $\mathcal{F}(f+g) = \mathcal{F}(f) + \mathcal{F}(g)$ for all $f, g \in \text{Mor}(A, B)$ and all $A, B \in \mathcal{O}$.

Remark. Subsequently, the term "functor" will mean covariant functor.

Examples

It is left to the reader to determine which of the following categories are additive and which of the following covariant/contravariant functors are additive.

10. **Identity Functors.** If $\mathcal{C}$ is a category, then the *identity functor* $\text{Id}_{\mathcal{C}} : \mathcal{C} \to \mathcal{C}$ is given by $\text{Id}_{\mathcal{C}}(A) = A$ for each object A of $\mathcal{C}$ and if $f : A \to B$ is a morphism in $\mathcal{C}$, then $\text{Id}_{\mathcal{C}}(f) : \text{Id}_{\mathcal{C}}(A) \to \text{Id}_{\mathcal{C}}(B)$ is the morphism $f : A \to B$.

11. **Embedding Functors.** If $\mathcal{C}$ is a subcategory of $\mathcal{D}$ and $\mathcal{F} : \mathcal{C} \to \mathcal{D}$ is such that $\mathcal{F}(A) = A$ and $\mathcal{F}(f) = f$ for each object A and each morphism f of $\mathcal{C}$, then $\mathcal{F}$ is the *canonical embedding functor*.

12. **Forgetful Functors.** If $\mathcal{F} : \text{Mod}_R \to \text{Ab}$ is such that $\mathcal{F}(M)$ is the additive abelian group underlying the R-module structure on M and if $\mathcal{F}(f) : \mathcal{F}(M) \to \mathcal{F}(N)$ is the group homomorphism underlying the R-linear map $f : M \to N$, then $\mathcal{F}$ is a *forgetful functor*. Thus, $\mathcal{F}$ strips the module structure from M and leaves M as an additive abelian group and $\mathcal{F}(f)$ as a group homomorphism.

13. **The Contravariant Functor $\text{Hom}_R(-, X)$.** Let X be a fixed R-module and suppose that $\text{Hom}_R(-, X) : \text{Mod}_R \to \text{Ab}$ is such that for each R-module M, $\text{Hom}_R(-, X)$ at M is $\text{Hom}_R(M, X)$. If $f : M \to N$ is a morphism in Mod_R, let $\text{Hom}_R(f, X) = f^*$, where

$$f^* : \text{Hom}_R(N, X) \to \text{Hom}_R(M, X) \quad \text{is given by } f^*(h) = hf.$$

Then $\text{Hom}_R(-, X)$ is a contravariant functor from Mod_R to Ab.

14. **The Functor $\text{Hom}_R(X, -)$.** Let X be a fixed R-module and suppose that $\text{Hom}_R(X, -) : \mathbf{Mod}_R \to \mathbf{Ab}$ is such that for each R-module M, $\text{Hom}_R(X, -)$ at M is $\text{Hom}_R(X, M)$. If $f : M \to N$ is a morphism in $\mathbf{Mod}_R$, let $\text{Hom}_R(X, f) = f_*$, where

$$f_* : \text{Hom}_R(X, M) \to \text{Hom}_R(X, N) \quad \text{is defined by } f_*(h) = fh.$$

Then $\text{Hom}_R(X, -)$ is a functor from $\mathbf{Mod}_R$ to $\mathbf{Ab}$.

15. **Composition of Functors.** If $\mathcal{F} : \mathcal{C} \to \mathcal{D}$ and $\mathcal{G} : \mathcal{D} \to \mathcal{E}$ are functors, then $\mathcal{G}\mathcal{F} : \mathcal{C} \to \mathcal{E}$, the composition of $\mathcal{F}$ and $\mathcal{G}$, is a functor if $\mathcal{G}\mathcal{F}$ is defined in the obvious way.

Remark. For the contravariant functor $\text{Hom}_R(-, X)$ of Example 13, the notation $f^*(h) = hf$ means that f is applied first. For the functor $\text{Hom}_R(X, -)$ of Example 14, the notation $f_*(h) = fh$ indicates that f is applied second. Thus, the upper star on f means we are working with the first variable of $\text{Hom}_R(-, X)$ with f applied first, while the lower star on f means that we are working in the second variable of $\text{Hom}_R(X, -)$ with f applied second.

Properties of Morphisms

In the category **Set**, an injective function is often defined as a function $f : B \to C$ such that if $f(x) = f(y)$, then $x = y$. A function $f : B \to C$ is also injective if and only if f is *left cancellable*, that is, if $g, h : A \to B$ are functions such that $fg = fh$, then $g = h$. Indeed, if f is injective, let $g, h : A \to B$ be such that $fg = fh$. Then for any $a \in A$ we see that $fg(a) = fh(a)$, so $f(g(a)) = f(h(a))$ gives $g(a) = h(a)$. Hence, $g = h$. Conversely, suppose that f is left cancellable and let $x, y \in B$ be such that $f(x) = f(y)$. Next, let $A = \{a\}$ and define $g, h : \{a\} \to B$ by $g(a) = x$ and $h(a) = y$. Since $fg = fh$ implies that $g = h$, we see that $fg(a) = f(x) = f(y) = fh(a)$ gives $x = g(a) = h(a) = y$ and so f is injective.

One can also show that a morphism $f : A \to B$ in **Set** is a surjective function if and only if whenever $g, h : B \to C$ are functions such that $gf = hf$, then $g = h$, that is, if f is *right cancellable*. These observations lead to the following definition.

Definition 3.1.5. A morphism f in a category $\mathcal{C}$ is said to be *monic* if f is left cancellable. Likewise, f is said to be *epic* if f is *right cancellable*. A *bimorphism* in $\mathcal{C}$ is a morphism in $\mathcal{C}$ that is monic and epic.

We will refer to a category whose objects are sets as a *concrete category*. Often the sets in a concrete category will have additional structure and its morphisms will, in some sense, be structure preserving. For example, **Set**, **Ab** and $\mathbf{Mod}_R$ are examples of concrete categories. A monic (an epic, a bimorphism) morphism may not be as "well behaved" in all concrete categories as they are in **Set**. The following two examples illustrate that fact that a monic (an epic, a bimorphism) morphism in a concrete category may fail to be an injection (a surjection, a bijection).

Examples

16. Consider the natural group homomorphism $\eta : \mathbb{Q} \to \mathbb{Q}/\mathbb{Z}$ in the category **DivAb** of Example 4. We claim that this morphism is monic in **DivAb**. For this, suppose that G is a divisible abelian group and let $g, h : G \to \mathbb{Q}$ be group homomorphisms such that $g \neq h$. Then there is an $x \in G$ such that $g(x) \neq h(x)$. This means that there is an m/n in $\mathbb{Q}$, $m/n \neq 0$, such that $g(x) - h(x) = m/n$ and m/n can be chosen so that $n \neq \pm 1$. Since G is divisible, there is a $y \in G$ such that $x = my$. Thus,

$$m[g(y) - h(y)] = mg(y) - mh(y)$$
$$= g(my) - h(my)$$
$$= g(x) - h(x) = m/n.$$

Hence, it follows that $g(y) - h(y) = 1/n$, so $g(y) + \mathbb{Z} \neq h(y) + \mathbb{Z}$ in $\mathbb{Q}/\mathbb{Z}$. Thus, $g \neq h$ gives $\eta g \neq \eta h$ and so $\eta g = \eta h$ implies that $g = h$. Therefore, η is monic. But η is clearly not injective, so in general, monic $\not\Rightarrow$ injection.

17. In the category **Ring** of Example 6, consider the canonical injection $j : \mathbb{Z} \to \mathbb{Q}$. We claim that j is epic. Suppose that $g, h : \mathbb{Q} \to \mathbb{Q}$ are such that $gj = hj$. Then $g(n) = h(n)$ for each $n \in \mathbb{Z}$ and so for each m/n in $\mathbb{Q}$ we see that

$$g(m/n) = g(m \cdot n^{-1}) = g(m)g(n)^{-1}$$
$$= h(m)h(n)^{-1} = h(m \cdot n^{-1}) = h(m/n).$$

Hence, $g = h$ and so j is epic. However, j is clearly not a surjective ring homomorphism, so in general, epic $\not\Rightarrow$ surjection. It follows easily that j is also monic, so j is a bimorphism in **Ring**. Hence, in general, bimorphism $\not\Rightarrow$ bijection.

Definition 3.1.6. A morphism $f : A \to B$ in a category $\mathcal{C}$ is said to be a *section* (*retraction*) if there is a morphism $g : B \to A$ in $\mathcal{C}$ such that $gf = \mathrm{id}_A$ ($fg = \mathrm{id}_B$). An *isomorphism* is a morphism that is a section and a retraction.

If $f : A \to B$ is a section (retraction), then it is easy to show that f is monic (epic). However, the converse does not hold, as is shown by the following two examples.

Examples

18. Consider the mapping $f : \mathbb{Z} \to \mathbb{Z}$ in $\mathbf{Mod}_{\mathbb{Z}}$ given by $f(n) = 2n$. It follows easily that f is monic, however, f is not a section. To see that f cannot be a section, suppose that there is a morphism $g : \mathbb{Z} \to \mathbb{Z}$ in $\mathbf{Mod}_{\mathbb{Z}}$ such that $gf = \mathrm{id}_{\mathbb{Z}}$. Then for any integer n, $gf(n) = n$, so $g(2n) = n$. But g is $\mathbb{Z}$-linear, so $g(2n) = 2g(n)$. Hence, $2g(n) = n$ and, in particular, $2g(1) = 1$. But such an integer $g(1)$ cannot exist, so such a g cannot exist. Consequently, in general, monic $\not\Rightarrow$ section.

19. Let p be a prime number and consider the abelian group

$$\mathbb{Q}_p = \{x \in \mathbb{Q} \mid x = mp^{-k} \text{ for some } m \in \mathbb{Z} \text{ and } k \in \mathbb{N}_0\}$$

in $\mathbf{Mod}_{\mathbb{Z}}$. Now $\mathbb{Z}$ is a subgroup of $\mathbb{Q}_p$, so consider the $\mathbb{Z}$-linear map $f : \mathbb{Q}_p/\mathbb{Z} \to \mathbb{Q}_p/\mathbb{Z}$ defined by $f(x + \mathbb{Z}) = px + \mathbb{Z}$. Since $f(mp^{-(k+1)} + \mathbb{Z}) = p(mp^{-(k+1)} + \mathbb{Z} = mp^{-k} + \mathbb{Z}$, it follows that f is epic. We claim that f is not a retraction. Suppose that a morphism $g : \mathbb{Q}_p/\mathbb{Z} \to \mathbb{Q}_p/\mathbb{Z}$ exists in $\mathbf{Mod}_{\mathbb{Z}}$ such that $fg = \mathrm{id}_{\mathbb{Q}_p/\mathbb{Z}}$. Then

$$
\begin{aligned}
p^{-1} + \mathbb{Z} = fg(p^{-1} + \mathbb{Z}) &= pg(p^{-1} + \mathbb{Z}) \\
&= g(pp^{-1} + \mathbb{Z}) = g(1 + \mathbb{Z}) \\
&= g(\mathbb{Z}) = 0
\end{aligned}
$$

which implies that $p^{-1} \in \mathbb{Z}$, a contradiction. Hence, such a g cannot exist, so f is not a retraction. Thus, in general, epic $\not\Rightarrow$ retraction.

Problem Set 3.1

1. Let R be a commutative ring. Verify that we can form a category $\mathcal{C}$ by letting the objects of $\mathcal{C}$ be the positive integers ≥ 2 and by letting $\mathrm{Mor}(m, n)$ be the set of $n \times m$ matrices over R. The rule of composition is matrix multiplication.

2. Show that $\varnothing$ is an initial object and $\{x\}$ is a terminal object in the category **Set**.

3. Prove Proposition 3.1.3.

4. Let $\mathcal{F} : \mathcal{C} \to \mathcal{D}$ and $\mathcal{G} : \mathcal{D} \to \mathcal{E}$ be functors.

    (a) Give an appropriate definition of the composition $\mathcal{G}\mathcal{F} : \mathcal{C} \to \mathcal{E}$ of two functors $\mathcal{F}$ and $\mathcal{G}$ and show that $\mathcal{G}\mathcal{F}$ is a functor.

    (b) If one of the functors $\mathcal{F}$ and $\mathcal{G}$ is covariant and the other is contravariant, then is $\mathcal{G}\mathcal{F}$ covariant or contravariant?

    (c) If $\mathcal{F}$ and $\mathcal{G}$ are contravariant, then is $\mathcal{G}\mathcal{F}$ covariant or contravariant?

5. Consider the category **Set** and for each object A of **Set**, let $\mathcal{F}(A) = \wp(A)$, the power set of A. If $f : A \to B$ is a morphism in **Set**, let $\mathcal{F}(f) = f^{\#}$, where $f^{\#} : \wp(A) \to \wp(B)$ is given by $f^{\#}(X) = f(X)$ for all $X \in \wp(A)$. Show that $\mathcal{F} : \mathbf{Set} \to \mathbf{Set}$ is a functor.

6. Consider the category **Set** and for each object A of **Set**, let $\mathcal{F}(A) = \wp(A)$. If $f : A \to B$ is a morphism in **Set**, let $\mathcal{F}(f) = f^{\#\#}$, where $f^{\#\#} : \wp(B) \to \wp(A)$ is given by $f^{\#\#}(Y) = f^{-1}(Y)$ for all $Y \in \wp(B)$. Show that $\mathcal{F} : \mathbf{Set} \to \mathbf{Set}$ is a contravariant functor.

7. Let $\mathcal{C}$ and $\mathcal{D}$ be additive categories and suppose that $\mathcal{F} : \mathcal{C} \to \mathcal{D}$ is an additive functor.

(a) If $A, B \in \mathcal{O}_{\mathcal{C}}$, let 0_{AB} denote the zero morphism from A to B. That is, 0_{AB} is the additive identity of the abelian group $\mathrm{Mor}_{\mathcal{C}}(A, B)$. Show that $\mathcal{F}(0_{AB})$ is the additive identity of $\mathrm{Mor}_{\mathcal{D}}(\mathcal{F}(A), \mathcal{F}(B))$. Conclude that $\mathcal{F}(0_{AB}) = 0_{\mathcal{F}(A)\mathcal{F}(B)}$ for all $A, B \in \mathcal{O}_{\mathcal{C}}$.

(b) If $f \in \mathrm{Mor}_{\mathcal{C}}(A, B)$, show that $\mathcal{F}(-f)$ is the additive inverse of $\mathcal{F}(f)$ in $\mathrm{Mor}_{\mathcal{D}}(\mathcal{F}(A), \mathcal{F}(B))$. Conclude that $\mathcal{F}(-f) = -\mathcal{F}(f)$ for every morphism f of $\mathcal{C}$.

8. If $\mathcal{C}$ and $\mathcal{D}$ are categories, let $\mathcal{C} \times \mathcal{D}$ be the *product category* defined as follows: The objects of $\mathcal{C} \times \mathcal{D}$ are ordered pairs (A, B), where A is an object of $\mathcal{C}$ and B is an object of $\mathcal{D}$. A morphism in $\mathrm{Mor}_{\mathcal{C} \times \mathcal{D}}((A_1, B_1), (A_2, B_2))$ is a pair (f, g), where $f : A_1 \to A_2$ is a morphism in $\mathcal{C}$ and $g : B_1 \to B_2$ is a morphism in $\mathcal{D}$. Composition of morphisms is given by $(f_2, g_2)(f_1, g_1) = (f_2 f_1, g_2 g_1)$ when $f_2 f_1$ is defined in $\mathcal{C}$ and $g_2 g_1$ is defined in $\mathcal{D}$.

(a) Verify that $\mathcal{C} \times \mathcal{D}$ is a category.

(b) If $\mathcal{C}$ and $\mathcal{D}$ are additive categories, then is $\mathcal{C} \times \mathcal{D}$ additive?

(c) If $\mathcal{F} : \mathcal{C}_1 \to \mathcal{C}_2$ and $\mathcal{G} : \mathcal{D}_1 \to \mathcal{D}_2$ are functors, show that a functor $\mathcal{F} \times \mathcal{G} : \mathcal{C}_1 \times \mathcal{D}_1 \to \mathcal{C}_2 \times \mathcal{D}_2$ can be defined in an obvious way.

9. Let $\mathcal{C}$, $\mathcal{D}$ and $\mathcal{E}$ be categories and suppose that $\mathcal{F} : \mathcal{C} \times \mathcal{D} \to \mathcal{E}$ is a functor. We refer to such a two variable functor $\mathcal{F}$ as a *bifunctor*. Consider the bifunctor $\mathrm{Hom}_R(-, -) : \mathbf{Mod}_R^{\mathrm{op}} \times \mathbf{Mod}_R \to \mathbf{Ab}$ defined as follows: If (M, N) is an object of $\mathbf{Mod}_R^{\mathrm{op}} \times \mathbf{Mod}_R$, then $\mathrm{Hom}_R(M, N)$ is the additive abelian group of **Ab** defined by addition of R-linear mappings. Also if $(f^{\mathrm{op}}, g) : (M, N) \to (M', N')$ is a morphism in $\mathbf{Mod}_R^{\mathrm{op}} \times \mathbf{Mod}_R$, then

$$\mathrm{Hom}_R(f^{\mathrm{op}}, g) : \mathrm{Hom}_R(M, N) \to \mathrm{Hom}_R(M', N')$$

is defined by $\mathrm{Hom}_R(f^{\mathrm{op}}, g)(h) = ghf = g_* f^*(h)$. Note that since $f^{\mathrm{op}} : M \to M'$ is a morphism in $\mathbf{Mod}_R^{\mathrm{op}}$, $f : M' \to M$ in $\mathbf{Mod}_R$.

(a) If $N = N' = X$ is a fixed R-module and $g = \mathrm{id}_X$ and if we write $\mathrm{Hom}_R(f^{\mathrm{op}}, X) = \mathrm{Hom}_R(f^{\mathrm{op}}, \mathrm{id}_X)$ whenever $(f^{\mathrm{op}}, \mathrm{id}_X) : (M, X) \to (M', X)$

is a morphism in $\mathbf{Mod}_R^{\mathrm{op}} \times \mathbf{Mod}_R$, show that $\mathrm{Hom}_R(-, X)$ gives the contravariant functor of Example 13.

(b) If $M = M' = X$ is a fixed R-module and $f^{\mathrm{op}} = \mathrm{id}_X$ and if we write $\mathrm{Hom}_R(X, g) = \mathrm{Hom}_R(\mathrm{id}_X, g)$ when $(\mathrm{id}_X, g) : (X, N) \to (X, N')$ is a morphism in $\mathbf{Mod}_R^{\mathrm{op}} \times \mathbf{Mod}_R$, show that $\mathrm{Hom}_R(X, -)$ is the functor of Example 14. The functors $\mathrm{Hom}_R(-, X)$ and $\mathrm{Hom}_R(X, -)$ are said to be *functors induced by the bifunctor* $\mathrm{Hom}_R(-, -)$.

10. Consider the bifunctor $- \otimes_R - : \mathbf{Mod}_R \times {}_R\mathbf{Mod} \to \mathbf{Ab}$ defined as follows: If (M, N) is an object in $\mathbf{Mod}_R \times {}_R\mathbf{Mod}$, then $M \otimes_R N$ is the additive abelian group formed by taking the tensor product of M and N. If $(f, g) : (M, N) \to (M', N')$ is a morphism in $\mathbf{Mod}_R \times {}_R\mathbf{Mod}$, then $f \otimes g : M \otimes_R N \to M' \otimes_R N'$ is the corresponding morphism in $\mathbf{Ab}$. If $N = N' = X$ is fixed, show that $- \otimes_R -$ induces a functor $- \otimes_R X$ from $\mathbf{Mod}_R$ to $\mathbf{Ab}$. Likewise, if $M = M' = X$ is fixed, show that $- \otimes_R -$ induces a functor $X \otimes_R -$ from ${}_R\mathbf{Mod}$ to $\mathbf{Ab}$.

11. (a) Prove in the category $\mathbf{Set}$ that a morphism is surjective if and only if it right cancellable.

(b) If an arrow indicates implication, verify that each implication given in the table below holds in a concrete category.

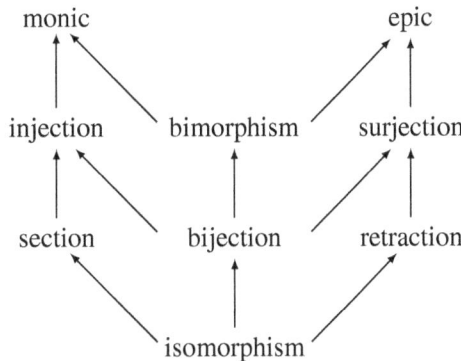

12. Prove each of the following for morphisms f and g in a category $\mathcal{C}$ such that fg is defined.
    (a) If f and g are monic (epic), then fg is monic (epic).
    (b) If fg is monic, then g is monic.
    (c) If fg is epic, then f is epic.

13. If f is a morphism in a category $\mathcal{C}$, prove that the following are equivalent.
    (a) f is monic and a retraction.
    (b) f is epic and a section.
    (c) f is an isomorphism.

14. Prove that each of the following hold in $\mathbf{Mod}_R$.

(a) Every monic morphism is an injection.

(b) Every epic morphism is a surjection.

15. A category $\mathcal{C}$ is said to be *balanced* if every bimorphism is an isomorphism. Prove that $\mathbf{Mod}_R$ is balanced.

16. A functor $\mathcal{F} : \mathcal{C} \to \mathcal{D}$ is said to be *faithful* (*full*) if for all objects M and N of $\mathcal{C}$ the mapping $\mathrm{Mor}_{\mathcal{C}}(M, N) \to \mathrm{Mor}_{\mathcal{D}}(\mathcal{F}(M), \mathcal{F}(N))$ given by $f \mapsto \mathcal{F}(f)$ is injective (surjective).

(a) Show that the forgetful functor $\mathcal{F} : \mathbf{Ab} \to \mathbf{Set}$ is faithful but not full.

(b) Show that the functor $\mathcal{F} : \mathbf{Ab} \to \mathbf{Mod}_{\mathbb{Z}}$ is full and faithful.

A functor $\mathcal{F} : \mathcal{C} \to \mathcal{D}$ is said to *reflect monics* (*epics*) if $\mathcal{F}(f)$ monic (epic) in $\mathcal{D}$ implies that f is monic (epic) in $\mathcal{C}$.

(c) Prove that a faithful functor reflects monics and epics.

Conclude from (c) that if $\mathcal{F} : \mathbf{Mod}_R \to \mathbf{Mod}_S$ is a faithful functor and $0 \to \mathcal{F}(M_1) \to \mathcal{F}(M) \to \mathcal{F}(M_2) \to 0$ is exact in $\mathbf{Mod}_S$, then $0 \to M_1 \to M \to M_2 \to 0$ is exact in $\mathbf{Mod}_R$.

A functor $\mathcal{F} : \mathcal{C} \to \mathcal{D}$ is said to *reflect isomorphisms* if $\mathcal{F}(f)$ an isomorphism in $\mathcal{D}$ implies that f is an isomorphism in $\mathcal{C}$.

(d) Prove that a full and faithful functor reflects isomorphisms.

3.2   Exact Sequences in $\mathbf{Mod}_R$

Definition 3.2.1. A sequence $M_1 \xrightarrow{f} M \xrightarrow{g} M_2$ of R-modules and R-module homomorphisms is said to be *exact at* M if $\mathrm{Im}\, f = \mathrm{Ker}\, g$, while a sequence of the form

$$\mathbf{S} : \cdots \to M_{n-1} \xrightarrow{f_{n-1}} M_n \xrightarrow{f_n} M_{n+1} \to \cdots,$$

$n \in \mathbb{Z}$, is said to be an *exact sequence* if it is exact at M_n for each $n \in \mathbb{Z}$. A sequence such as

$$\mathbf{S} : 0 \to M_1 \xrightarrow{f} M \xrightarrow{g} M_2 \to 0$$

that is exact at M_1, at M and at M_2 is called a *short exact sequence*. A sequence will often be referred to by its prefix $\mathbf{S}$.

Remark. There is nothing special about considering sequences such as

$$\mathbf{S} : \cdots \to M_{n-1} \xrightarrow{f_{n-1}} M_n \xrightarrow{f_n} M_{n+1} \to \cdots$$

with increasing subscripts. We could just as well have considered sequences such as

$$\mathbf{S}: \cdots \to M_{n+1} \xrightarrow{f_{n+1}} M_n \xrightarrow{f_{n-1}} M_{n-1} \to \cdots$$

with decreasing subscripts. Both types of sequences will be considered later in Chapter 11 when chain and cochain complexes are discussed.

If $0 \xrightarrow{}_1 M_1 \xrightarrow{f} M \xrightarrow{g} M_2 \to 0$ is a short exact sequence, then it follows that f is a monomorphism and g is an epimorphism. If N is a submodule of M, then $0 \to N \xrightarrow{i} M \xrightarrow{\eta} M/N \to 0$ is a short exact sequence, where i is the canonical injection and η is the natural mapping. An R-linear map $f : M \to N$ gives rise to two canonical short exact sequences

$$0 \to \operatorname{Ker} f \to M \to f(M) \to 0 \quad \text{and}$$
$$0 \to f(M) \to N \to \operatorname{Coker} f \to 0,$$

where the maps are the obvious ones.

Definition 3.2.2. If $f : M_1 \to M$ is a monomorphism, then we say that f is a *split monomorphism* if there is an R-linear mapping $f' : M \to M_1$ such that $f' f = \operatorname{id}_{M_1}$. The map f' is said to be a *splitting map for f*. Likewise, if $g : M \to M_2$ is an epimorphism, then an R-linear mapping $g' : M_2 \to M$ such that $gg' = \operatorname{id}_{M_2}$ is a *splitting map for g* and g is called a *split epimorphism*. (See Definition 3.1.6.)

Note that if f is a split monomorphism with splitting map f', then f' is a split epimorphism and if g is a split epimorphism with splitting map g', then g' is a split monomorphism.

Proposition 3.2.3. If $M_1 \xrightarrow{f} M$ is a split monomorphism with splitting map $f' : M \to M_1$, then $M = \operatorname{Im} f \oplus \operatorname{Ker} f'$.

Proof. If $x \in M$, then $f'(x) \in M_1$, so $f(f'(x)) \in M$. If $z = x - f(f'(x))$, then $f'(z) = f'(x) - f'(f(f'(x))) = 0$ since $f' f = \operatorname{id}_{M_1}$. Thus, $z \in \operatorname{Ker} f'$ and $x = f(f'(x)) + z \in \operatorname{Im} f + \operatorname{Ker} f'$. Therefore, $M = \operatorname{Im} f + \operatorname{Ker} f'$. If $y \in \operatorname{Im} f \cap \operatorname{Ker} f'$, then $y = f(x)$ for some $x \in M_1$, so $0 = f'(y) = f' f(x) = x$. Hence, $y = 0$ and we have $M = \operatorname{Im} f \oplus \operatorname{Ker} f'$. $\square$

Proposition 3.2.3 also establishes the following proposition.

Proposition 3.2.4. If $g : M \to M_2$ is a split epimorphism and $g' : M_2 \to M$ is a splitting map for g, then $M = \operatorname{Im} g' \oplus \operatorname{Ker} g$.

Split Short Exact Sequences

If $0 \to M_1 \xrightarrow{f} M \xrightarrow{g} M_2 \to 0$ is a short exact sequence of R-modules and R-homomorphisms and f is a split monomorphism, then is g a split epimorphism? Conversely, if g is a split epimorphism, then is f a split monomorphism? These questions lead to the following definition and proposition.

Definition 3.2.5. If $S : 0 \to M_1 \xrightarrow{f} M \xrightarrow{g} M_2 \to 0$ is a short exact sequence of R-modules and R-module homomorphisms and f is a split monomorphism, then S is said to *split on the left*. If g is a split epimorphism, then S is said to *split on the right*.

The connection between a short exact sequence that splits on the left and one that splits on the right is given by the following proposition.

Proposition 3.2.6. *A short exact sequence of R-modules and R-module homomorphisms splits on the left if and only if it splits on the right.*

Proof. Suppose that $S : 0 \to M_1 \xrightarrow{f} M \xrightarrow{g} M_2 \to 0$ is a short exact sequence of R-modules and R-module homomorphisms that splits on the left and let $f' : M \to M_1$ be a splitting map such that $f'f = \mathrm{id}_{M_1}$. Then Proposition 3.2.3 shows that $M = \mathrm{Im}\, f \oplus \mathrm{Ker}\, f'$. Next, note that

$$f'(x - f(f'(x))) = f'(x) - f'(f(f'(x))) = f'(x) - f'(x) = 0,$$

so $x - f(f'(x)) \in \mathrm{Ker}\, f'$ for each $x \in M$. Now define $g' : M_2 \to M$ by $g'(y) = x - f(f'(x))$, where $x \in M$ is such that $g(x) = y$. Such an x exists since g is an epimorphism, but there may be more than one such x. Nevertheless, we claim that g' is well defined. Suppose that $x' \in M$ is also such that $g(x') = y$. Then $x - x' \in \mathrm{Ker}\, g = \mathrm{Im}\, f$, so

$$(x - f(f'(x))) - (x' - f(f'(x'))) = (x - x') - (f(f'(x)) - f(f'(x')))$$
$$= (x - x') - f(f'(x - x'))$$
$$\in \mathrm{Ker}\, f' \cap \mathrm{Im}\, f = 0.$$

Thus, it follows that g' is well defined. If $y \in M_2$ and $g'(y) = x - f(f'(x))$, where $x \in M$ is such that $g(x) = y$, then $g(g'(y)) = g(x - f(f'(x))) = g(x) - g(f(f'(x))) = g(x) = y$ since $gf = 0$. Hence, $gg' = \mathrm{id}_{M_2}$, so S splits on the right. The converse has a similar proof.   □

Because of Proposition 3.2.6, if a short exact sequence splits on either the left or the right, then we simply say that the *sequence splits* or that it is *split exact*. The following proposition gives additional information on split short exact sequences.

Proposition 3.2.7. *A short exact sequence* $0 \to M_1 \xrightarrow{f} M \xrightarrow{g} M_2 \to 0$ *splits if and only if one of the following three equivalent conditions holds.*

(1) Im f *is a direct summand of* M.

(2) Ker g *is a direct summand of* M.

(3) $M \cong M_1 \oplus M_2$.

Proof. We first show that the three conditions are equivalent. Note that (1) clearly implies (2) since Im f = Ker g. If Ker g is a direct summand of M, let N be a submodule of M such that M = Ker $g \oplus N$. Then $M_2 \cong M/\mathrm{Ker}\, g \cong N$ and Ker g = Im $f \cong M_1$ and so we have $M \cong M_1 \oplus M_2$. Thus, (2) implies (3).

Next, suppose that (3) holds. Since Im $f \cong M_1$, we have $M \cong \mathrm{Im}\, f \oplus M_2$, so Proposition 2.1.11 shows that there are submodules N_1 and N_2 of M such that $M = N_1 \oplus N_2$ with Im $f \cong N_1$ and $M_2 \cong N_2$. We claim that $M = \mathrm{Im}\, f \oplus N_2$. If $z \in M$, then there are $x \in N_1$ and $y \in N_2$ such that $z = x + y$. If $w \in M_1$ is such that $f(w) = x$, then $z = f(w) + y$, so $M = \mathrm{Im}\, f + N_2$. It follows that Im $f \cap N_2 = 0$ and so Im f is a direct summand of M. Thus, (3) implies (1).

Finally, suppose that $M_1 \to M_1 \xrightarrow{f} M \xrightarrow{g} M_2 \to 0$ splits. Then Proposition 3.2.3 gives $M = \mathrm{Im}\, f \oplus \mathrm{Ker}\, f'$, where f' is a splitting map for f. Hence, (1) holds. Conversely, suppose that (1) holds and let N be a submodule of M such that $M = \mathrm{Im}\, f \oplus N$. If $f' : \mathrm{Im}\, f \oplus N \to M_1$ is such that $f'(f(x) + y) = x$, then f' is a splitting map for f, so $0 \to M_1 \xrightarrow{f} M \xrightarrow{g} M_2 \to 0$ splits. $\square$

A category $\mathcal{C}$ can be formed by letting the class $\mathcal{SE}$ of short exact sequences in **Mod**$_R$ be the objects of $\mathcal{C}$. If $\mathbf{S}_1 : 0 \to M_1 \to M_1 \to M_2 \to 0$ and $\mathbf{S}_2 : 0 \to N_1 \to N \to N_2 \to 0$ are objects of $\mathcal{C}$, then Mor($\mathbf{S}_1, \mathbf{S}_2$) is the set of all triples $\delta = (\alpha, \beta, \gamma)$ of R-module homomorphisms such that the diagram

$$
\begin{array}{ccccccccc}
0 & \longrightarrow & M_1 & \longrightarrow & M & \longrightarrow & M_2 & \longrightarrow & 0 \\
 & & \downarrow{\scriptstyle \alpha} & & \downarrow{\scriptstyle \beta} & & \downarrow{\scriptstyle \gamma} & & \\
0 & \longrightarrow & N_1 & \longrightarrow & N & \longrightarrow & N_2 & \longrightarrow & 0
\end{array}
$$

is commutative. If $\delta_1 : \mathbf{S}_1 \to \mathbf{S}_2$ and $\delta_2 : \mathbf{S}_2 \to \mathbf{S}_3$ are morphisms in $\mathcal{C}$, then the rule of composition is given by $\delta_2\delta_1 = (\alpha_2, \beta_2, \gamma_2)(\alpha_1, \beta_1, \gamma_1) = (\alpha_2\alpha_1, \beta_2\beta_1, \gamma_2\gamma_1)$. We call a morphism $\delta = (\alpha, \beta, \gamma)$ in $\mathcal{C}$ a monomorphism (an epimorphism, an isomorphism) if α, β and γ are monomorphisms (epimorphisms, isomorphisms) in **Mod**$_R$. The Short Five Lemma (See Exercise 9.) shows that a morphism $\delta = (\alpha, \beta, \gamma)$ in $\mathcal{C}$ is a monomorphism (an epimorphism, an isomorphism) when α and γ are monomorphisms (epimorphisms, isomorphisms) in **Mod**$_R$. If the relation $\sim$ is defined on $\mathcal{SE}$ by $S_1 \sim S_2$ if and only if there is an isomorphism $\delta : S_1 \to S_2$, then $\sim$ is an equivalence relation on $\mathcal{SE}$. If $[S]$ is an equivalence class of short exact sequences in $\mathcal{SE}$, then one can show that a sequence in $[S]$ splits if and only if every sequence in $[S]$ splits.

Problem Set 3.2

1. Let $f : M \to N$ be an R-linear mapping. Identify each R-module homomorphism in each of the following sequences and then show that each sequence is exact.

   (a) $0 \to \operatorname{Ker} f \to M \to f(M) \to 0$

   (b) $0 \to f(M) \to N \to \operatorname{Coker} f \to 0$

   (c) $0 \to \operatorname{Ker} f \to M \to N \to \operatorname{Coker} f \to 0$

2. (a) Let

$$S_1 : 0 \to M_1 \to M \xrightarrow{f} N \to 0 \quad \text{and}$$

$$S_2 : 0 \to N \xrightarrow{g} M_2 \to M_3 \to 0$$

   be short exact sequences. Prove that

$$S_3 : 0 \to M_1 \to M \xrightarrow{gf} M_2 \to M_3 \to 0$$

   is exact. The sequences S_1 and S_2 are said to be *spliced together* to form S_3.

   (b) Prove that any exact sequence can be obtained by splicing together appropriately chosen short exact sequences.

3. (a) Show that if f is a split monomorphism with splitting map f', then f' is a split epimorphism.

   (b) Prove that if g is a split epimorphism with splitting map g', then g' is a split monomorphism.

   (c) Prove Proposition 3.2.4.

4. Complete the proof of Proposition 3.2.6.

5. If $f : M \to M$ is an R-linear mapping such that $f^2 = f$, then f said to be an *idempotent endomorphism* in the ring $\operatorname{End}_R(M)$.

   (a) Suppose also that $M = N_1 \oplus N_2$, where N_1 and N_2 are submodules of M and suppose that $f_1, f_2 : M \to M$ are such that $f_1 = i_1\pi_1$ and $f_2 = i_2\pi_2$, where i_k and π_k are the canonical injections and projections for $k = 1, 2$, respectively. Show that f_1 and f_2 are idempotent endomorphisms of M.

(b) Prove that $M = \text{Im } f \oplus \text{Ker } f = \text{Im } f \oplus \text{Im}(\text{id}_M - f)$ for every idempotent endomorphism f of $\text{End}_R(M)$. Conclude from (a) and (b) that if $M = N_1 \oplus N_2$, then there is an idempotent endomorphism $f \in \text{End}_R(M)$ such that $\text{Im } f = N_1$ and $\text{Ker}(\text{id}_M - f) = N_2$.

(c) If $M = \bigoplus_\Delta M_\alpha$, prove that there is a set of orthogonal idempotents $\{f_\alpha\}_\alpha$ of the ring $\text{End}_R(M)$ such that $f_\alpha(M) = M_\alpha$ for each $\alpha \in \Delta$. [Hint: If $N_\alpha = \sum_{\beta \neq \alpha} M_\beta$, then $M = M_\alpha \oplus N_\alpha$ for each $\alpha \in \Delta$.]

6. If $0 \to M_1 \xrightarrow{f} M \xrightarrow{g} M_2 \to 0$ is a split short exact sequence and f' and g' are splitting maps for f and g, respectively, show that $0 \to M_2 \xrightarrow{g'} M \xrightarrow{f'} M_1 \to 0$ is a split short exact sequence.

7. If N_1 and N_2 are submodules of an R-module M, prove that

$$0 \to N_1 \cap N_2 \xrightarrow{f} N_1 \oplus N_2 \xrightarrow{g} N_1 + N_2 \to 0$$

is exact if $f(x) = (x, x)$ and $g(x, y) = x - y$.

8. If R be an integral domain, let $t(M)$ denote the torsion submodule of M.

(a) If $f : M \to N$ is an R-linear mapping, show that there is an R-linear mapping $t(f) : t(M) \to t(N)$.

(b) Prove that $t : \mathbf{Mod}_R \to \mathbf{Mod}_R$ is a functor.

(c) If $0 \to M_1 \xrightarrow{f} M \xrightarrow{g} M_2 \to 0$ is an exact sequence of R-modules and R-module homomorphisms, prove that $0 \to t(M_1) \xrightarrow{t(f)} t(M) \xrightarrow{t(g)} t(M_2)$ is exact.

(d) Show by example that if $g : M \to N$ is an epimorphism, then $t(g) : t(M) \to t(N)$ need not be an epimorphism. Conclude that the functor t does not preserve short exact sequences.

9. **The Short Five Lemma.** Let

$$
\begin{array}{ccccccccc}
0 & \longrightarrow & M_1 & \longrightarrow & M & \longrightarrow & M_2 & \longrightarrow & 0 \\
& & \downarrow{\alpha} & & \downarrow{\beta} & & \downarrow{\gamma} & & \\
0 & \longrightarrow & N_1 & \longrightarrow & N & \longrightarrow & N_2 & \longrightarrow & 0
\end{array}
$$

be a row exact commutative diagram of R-modules and R-module homomorphisms. Prove each of the following by chasing the diagram.

(a) If α and γ are monomorphisms, then β is a monomorphism.

(b) If α and γ are epimorphisms, then β is an epimorphism.

Conclude that if α and γ are isomorphisms, then β is an isomorphism.

10. Let $\mathcal{C}$ be the structure described in the paragraph immediately preceding this problem set and let $\sim$ be the relation defined on $\mathcal{C}$.

(a) Show that $\mathcal{C}$ is a category.

(b) Verify that $\sim$ is an equivalence relation on $\mathcal{SE}$.

(c) Prove that a sequence in an equivalence class $[\mathbf{S}]$ splits if and only if every sequence in $[\mathbf{S}]$ splits.

11. **The Five Lemma.** Let

$$
\begin{array}{ccccccccc}
M_1 & \longrightarrow & M_2 & \longrightarrow & M_3 & \longrightarrow & M_4 & \longrightarrow & M_5 \\
\downarrow{\scriptstyle\alpha} & & \downarrow{\scriptstyle\beta} & & \downarrow{\scriptstyle\gamma} & & \downarrow{\scriptstyle\delta} & & \downarrow{\scriptstyle\varepsilon} \\
N_1 & \longrightarrow & N_2 & \longrightarrow & N_3 & \longrightarrow & N_4 & \longrightarrow & N_5
\end{array}
$$

be a row exact commutative diagram of R-modules and R-module homomorphisms. Prove each of the following by chasing the diagram.

(a) If α is an epimorphism and β and δ are monomorphisms, then γ is a monomorphism.

(b) If ε is a monomorphism and β and δ are epimorphisms, then γ is an epimorphism.

(c) If α, β, δ and ε are isomorphisms, then γ is an isomorphism.

12. Consider the commutative row exact

$$
\begin{array}{ccccc}
M_1 & \xrightarrow{f} & M & \xrightarrow{g} & M_2 \\
\downarrow{\scriptstyle\alpha} & & \downarrow{\scriptstyle\beta} & & \downarrow{\scriptstyle\gamma} \\
N_1 & \xrightarrow{f'} & N & \xrightarrow{g'} & N_2
\end{array}
$$

of R-modules and R-module homomorphisms. Prove each of the following by chasing the diagram.

(a) If β is a monomorphism and α and g are epimorphisms, then γ is a monomorphism.

(b) If β is an epimorphism and f' and γ are monomorphisms, then α is an epimorphism.

13. Suppose that the diagram

$$
\begin{array}{ccccccccc}
& & 0 & & 0 & & 0 & & \\
& & \downarrow & & \downarrow & & \downarrow & & \\
0 & \longrightarrow & L_1 & \xrightarrow{\;f'\;} & M_1 & \xrightarrow{\;g'\;} & N_1 & \longrightarrow & 0 \\
& & \alpha_1 \downarrow & & \beta_1 \downarrow & & \gamma_1 \downarrow & & \\
0 & \longrightarrow & L & \xrightarrow{\;f\;} & M & \xrightarrow{\;g\;} & N & \longrightarrow & 0 \\
& & \alpha_2 \downarrow & & \beta_2 \downarrow & & \gamma_2 \downarrow & & \\
& & L_2 & & M_2 & & N_2 & & \\
& & \downarrow & & \downarrow & & \downarrow & & \\
& & 0 & & 0 & & 0 & &
\end{array}
$$

of R-modules and R-module homomorphisms is commutative and that the columns and rows are exact. Prove that there is a monomorphism $f'' : L_2 \to M_2$ and an epimorphism $g'' : M_2 \to N_2$ such that the bottom row is exact and such that the completed diagram is commutative. [Hint: Observe that $\beta_2 f \alpha_1 = 0$ gives $\operatorname{Ker} \alpha_2 = \operatorname{Im} \alpha_1 \subseteq \operatorname{Ker} \beta_2 f$, so use this to induce a map $f'' : L_2 \to M_2$.]

14. A category $\mathcal{C}$ can be formed by letting sequences

$$\mathbf{S}: \cdots \to M_{n-2} \xrightarrow{f_{n-2}} M_{n-1} \xrightarrow{f_{n-1}} M_n \xrightarrow{f_n} M_{n+1} \xrightarrow{f_{n+1}} M_{n+2} \to \cdots$$

of R-modules and R-module homomorphisms be the objects of $\mathcal{C}$. If $\mathbf{S}$ and $\mathbf{T}$ are objects of $\mathcal{C}$, then each morphism in $\operatorname{Mor}(\mathbf{S}, \mathbf{T})$ is a family $\delta = \{\alpha_n\}_{\mathbb{Z}}$ of R-linear mappings such that the diagram

$$
\begin{array}{ccccccccc}
\mathbf{S}: \cdots \longrightarrow & M_{n-2} & \xrightarrow{f_{n-2}} & M_{n-1} & \xrightarrow{f_{n-1}} & M_n & \xrightarrow{f_n} & M_{n+1} & \xrightarrow{f_{n+1}} & M_{n+2} \longrightarrow \cdots \\
& \alpha_{n-2} \downarrow & & \alpha_{n-1} \downarrow & & \alpha_n \downarrow & & \alpha_{n+1} \downarrow & & \alpha_{n+2} \downarrow \\
\mathbf{T}: \cdots \longrightarrow & N_{n-2} & \xrightarrow{g_{n-2}} & N_{n-1} & \xrightarrow{g_{n-1}} & N_n & \xrightarrow{g_n} & N_{n+1} & \xrightarrow{g_{n+1}} & N_{n+2} \longrightarrow \cdots
\end{array}
$$

is commutative. Composition of morphisms is defined in the obvious way. A morphism $\delta : \mathbf{S} \to \mathbf{T}$ is said to be a monomorphism (an epimorphism, an isomorphism) if each α_n in δ is a monomorphism (an epimorphism, an isomorphism). Prove that $\mathcal{C}$ is actually a category and let the relation $\sim$ be defined on the objects of $\mathcal{C}$ by $\mathbf{S} \sim \mathbf{T}$ if there is an isomorphism $\delta : \mathbf{S} \to \mathbf{T}$. Prove that $\sim$ is an equivalence relation on the objects of $\mathcal{C}$ and show that if $[\mathbf{S}]$ is an equivalence class determined by an object $\mathbf{S}$ of $\mathcal{C}$, then $\mathbf{S}$ is exact if and only if every $\mathbf{T} \in [\mathbf{S}]$ is exact.

3.3   Hom and $\otimes$ as Functors

We now investigate two important bifunctors both of which are additive in each variable. These functors play an important role in homological algebra, a topic to be
introduced in a subsequent chapter. For the remainder of the text, we adopt the notation Hom and $\otimes$ for the bifunctors

$$\mathrm{Hom}_R(-,-) : \mathbf{Mod}_R^{\mathrm{op}} \times \mathbf{Mod}_R \to \mathbf{Ab} \quad \text{and}$$
$$- \otimes_R - : \mathbf{Mod}_R \times_R \mathbf{Mod} \to \mathbf{Ab},$$

respectively.

Properties of Hom

We have seen in Examples 13 and 14 in Section 3.1 that for any fixed R-module
X, Hom can be used to define a contravariant functor $\mathrm{Hom}_R(-, X)$ and a covariant
functor $\mathrm{Hom}_R(X, -)$ from $\mathbf{Mod}_R$ to $\mathbf{Ab}$. If X is a fixed R-module and $f : M \to N$
is an R-linear mapping, then the contravariant functor

$$\mathrm{Hom}_R(-, X) : \mathbf{Mod}_R \to \mathbf{Ab}$$

is such that

$$\mathrm{Hom}_R(f, X) = f^* : \mathrm{Hom}_R(N, X) \to \mathrm{Hom}_R(M, X), \quad \text{where } f^*(h) = hf.$$

For the covariant functor

$$\mathrm{Hom}_R(X, -) : \mathbf{Mod}_R \to \mathbf{Ab}$$

we have

$$\mathrm{Hom}_R(X, f) = f_* : \mathrm{Hom}_R(X, M) \to \mathrm{Hom}_R(X, N), \quad \text{where } f_*(h) = fh.$$

Thus, if

$$M_1 \xrightarrow{f} M \xrightarrow{g} M_2$$

is a sequence of R-modules and R-module homomorphisms, then for any R-mod-
ule X

$$\mathrm{Hom}_R(M_2, X) \xrightarrow{g^*} \mathrm{Hom}_R(M, X) \xrightarrow{f^*} \mathrm{Hom}_R(M_1, X) \quad \text{and}$$

$$\mathrm{Hom}_R(X, M_1) \xrightarrow{f_*} \mathrm{Hom}_R(X, M) \xrightarrow{g_*} \mathrm{Hom}_R(X, M_2)$$

are sequences of additive abelian groups and group homomorphisms.

One of the first questions that needs to be addressed concerning Hom is, if

$$0 \to M_1 \xrightarrow{f} M \xrightarrow{g} M_2 \to 0$$

is a short exact sequence in $\mathbf{Mod}_R$, then are

$$0 \to \operatorname{Hom}_R(M_2, X) \xrightarrow{g^*} \operatorname{Hom}_R(M, X) \xrightarrow{f^*} \operatorname{Hom}_R(M_1, X) \to 0 \quad \text{and}$$

$$0 \to \operatorname{Hom}_R(X, M_1) \xrightarrow{f_*} \operatorname{Hom}_R(X, M) \xrightarrow{g_*} \operatorname{Hom}_R(X, M_2) \to 0$$

exact sequences in $\mathbf{Ab}$? The following definition will help clarify our discussion.

Definition 3.3.1. Let R and S be rings. A functor $\mathcal{F} : \mathbf{Mod}_R \to \mathbf{Mod}_S$ is said to be *left exact* if

$$0 \to \mathcal{F}(M_1) \xrightarrow{\mathcal{F}(f)} \mathcal{F}(M) \xrightarrow{\mathcal{F}(g)} \mathcal{F}(M_2)$$

is an exact sequence in $\mathbf{Mod}_S$ whenever

$$0 \to M_1 \xrightarrow{f} M \xrightarrow{g} M_2$$

is exact in $\mathbf{Mod}_R$. Likewise, if

$$\mathcal{F}(M_1) \xrightarrow{\mathcal{F}(f)} \mathcal{F}(M) \xrightarrow{\mathcal{F}(g)} \mathcal{F}(M_2) \to 0$$

is exact in $\mathbf{Mod}_S$ for each exact sequence

$$M_1 \xrightarrow{f} M \xrightarrow{g} M_2 \to 0$$

$\mathbf{Mod}_R$, then $\mathcal{F}$ is said to be *right exact*. If $\mathcal{F}$ is left and right exact, then $\mathcal{F}$ is said to be an *exact functor*. If $\mathcal{F} : \mathbf{Mod}_R \to \mathbf{Mod}_S$ is contravariant and

$$0 \to \mathcal{F}(M_2) \xrightarrow{\mathcal{F}(g)} \mathcal{F}(M) \xrightarrow{\mathcal{F}(f)} \mathcal{F}(M_1)$$

is exact in $\mathbf{Mod}_S$ whenever

$$M_1 \xrightarrow{f} M \xrightarrow{g} M_2 \to 0$$

is exact in $\mathbf{Mod}_R$, then $\mathcal{F}$ is said to be *left exact*. If

$$\mathcal{F}(M_2) \xrightarrow{\mathcal{F}(g)} \mathcal{F}(M) \xrightarrow{\mathcal{F}(f)} \mathcal{F}(M_1) \to 0$$

is exact in $\mathbf{Mod}_S$ for every exact sequence

$$0 \to M_1 \xrightarrow{f} M \xrightarrow{g} M_2$$

in $\mathbf{Mod}_R$, then $\mathcal{F}$ is *right exact*. If $\mathcal{F}$ is left and right exact, then $\mathcal{F}$ is an *exact contravariant functor*.

Thus, exact functors, either covariant or contravariant, are precisely those functors that send short exact sequences to short exact sequences.

Proposition 3.3.2. *For any R-module X, $\mathrm{Hom}_R(-, X)$: $\mathbf{Mod}_R \to \mathbf{Ab}$ and $\mathrm{Hom}_R(X, -)$: $\mathbf{Mod}_R \to \mathbf{Ab}$ are left exact.*

Proof. If $0 \to M_1 \xrightarrow{f} M \xrightarrow{g} M_2$ is an exact sequence in $\mathbf{Mod}_R$, we will show that

$$0 \to \mathrm{Hom}_R(X, M_1) \xrightarrow{f_*} \mathrm{Hom}_R(X, M) \xrightarrow{g_*} \mathrm{Hom}_R(X, M_2)$$

is exact. It is clear that f_* and g_* are group homomorphisms, so we show first that f_* is injective. If $h \in \mathrm{Hom}_R(X, M_1)$ is such that $f_*(h) = 0$, then $fh = 0$. Hence, $fh(x) = 0$ for all $x \in X$ implies that $h(x) = 0$ for all $x \in X$ since f is an injection. Thus, $h = 0$, so f_* is injective. Next, note that if $h \in \mathrm{Im}\, f_*$, then there is an $h' \in \mathrm{Hom}_R(X, M_1)$ such that $h = f_*(h')$. This gives $g_*(h) = g_*f_*(h') = gfh' = 0$ since $gf = 0$. Therefore, $h \in \mathrm{Ker}\, g_*$ and so $\mathrm{Im}\, f_* \subseteq \mathrm{Ker}\, g_*$. Finally, if $h \in \mathrm{Ker}\, g_*$, then $gh = g_*(h) = 0$. Hence, if $x \in X$, then $gh(x) = 0$, so $h(x) \in \mathrm{Ker}\, g = \mathrm{Im}\, f$. Since f is an injection, there is a unique $y \in M_1$ such that $f(y) = h(x)$. Define $h'' : X \to M_1$ by $h''(x) = y$. Then $h'' \in \mathrm{Hom}_R(X, M_1)$ and $fh''(x) = f(y) = h(x)$. Thus, $f_*(h'') = h$, so $h \in \mathrm{Im}\, f_*$ and we have $\mathrm{Ker}\, g_* \subseteq \mathrm{Im}\, f_*$. Hence, $\mathrm{Im}\, f_* = \mathrm{Ker}\, g_*$ and we have that

$$0 \to \mathrm{Hom}_R(X, M_1) \xrightarrow{f_*} \mathrm{Hom}_R(X, M) \xrightarrow{g_*} \mathrm{Hom}_R(X, M_2)$$

is exact. The proof that $\mathrm{Hom}_R(-, X)$: $\mathbf{Mod}_R \to \mathbf{Ab}$ is left exact is similar. $\square$

Because of Proposition 3.3.2 we see that Hom is left exact in each variable. The following two examples show that, in general, Hom is not right exact in either variable.

Examples

In each of the following examples, we consider the short exact sequence $0 \to \mathbb{Z} \xrightarrow{f} \mathbb{Z} \xrightarrow{\eta} \mathbb{Z}_6 \to 0$, where $f(x) = 6x$ for all $x \in \mathbb{Z}$ and η is the natural map.

1. **$\mathrm{Hom}_{\mathbb{Z}}(-, \mathbb{Z}_6)$ Is Not Right Exact** Consider the exact sequence

$$0 \to \mathrm{Hom}_{\mathbb{Z}}(\mathbb{Z}_6, \mathbb{Z}_6) \xrightarrow{\eta^*} \mathrm{Hom}_{\mathbb{Z}}(\mathbb{Z}, \mathbb{Z}_6) \xrightarrow{f^*} \mathrm{Hom}_{\mathbb{Z}}(\mathbb{Z}, \mathbb{Z}_6).$$

If $g \in \mathrm{Hom}_{\mathbb{Z}}(\mathbb{Z}, \mathbb{Z}_6)$ and $x \in \mathbb{Z}$, then $(f^*g)(x) = gf(x) = g(6x) = 6g(x) = 0$. Hence, $f^*(g) = 0$ and so $f^* = 0$. But $\mathrm{Hom}_{\mathbb{Z}}(\mathbb{Z}, \mathbb{Z}_6) \neq 0$, so f^* cannot be an epimorphism. Consequently, $\mathrm{Hom}_{\mathbb{Z}}(-, \mathbb{Z}_6)$ is not right exact.

2. **$\mathrm{Hom}_{\mathbb{Z}}(\mathbb{Z}_6, -)$ Is Not Right Exact** First, note that if $h \in \mathrm{Hom}_{\mathbb{Z}}(\mathbb{Z}_6, \mathbb{Z})$, $h \neq 0$, then there must be an $[x] \in \mathbb{Z}_6$ such that $h([x]) \neq 0$. Thus, $h([x])6$ is not zero in $\mathbb{Z}$. But $h([x])6 = h([x]6) = h([0]) = 0$ and we have a contradiction. Hence, $\mathrm{Hom}_{\mathbb{Z}}(\mathbb{Z}_6, \mathbb{Z}) = 0$. Therefore, the exact sequence

$$0 \to \mathrm{Hom}_{\mathbb{Z}}(\mathbb{Z}_6, \mathbb{Z}) \xrightarrow{f_*} \mathrm{Hom}_{\mathbb{Z}}(\mathbb{Z}_6, \mathbb{Z}) \xrightarrow{\eta_*} \mathrm{Hom}_{\mathbb{Z}}(\mathbb{Z}_6, \mathbb{Z}_6)$$

reduces to

$$0 \to 0 \xrightarrow{f_*} 0 \xrightarrow{\eta_*} \mathrm{Hom}_{\mathbb{Z}}(\mathbb{Z}_6, \mathbb{Z}_6) \neq 0,$$

so η_* is not an epimorphism.

Even though Hom is not an exact functor in either variable, there are short exact sequences that are preserved by Hom in both variables.

Proposition 3.3.3. *If* $0 \to M_1 \xrightarrow{f} M \xrightarrow{g} M_2 \to 0$ *is a split short exact sequence in* $\mathbf{Mod}_R$, *then for any R-module* X

(1) $0 \to \mathrm{Hom}_R(M_2, X) \xrightarrow{g^*} \mathrm{Hom}_R(M, X) \xrightarrow{f^*} \mathrm{Hom}_R(M_1, X) \to 0$ *and*

(2) $0 \to \mathrm{Hom}_R(X, M_1) \xrightarrow{f_*} \mathrm{Hom}_R(X, M) \xrightarrow{g_*} \mathrm{Hom}_R(X, M_2) \to 0$

are split short exact sequences in **Ab**.

Proof. We prove (1) and leave the proof of (2) as an exercise. Proposition 3.3.2 shows that

$$0 \to \mathrm{Hom}_R(M_2, X) \xrightarrow{g^*} \mathrm{Hom}_R(M, X) \xrightarrow{f^*} \mathrm{Hom}_R(M_1, X)$$

is exact, so we are only required to show that f^* is an epimorphism and that one of f^* and g^* has a splitting map. Since $0 \to M_1 \xrightarrow{f} M \xrightarrow{g} M_2 \to 0$ is a split short exact sequence, there is an R-linear mapping $f' : M \to M_1$ such that $f'f = \mathrm{id}_{M_1}$. If $h \in \mathrm{Hom}_R(M_1, X)$, then $hf' \in \mathrm{Hom}_R(M, X)$ and $f^*(hf') = hf'f = h$, so f^* is an epimorphism. Finally, if $f' : M \to M_1$ is a splitting map for f, then $f'^* : \mathrm{Hom}_R(M_1, X) \to \mathrm{Hom}_R(M, X)$, so if $h \in \mathrm{Hom}_R(M_1, X)$, then $f^* f'^*(h) = f^*(hf') = hf'f = h\mathrm{id}_{M_1} = h$. Hence, $f^* f'^* = \mathrm{id}_{\mathrm{Hom}_R(M_1, X)}$ which shows that f'^* is a splitting map for f^*. Thus,

$$0 \to \mathrm{Hom}_R(M_2, X) \xrightarrow{g^*} \mathrm{Hom}_R(M, X) \xrightarrow{f^*} \mathrm{Hom}_R(M_1, X) \to 0$$

is a split short exact sequence in **Ab**. $\square$

Properties of Tensor Products

Tensor product can be used to define a functor from $\mathbf{Mod}_R$ to $\mathbf{Ab}$ and a functor from $_R\mathbf{Mod}$ to $\mathbf{Ab}$. If X is a fixed left R-module, let

$$- \otimes_R X : \mathbf{Mod}_R \to \mathbf{Ab}$$

be such that $(- \otimes_R X)(M) = M \otimes_R X$ and $(- \otimes_R X)(f) = f \otimes \mathrm{id}_X : M \otimes_R X \to N \otimes_R X$ for each R-linear mapping $f : M \to N$ in $\mathbf{Mod}_R$. Then $- \otimes_R X$ is a functor from $\mathbf{Mod}_R$ to $\mathbf{Ab}$. Similarly, if X is a fixed R-module, then

$$X \otimes_R - : {}_R\mathbf{Mod} \to \mathbf{Ab}$$

is a functor from $_R\mathbf{Mod}$ to $\mathbf{Ab}$.

We have seen that the bifunctor $\mathrm{Hom}_R(-, -)$ is left exact in both variables but, in general, is not right exact in either variable. This is in contrast to the bifunctor $- \otimes_R -$ which is right exact in both variables but, in general, is not left exact in either variable.

Proposition 3.3.4. *If $M_1 \xrightarrow{f} M \xrightarrow{g} M_2 \to 0$ is an exact sequence in $\mathbf{Mod}_R$, then for any left R-module X the sequence*

$$(1)\ M_1 \otimes_R X \xrightarrow{f \otimes \mathrm{id}_X} M \otimes_R X \xrightarrow{g \otimes \mathrm{id}_X} M_2 \otimes_R X \to 0$$

is exact in $\mathbf{Ab}$. Similarly, if $M_1 \xrightarrow{f} M \xrightarrow{g} M_2 \to 0$ is an exact sequence in $_R\mathbf{Mod}$, then for any R-module X, the sequence

$$(2)\ X \otimes_R M_1 \xrightarrow{\mathrm{id}_X \otimes f} X \otimes_R M \xrightarrow{\mathrm{id}_X \otimes g} X \otimes_R M_2 \to 0$$

is exact in $\mathbf{Ab}$.

Proof. To show that (1) is exact, we need to show that $\mathrm{Im}(f \otimes \mathrm{id}_X) = \mathrm{Ker}(g \otimes \mathrm{id}_X)$ and that $g \otimes \mathrm{id}_X$ is an epimorphism. Since g and id_X are epimorphisms, it follows easily that $g \otimes \mathrm{id}_X$ is an epimorphism. Note also that $gf = 0$ gives $\mathrm{Im}(f \otimes \mathrm{id}_X) \subseteq \mathrm{Ker}(g \otimes \mathrm{id}_X)$, so there is an induced mapping

$$h : (M \otimes_R X)/\mathrm{Im}(f \otimes \mathrm{id}_X) \to M_2 \otimes_R X$$

such that

$$h(y \otimes x + \mathrm{Im}(f \otimes \mathrm{id}_X)) = g(y) \otimes x.$$

Hence, we have a row exact commutative diagram

The map $\rho : M_2 \times X \to (M \otimes_R X)/\operatorname{Im}(f \otimes \operatorname{id}_X)$ defined by $\rho(z, x) = y \otimes x + \operatorname{Im}(f \otimes \operatorname{id}_X)$, where $y \in M$ is such that $g(y) = z$, is R-balanced, but there may be more than one such y, so we must show that ρ is independent of the choice of y. If $y' \in M$ is also such that $g(y') = z$, then $y - y' \in \operatorname{Ker} g = \operatorname{Im} f$. Let $u \in M_1$ be such that $f(u) = y - y'$. Then $(f \otimes \operatorname{id}_X)(u \otimes x) = f(u) \otimes x = (y - y') \otimes x$, so $(y - y') \otimes x$ is in $\operatorname{Im}(f \otimes \operatorname{id}_X)$. Hence, $y \otimes x + \operatorname{Im}(f \otimes \operatorname{id}_X) = y' \otimes x + \operatorname{Im}(f \otimes \operatorname{id}_X)$ which shows that ρ is well defined. Since ρ is R-balanced, there is a unique group homomorphism

$$\bar{h} : M_2 \otimes_R X \to (M \otimes_R X)/\operatorname{Im}(f \otimes \operatorname{id}_X)$$

such that

$$\bar{h}(z \otimes x) = y \otimes x + \operatorname{Im}(f \otimes \operatorname{id}_X),$$

where again we pick $y \in M$ to be such that $g(y) = z$. One can easily verify that $\bar{h} = h^{-1}$, so h is a group isomorphism. Hence, $(M \otimes_R X)/\operatorname{Ker}(g \otimes \operatorname{id}_X) \cong (M \otimes_R X)/\operatorname{Im}(f \otimes \operatorname{id}_X)$ and it follows that $\operatorname{Im}(f \otimes \operatorname{id}_X) = \operatorname{Ker}(g \otimes \operatorname{id}_X)$. The proof of (2) follows by symmetry. $\square$

The following example shows that, in general, $\otimes$ is not left exact in either variable.

Example

3. $-\otimes_{\mathbb{Z}} \mathbb{Z}_6$ **Is Not Left Exact.** The sequence $0 \to \mathbb{Z} \xrightarrow{i} \mathbb{Q} \xrightarrow{\eta} \mathbb{Q}/\mathbb{Z} \to 0$ is exact, where i is the canonical injection and η is the natural mapping. Furthermore, the sequence

$$\mathbb{Z} \otimes_{\mathbb{Z}} \mathbb{Z}_6 \xrightarrow{i \otimes \operatorname{id}_{\mathbb{Z}_6}} \mathbb{Q} \otimes_{\mathbb{Z}} \mathbb{Z}_6 \xrightarrow{\eta \otimes \operatorname{id}_{\mathbb{Z}_6}} (\mathbb{Q}/\mathbb{Z}) \otimes_{\mathbb{Z}} \mathbb{Z}_6 \to 0$$

is exact, but $i \otimes \operatorname{id}_{\mathbb{Z}_6}$ is not an injection. Surely $\mathbb{Q} \otimes_{\mathbb{Z}} \mathbb{Z}_6 = 0$, since for any $\frac{p}{q} \otimes [n] \in \mathbb{Q} \otimes_{\mathbb{Z}} \mathbb{Z}_6$,

$$\frac{p}{q} \otimes [n] = \frac{p}{6q} \otimes 6[n] = \frac{p}{6q} \otimes 0 = 0.$$

But $\mathbb{Z} \otimes_{\mathbb{Z}} \mathbb{Z}_6 \cong \mathbb{Z}_6 \neq 0$, so $i \otimes \operatorname{id}_{\mathbb{Z}_6}$ cannot be an injective mapping. Hence, $\otimes_{\mathbb{Z}}$ is not left exact in the first variable and, by symmetry, $\otimes_{\mathbb{Z}}$ fails to be left exact in the second variable as well.

Problem Set 3.3

1. Complete the proof of Proposition 3.3.2.

2. Prove (2) of Proposition 3.3.3.

3. Show that each of the bifunctors Hom and $\otimes$ is additive in each variable.

4. (a) Let R and S be rings and suppose that $\mathcal{F} : \mathbf{Mod}_R \to \mathbf{Mod}_S$ is a functor. If $f \in \mathrm{Hom}_R(M, N)$ has a left (right) inverse in $\mathrm{Hom}_R(N, M)$, then does $\mathcal{F}(f)$ have a left (right) inverse in $\mathrm{Hom}_S(\mathcal{F}(N), \mathcal{F}(M))$?

   (b) Answer questions similar to those posed in (a) but for a contravariant functor $\mathcal{F}$.

5. Answer each of the following questions for an additive functor $\mathcal{F} : \mathbf{Mod}_R \to \mathbf{Mod}_{R'}$.

   (a) If $(P, p_\alpha)_\Delta$ is a direct product of a family $\{M_\alpha\}_\Delta$ of R-modules, then is $(\mathcal{F}(P), \mathcal{F}(p_\alpha))_\Delta$ a direct product of the family $\{\mathcal{F}(M_\alpha)\}_\Delta$ of R'-modules?

   (b) If $(S, u_\alpha)_\Delta$ is a direct sum of a family $\{M_\alpha\}_\Delta$ of R-modules, then is $(\mathcal{F}(S), \mathcal{F}(u_\alpha))_\Delta$ a direct sum of the family $\{\mathcal{F}(M_\alpha)\}_\Delta$ of R'-modules?

   (c) Does the functor $\mathrm{Hom}_R(M, -)$ preserve direct products and direct sums?

6. (a) Let $0 \to M_1 \xrightarrow{f} M \xrightarrow{g} M_2 \to 0$ be a split short exact sequence of R-modules and R-module homomorphisms. If $f' : M \to M_1$ and $g' : M_2 \to M$ are splitting maps for f and g, respectively, prove that

$$0 \to \mathrm{Hom}_R(M_1, X) \xrightarrow{f'^*} \mathrm{Hom}_R(M, X) \xrightarrow{g'^*} \mathrm{Hom}_R(M_2, X) \to 0 \quad \text{and}$$

$$0 \to \mathrm{Hom}_R(X, M_2) \xrightarrow{g'_*} \mathrm{Hom}_R(X, M) \xrightarrow{f'_*} \mathrm{Hom}_R(X, M_1) \to 0$$

are split short exact sequences in **Ab**.

7. (a) If $0 \to M_1 \xrightarrow{f} M \xrightarrow{g} M_2 \to 0$ is a short exact sequence of R-modules and R-module homomorphisms, prove that

$$0 \to \mathrm{Hom}_R(R^{(n)}, M_1) \xrightarrow{f_*} \mathrm{Hom}_R(R^{(n)}, M) \xrightarrow{g_*} \mathrm{Hom}_R(R^{(n)}, M_2) \to 0$$

is exact for any integer $n \geq 1$.

   (b) Does the conclusion of (a) remain valid if "for any integer $n \geq 1$" is replaced by "for a set Δ" in the statement of Part (a) and $R^{(n)}$ is replaced by $R^{(\Delta)}$?

8. If $0 \to N \to M$ is an exact sequence of left R-modules, prove that $0 \to R^{(\Delta)} \otimes_R N \to R^{(\Delta)} \otimes_R M$ is exact for any set Δ. Conclude that

$$R^{(\Delta)} \otimes_R - :_R \mathbf{Mod} \to \mathbf{Ab}$$

is an exact functor for any nonempty set Δ with a similar observation holding for

$$- \otimes_R R^{(\Delta)} : \mathbf{Mod}_R \to \mathbf{Ab}.$$

9. Given modules L_R, $_RM_S$ and $_SN$, form $L \otimes_R M \otimes_S N$. Show that by fixing any two of the modules and allowing the other to vary we obtain a right exact functor. Give the category that serves as the "domain" of each functor.

10. (a) Recall that a sequence

$$\mathbf{S} : \cdots \to M_{n-1} \to M_n \to M_{n+1} \to \cdots$$

of R-modules and R-module homomorphisms is exact if it is exact at M_n for each $n \in \mathbb{Z}$. If R and S are rings, prove that $\mathscr{F} : \mathbf{Mod}_R \to \mathbf{Mod}_S$ is an exact functor, (that is, $\mathscr{F}$ preserves short exact sequences) if and only if for every exact sequence $\mathbf{S}$ in $\mathbf{Mod}_R$, the sequence

$$\mathscr{F}(\mathbf{S}) : \cdots \to \mathscr{F}(M_{n-1}) \to \mathscr{F}(M_n) \to \mathscr{F}(M_{n+1}) \to \cdots$$

is exact in $\mathbf{Mod}_S$. Make and prove a similar statement for a contravariant functor $\mathscr{F}$. [Hint: Exercise 2 in Problem Set 3.2. Consider the short exact sequences

$$0 \to \mathrm{Ker}\, f_{n-1} \to M_{n-1} \to \mathrm{Ker}\, f_n \to 0 \quad \text{and}$$
$$0 \to \mathrm{Ker}\, f_n \to M_n \to \mathrm{Ker}\, f_{n+1} \to 0,$$

where $\cdots \to M_{n-1} \xrightarrow{f_{n-1}} M_n \xrightarrow{f_n} M_{n+1} \to \cdots$ is an exact sequence in $\mathbf{S}$.]

(b) If $\mathscr{F} : \mathbf{Mod}_R \to \mathbf{Mod}_S$ is an exact functor and $0 \to M_1 \to M \to M_2 \to 0$ is a split short exact sequence in $\mathbf{Mod}_R$, then does $0 \to \mathscr{F}(M_1) \to \mathscr{F}(M) \to \mathscr{F}(M_2) \to 0$ split in $\mathbf{Mod}_S$?

3.4   Equivalent Categories and Adjoint Functors

Two R-modules M and N are isomorphic if there are R-linear mappings $f : M \to N$ and $g : N \to M$ such that $fg = \mathrm{id}_N$ and $gf = \mathrm{id}_M$. There is an analogous concept for categories.

Definition 3.4.1. A functor $\mathscr{F} : \mathscr{C} \to \mathscr{D}$ is said to be a *category isomorphism* if there is a functor $\mathscr{G} : \mathscr{D} \to \mathscr{C}$ such that $\mathscr{F}\mathscr{G} = \mathbf{Id}_{\mathscr{D}}$ and $\mathscr{G}\mathscr{F} = \mathbf{Id}_{\mathscr{C}}$. When such a functor $\mathscr{F}$ exists, $\mathscr{C}$ and $\mathscr{D}$ are said to be *isomorphic categories*. The notation $\mathscr{C} \cong \mathscr{D}$ will indicate that $\mathscr{C}$ and $\mathscr{D}$ are isomorphic.

Examples

1. The identity functor $\mathbf{Id}_{\mathscr{C}} : \mathscr{C} \to \mathscr{C}$ is a category isomorphism.

2. Let $\mathscr{C}$ and $\mathscr{D}$ be categories and form the product category $\mathscr{C} \times \mathscr{D}$. Suppose that $\mathscr{F} : \mathscr{C} \times \mathscr{D} \to \mathscr{D} \times \mathscr{C}$ is such that $\mathscr{F}((A, B)) = (B, A)$ and $\mathscr{F}((f, g)) = (g, f)$. Next, let $\mathscr{G} : \mathscr{D} \times \mathscr{C} \to \mathscr{C} \times \mathscr{D}$ be defined by $\mathscr{G}((B, A)) = (A, B)$ and $\mathscr{G}((g, f)) = (f, g)$. Then $\mathscr{F}\mathscr{G} = \mathbf{Id}_{\mathscr{D} \times \mathscr{C}}$ and $\mathscr{G}\mathscr{F} = \mathbf{Id}_{\mathscr{C} \times \mathscr{D}}$, so $\mathscr{C} \times \mathscr{D} \cong \mathscr{D} \times \mathscr{C}$.

Nontrivial isomorphic categories do not occur very often. A more commonly occurring situation is described by the following definition and examples.

Definition 3.4.2. Let $\mathcal{F}, \mathcal{G} : \mathcal{C} \to \mathcal{D}$ be functors and suppose that for each object A of $\mathcal{C}$ there is a morphism $\eta_A : \mathcal{F}(A) \to \mathcal{G}(A)$ in $\mathcal{D}$ such that for each morphism $f : A \to B$ in $\mathcal{C}$ the diagram

$$
\begin{array}{ccc}
\mathcal{F}(A) & \xrightarrow{\ \eta_A\ } & \mathcal{G}(A) \\
{\scriptstyle \mathcal{F}(f)}\Big\downarrow & & \Big\downarrow{\scriptstyle \mathcal{G}(f)} \\
\mathcal{F}(B) & \xrightarrow{\ \eta_B\ } & \mathcal{G}(B)
\end{array}
$$

is commutative. Then the class of morphisms $\eta = \{\eta_A : \mathcal{F}(A) \to \mathcal{G}(A)\}$, indexed over the objects of $\mathcal{C}$, is said to be a *natural transformation* from $\mathcal{F}$ to $\mathcal{G}$. Such a transformation, also called a *functorial morphism*, will be denoted by $\eta : \mathcal{F} \to \mathcal{G}$. If η_A is an isomorphism in $\mathcal{D}$ for each object A of $\mathcal{C}$, then η is said to be a *natural isomorphism* and $\mathcal{F}$ and $\mathcal{G}$ are said to be *naturally equivalent functors*, denoted by $\mathcal{F} \approx \mathcal{G}$.

A *composition of natural transformations* can also be defined. Let $\mathcal{F}, \mathcal{G}, \mathcal{H} : \mathcal{C} \to \mathcal{D}$ be functors and suppose that $\zeta : \mathcal{F} \to \mathcal{G}$ and $\eta : \mathcal{G} \to \mathcal{H}$ are natural transformations. Then $\eta\zeta : \mathcal{F} \to \mathcal{H}$ is the natural transformation defined by $(\eta\zeta)_A = \eta_A \zeta_A : \mathcal{F}(A) \to \mathcal{H}(A)$ for each object A of $\mathcal{C}$.

If $\mathcal{F} : \mathcal{C} \to \mathcal{D}$ and $\mathcal{G} : \mathcal{D} \to \mathcal{C}$ are functors such that $\mathcal{G}\mathcal{F} \approx \mathbf{Id}_{\mathcal{C}}$ and $\mathcal{F}\mathcal{G} \approx \mathbf{Id}_{\mathcal{D}}$, then $\mathcal{C}$ and $\mathcal{D}$ are said to be *equivalent categories*, denoted by $\mathcal{C} \approx \mathcal{D}$, and we say that the pair $(\mathcal{F}, \mathcal{G})$ gives a *category equivalence* between $\mathcal{C}$ and $\mathcal{D}$. If $\mathcal{F}$ and $\mathcal{G}$ are contravariant functors such that $\mathcal{G}\mathcal{F} \approx \mathbf{Id}_{\mathcal{C}}$ and $\mathcal{F}\mathcal{G} \approx \mathbf{Id}_{\mathcal{D}}$, then the pair $(\mathcal{F}, \mathcal{G})$ is said to be give a *duality* between $\mathcal{C}$ and $\mathcal{D}$ and we say that $\mathcal{C}$ and $\mathcal{D}$ are *dual categories*. We write $\mathcal{C} \approx \mathcal{D}^{\mathrm{op}}$ to indicate that $\mathcal{C}$ and $\mathcal{D}$ are dual categories.

Examples

3. If $\mathcal{F} : \mathcal{C} \to \mathcal{D}$ is a functor, then $\eta : \mathcal{F} \to \mathcal{F}$ such that $\eta_A : \mathcal{F}(A) \to \mathcal{F}(A)$ is the identity morphism $\mathrm{id}_{\mathcal{F}(A)}$ for each object A of $\mathcal{C}$ is a natural transformation called the *identity natural transformation*.

4. For a given set X, let $\mathcal{F}(X)$ be the free R-module on X. In view of the first Remark of Section 2.2, each element of $\mathcal{F}(X)$ can be written as $\sum_X x a_x$, where $a_x = 0$ for almost all $x \in X$, and X is a basis for $\mathcal{F}(X)$. If $f : X \to Y$ is a function, then we have an R-linear map $\mathcal{F}(f) : \mathcal{F}(X) \to \mathcal{F}(Y)$ defined by $\mathcal{F}(f)(\sum_X x a_x) = \sum_X f(x) a_x$. This gives a functor $\mathcal{F} : \mathbf{Set} \to \mathbf{Mod}_R$ called the *free module functor*. Next, for each R-module M strip the module structure from M and consider M simply as a set. With this done, if $f : M \to N$ is an R-linear map, then f is now just a mapping from the set M to

the set N. Thus, $\mathcal{F}(M)$ is a free module on M and $\mathcal{F}(f) : \mathcal{F}(M) \to \mathcal{F}(N)$ is an R-linear mapping. Consequently, we have a functor $\mathcal{F} : \mathbf{Mod}_R \to \mathbf{Mod}_R$. If $\eta_M : M \to \mathcal{F}(M)$ is defined by $\eta_M(x) = x$, for each R-module M, then the diagram

$$\mathbf{Id}_{\mathbf{Mod}_R}(M) \xrightarrow{\eta_M} \mathcal{F}(M)$$

$$\mathbf{Id}_{\mathbf{Mod}_R}(f) \Big\downarrow \qquad \qquad \Big\downarrow \mathcal{F}(f)$$

$$\mathbf{Id}_{\mathbf{Mod}_R}(N) \xrightarrow{\eta_N} \mathcal{F}(N)$$

is commutative and η is a natural transformation from the functor $\mathbf{Id}_{\mathbf{Mod}_R} : \mathbf{Mod}_R \to \mathbf{Mod}_R$ to the functor $\mathcal{F} : \mathbf{Mod}_R \to \mathbf{Mod}_R$.

5. Let M be a fixed R-module and suppose that

$$\mathcal{F}^M, \mathcal{F}_M : \mathbf{Mod}_R \to \mathbf{Mod}_R$$

are such that:

(a) $\mathcal{F}^M(N) = N \oplus M$ for each R-module N.

(b) If $f : N_1 \to N_2$ is an R-linear mapping, then $\mathcal{F}^M(f) : N_1 \oplus M \to N_2 \oplus M$ is such that $\mathcal{F}^M(f)((x, y)) = (f(x), y)$.

(c) $\mathcal{F}_M(N) = M \oplus N$ for each R-module N.

(d) If $f : N_1 \to N_2$ is an R-linear mapping, then $\mathcal{F}_M(f) : M \oplus N_1 \to M \oplus N_2$ is such that $\mathcal{F}_M(f)((y, x)) = (y, f(x))$.

Now let $\eta : \mathcal{F}^M \to \mathcal{F}_M$ be the natural transformation such that for each R-module N, $\eta_N : \mathcal{F}^M(N) \to \mathcal{F}_M(N)$ is given by $\eta_N((x, y)) = (y, x)$. Then η is a natural isomorphism and we have $\mathcal{F}^M \approx \mathcal{F}_M$.

The proof of the following proposition is left to the reader.

Proposition 3.4.3. *If $\mathcal{F}, \mathcal{G} : \mathcal{C} \to \mathcal{D}$ are functors, then a natural transformation $\eta : \mathcal{F} \to \mathcal{G}$ is a natural isomorphism if and only if there is a natural transformation $\eta^{-1} : \mathcal{G} \to \mathcal{F}$ such that $\eta_A^{-1}\eta_A = \mathrm{id}_{\mathcal{F}(A)}$ and $\eta_A\eta_A^{-1} = \mathrm{id}_{\mathcal{G}(A)}$ for each object A of $\mathcal{C}$.*

Adjoints

There is also an adjoint connection that may occur between two functors. Such a relation holds between the functors $- \otimes_R X : \mathbf{Mod}_R \to \mathbf{Mod}_S$ and $\mathrm{Hom}_S(X, -) : \mathbf{Mod}_S \to \mathbf{Mod}_R$, where $_R X_S$ is an (R, S)-bimodule.

Definition 3.4.4. If $\mathcal{F} : \mathcal{C} \to \mathcal{D}$ and $\mathcal{G} : \mathcal{D} \to \mathcal{C}$ are functors, then $\mathcal{F}$ is said to be a *left adjoint* of $\mathcal{G}$ and $\mathcal{G}$ is a *right adjoint* of $\mathcal{F}$ if for each object A of $\mathcal{C}$ and each object B of $\mathcal{D}$ there is a bijection

$$\eta_{AB} : \mathrm{Mor}_{\mathcal{D}}(\mathcal{F}(A), B) \to \mathrm{Mor}_{\mathcal{C}}(A, \mathcal{G}(B))$$

that is a natural transformation in A and in B. We will refer to the family of mappings $\eta = \{\eta_{AB}\}_{(A,B)\in\mathcal{C}\times\mathcal{D}}$ as an *adjoint transformation*. Such transformation is said to be *natural in A* if when B is fixed and A is allowed to vary, the diagram

$$
\begin{array}{ccc}
\mathrm{Mor}_{\mathcal{D}}(\mathcal{F}(A'), B) & \xrightarrow{\eta_{A'B}} & \mathrm{Mor}_{\mathcal{C}}(A', \mathcal{G}(B)) \\
\downarrow{\scriptstyle \mathcal{F}(f)^*} & & \downarrow{\scriptstyle f^*} \\
\mathrm{Mor}_{\mathcal{D}}(\mathcal{F}(A), B) & \xrightarrow{\eta_{AB}} & \mathrm{Mor}_{\mathcal{C}}(A, \mathcal{G}(B))
\end{array}
$$

is commutative for each morphism $f : A \to A'$ in $\mathcal{C}$. Likewise, such a transformation is said to be *natural in B* if whenever A is fixed and B is allowed to vary, the diagram

$$
\begin{array}{ccc}
\mathrm{Mor}_{\mathcal{D}}(\mathcal{F}(A), B) & \xrightarrow{\eta_{AB}} & \mathrm{Mor}_{\mathcal{C}}(A, \mathcal{G}(B)) \\
\downarrow{\scriptstyle g_*} & & \downarrow{\scriptstyle \mathcal{G}(g)_*} \\
\mathrm{Mor}_{\mathcal{D}}(\mathcal{F}(A), B') & \xrightarrow{\eta_{AB'}} & \mathrm{Mor}_{\mathcal{C}}(A, \mathcal{G}(B'))
\end{array}
$$

is commutative for each morphism $g : B \to B'$ in $\mathcal{D}$. If there is an adjoint transformation $\eta : \mathcal{F} \to \mathcal{G}$, then $(\mathcal{F}, \mathcal{G})$ is said to be an *adjoint pair*.

Example

6. Let $\mathcal{F} : \mathbf{Set} \to \mathbf{Mod}_R$ be the free module functor of Example 4 and let $\mathcal{G} : \mathbf{Mod}_R \to \mathbf{Set}$ be the forgetful functor which takes each R-module to its underlying set and each R-linear mapping to its underlying function. For each set X and each R-module M, each function $f : X \to \mathcal{G}(M)$ can be extended linearly to an R-linear mapping $\bar{f} : \mathcal{F}(M) \to M$. The correspondence $\eta_{XM}(f) = f|_X$ has an inverse given by $\eta_{XM}^{-1}(f) = \bar{f}$ and this establishes a bijection

$$\eta_{XM} : \mathrm{Hom}_R(\mathcal{F}(X), M) \to \mathrm{Mor}_{\mathbf{Set}}(X, \mathcal{G}(M)).$$

If $\eta = \{\eta_{XM}\}_{(X,M)\in\mathbf{Set}\times\mathbf{Mod}_R}$, then $\eta : \mathcal{F} \to \mathcal{G}$ is an adjoint transformation and $(\mathcal{F}, \mathcal{G})$ is an adjoint pair.

Now let R and S be rings, suppose that $_R X_S$ is an (R, S)-bimodule and consider the following:

A. Define $\mathscr{F} : \mathbf{Mod}_R \to \mathbf{Mod}_S$ as follows: Set $\mathscr{F}(M) = M \otimes_R X$ for each R-module M and note that $M \otimes X$ is an S-module if we set $(x \otimes y)s = x \otimes ys$ for $x \otimes y \in M \otimes_R X$ and $s \in S$. Also if $f : M \to M'$ is an R-linear mapping, then $(f \otimes \mathrm{id}_X)((x \otimes y)s) = (f \otimes \mathrm{id}_X)(x \otimes ys) = f(x) \otimes ys = (f(x) \otimes y)s = (f \otimes \mathrm{id}_X)(x \otimes y)s$. Hence, if $\mathscr{F}(f) = f \otimes \mathrm{id}_X$, then it follows that $F(f) : M \otimes_R X \to M \otimes_R X$ is S-linear. Consequently, $\mathscr{F}$ is a functor from $\mathbf{Mod}_R$ to $\mathbf{Mod}_S$.

B. Let $\mathscr{G} : \mathbf{Mod}_S \to \mathbf{Mod}_R$ be such that $\mathscr{G}(N) = \mathrm{Hom}_S(X, N)$. Then $\mathrm{Hom}_S(X, N)$ is an R-module if we set $(ga)(x) = g(ax)$ for each $g \in \mathrm{Hom}_S(X, N), a \in R$ and $x \in X$. Moreover, if $f : N \to N'$ is an S-linear map, then $\mathscr{G}(f) = f_* : \mathrm{Hom}_S(X, N) \to \mathrm{Hom}_S(X, N')$ is an R-linear mapping since $f_*(ga)(x) = f(ga)(x) = (fg)(ax) = ((fg)a)(x) = (f_*(g)a)(x)$, so $f_*(ga) = f_*(g)a$. It follows that $\mathscr{G}$ is a functor from $\mathbf{Mod}_S$ to $\mathbf{Mod}_R$.

The following proposition establishes an important connection between the functors $\mathscr{F}(-) = - \otimes_R X$ and $\mathscr{G}(-) = \mathrm{Hom}_S(X, -)$ described in A and B above.

Proposition 3.4.5 (Adjoint Associativity). *If R and S are rings and $_R X_S$ is an (R, S)-bimodule, then $(- \otimes_R X, \mathrm{Hom}_S(X, -))$ is an adjoint pair and the adjoint transformation $\eta = \{\eta_{MN}\}_{(M,N) \in \mathbf{Mod}_R \times \mathbf{Mod}_S}$ gives a natural isomorphism of abelian groups*

$$\eta_{MN} : \mathrm{Hom}_S(M \otimes_R X, N) \to \mathrm{Hom}_R(M, \mathrm{Hom}_S(X, N))$$

for each object (M, N) in $\mathbf{Mod}_R \times \mathbf{Mod}_S$.

Proof. Suppose that M is an R-module and that N is an S-module. If $f \in \mathrm{Hom}_S(M \otimes_R X, N)$, let $\eta_{MN}(f) : M \to \mathrm{Hom}_S(X, N)$ be defined by $\eta_{MN}(f)(x) = f_x$, where $f_x \in \mathrm{Hom}_S(X, N)$ is such that $f_x(y) = f(x \otimes y)$. Then f_x is clearly an S-linear mapping since f is. Note also that since X is a left R-module, $\mathrm{Hom}_S(X, N)$ is an R-module, so $(f_x a)(y) = f_x(ay) = f(x \otimes ay) = f(xa \otimes y) = f_{xa}(y)$. This gives $f_x a = f_{xa}$ and we see that $\eta_{MN}(f)(xa) = f_{xa} = f_x a = \eta_{MN}(f)(x)a$. Therefore, $\eta_{MN}(f) \in \mathrm{Hom}_R(M, \mathrm{Hom}_S(X, N))$ as required. If f and g are in $\mathrm{Hom}_S(M \otimes_R X, N)$, then a direct computation shows that $\eta_{MN}(f + g) = \eta_{MN}(f) + \eta_{MN}(g)$, so we have established a group homomorphism

$$\eta_{MN} : \mathrm{Hom}_S(M \otimes_R X, N) \to \mathrm{Hom}_R(M, \mathrm{Hom}_S(X, N)).$$

To show that η_{MN} is an isomorphism, we will show that η_{MN} has an inverse. Let $f \in \mathrm{Hom}_R(M, \mathrm{Hom}_S(X, N))$. Since $f(x) \in \mathrm{Hom}_S(X, N)$ for each $x \in M$, this gives a mapping $M \to \mathrm{Hom}_S(X, N)$ defined by $x \mapsto f_x$, where $f(x) = f_x$. Moreover, $f_x(y) \in N$ for each $y \in X$. If $\bar{f}$ is defined on the set $M \times X$ by $\bar{f} : M \times X \to N$, where $\bar{f}((x, y)) = f_x(y)$, then $\bar{f}(xa, y) = f_{xa}(y) =$

$[f(xa)](y) = [f(x)](ay) = f_x(ay) = \bar{f}(x, ay)$ for each $a \in R$ and it follows that $\bar{f}$ is R-balanced. Thus, by the definition of a tensor product, there is a unique group homomorphism $g : M \otimes_R X \to N$ such that $g(x \otimes y) = f_x(y)$ which is easily seen to be S-linear. If we now define

$$\eta_{MN}^{-1} : \mathrm{Hom}_R(M, \mathrm{Hom}_S(X, N)) \to \mathrm{Hom}_S(M \otimes_R X, N),$$

by $\eta_{MN}^{-1}(f) = g$, then η_{MN}^{-1} is an inverse function for η_{MN}, so η_{MN} is a group isomorphism.

If we fix N and allow M to vary, then we get a commutative diagram,

$$
\begin{array}{ccc}
\mathrm{Hom}_S(M' \otimes_R X, N) & \xrightarrow{\eta_{M'N}} & \mathrm{Hom}_R(M', \mathrm{Hom}_S(X, N)) \\
\downarrow{\scriptstyle (f \otimes \mathrm{id}_X)^*} & & \downarrow{\scriptstyle f^*} \\
\mathrm{Hom}_S(M \otimes_R X, N) & \xrightarrow{\eta_{MN}} & \mathrm{Hom}_R(M, \mathrm{Hom}_S(X, N))
\end{array}
$$

where $f : M \to M'$. Thus, η_{MN} is natural in M. Similarly, η_{MN} is natural in N, so η_{MN} is a natural group isomorphism.   □

There is another form of Proposition 3.4.5. The proof is an exercise.

Proposition 3.4.6 (Adjoint Associativity). *If R and S are rings and $_R M$, $_S X_R$ and $_S N$ are modules, then there is a natural abelian group isomorphism*

$$\eta_{MN} : \mathrm{Hom}_S(X \otimes_R M, N) \to \mathrm{Hom}_R(M, \mathrm{Hom}_S(X, N))$$

and $(X \otimes_R -, \mathrm{Hom}_S(X, -))$ is an adjoint pair.

Problem Set 3.4

1. Verify that the transformation η of Example 5 is a natural isomorphism.

2. Prove Proposition 3.4.3.

3. If R is a commutative ring, then are the functors

   $$- \otimes_R M : \mathbf{Mod}_R \to \mathbf{Mod}_R \quad \text{and} \quad M \otimes_R - : \mathbf{Mod}_R \to \mathbf{Mod}_R$$

   naturally equivalent?

4. Verify the details of Example 6.

5. Prove that the function η_{MN}^{-1} given in the proof of Proposition 3.4.5 is an inverse function for η_{MN}.

6. Show that the diagram given in the proof of Proposition 3.4.5 is commutative.

7. Prove that the transformation η_{MN} established in the proof of Proposition 3.4.5 is natural in N.

8. Prove Proposition 3.4.6. [Hint: Make the necessary changes to the proof of Proposition 3.4.5.]

9. Let $M' \to M \to M'' \to 0$ be a sequence of R-modules and R-module homomorphisms. If $0 \to \operatorname{Hom}_R(M'', N) \to \operatorname{Hom}_R(M, N) \to \operatorname{Hom}_R(M', N)$ is exact in **Ab** for every R-module N, prove that $M' \to M \to M'' \to 0$ is exact.

10. Let $\mathscr{F} : \mathbf{Mod}_R \to \mathbf{Mod}_S$ and $\mathscr{G} : \mathbf{Mod}_S \to \mathbf{Mod}_R$ be functors. If $(\mathscr{F}, \mathscr{G})$ is an adjoint pair, prove that $\mathscr{F}$ is right exact and $\mathscr{G}$ is left exact. [Hint: If $M' \to M \to M'' \to 0$ is exact, then we need to show that $\mathscr{F}(M') \to \mathscr{F}(M) \to \mathscr{F}(M'') \to 0$ is exact. Consider the commutative diagram

$$0 \longrightarrow \operatorname{Hom}_S(\mathscr{F}(M''), N) \longrightarrow \operatorname{Hom}_S(\mathscr{F}(M), N) \longrightarrow \operatorname{Hom}_S(\mathscr{F}(M'), N)$$

$$\downarrow \qquad\qquad\qquad \downarrow \qquad\qquad\qquad \downarrow$$

$$0 \longrightarrow \operatorname{Hom}_R(M'', \mathscr{G}(N)) \longrightarrow \operatorname{Hom}_R(M, \mathscr{G}(N)) \longrightarrow \operatorname{Hom}_R M', \mathscr{G}(N)),$$

where the down arrows are natural isomorphisms and N is any S-module. Show that the top row is exact and consider Exercise 9.] The fact that $\mathscr{G}$ is left exact has a similar proof.

Observe that if $(\mathscr{F}, \mathscr{G})$ is an adjoint pair, then $\mathscr{F}$ may not be left exact and $\mathscr{G}$ need not be right exact as is pointed out by the adjoint pair $(- \otimes_R X, \operatorname{Hom}_S(X, -))$.

11. Let $\mathscr{F} : \mathscr{C} \to \mathscr{D}$ be a functor that has a right adjoint $\mathscr{G} : \mathscr{D} \to \mathscr{C}$. Prove that $\mathscr{G}$ is unique up to natural isomorphism. Conclude that a similar proof will show that if $\mathscr{G} : \mathscr{D} \to \mathscr{C}$ is a functor that has a left adjoint $\mathscr{F} : \mathscr{C} \to \mathscr{D}$, then $\mathscr{F}$ is unique up to natural isomorphism.

Chapter 4
Chain Conditions

We begin with a discussion of the generation and cogeneration of modules which is followed by an introduction to chain conditions on modules. Chain conditions play an important role in the classification of rings.

4.1 Generating and Cogenerating Classes

Recall that a (finite) subset X of an R-module M is a (finite) *set of generators* of M if $M = \sum_X xR$. If $\mathcal{S}$ is a set of submodules of M such that $M = \sum_{\mathcal{S}} N$, then $\mathcal{S}$ is said to *span* M. Recall also that every module M has at least one set X of generators, namely the set $X = M$.

Proposition 4.1.1. *The following are equivalent for an R-module M.*

(1) *For every family $\{M_\alpha\}_\Delta$ of submodules of M such that $\{M_\alpha\}_\Delta$ spans M there is a finite subset $F \subseteq \Delta$ such that $\{M_\alpha\}_F$ spans M.*

(2) *If $\{M_\alpha\}_\Delta$ is any family of R-modules, then for every set of R-linear mappings $\{f_\alpha : M_\alpha \to M\}_\Delta$ such that $\{\operatorname{Im} f_\alpha\}_\Delta$ spans M there is a finite subset $F \subseteq \Delta$ such that $\{\operatorname{Im} f_\alpha\}_F$ spans M.*

(3) *For every family $\{M_\alpha\}_\Delta$ of R-modules for which there is an epimorphism $\bigoplus_\Delta M_\alpha \to M$ there is a finite subset $F \subseteq \Delta$ and an epimorphism $\bigoplus_F M_\alpha \to M$.*

(4) *M is finitely generated.*

Proof. (1) $\Rightarrow$ (2) is obvious.

(2) $\Rightarrow$ (3). If $\varphi : \bigoplus_\Delta M_\alpha \to M$ is an epimorphism and $i_\alpha : M_\alpha \to \bigoplus_\Delta M_\alpha$ is the αth canonical injection, then $f_\alpha = \varphi i_\alpha : M_\alpha \to M$ for each $\alpha \in \Delta$ is such that $\{\operatorname{Im} f_\alpha\}_\Delta$ spans M. Thus, by (2) there is a finite subset $F \subseteq \Delta$ such that $\{\operatorname{Im} f_\alpha\}_F$ spans M and this gives an epimorphism $\bigoplus_F M_\alpha \to M$.

(3) $\Rightarrow$ (4). Every R-module is the homomorphic image of a free R-module, so there is a set Δ and an epimorphism $R^{(\Delta)} \to M$. Thus, by (3) there is a finite set $F \subseteq \Delta$ and an epimorphism $\varphi : R^{(F)} \to M$. If we let $F = \{1, 2, \ldots n\}$ and if $\{e_i\}_{i=1}^n$ is the canonical basis for the free R-module $R^{(n)}$, then the finite set $X = \{\varphi(e_i)\}_{i=1}^n$ will generate M.

(4) $\Rightarrow$ (1). Let $\{M_\alpha\}_\Delta$ be a family of submodules of M that spans M. If $X = \{x_1, x_2, \ldots, x_n\}$ is a finite set of generators of M, then $M = \sum_{i=1}^n x_i R = \sum_\Delta M_\alpha$.

Thus, for each i, there is a finite set $F_i \subseteq \Delta$ such that $x_i \in \sum_{F_i} M_\alpha$. If $F = \bigcup_{i=1}^n F_i$, then $F \subseteq \Delta$ and $\{M_\alpha\}_F$ spans M. Hence, we have (1).   □

Because of the equivalence of (3) and (4) in the previous proposition, we can now make a more general observation regarding what it means for a module to be generated by a set of R-modules.

Definition 4.1.2. An R-module M is said to be *generated by a set* $\{M_\alpha\}_\Delta$ of R-modules if there is an epimorphism $\bigoplus_\Delta M_\alpha \rightarrow M$. A class $\mathcal{C}$ of R-modules is said to *generate* $\mathbf{Mod}_R$ if every R-module M is generated by a set $\{M_\alpha\}_\Delta$ of modules in $\mathcal{C}$. In this case, we also say that $\mathcal{C}$ is a *class of generators* for $\mathbf{Mod}_R$. An R-module M is said to generate an R-module N if there is an epimorphism $M^{(\Delta)} \rightarrow N$ for some set Δ. If $\mathcal{C} = \{M\}$ generates $\mathbf{Mod}_R$, then we will simply say that M generates $\mathbf{Mod}_R$.

It is obvious that $\mathbf{Mod}_R$ has the class of all R-modules as a class of generators. Moreover, if $f : M \rightarrow N$ is an epimorphism, then any set of modules that generates M will also generate N. Consequently, if M is finitely generated, then so is N. Note also that every module is the homomorphic image of a free R-module, so R generates $\mathbf{Mod}_R$.

Due to Proposition 4.1.1 an R-module M is finitely generated if and only if for each set of submodules $\{M_\alpha\}_\Delta$ such that $M = \sum_\Delta M_\alpha$, there is a finite subset $F \subseteq \Delta$ such that $M = \sum_F M_\alpha$. This leads to the following definition of a finitely cogenerated module.

Definition 4.1.3. An R-module M is said to be *finitely cogenerated* if whenever $\{M_\alpha\}_\Delta$ is a set of submodules of M such that $\bigcap_\Delta M_\alpha = 0$ there is a finite subset $F \subseteq \Delta$ such that $\bigcap_F M_\alpha = 0$.

Proposition 4.1.4. *The following are equivalent for an R-module M.*

(1) *M is finitely cogenerated.*

(2) *For every family $\{M_\alpha\}_\Delta$ of R-modules and each family $\{f_\alpha : M \rightarrow M_\alpha\}_\Delta$ of R-linear mapping with $\bigcap_\Delta \mathrm{Ker}\, f_\alpha = 0$ there is a finite subset F of Δ such that $\bigcap_F \mathrm{Ker}\, f_\alpha = 0$.*

(3) *For every family $\{M_\alpha\}_\Delta$ of R-modules for which there is a monomorphism $M \rightarrow \prod_\Delta M_\alpha$ there is a finite subset $F \subseteq \Delta$ and a monomorphism $M \rightarrow \prod_F M_\alpha$.*

Proof. (1) $\Rightarrow$ (2) is clear.

(2) $\Rightarrow$ (3). If $\phi : M \rightarrow \prod_\Delta M_\alpha$ is a monomorphism, then $\bigcap_\Delta \mathrm{Ker}(\pi_\alpha \phi) = 0$, where $\pi_\alpha : \prod_\Delta M_\alpha \rightarrow M_\alpha$ is the αth canonical projection for each $\alpha \in \Delta$. By assumption there is a finite subset $F \subseteq \Delta$ such that $\bigcap_F \mathrm{Ker}(\pi_\alpha \phi) = 0$ and this gives a monomorphism $M \rightarrow \prod_F M_\alpha$.

(3) $\Rightarrow$ (1). Let $\{M_\alpha\}_\Delta$ be a family of submodules of M such that $\bigcap_\Delta M_\alpha = 0$. If $\eta_\alpha : M \to M/M_\alpha$ is the natural mapping for each $\alpha \in \Delta$, then $\phi : M \to \prod_\Delta M/M_\alpha$ defined by $\phi(x) = (\eta_\alpha(x))$ is such that $\mathrm{Ker}\,\phi = \bigcap_\Delta M_\alpha = 0$, so ϕ is a monomorphism. But (3) gives a finite subset $F \subseteq \Delta$ and a monomorphism $M \to \prod_F M/M_\alpha$. Thus, $\bigcap_F M_\alpha = 0$ and so we have (1). $\qquad\qquad\square$

Definition 4.1.5. An R-module M is said to be *cogenerated by a set* $\{M_\alpha\}_\Delta$ of R-modules if there is a monomorphism $M \to \prod_\Delta M_\alpha$. A class $\mathcal{C}$ of R-modules is said to be a *class of cogenerators for* $\mathbf{Mod}_R$ if every R-module is cogenerated by a set of modules in $\mathcal{C}$. In this case, we will say that $\mathcal{C}$ *cogenerates* $\mathbf{Mod}_R$. If $\mathcal{C} = \{M\}$, then an R-module N is cogenerated by M if there is a set Δ and a monomorphism $N \to M^\Delta$. If $\mathcal{C} = \{M\}$ and $\mathcal{C}$ cogenerates $\mathbf{Mod}_R$, then we say that M *cogenerates* $\mathbf{Mod}_R$.

It is clear that $\mathbf{Mod}_R$ has the class of all R-modules as a class of cogenerators. Furthermore, if an R-module M is (finitely) cogenerated by $\{M_\alpha\}_\Delta$ and if $f : N \to M$ is a monomorphism, then $\{M_\alpha\}_\Delta$ also (finitely) cogenerates N.

Problem Set 4.1

1. If M and N are R-modules, prove that M (finitely) generates N if an only if there is a (finite) subset $H \subseteq \mathrm{Hom}_R(M, N)$ such that $N = \sum_H \mathrm{Im}\, f$.

2. Let M and N be R-modules. Prove that M generates N if and only if for each nonzero R-linear mapping $f : N \to N'$ there is an R-linear mapping $h : M \to N$ such that $fh \neq 0$.

3. Prove that the following are equivalent for R-modules M and N.

   (a) N is cogenerated by M.

   (b) For each $x \in N$, $x \neq 0$, there is an R-linear map $f : N \to M$ such that $f(x) \neq 0$.

   Conclude that M is a cogenerator for $\mathbf{Mod}_R$ if and only if for each R-module N and each $x \in N$, $x \neq 0$, there is an R-linear mapping $f : N \to M$ such that $f(x) \neq 0$.

4. Let M and N be R-modules. Prove that M cogenerates N if and only if for each nonzero R-linear mapping $f : N' \to N$ there is an R-linear mapping $h : N \to M$ such that $hf \neq 0$.

5. (a) Let $\mathcal{C}_1$ and $\mathcal{C}_2$ be classes of R-modules. If every module in $\mathcal{C}_2$ is generated by a set of modules in $\mathcal{C}_1$ and $\mathcal{C}_2$ generates $\mathbf{Mod}_R$, prove that $\mathcal{C}_1$ also generates $\mathbf{Mod}_R$.

   (b) Show that a class $\mathcal{C}$ generates $\mathbf{Mod}_R$ if and only if R is generated by a set of modules in $\mathcal{C}$.

6. Let $\mathcal{C}$ be a class of R-modules and suppose that $\{M_\alpha\}_\Delta \subseteq \mathcal{C}$ is a set such that $M_\alpha \ncong M_\beta$ when $\alpha \neq \beta$. If each module in $\mathcal{C}$ is isomorphic to an $M_\alpha \in \{M_\alpha\}_\Delta$, then $\{M_\alpha\}_\Delta$ is said to be a *complete set of representatives* of the modules in $\mathcal{C}$.

(a) If $\mathcal{C}$ is a class of generators for $\mathbf{Mod}_R$ and if $\{M_\alpha\}_\Delta$ is a complete set of representatives of $\mathcal{C}$, prove that $\bigoplus_\Delta M_\alpha$ generates $\mathbf{Mod}_R$.

(b) If $\mathcal{C}$ is a class of cogenerators for $\mathbf{Mod}_R$ and if $\{M_\alpha\}_\Delta$ is a complete set of representatives of $\mathcal{C}$, prove that $\prod_\Delta M_\alpha$ cogenerates $\mathbf{Mod}_R$.

7. (a) Prove that an R-module N is cogenerated by a set $\{M_\alpha\}_\Delta$ of R-modules if and only if there is a set $\mathcal{N}$ of submodules of N such that N/N' embeds in some $M_\beta \in \{M_\alpha\}_\Delta$ for each $N' \in \mathcal{N}$ and $\bigcap_\mathcal{N} N' = 0$.

(b) Let $\mathcal{C}_1$ and $\mathcal{C}_2$ be classes of R-modules. If every module in $\mathcal{C}_2$ is cogenerated by a set of modules in $\mathcal{C}_1$ and $\mathcal{C}_2$ cogenerates $\mathbf{Mod}_R$, prove that $\mathcal{C}_1$ also cogenerates $\mathbf{Mod}_R$.

8. If M and N are R-modules, prove that M (finitely) cogenerates N if and only if there is a (finite) subset $H \subseteq \mathrm{Hom}_R(N, M)$ such that $\bigcap_H \mathrm{Ker}\, f = 0$.

4.2 Noetherian and Artinian Modules

In this section we investigate the ascending and descending chain conditions on modules. As we will see, there is a connection between modules that satisfy the ascending (descending) chain condition and modules that are finitely generated (finitely cogenerated).

Definition 4.2.1. An R-module M is said to satisfy the *ascending chain condition* if whenever

$$M_1 \subseteq M_2 \subseteq M_3 \subseteq \cdots$$

is an ascending chain of submodules of M, there is a positive integer n such that $M_i = M_n$ for all $i \geq n$. In this case, we say that the *chain terminates at M_n* or simply that the chain terminates. An R-module that satisfies the ascending chain condition is called *noetherian*. If the ring R is noetherian when viewed as an R-module, then R is said to be a *right noetherian ring*. Noetherian left R-modules and left noetherian rings are defined in the obvious way. A ring R that is left and right noetherian is a *noetherian ring*.

Examples

1. **The Ring of Integers.** The ring of integers $\mathbb{Z}$ is noetherian and, in fact, every principal ideal ring is noetherian (Corollary 4.2.12).

2. **Finite R-modules.** Any finite R-module is noetherian and if M is an R-module with the property that every submodule of M is finitely generated, then, as we will see, M is noetherian.

3. **Finite Dimensional Vector Spaces.** Any finite dimensional vector space over a division ring D is noetherian while an infinite dimensional vector space over D is not. If V is infinite dimensional with basis $\{x_\alpha\}_\Delta$, then $\text{card}(\Delta) \geq \aleph_0$. If $\Gamma = \{\alpha_1, \alpha_2, \alpha_3, \ldots\} \subseteq \Delta$, then $V_n = \bigoplus_{i=1}^n x_{\alpha_i} D$ is a subspace of V for each $n \geq 1$ and $V_1 \subseteq V_2 \subseteq V_3 \subseteq \cdots$ is an ascending chain of subspaces of V that fails to terminate.

4. **Right noetherian but not left noetherian.** There are rings that are right noetherian but not left noetherian and conversely. For example, the matrix ring

$$R = \begin{pmatrix} \mathbb{Z} & \mathbb{Q} \\ 0 & \mathbb{Q} \end{pmatrix} = \left\{ \begin{pmatrix} a_{11} & a_{12} \\ 0 & a_{22} \end{pmatrix} \,\middle|\, a_{11} \in \mathbb{Z}, a_{12}, a_{22} \in \mathbb{Q} \right\}$$

is right noetherian but not left noetherian. Indeed, the only ascending chains of right ideals of R are of the form

$$\begin{pmatrix} I_1 & \mathbb{Q} \\ 0 & 0 \end{pmatrix} \subseteq \begin{pmatrix} I_2 & \mathbb{Q} \\ 0 & 0 \end{pmatrix} \subseteq \begin{pmatrix} I_3 & \mathbb{Q} \\ 0 & 0 \end{pmatrix} \subseteq \cdots \text{ and}$$

$$\begin{pmatrix} I_1 & \mathbb{Q} \\ 0 & \mathbb{Q} \end{pmatrix} \subseteq \begin{pmatrix} I_2 & \mathbb{Q} \\ 0 & \mathbb{Q} \end{pmatrix} \subseteq \begin{pmatrix} I_3 & \mathbb{Q} \\ 0 & \mathbb{Q} \end{pmatrix} \subseteq \cdots$$

where $I_1 \subseteq I_2 \subseteq I_3 \subseteq \cdots$ is an ascending chain of ideals of $\mathbb{Z}$. But $\mathbb{Z}$ is noetherian, so any such chain in R must terminate since $I_1 \subseteq I_2 \subseteq I_3 \subseteq \cdots$ terminates. On the other hand, $\mathbb{Q}$ is not noetherian when viewed as a $\mathbb{Z}$-module. Hence, if $S_1 \subseteq S_2 \subseteq S_3 \subseteq \cdots$ is a nonterminating chain of $\mathbb{Z}$-submodules of $\mathbb{Q}$, then

$$\begin{pmatrix} 0 & S_1 \\ 0 & 0 \end{pmatrix} \subseteq \begin{pmatrix} 0 & S_2 \\ 0 & 0 \end{pmatrix} \subseteq \begin{pmatrix} 0 & S_3 \\ 0 & 0 \end{pmatrix} \subseteq \cdots$$

is a nonterminating chain of left ideals of R.

Definition 4.2.2. An R-module M is said to satisfy the *descending chain condition* if whenever

$$M_1 \supseteq M_2 \supseteq M_3 \supseteq \cdots$$

is a descending chain of submodules of M, there is a positive integer n such that $M_i = M_n$ for all $i \geq n$. In this case, we say that the *chain terminates at M_n*. An R-module that satisfies the descending chain condition is called *artinian*. If the ring R is artinian when considered as an R-module, then R is said to be a *right artinian ring*. Artinian left R-modules and left artinian rings are defined analogously. A ring R that is left and right artinian is said to be an *artinian ring*.

Examples

5. **Noetherian but not artinian.**   We have seen in Example 1 that the ring of integers $\mathbb{Z}$ is noetherian. However $\mathbb{Z}$ is not artinian since, for example,

$$(2) \supseteq (4) \supseteq (8) \supseteq \cdots \supseteq (2^n) \supseteq \cdots$$

is a chain of ideals of $\mathbb{Z}$ that does not terminate. Consequently, there are rings that are noetherian but not artinian. The converse is false. We will see in a later chapter that every right (left) artinian ring is right (left) noetherian.

6. **Right artinian but not left artinian.**   The matrix ring

$$\begin{pmatrix} \mathbb{Q} & \mathbb{R} \\ 0 & \mathbb{R} \end{pmatrix} = \left\{ \begin{pmatrix} a_{11} & a_{12} \\ 0 & a_{22} \end{pmatrix} \,\middle|\, a_{11} \in \mathbb{Q}, a_{12}, a_{22} \in \mathbb{R} \right\}$$

is right artinian but not left artinian. Thus, there are right artinian rings that are not left artinian. Similarly, there are left artinian rings that are not right artinian.

7. **Division Rings.**   A division ring D is artinian and noetherian since the only right or left ideals of D are 0 and D. Furthermore, the $n \times n$ matrix ring $\mathbb{M}_n(D)$ is artinian and noetherian.

8. **Direct Products.**   Let $\{M_n\}_{\mathbb{N}}$ be a family of nonzero R-modules. Then the direct product $\prod_{\mathbb{N}} M_n$ is neither noetherian nor artinian. If

$$N_n = M_1 \times M_2 \times \cdots \times M_n \times 0 \times 0 \times \cdots$$

for each $n \geq 1$, then

$$N_1 \subseteq N_2 \subseteq N_3 \subseteq \cdots$$

is an ascending chain of submodules of $\prod_{\mathbb{N}} M_n$ that fails to terminate. Likewise, if

$$N_n = 0 \times 0 \times \cdots \times 0 \times M_n \times M_{n+1} \times \cdots,$$

then

$$N_1 \supseteq N_2 \supseteq N_3 \supseteq \cdots$$

is a decreasing chain of submodules of $\prod_{\mathbb{N}} M_n$ that does not terminate.

Proposition 4.2.3. *The following are equivalent for an R-module M.*

(1) *M is noetherian.*

(2) *Every nonempty collection of submodules of M, when ordered by inclusion, has a maximal element.*

(3) *Every submodule of M is finitely generated.*

Proof. (1) $\Rightarrow$ (2). Suppose that $\mathcal{S}$ is a nonempty collection of submodules of M. If $M_1 \in \mathcal{S}$ and M_1 is maximal in $\mathcal{S}$, then we are finished. If M_1 is not maximal in $\mathcal{S}$, then there is an $M_2 \in \mathcal{S}$ such that $M_1 \subsetneq M_2$. If M_2 is maximal in $\mathcal{S}$, then the proof is complete. If M_2 is not maximal in $\mathcal{S}$, then there is an $M_3 \in \mathcal{S}$ such that $M_1 \subsetneq M_2 \subsetneq M_3$. Continuing in this way, we see that if $\mathcal{S}$ fails to have a maximal element, then we can obtain an ascending chain $M_1 \subsetneq M_2 \subsetneq M_3 \subsetneq \cdots$ that does not terminate and so M is not noetherian. Hence, if M is noetherian, then $\mathcal{S}$ must have a maximal element.

(2) $\Rightarrow$ (3). Let N be a submodule of M and suppose that $\mathcal{S}$ is the set of all finitely generated submodules of N. Note that $\mathcal{S} \neq \emptyset$ since the zero submodule is in $\mathcal{S}$. Now let N^* be a maximal element of $\mathcal{S}$. If $N^* = N$, then N is finitely generated and the proof is complete. If $N^* \neq N$, let $x \in N - N^*$. Then $N^* + xR$ is a finitely generated submodule of N that properly contains N^*, a contradiction. Therefore, it must be the case that $N = N^*$.

(3) $\Rightarrow$ (1). If $M_1 \subseteq M_2 \subseteq M_3 \subseteq \cdots$ is an ascending chain of submodules of M, then $\bigcup_{i=1}^{\infty} M_i$ is a finitely generated submodule of M. Furthermore, if $\{x_1, x_2, \ldots, x_m\}$ is a set of generators of $\bigcup_{i=1}^{\infty} M_i$, then there is an integer $n \geq 1$ such that $\{x_1, x_2, \ldots, x_m\} \subseteq M_n$. Hence, $M_n = \bigcup_{i=1}^{\infty} M_i$, so $M_i = M_n$ for all $i \geq n$. Thus, M is noetherian. $\square$

The following proposition is dual to the previous proposition.

Proposition 4.2.4. *The following are equivalent for an R-module M.*

(1) *M is artinian.*

(2) *Every nonempty collection of submodules of M has a minimal element.*

(3) *Every factor module of M is finitely cogenerated.*

(4) *If $\{M_\alpha\}_\Delta$ is a family of submodules of M, there is a finite subset F of Δ such that $\bigcap_F M_\alpha = \bigcap_\Delta M_\alpha$.*

Proof. (1) $\Rightarrow$ (2). Let $\mathcal{S}$ be a nonempty collection of submodules of M and suppose that $\mathcal{S}$ does not have a minimal element. If M_1 is a submodule of M that is in $\mathcal{S}$, then M_1 must have proper submodules that are in $\mathcal{S}$ for otherwise M_1 would be a minimal submodule of $\mathcal{S}$. If $M_2 \in \mathcal{S}$ is a proper submodule of M_1, then the same reasoning applied to M_2 shows that there is an $M_3 \in \mathcal{S}$ that is properly contained in M_2. If this is continued, then we have a decreasing chain $M_1 \supsetneq M_2 \supsetneq M_3 \supsetneq \cdots$ of submodules of M that does not terminate. Thus, M is not artinian. Hence, if M is artinian, then (2) must hold.

(2) $\Rightarrow$ (3). Let N be a submodule of M and suppose that $\{M_\alpha/N\}_\Delta$ is a family of submodules of M/N such that $\bigcap_\Delta (M_\alpha/N) = 0$. If

$$\mathcal{S} = \left\{ \bigcap_\Gamma M_\alpha \mid \Gamma \text{ a finite subset of } \Delta \right\},$$

choose $F \subseteq \Delta$ to be such that $\bigcap_F M_\alpha$ is minimal in $\mathcal{S}$. If $\bigcap_F M_\alpha \neq N$, then $\bigcap_F M_\alpha \not\supseteq \bigcap_\Delta M_\alpha = N$, so we can find a $\beta \in \Delta$ such that $\bigcap_F M_\alpha \not\supseteq \bigcap_{F \cup \{\beta\}} M_\alpha$. But this would mean that $\bigcap_F M_\alpha$ is not a minimal element of $\mathcal{S}$. Thus, $\bigcap_F M_\alpha = N$ and this gives $\bigcap_F (M_\alpha/N) = 0$. Hence, M/N is finitely cogenerated.

(3) $\Rightarrow$ (1). Suppose that $M_1 \supseteq M_2 \supseteq \cdots$ is a decreasing chain of submodules of M and let $N = \bigcap_{\mathbb{N}} M_i$. Then $M_1/N \supseteq M_2/N \supseteq \cdots$ is a decreasing chain of submodules of M/N and $\bigcap_{\mathbb{N}} (M_i/N) = 0$. Thus, there is a finite subset F of $\mathbb{N}$ such that $\bigcap_F (M_i/N) = 0$. Hence, $M_n = \bigcap_F M_i = N$ for some integer $n \geq 1$. If $k \geq n$, then $M_k \supseteq N = M_n$, so $M_k = M_n$ for all $k \geq n$. Thus, M is artinian.

(1) $\Leftrightarrow$ (4) is Exercise 8. $\square$

Remark. If $\mathcal{S}$ is a nonempty collection of submodules of an R-module M, then we say that *the submodules in $\mathcal{S}$ satisfy the ascending (descending) chain condition* if every ascending (descending) chain $M_1 \subseteq M_2 \subseteq M_3 \subseteq \cdots (M_1 \supseteq M_2 \supseteq M_3 \supseteq \cdots)$ of submodules in $\mathcal{S}$ terminates. If the submodules in $\mathcal{S}$ satisfy the ascending (descending) chain condition, even though M may not be noetherian (artinian), it is not difficult to show that there is a submodule in $\mathcal{S}$ that is maximal (minimal) among the submodules in $\mathcal{S}$.

The following two propositions hold for artinian and for noetherian modules. A proof of the noetherian case is given for each proposition with the artinian case left as an exercise. Before beginning, we point out the obvious fact that if $f : M \to N$ is an isomorphism, then M is noetherian (artinian) if and only if N is noetherian (artinian).

Proposition 4.2.5. *Let N be a submodule of M. Then M is noetherian (artinian) if and only if N and M/N are noetherian (artinian).*

Proof. Suppose that M is noetherian. If N is a submodule of M, then any ascending chain of submodules of N is an ascending chain of submodules of M and so must terminate. Hence, N is noetherian. Next, note that any ascending chain of submodules of M/N is of the form

$$M_1/N \subseteq M_2/N \subseteq M_3/N \subseteq \cdots, \tag{4.1}$$

where

$$M_1 \subseteq M_2 \subseteq M_3 \subseteq \cdots \tag{4.2}$$

is an ascending chain of submodules of M with each module in the chain containing N. Since the chain (4.2) terminates, the chain (4.1) must also terminate, so M/N is noetherian.

Conversely, suppose that N and M/N are noetherian. Let

$$M_1 \subseteq M_2 \subseteq M_3 \subseteq \cdots$$

be an ascending chain of submodules of M. Then

$$M_1 \cap N \subseteq M_2 \cap N \subseteq M_3 \cap N \subseteq \cdots$$

is an ascending chain of submodules of N, so there is an integer $n_1 \geq 1$ such that $M_i \cap N = M_{n_1} \cap N$ for all $i \geq n_1$. Furthermore,

$$(M_1 + N)/N \subseteq (M_2 + N)/N \subseteq (M_3 + N)/N \subseteq \cdots$$

is an ascending chain of submodules of M/N and there is an integer $n_2 \geq 1$ such that $(M_i + N)/N = (M_{n_2} + N)/N$ for all $i \geq n_2$. Let $n = \max(n_1, n_2)$. Then $M_i \cap N = M_n \cap N$ and $(M_i + N)/N = (M_n + N)/N$ for all $i \geq n$. If $i \geq n$ and $x \in M_i$, then $x + N \in (M_i + N)/N = (M_n + N)/N$, so there is a $y \in M_n$ such that $x + N = y + N$. This gives $x - y \in N$ and since $M_n \subseteq M_i$ we have $x - y \in M_i \cap N = M_n \cap N$ when $i \geq n$. If $x - y = z \in M_n \cap N$, then $x = y + z \in M_n$, so $M_i \subseteq M_n$. Hence, $M_i = M_n$ whenever $i \geq n$, so M is noetherian. $\qquad \square$

Corollary 4.2.6. *If $0 \rightarrow M_1 \rightarrow M \rightarrow M_2 \rightarrow 0$ is a short exact sequence of R-modules and R-module homomorphisms, then M is noetherian (artinian) if and only if M_1 and M_2 are noetherian (artinian).*

Example 8 shows that an infinite direct product of noetherian (artinian) R-modules need not be noetherian (artinian). The situation changes when the indexing set is finite.

Proposition 4.2.7. *For any positive integer n, a direct sum $\bigoplus_{i=1}^{n} M_i$ of R-modules is noetherian (artinian) if and only if each M_i is noetherian (artinian).*

Proof. Suppose that $\bigoplus_{i=1}^{n} M_i$ is noetherian. Since submodules of noetherian modules are noetherian and since each M_i is isomorphic to a submodule of $\bigoplus_{i=1}^{n} M_i$, it follows that each M_i is noetherian.

Conversely, suppose that each M_i is noetherian. Suppose also that $\bigoplus_{i=1}^{m} M_i$ is noetherian for each integer m such that $1 \leq m < n$. Then the short exact sequence $0 \rightarrow \bigoplus_{i=1}^{n-1} M_i \rightarrow \bigoplus_{i=1}^{n} M_i \rightarrow M_n \rightarrow 0$ and Corollary 4.2.6 show that $\bigoplus_{i=1}^{n} M_i$ is noetherian. The result follows by induction. $\qquad \square$

Corollary 4.2.8. *A ring R is right noetherian (artinian) if and only if the free R-module $R^{(n)}$ is noetherian (artinian) for each integer $n \geq 1$.*

Definition 4.2.9. A nonzero R-module M is said to be *decomposable* if there are nonzero submodules N_1 and N_2 of M such that $M = N_1 \oplus N_2$; otherwise M is said to be *indecomposable*.

We now prove a property of artinian and noetherian modules that will be useful in a subsequent chapter.

Proposition 4.2.10. *If M is a nonzero noetherian or a nonzero artinian R-module, then M is a finite direct sum*

$$M = M_1 \oplus M_2 \oplus \cdots \oplus M_n$$

of indecomposable R-modules.

Proof. Let M be a nonzero noetherian R-module. If M is indecomposable, then we are done, so suppose that M is not indecomposable and that M fails to have such a decomposition. Since M is not indecomposable, we may write $M = X \oplus Y$. At least one of X and Y cannot be a finite direct sum of its indecomposable submodules. Suppose this is the case for X. Then X is not indecomposable, and we can write $X = X' \oplus Y'$ and at least one of X' and Y' cannot be a finite direct sum of its indecomposable submodules. If this is X', then X' cannot be indecomposable and we can write $X' = X'' \oplus Y''$. Continuing in this way we obtain as ascending chain $Y \subseteq Y \oplus Y' \subseteq Y \oplus Y' \oplus Y'' \subseteq \cdots$ of submodules of M that fails to terminate. This contradicts the fact that M is noetherian, so it must be the case that M can be expressed as a finite direct sum of indecomposable submodules. In the artinian case, $X \supseteq X' \supseteq X'' \supseteq \cdots$ fails to terminate.   □

Proposition 4.2.11. *The following are equivalent for a ring R.*

(1) *R is right noetherian.*

(2) *Every finitely generated R-module is noetherian.*

(3) *Every right ideal of R is finitely generated.*

Proof. (1) $\Rightarrow$ (2). If R is right noetherian, then by considering Corollary 4.2.8 we see that $R^{(n)}$ is a noetherian R-module for each integer $n \geq 1$. Also if M is a finitely generated R-module, then M is the homomorphic image of $R^{(n)}$ for some integer $n \geq 1$. Hence, M is noetherian, since Corollary 4.2.6 shows that homomorphic images of noetherian modules are noetherian.

(2) $\Rightarrow$ (3). R is generated by 1, so R_R is noetherian. Proposition 4.2.3 gives the result.

(3) $\Rightarrow$ (1). This follows immediately, again from Proposition 4.2.3.   □

Corollary 4.2.12. *Every principal ideal ring is noetherian.*

Example

9. If K is a field and I is an ideal of $K[X]$, then $I = (p(X))$, where $p(X)$ is a monic polynomial in I of minimal degree. Thus, $K[X]$ is a principal ideal domain and so Corollary 4.2.12 shows that $K[X]$ is noetherian.

We now investigate a concept possessed by modules that are artinian and noetherian.

Definition 4.2.13. A decreasing chain

$$M = M_0 \supsetneq M_1 \supsetneq M_2 \supsetneq \cdots \supsetneq M_n = 0$$

of submodules of a nonzero R-module M is said to be a *composition series* of M if M_i is a maximal submodule of M_{i-1} for $i = 1, 2, \ldots, n$. The simple R-modules M_{i-1}/M_i are said to be the *composition factors* of the composition series and the integer n is the *length* of the composition series. If

(1) $$M = M_0 \supsetneq M_1 \supsetneq M_2 \supsetneq \cdots \supsetneq M_n = 0 \quad \text{and}$$

(2) $$M = N_0 \supsetneq N_1 \supsetneq N_2 \supsetneq \cdots \supsetneq N_m = 0$$

are composition series for M, then (1) and (2) are said to be *equivalent composition series* if $m = n$ and there is a permutation

$$\sigma : \{1, 2, \ldots, n\} \to \{1, 2, \ldots, n\}$$

such that $M_{i-1}/M_i \cong N_{\sigma(i)-1}/N_{\sigma(i)}$ for $i = 1, 2, \ldots, n$.

We see from the definition that equivalent composition series determine, up to isomorphism, the same set of composition factors.

Proposition 4.2.14. *A nonzero R-module M has a composition series if and only if it is artinian and noetherian.*

Proof. If M has a composition series, then among all the composition series of M there is one of minimal length, say n. If $n = 1$, then M is a simple R-module, so M is clearly artinian and noetherian. Now suppose that any module with a composition series of length less than n is artinian and noetherian and let $M = M_0 \supsetneq M_1 \supsetneq \cdots \supsetneq M_n = 0$ be a composition series of M of length n. Then $M_1 \supsetneq \cdots \supsetneq M_n = 0$ is a composition series of M_1 of length less than n, so M_1 is artinian and noetherian. Since M/M_1 is a simple R-module, M/M_1 is also artinian and noetherian, so Proposition 4.2.5 shows that M is artinian and noetherian. It follows by induction that any module with a composition series must be artinian and noetherian.

Conversely, suppose that M is artinian and noetherian. Then in view of Proposition 4.2.3, M has a maximal submodule M_1. But Proposition 4.2.5 shows that submodules of noetherian modules are noetherian, so let M_2 be a maximal submodule of M_1. At this point, we have $M = M_0 \supsetneq M_1 \supsetneq M_2$, where M_1 is a maximal submodule of M and M_2 is a maximal submodule of M_1. Continuing in this way, we obtain a decreasing chain $M = M_0 \supsetneq M_1 \supsetneq M_2 \supsetneq \cdots$ of submodules of M such that M_i is a maximal submodule of M_{i-1} for $i = 1, 2, 3, \ldots$. Since M is artinian, this chain must terminate at some M_n and we have constructed a composition series for M.   □

Corollary 4.2.15. *If* $0 \to M_1 \to M \to M_2 \to 0$ *is an exact sequence of nonzero R-modules and R-module homomorphisms, then M has a composition series if and only if M_1 and M_2 have a composition series.*

Proof. This follows immediately from the proposition and Corollary 4.2.6. □

The zero module is artinian and noetherian, so in order to remain consistent with Proposition 4.2.14, we will say that the zero module has a composition series by definition.

The following proposition is well known. It establishes the fact that the length of a composition series of an R-module M is an invariant and that, after a suitable reordering, the composition factors of any two composition series of the module are pairwise isomorphic.

Proposition 4.2.16 (Jordan–Hölder). *If a nonzero R-module M has a composition series, then any two composition series of M are equivalent.*

Proof. Let M be a nonzero R-module. If M has composition series, then among all the composition series of M there is one of minimal length . The proof proceeds by induction on the length of this "minimal" composition series . If an R-module M has a composition series $M = M_0 \supsetneq M_1 = 0$ of minimal length 1, then 0 is a maximal submodule of M, so M is a simple R-module. It is immediate from this observation that if an R-module has a composition series with minimal length 1, then all of its composition series are equivalent . Next, make the induction hypothesis that if an R-module has a composition series of minimal length less than n, then all of its composition series are equivalent. Let M be an R-module with composition series

$$M = M_0 \supsetneq M_1 \supsetneq M_2 \supsetneq \cdots \supsetneq M_n = 0 \tag{4.3}$$

of minimal length n and suppose that

$$M = N_0 \supsetneq N_1 \supsetneq N_2 \supsetneq \cdots \supsetneq N_m = 0 \tag{4.4}$$

is also a composition series of M. We will show that (4.3) and (4.4) are equivalent. Consider

$$M_1 \supsetneq M_2 \supsetneq \cdots \supsetneq M_n = 0 \quad \text{and} \tag{4.5}$$
$$N_1 \supsetneq N_2 \supsetneq \cdots \supsetneq N_m = 0 \tag{4.6}$$

which are composition series of M_1 and N_1. Note that the length of (4.5) is less than n. Moreover, the length of (4.5) must be minimal, for if not, then the length of (4.3) is not minimal. If $M_1 = N_1$, then (4.5) and (4.6) are composition series of M_1, so the induction hypothesis shows that (4.5) and (4.6) are equivalent and this in turn renders (4.3) and (4.4) equivalent. If $M_1 \neq N_1$, then $M_1 + N_1 = M$ since N_1 is a maximal submodule of M. Therefore,

$$M/M_1 = (M_1 + N_1)/M_1 \cong N_1/(M_1 \cap N_1) \quad \text{and} \qquad (4.7)$$

$$M/N_1 = (M_1 + N_1)/N_1 \cong M_1/(M_1 \cap N_1), \qquad (4.8)$$

so $M_1 \cap N_1$ is a maximal submodule of M_1 and of N_1, since M/M_1 and M/N_1 are simple R-modules. Using Proposition 4.2.14 we see that M is artinian and noetherian and Proposition 4.2.5 indicates that $M_1 \cap N_1$ is artinian and noetherian. Thus, $M_1 \cap N_1$ has a composition series

$$M_1 \cap N_1 = X_0 \supsetneq X_1 \supsetneq X_2 \supsetneq \cdots \supsetneq X_s = 0, \quad \text{so}$$

$$M_1 \supsetneq X_0 \supsetneq X_1 \supsetneq X_2 \supsetneq \cdots \supsetneq X_s = 0 \quad \text{and} \qquad (4.9)$$

$$N_1 \supsetneq X_0 \supsetneq X_1 \supsetneq X_2 \supsetneq \cdots \supsetneq X_s = 0 \qquad (4.10)$$

are composition series of M_1 and N_1, respectively. But M_1 has a composition series (4.5) of minimal length less than n and so by the induction hypothesis, (4.5) and (4.9) are equivalent. Hence, the composition series

$$M = M_0 \supsetneq M_1 \supsetneq M_2 \supsetneq \cdots \supsetneq M_n = 0 \quad \text{and} \qquad (4.3)$$

$$M = M_0 \supsetneq M_1 \supsetneq X_0 \supsetneq X_1 \supsetneq X_2 \supsetneq \cdots \supsetneq X_s = 0 \qquad (4.11)$$

of M are equivalent. Since the equivalence of (4.3) and (4.11) gives $s < n - 1$, N_1 has a composition series (4.10) of length less than n, so N_1 has a composition series of minimal length which is less than n. Therefore, the induction hypothesis shows that any two composition series of N_1 are equivalent. Hence, we see that

$$M = N_0 \supsetneq N_1 \supsetneq N_2 \supsetneq \cdots \supsetneq N_m = 0 \quad \text{and} \qquad (4.4)$$

$$M = N_0 \supsetneq N_1 \supsetneq X_0 \supsetneq X_1 \supsetneq X_2 \supsetneq \cdots \supsetneq X_s = 0 \qquad (4.12)$$

are equivalent. Since (4.7) and (4.8) show that $M/M_1 \cong N_1/X_0$ and $M/N_1 \cong M_1/X_0$, respectively, we see that (4.11) and (4.12) are equivalent. Therefore, we have that (4.3) is equivalent to (4.11), that (4.11) is equivalent to (4.12) and that (4.12) is equivalent to (4.4). Thus, (4.3) and (4.4) are equivalent.   □

We saw in Example 9 that if K is a field, then $K[X]$ is a noetherian ring. We close this section with a more general result due to Hilbert.

Proposition 4.2.17 (Hilbert's Basis Theorem). *If R is a right (left) noetherian ring, then so is the polynomial ring $R[X]$.*

Proof. Let R be a right noetherian ring, suppose that A is a right ideal in $R[X]$ and let A_n be the set of all $a \in R$ such that there is a polynomial in A of degree n with leading coefficient a. Then A_n together with the zero polynomial is easily seen to be a right ideal of R. If $a \in A_n$ and $p(X) \in A$ has degree n and a as its leading coefficient, then $p(X)X \in A$ and $p(X)X$ is of degree $n + 1$ with leading coefficient a. Hence, $a \in A_{n+1}$, so $A_n \subseteq A_{n+1}$ for all $n \geq 0$. Therefore, $A_0 \subseteq A_1 \subseteq A_2 \subseteq \cdots$ is an ascending chain of right ideals of R. But R is noetherian, so there is a nonnegative integer n_0 such that $A_i = A_{n_0}$ for all $i \geq n_0$. Furthermore, each A_i is finitely generated, so for each i with $0 \leq i \leq n_0$, let $\{a_{ij}\}_{j=1}^{k(i)}$ be a finite set of generators of A_i, with k depending on i. Next for $i = 0, 1, 2, \ldots, n$, let $p_{ij}(X) = X^i a_{ij} + \cdots$ be a polynomial in A of degree i with leading coefficient a_{ij}. We claim that the finite set of polynomials $\{p_{ij}(X)\}_{i=0}^{n_0}$ is a set of generators for A.

Among the polynomials in A that cannot be written as a finite linear combination of the $p_{ij}(X)$, choose a polynomial $q(X)$ of minimal degree. If $q(X) = X^m b +$ lower terms, then $b \in A_m$ and if $m \leq n_0$, then A_m is generated by the finite set $\{a_{mj}\}_{j=1}^{k(m)}$. If $b = \sum_{j=1}^{k(m)} a_{mj} b_j$, then

$$
q(X) - \sum_{j=1}^{k(m)} p_{mj}(X)b_j
$$

$$
= q(X) - (p_{m1}(X)b_1 + p_{m2}(X)b_2 + \cdots + p_{mk(m)}(X)b_{k(m)})
$$

$$
= (X^m b + \text{ lower terms})
$$
$$
- (X^m a_{m1} b_1 + \text{lower terms} + X^m a_{m2} b_2 + \text{lower terms}
$$
$$
+ \cdots + X^m a_{mk(m)} b_{k(m)} + \text{lower terms})
$$

$$
= (X^m b + \text{ lower terms})
$$
$$
- (X^m (a_{m1} b_1 + a_{m2} b_2 + \cdots + a_{mk(m)} b_{k(m)}) + \text{lower terms})
$$

$$
= (X^m b + \text{ lower terms})
$$
$$
- (X^m b + \text{ lower terms})
$$

$$
= \text{a polynomial with degree} < m.
$$

Thus, $q(X) - \sum_{j=1}^{k(m)} p_{mj}(X)b_j$ is a linear combination of the $p_{ij}(X)$. But this implies that $q(X)$ is a linear combination of the $p_{ij}(X)$, a contradiction.

If $m > n_0$, then $A_m = A_{n_0}$, so A_m is generated by $\{a_{n_0 j}\}_{j=1}^{k(n_0)}$ and we can write $b = \sum_{j=1}^{k(n_0)} a_{n_0 j} b_j$. In this case, we have a polynomial

$$
q(X) - \sum_{j=1}^{k(n_0)} p_{n_0 j}(X) X^{m-n_0} b_j
$$

that, when expanded as above, is seen to have degree less than m. This again leads to a contradiction, so the set of polynomials in A that cannot be written as a finite linear combination of the $p_{ij}(X)$ is empty. Hence, A is finitely generated by the set of polynomials $\{p_{ij}(X)\}_{i=0}^{n_0}$, so $R[X]$ is a right noetherian ring.   □

Problem Set 4.2

1. Prove that a ring R is right noetherian if and only if every submodule of every finitely generated R-module is finitely generated.

2. Prove that an R-module M is noetherian if and only if M satisfies the ascending chain condition on finitely generated submodules.

3. Suppose that $\{M_\alpha\}_\Delta$ is a family of submodules of an R-module M. If Δ is well ordered, then $\{M_\alpha\}_\Delta$ is said to be a *increasing chain (decreasing chain)* of submodules of M if $M_\alpha \subseteq M_\beta (M_\alpha \supseteq M_\beta)$ whenever $\alpha \leq \beta$. Show that M is noetherian (artinian) if and only if every increasing (decreasing) chain of submodules of M indexed over a well-ordered set terminates. [Hint: If M is noetherian (artinian), then every nonempty collection of submodules of M has a maximal (minimal) element.]

4. Prove Proposition 4.2.5 for the artinian case.

5. Prove Proposition 4.2.7 for the artinian case.

6. Verify Example 6.

7. Prove that every artinian R-module has at least one simple submodule.

8. Prove that M is an artinian R-module if and only if whenever $\{M_\alpha\}_\Delta$ is a family of submodules of M, there is a finite subset F of Δ such that $\bigcap_F M_\alpha = \bigcap_\Delta M_\alpha$.

9. (a) If R is a right artinian ring, prove that if $a, b \in R$ are such that $ab = 1$, then $ba = 1$. [Hint: Consider $bR \supseteq b^2 R \supseteq b^2 R \supseteq \cdots$.]

   (b) Prove that if R is a right artinian ring without zero divisors, then R is a division ring. [Hint: Consider $aR \supseteq a^2 R \supseteq a^3 R \supseteq \cdots$.]

10. Let $f : R \to S$ be a ring homomorphism and suppose that M is an S-module. Make M into an R-module by pullback along f. Prove that if M is noetherian (artinian) as an R-module, then it is noetherian (artinian) as an S-module.

11. (a) If R is a right noetherian ring, show that the polynomial ring $R[X_1, X_2, \ldots, X_n]$ in n commuting indeterminates $X_1, X_2, \ldots, X_n$ is also right noetherian.

    (b) If R is right noetherian and A is an ideal of R, prove that R/A is a right noetherian ring.

12. (a) Show that if R is right artinian, then $R[X]$ need not be right artinian.

    (b) If R is right artinian and A is an ideal of R, then is the ring R/A right artinian?

13. Prove or give a counterexample to the following statement. A ring R is right artinian if and only if every finitely generated R-module is artinian.

14. Prove that an endomorphism $f : M \to M$ of a noetherian (artinian) R-module M is an isomorphism if and only if it is an epimorphism (a monomorphism). [Hint: Consider the chain $\operatorname{Ker} f \subseteq \operatorname{Ker} f^2 \subseteq \operatorname{Ker} f^3 \subseteq \cdots$ ($M \supseteq \operatorname{Im} f \supseteq \operatorname{Im} f^2 \supseteq \operatorname{Im} f^3 \supseteq \cdots$).]

15. Let M be an artinian and a noetherian R-module.

    (a) **Fitting's lemma.** Prove that if $f : M \to M$ is R-linear, then $M = \operatorname{Im} f^n \oplus \operatorname{Ker} f^n$ for some positive integer n. [Hint: $\operatorname{Im} f^n = \operatorname{Im} f^{n+1}$ for some positive integer n, so f^n induces an epimorphism on the noetherian module $\operatorname{Im} f^n$ which, by Exercise 14, must be an isomorphism. Hence, $\operatorname{Im} f^n \cap \operatorname{Ker} f^n = 0$. So if $a \in M$, then $f^n(a) = f^{n+1}(b)$ for some $b \in M$.]

    (b) If M is indecomposable, prove that an R-linear mapping $f : M \to M$ is either nilpotent or an isomorphism.

16. Let N be a submodule of an R-module M. A decreasing chain

$$M = M_0 \supsetneq M_1 \supsetneq M_2 \supsetneq \cdots \supsetneq M_n = N$$

of submodules of an R-module M is said to be a *composition series from M to N* if M_i is a maximal submodule of M_{i-1} for $i = 1, 2, \ldots, n$. The simple R-modules M_{i-1}/M_i are said to be the *composition factors* of the composition series from M to N. If

(1) $\qquad M = M_0 \supsetneq M_1 \supsetneq M_2 \supsetneq \cdots \supsetneq M_n = N$ and

(2) $\qquad M = N_0 \supsetneq N_1 \supsetneq N_2 \supsetneq \cdots \supsetneq N_m = N$

are composition series from M to N, then (1) and (2) are said to be *equivalent composition series from M to N* if $m = n$ and there is a permutation

$$\sigma : \{1, 2, \ldots, n\} \to \{1, 2, \ldots, n\}$$

such that $M_{i-1}/M_i \cong N_{\sigma(i)-1}/N_{\sigma(i)}$ for $i = 1, 2, \ldots, n$.

    (a) Show that there is a composition series from M to N if and only if M/N has a composition series. Conclude that there is a composition series from M to N if and only if M/N is artinian and noetherian.

    (b) Show that any two composition series from M to N are equivalent.

    (c) Let $\ell(M)$ denote the length of a composition series of M. If M has a composition series and N is a submodule of M, then N and M/N have a composition series. Prove that $\ell(M) = \ell(M/N) + \ell(N)$. Conclude that if $0 \to M_1 \to$

$M \to M_2 \to 0$ is an exact sequence of R-modules and M has a composition series, then M_1 and M_2 have composition series and $\ell(M) = \ell(M_1) + \ell(M_2)$.

(d) If M has a composition series and $M = \bigoplus_{i=1}^{n} M_i$, prove that $\ell(M) = \sum_{i=1}^{n} \ell(M_i)$.

4.3   Modules over Principal Ideal Domains

Principal ideal domains are noetherian and, as a consequence, every finitely generated module over such a ring is noetherian. In this section we investigate the additional effects that a principal ideal domain has on its finitely generated modules. The two main results, Proposition 4.3.21 (Invariant Factors) and Proposition 4.3.25 (Elementary Divisors), give the usual decompositions of finitely generated abelian groups when R is the ring $\mathbb{Z}$ of integers. For example, see [10] or [20].

Before investigating these modules, we need to establish the "mathematical machinery" that will be required.

Throughout this section R will denote a commutative ring.

Definition 4.3.1. If $a, b \in R$, then we say that b *divides* a if there is a $c \in R$ such that $a = bc$. We write $b \mid a$ when b divides a and $b \nmid a$ will indicate that b does not divide a.

Definition 4.3.2. If $a_1, a_2, \ldots, a_n$ are nonzero elements of R, then a nonzero element $d \in R$ is said to be a *greatest common divisor* of $a_1, a_2, \ldots, a_n$ if the following conditions are satisfied:

(1) $d \mid a_i$ for $i = 1, 2, \ldots, n$.

(2) If $b \in R$ and $b \mid a_i$ for each i, then $b \mid d$.

The notation $\gcd(a_1, a_2, \ldots, a_n)$ will be used for a greatest common divisor of $a_1, a_2, \ldots, a_n$ when it can be shown to exist. If $a, b \in R$ and $\gcd(a, b) = 1$, then a and b are said to be *relatively prime*.

A greatest common divisor of a set of nonzero elements of R may fail to exist. The case is different for a principal ideal ring.

Proposition 4.3.3. *Let R be a principal ideal ring. If $a_1, a_2, \ldots, a_n \in R$ are nonzero, then the elements $a_1, a_2, \ldots, a_n$ have a greatest common divisor. Furthermore, if $\gcd(a_1, a_2, \ldots, a_n) = d$, then there are elements $s_1, s_2, \ldots, s_n$ in R such that $d = a_1 s_1 + a_2 s_2 + \cdots + a_n s_n$.*

Proof. If $(d) = a_1 R + a_2 R + \cdots + a_n R$, then each a_i is in (d), so there is a $b_i \in R$ such that $a_i = db_i$. Hence, $d \mid a_i$ for $i = 1, 2, \ldots, n$. Now $d \in a_1 R + a_2 R + \cdots + a_n R$, so d has a representation as $d = a_1 s_1 + a_2 s_2 + \cdots + a_n s_n$ for suitably chosen s_i in R. Finally, suppose that $c \mid a_i$ for $i = 1, 2, \ldots, n$ and let $d_i \in R$ be such that $a_i = cd_i$ for $i = 1, 2, \ldots, n$. This gives $d = (d_1s_1 + d_2s_2 + \cdots + d_ns_n)c$, so $c \mid d$. Thus, $d = \gcd(a_1, a_2, \ldots, a_n)$. $\square$

Definition 4.3.4. A nonzero nonunit element $a \in R$ is said to be *irreducible* if whenever $a = bc$, then either b or c is a unit. A nonzero nonunit element $p \in R$ is *prime* if $p \mid ab$ implies that $p \mid a$ or $p \mid b$. Two elements $a, b \in R$ are *associates* if there is a unit $u \in R$ such that $a = bu$.

In general, the greatest common divisor of a set of elements in R is not unique. However, if R is an integral domain and if d and d' are greatest common divisors of a set of nonzero elements of R, then d and d' are associates. Conversely, if d and d' are associates and if d is a greatest common divisor of a set of nonzero elements in an integral domain R, then so is d'.

Examples

1. It is easy to check that $[2]$ is prime in $\mathbb{Z}_6$. However, $[2] = [2][4]$ and neither $[2]$ nor $[4]$ is a unit in $\mathbb{Z}_6$. Hence, $[2]$ is prime but not irreducible. Thus, in a commutative ring, a prime element may not be irreducible.

2. Let $R = \{a + b\sqrt{5}i \mid a, b \in \mathbb{Z}\}$, where $i = \sqrt{-1}$. The mapping

$$N : R \to \mathbb{Z} \quad \text{such that } N(a + b\sqrt{5}i) = a^2 + 5b^2,$$

   is such that $N(xy) = N(x)N(y)$ for all $x, y \in R$, u is a unit in R if and only if $N(u) = 1$, and $N(x) = 0$ if and only if $x = 0$. It follows that 3 is an irreducible element of R. Now $3 \mid 6$ and $6 = (1 + \sqrt{5}i)(1 - \sqrt{5}i)$, so $3 \mid (1 + \sqrt{5}i)(1 - \sqrt{5}i)$. One can show that $3 \nmid (1 + \sqrt{5}i)$ and $3 \nmid (1 - \sqrt{5}i)$, so in a commutative ring an irreducible element need not be prime.

3. If we remove the requirement that a ring has an identity and apply Definition 4.3.2 to the ring $2\mathbb{Z}$, then two or more elements may fail to have a greatest common divisor. For example, 2 has no divisors in $2\mathbb{Z}$. Hence, if $a \in 2\mathbb{Z}$, then a and 2 do not have a common divisor much less a greatest common divisor.

Examples 1 and 2 show that in a commutative ring, prime and irreducible elements may be distinct. However, if the ring is an integral domain, then prime and irreducible elements take on characteristics enjoyed by a prime number in $\mathbb{Z}$.

Proposition 4.3.5. *The following hold in any integral domain.*

(1) *If $a, b \in R$, then $(a) = (b)$ if and only if a and b are associates.*

(2) *An element $p \in R$ is prime if and only if (p) is a prime ideal of R.*

(3) *An element $a \in R$ is irreducible if and only if (a) is a maximal ideal of R.*

(4) *Every prime element of R is irreducible.*

(5) *The only divisors of an irreducible element of R are its associates and the units of R.*

(6) *Every associate of a prime (irreducible) element of R is a prime (irreducible) element of R.*

Proof. We only prove part (4) of the proposition. The proofs the other parts can be found in almost every undergraduate abstract algebra text. As a review, the reader can elect to prove any of these that might be required to refresh his or her memory. So let p be a prime element of R and suppose that $p = ab$. Then either $p \mid a$ or $p \mid b$. If $p \mid a$, then there is an element $c \in R$ such that $a = pc$. Thus, $a = abc$ and so $bc = 1$. Thus, b is a unit in R and so p is irreducible.   □

The converse of part (4) of Proposition 4.3.5 holds if the ring is a principal ideal domain.

Proposition 4.3.6. *If R is a principal ideal domain, then an irreducible element of R is prime.*

Proof. Suppose that p is irreducible and that $p \mid ab$. Then there is a $c \in R$ such that $pc = ab$, so let $(d) = pR + aR$. Since $p \in (d)$, there is an $r \in R$ such that $p = dr$. But p is irreducible, so either d or r is a unit in R. If d is a unit, then $pR + aR = R$, so there are $s, t \in R$ such that $1 = ps + at$. Thus, $b = bps + bat = bps + ptc = p(bs + tc)$. Hence, $p \mid b$. If r is a unit in R, then $d = pr^{-1} \in pR$ and so $pR + aR = (d) \subseteq pR$. Thus, $a \in pR$, so $p \mid a$. Hence if $p \mid ab$, then $p \mid a$ or $p \mid b$, so p is prime.   □

Definition 4.3.7. Let R be an integral domain. Then R is said to be a *factorization domain* if each nonzero nonunit $a \in R$ can be expressed as $a = p_1 p_2 \cdots p_n$, where the p_i are not necessarily distinct irreducible elements of R. Each p_i is a *factor* of a and $a = p_1 p_2 \cdots p_n$ is said to be a *factorization of a*(as a product of irreducible elements). If $q_1, q_2, \ldots, q_k$ are the distinct irreducible elements in the factorization $a = p_1 p_2 \cdots p_n$ of a, then we can write $a = q_1^{n_1} q_2^{n_2} \cdots q_k^{n_k}$, where each n_i is a positive integer and $n = n_1 + n_2 + \cdots + n_k$. If a itself is irreducible, then we consider a to be a product of irreducible elements with one factor. If $a = p_1 p_2 \cdots p_n = q_1 q_2 \cdots q_m$ are two factorizations of a, then we say that the factorization is unique up to the order of the factors and associates, if $m = n$ and there is a permutation $\sigma :$ $\{1, 2, \ldots, n\} \to \{1, 2, \ldots, n\}$ such that p_i and $q_{\sigma(i)}$ are associates for $i = 1, 2, \ldots, n$. When this holds, we simplify terminology and simply say that the *factorization of a is unique*. If each nonzero nonunit of R has a unique factorization, then R is said to be a *unique factorization domain*.

Proposition 4.3.8. *Every principal ideal domain is a unique factorization domain.*

Proof. If a is a nonzero nonunit of R, then (a) is contained in some maximal ideal $\mathfrak{m}$. But (3) of Proposition 4.3.5 indicates that $\mathfrak{m} = (p_1)$ for some irreducible element $p_1 \in R$. Hence, $a \in (p_1)$, so there is a nonzero $a_1 \in R$ such that $a = p_1 a_1$. Thus $(a) \subseteq (a_1)$. If $(a) = (a_1)$, then $a_1 = ab$ for some $b \in R$. But this gives $a = p_1 a_1 = p_1 ab$ and so $p_1 b = 1$. Hence, p_1 is a unit in R, a contradiction. Therefore, $(a) \subsetneqq (a_1)$. Repeating this procedure, starting with a_1, gives an element $a_2 \in R$ such that $(a) \subsetneqq (a_1) \subsetneqq (a_2)$ and an irreducible element $p_2 \in R$ such that $a_1 = p_2 a_2$. Continuing we obtain an ascending chain of principal ideals $(a) \subsetneqq (a_1) \subsetneqq (a_2) \subsetneqq \cdots \subsetneqq (a_n) \subsetneqq \cdots$, where $a_n = p_n a_n$ for some irreducible $p_n \in R$. Now a principal ideal domain is noetherian, so this chain terminates. If the chain terminates at (a_n), then a_n is a unit in R and it follows that $a = p_1 p_2 \cdots p_{n-1} p'_n$, where $p'_n = p_n a_n$ and p' is an irreducible element that is an associate of p_n.

Finally, suppose that $a = p_1 p_2 \cdots p_n = q_1 q_2 \cdots q_m$ are two factorizations of a with $n \leq m$. Now $p_1 \mid (q_1 q_2 \cdots q_m)$, so since R is a principal ideal domain p_1 is prime, so p_1 must divide some q_i. Suppose that our notation is chosen so that this is q_1. Then $q_1 = p_1 u_1$ for some unit $u_1 \in R$. Hence, we have $p_2 p_3 \cdots p_n = u q_2 q_3 \cdots q_m$. Continuing in this way, after n steps we arrive at $1 = u_1 u_2 \cdots u_n \times q_{n+1} \cdots q_m$. Since the q_i are not units, this implies that $m = n$ and so each p_i is an associate of some q_j for $1 \leq i, j \leq n$.   □

Recall that if M is an R-module over an integral domain, then an element $x \in M$ is torsion if there is an $a \in R$, $a \neq 0$, such that $xa = 0$. Recall also that if zero is the only torsion element of M, then M is said to be torsion free.

Definition 4.3.9. Let M be an R-module, suppose that $x \in M$ and let $a \in R$. Then we say that a *divides* x if there is a $y \in M$ such that $x = ya$. If the only divisors of x are units of R, then x is a *primitive element* of M.

Note that if $x \in M$ and a is a unit in R, then $x = ya$, where $y = xa^{-1}$. Thus, x is divisible by every unit of R.

Lemma 4.3.10. *Let F be a finitely generated free module over a principal ideal domain R. Then each $x \in F$, $x \neq 0$, may be written as $x = x'a$, where x' is a primitive element of F and $a \in R$.*

Proof. Let x be a nonzero element of F. If x is primitive, then there is nothing to prove since $x = x1$. If x is not primitive, then $x = x_1 a_1$, where $x_1 \in F$ and a_1 is not a unit in R. Thus, $xR \subseteq x_1 R$ and we claim that this containment is proper. If $xR = x_1 R$, then $x_1 = xb$ for some $b \in R$, so $x_1 = x_1 a_1 b$ and therefore $x_1(1 - a_1 b) = 0$. But a free module over an integral domain is torsion free. Indeed, let F be a free module over an integral domain with basis $\{x_\alpha\}_\Delta$ and suppose that x is a torsion element of F such that $x = \sum_\Delta x_\alpha a_\alpha$. If $b \in R$, $b \neq 0$, is such that $xb = 0$, then

$\sum_{\Delta} x_{\alpha}(a_{\alpha}b) = 0$, so $a_{\alpha}b = 0$ for each $\alpha \in \Delta$. Hence, each a_{α} is zero and so $x = 0$ as asserted. Consequently, $x_1(1 - a_1b) = 0$ gives $a_1b = 1$ which indicates that a_1 is a unit in R. Therefore, it must be the case that $xR \subsetneqq x_1R$. If x_1 is primitive, then $x = x_1a_1$ and we are done. If x_1 is not primitive, we can repeat the argument with x_1 to obtain $x_1 = x_2a_2$, where a_2 is not a unit in R. If x_2 is primitive, then $x = x_2a_1a_2$ and we are done. If x_2 is not primitive, then $x_1R \subsetneqq x_2R$. Continuing in this way, we obtain an increasing chain $xR \subsetneqq x_1R \subsetneqq x_2R \subsetneqq \cdots$ of submodules of F. Now R is noetherian and since F is finitely generated, Proposition 4.2.11 indicates that F is noetherian as well. Thus, the chain $xR \subsetneqq x_1R \subsetneqq x_2R \subsetneqq \cdots$ must terminate. If the chain terminates at x_nR, then x_n is primitive and $x = x_na$, where $a = a_1a_2 \cdots a_n$.   □

Free Modules over a Principal Ideal Domain

The necessary "ground work" has now been established and we can investigate the structure of modules over principal ideal domains. The first step is to consider free modules over these rings. One basic result is given by the following proposition. The proof is delayed until Chapter 5, where it is part of Proposition 5.2.16.

Proposition 4.3.11. *If R is a principal ideal domain and M is a submodule of a free R-module F, then M is free and* $\operatorname{rank}(M) \leq \operatorname{rank}(F)$.

Lemma 4.3.12. *Let R be a principal ideal domain and suppose that F is a free R-module with basis $\{x_{\alpha}\}_{\Delta}$. If $x \in F$, $x \neq 0$, and $x = \sum_{\Delta} x_{\alpha}a_{\alpha}$, where $a_{\alpha} = 0$ for almost all $\alpha \in \Delta$, then x is primitive if and only if* $\gcd\{a_{\alpha} \mid a_{\alpha} \neq 0\} = 1$.

Proof. Suppose that $x \in F$ is primitive and let $x = \sum_{\Delta} x_{\alpha}a_{\alpha}$, where $a_{\alpha} = 0$ for almost all $\alpha \in \Delta$. If $d = \gcd\{a_{\alpha} \mid a_{\alpha} \neq 0\}$, then $x = d(\sum_{\Delta} x_{\alpha}b_{\alpha})$, where $a_{\alpha} = db_{\alpha}$ for all $\alpha \in \Delta$. Since x is primitive, d is a unit, so d and 1 are associates. Thus, 1 is a greatest common divisor of $\{a_{\alpha} \mid a_{\alpha} \neq 0\}$. Conversely, suppose that $\gcd\{a_{\alpha} \mid a_{\alpha} \neq 0\} = 1$ and let $x = ya$. If $y = \sum_{\Delta} x_{\alpha}c_{\alpha}$ with $c_{\alpha} = 0$ for almost all $\alpha \in \Delta$, then $\sum_{\Delta} x_{\alpha}a_{\alpha} = x = ya = \sum_{\Delta} x_{\alpha}(c_{\alpha}a)$. But $\{x_{\alpha}\}_{\Delta}$ is a basis for F, so $a_{\alpha} = c_{\alpha}a$ for all $\alpha \in \Delta$. Hence, $a \mid a_{\alpha}$ for each $\alpha \in \Delta$, so $a \mid (\gcd\{a_{\alpha} \mid a_{\alpha} \neq 0\} = 1)$. Thus, a is a unit, so x is primitive.   □

Given a primitive element x of a free module F over an principal ideal domain, we now show that a basis of F can always be found that contains x. The proof of this result is by induction, so we first prove this fact for a free module of rank 2.

Lemma 4.3.13. *Let R be a principal ideal domain and suppose that F is a free R-module of rank 2. If x is a primitive element of F, then there is a basis of F containing x.*

Proof. Let $\{x_1, x_2\}$ be a basis for F and suppose that $x = x_1a_1 + x_2a_2$. If x is primitive, then $\gcd\{a_1, a_2\} = 1$, so there are $s_1, s_2 \in R$ such that $a_1s_1 + a_2s_2 = 1$. If $x_2' = -x_1s_2 + x_2s_1$, then a routine calculation shows that

$$x_1 = xs_1 - x_2'a_2 \quad \text{and} \quad x_2 = xs_2 + x_2'a_1.$$

But $\{x_1, x_2\}$ generates F, so it follows that $\{x, x_2'\}$ generates F. Finally, suppose that $xb_1 + x_2'b_2 = 0$. Then

$$(x_1a_1 + x_2a_2)b_1 + (-x_1s_2 + x_2s_1)b_2 = 0 \quad \text{or}$$
$$x_1(a_1b_1 - s_2b_2) + x_2(a_2b_1 + s_1b_2) = 0.$$

Thus,

$$a_1b_1 - s_2b_2 = 0 \quad \text{and} \quad a_2b_1 + s_1b_2 = 0.$$

Multiplying the first of the last pair of equations by s_1 and the second by s_2 and adding gives $b_1 = (a_1s_1 + a_2s_2)b_1 = 0$. Similarly, $b_2 = 0$, so x and x_2' are linearly independent. Consequently, $\{x, x_2'\}$ is a basis for F. □

Proposition 4.3.14. *Let R be a principal ideal domain and suppose that F is a free R-module. If x is a primitive element of F, then there is a basis of F containing x.*

Proof. Let x be a primitive element of F and suppose that the rank of F is finite. If $\{x_1\}$ is a basis for F, then there is an $a \in R$ such that $x = x_1a$. But x is primitive, so a is a unit in R. Hence, $xR = x_1R$, so $\{x\}$ is a basis of F. Now suppose that $\text{rank}(F) = n$ and make the induction hypothesis that the proposition is true for any free module with rank less than n. If $\{x_1, x_2, \ldots, x_n\}$ is a basis for F, then there are $a_i \in R$ such that $x = x_1a_1 + x_2a_2 + \cdots + x_na_n$. If $a_n = 0$, then $x \in M = x_1R \oplus x_2R \oplus \cdots \oplus x_{n-1}R$ and M is a free R-module of rank less that n. The induction hypothesis gives a basis $\{x, x_2', \ldots, x_{n-1}'\}$ of M and it follows that $\{x, x_2', \ldots, x_{n-1}', x_n\}$ is a basis of F that contains x. If $a_n \neq 0$, let $y = x_1a_1 + x_2a_2 + \cdots + x_{n-1}a_{n-1}$, so that $x = y + x_na_n$. If $y = 0$, then $x = x_na_n$, so a_n is a unit since x is primitive. Thus, $\{x_1, x_2, \ldots, x_{n-1}, x\}$ is a basis of F. If $y \neq 0$, then we can write $y = y'b$, where y' is a primitive element of F and b is a nonzero element of R. We claim that y' and x_n are linearly independent. If

$$y'c_1 + x_nc_2 = 0,$$

then $y'bc_1 + x_nbc_2 = 0$ gives $yc_1 + x_nbc_2 = 0$. Thus,

$$x_1a_1c_1 + x_2a_2c_1 + \cdots + x_{n-1}a_{n-1}c_2 + x_nbc_2 = 0$$

and so $a_1c_1 = a_2c_1 = \cdots = a_{n-1}c_1 = bc_2 = 0$. Since $b \neq 0, c_2 = 0$, so $y'c_1 = 0$. Now F is free, so F is torsion free and so since y' is primitive and therefore nonzero, it must be the case that $c_1 = 0$. Hence, $y'R \oplus x_n R$ is a rank 2 R-module and $x \in y'R \oplus x_n R$ since $x = y'b + x_n a_n$. Thus, Lemma 4.3.13 indicates that $y'R \oplus x_n R$ has a basis $\{x, y''\}$. It follows that $\{x, x_2', \ldots x_{n-1}', y''\}$ is a basis for F, so the proposition holds for all free R-modules of finite rank.

Finally, let F be a free R-module of infinite rank and suppose that $\{x_\alpha\}_\Delta$ is a basis for F. If $\{x_{\alpha_1}, x_{\alpha_2}, \ldots, x_{\alpha_n}\} \subseteq \{x_\alpha\}_\Delta$ and $a_1, a_2, \ldots, a_n \in R$ are such that $x = x_{\alpha_1} a_1 + x_{\alpha_2} a_2 + \cdots + x_{\alpha_n} a_n$, then $x \in F' = x_{\alpha_1} R \oplus x_{\alpha_2} R \oplus \cdots \oplus x_{\alpha_n} R$ and F' is a free module with finite rank. Hence, from what we have just proved in the previous paragraph, there is a basis $\mathcal{B}'$ of F' that contains x. It follows that $\mathcal{B} = \mathcal{B}' \cup (\{x_\alpha\}_\Delta - \{x_{\alpha_1}, x_{\alpha_2}, \ldots, x_{\alpha_n}\})$ is a basis of F that contains x, so the proof is complete.   □

The next proposition will play a central role when we develop the structure of finitely generated modules over a principal ideal domain. But first we need the following definition.

Definition 4.3.15. Let F be a free module over a principal ideal domain R. If $x \in F$ and $x = x'a$, where x' is a primitive element of F, then a is said to be the *content of* x, denoted by $c(x)$.

Let F be a free module over a principal ideal domain R and assume that $\{x_\alpha\}_\Delta$ is a basis for F. Let $x \in F$, $x \neq 0$, and suppose that $x = x'c(x)$, where x' is a primitive element of F. If $x' = \sum_\Delta x_\alpha a_\alpha$, where $a_\alpha = 0$ for almost all $\alpha \in \Delta$, then $x = \sum_\Delta x_\alpha (a_\alpha c(x)) = \sum_\Delta x_\alpha c_\alpha$, where $c_\alpha = a_\alpha c(x)$ for each $\alpha \in \Delta$. Since x' is primitive, Lemma 4.3.12 shows that $\gcd\{a_\alpha \mid a_\alpha \neq 0\} = 1$, so $c(x) = \gcd\{c_\alpha \mid c_\alpha \neq 0\}$.

Proposition 4.3.16. *Suppose that F is a free module over a principal ideal domain R. If M is a submodule of F of finite rank k (M is free by Proposition 4.3.11), then there is a basis $\mathcal{B}$ of F, a subset $\{x_1, x_2, \ldots, x_k\}$ of $\mathcal{B}$, and nonzero elements $a_1, a_2, \ldots, a_k \in R$ such that $\{x_1 a_1, x_2 a_2, \ldots, x_k a_k\}$ is a basis for M and $a_i \mid a_{i+1}$ for $i = 1, 2, \ldots, k-1$.*

Proof. If $M = 0$, there is nothing to prove, so assume that $M \neq 0$. We proceed by induction on k, the rank of M. If $k = 1$, then M has a basis $\{x\}$ for some $x \in M$. Thus, $x = x'c(x)$, where $x' \in F$ is primitive. Due to Proposition 4.3.14 there is a basis $\mathcal{B}$ of F that contains x', so if we let $a_1 = c(x)$ and $x' = x_1$, then $x = x_1 a_1$ and $\{x_1 a_1\}$ is a basis for M. Clearly, $a_1 \neq 0$ and, moreover, the divisibility condition vacuously holds.

Now assume that the proposition holds for any submodule of F of rank less than k and let $\mathcal{S} = \{(c(x)) \mid x \in M\}$. Since R is noetherian, Proposition 4.2.3 indicates that $\mathcal{S}$ has a maximal element, say $(c(x))$. Then $x \in M$ and $x = x'c(x)$, where $x' \in F$ is primitive. Let $x_1 = x'$, let $a_1 = c(x)$ and suppose that $\mathcal{B}$ is a basis of F that contains x_1. If $\mathcal{B}' = \mathcal{B} - \{x_1\}$ and F' is the submodule of F generated by $\mathcal{B}'$, then we claim that

$$M = x_1 a_1 R \oplus (F' \cap M).$$

Since $x_1 a_1 R \subseteq x_1 R$ and $(F' \cap M) \subseteq F'$, we immediately see that

$$x_1 a_1 R \cap (F' \cap M) \subseteq x_1 R \cap F' = 0.$$

Hence, we need to show that $M \subseteq x_1 a_1 R + (F' \cap M)$. If $y \in M$, then we can use the basis $\mathcal{B}$ to write

$$y = x_1 b + \sum_{\mathcal{B}'} z a_z,$$

where $a_z = 0$ for almost all $z \in \mathcal{B}'$. If $d = \gcd\{a_1, b\}$, then there are $s_1, s_2 \in R$ such that $d = a_1 s_1 + b s_2$. If $w = x s_1 + y s_2$, then

$$w = x_1 a_1 s_1 + \left(x_1 b + \sum_{\mathcal{B}'} z a_z\right) s_2$$

$$= x_1 (a_1 s_1 + b s_2) + \sum_{\mathcal{B}'} z (a_z s_2)$$

$$= x_1 d + \sum_{\mathcal{B}'} z (a_z s_2).$$

If $w = w' c(w)$, then, since w' can be expressed as a linear combination of elements of $\mathcal{B}$, we see $c(w)$ divides each coefficient in the expression $x_1 d + \sum_{\mathcal{B}'} z (a_z s_2)$, and therefore $c(w) \mid d$. Hence, $(a_1) \subseteq (d) \subseteq (c(w))$ and the maximality of (a_1) gives $(a_1) = (d) = (c(w))$. Thus, $d = a_1 b'$ for some $b' \in R$ and we have $y = (x_1 a_1) b' + \sum_{\mathcal{B}'} z a_z$. Now $y \in M$ and $(x_1 a_1) b' = x b' \in M$, so $\sum_{\mathcal{B}'} z a_z = y - (x_1 a_1) b' \in M$. Therefore,

$$y = (x_1 a_1) b' + \sum_{\mathcal{B}'} z a_z \in x_1 a_1 R + (F' \cap M)$$

and so

$$M = x_1 a_1 R \oplus (F' \cap M).$$

Hence,

$$\text{rank}(F' \cap M) = \text{rank}(M) - 1 = k - 1,$$

so by the induction hypothesis there is a basis $\mathcal{B}''$ of F' and a subset $\{x_2, x_3, \ldots, x_k\}$ of $\mathcal{B}''$ and nonzero elements $a_2, a_3, \ldots, a_k \in R$ such that $\{x_2 a_2, x_3 a_3, \ldots, x_k a_k\}$

is a basis for $F' \cap M$, where $a_i \mid a_{i+1}$ for $i = 2, 3, \ldots, k - 1$. It follows that $\{x_1 a_1, x_2 a_2, \ldots, x_k a_k\}$ is a basis for M, so the proof will be complete if we can show that $a_1 \mid a_2$. Let $y = x_1 a_1 + x_2 a_2$. If $y = y'c(y)$, where y' is primitive in F, then the observation immediately following Definition 4.3.15 shows that $\gcd\{a_1, a_2\} = c(y)$. Hence, $(a_1) \subseteq (c(y))$ and so the maximality of (a_1) gives $(a_1) = (c(y))$. Thus, $a_1 \mid a_2$ and we are done.   □

Finitely Generated Modules over a Principal Ideal Domain

We are now in a position to begin development of the structure of finitely generated modules over a principal ideal domain.

Proposition 4.3.17. *Let M be a finitely generated module over a principal ideal domain R. Then M is a direct sum of nonzero cyclic submodules*

$$M = x_1 R \oplus x_2 R \oplus \cdots \oplus x_n R$$

and $\operatorname{ann}(x_1) \supseteq \operatorname{ann}(x_2) \supseteq \cdots \supseteq \operatorname{ann}(x_n)$.

Proof. Let $\{z_1, z_2, \ldots, z_n\}$ be a minimal set of nonzero generators of M. Then there is an epimorphism $\varphi : R^{(n)} \to M$ defined by $\varphi((r_i)) = \sum_{i=1}^{n} z_i r_i$. If $K = \operatorname{Ker} \varphi$, then $R^{(n)}/K \cong M$ and since $R^{(n)}$ is a free R-module of rank n, Proposition 4.3.11 indicates that K is a free R-module of rank k with $k \leq n$. Hence, Proposition 4.3.16 gives a basis $\{b_1, b_2, \ldots, b_n\}$ of $R^{(n)}$ and nonzero elements $a_1, a_2, \ldots, a_k \in R$ such that $\{b_1 a_1, b_2 a_2, \ldots, b_k a_k\}$ is a basis of K and $a_i \mid a_{i+1}$ for $i = 1, 2, \ldots, k - 1$. If $\varphi(b_i) = x_i$ for $i = 1, 2, \ldots, n$, then $\{x_1, x_2, \ldots, x_n\}$ is a set of generators of M, so we have

$$M = x_1 R + x_2 R + \cdots + x_n R.$$

We claim that this sum is direct. To see this, it suffices to show that if $x_1 r_1 + x_2 r_2 + \cdots + x_n r_n = 0$ in M, then $x_1 r_1 = x_2 r_2 = \cdots = x_n r_n = 0$. From

$$\varphi(b_1 r_1 + b_2 r_2 + \cdots + b_n r_n) = \varphi(b_1)r_1 + \varphi(b_2)r_2 + \cdots + \varphi(b_n)r_n$$
$$= x_1 r_1 + x_2 r_2 + \cdots + x_n r_n$$

we see that

$$b_1 r_1 + b_2 r_2 + \cdots + b_n r_n \in K = b_1 a_1 R \oplus b_2 a_2 R \oplus \cdots \oplus b_k a_k R$$

when $x_1 r_1 + x_2 r_2 + \cdots + x_n r_n = 0$. Since $\{b_1, b_2, \ldots, b_n\}$ is a basis for $R^{(n)}$, it follows that $r_i = a_i s_i$ for some $s_i \in R$ for $i = 1, 2, \ldots, k$ and $r_i = 0$ for $i = k + 1, \ldots, n$. Now $x_i r_i = \varphi(b_i) a_i s_i = \varphi(b_i a_i s_i) = 0$ since $b_i a_i s_i \in K$ for

$i = 1, 2, \ldots, k$. Hence, $x_1 r_1 + x_2 r_2 + \cdots + x_n r_n = 0$ gives $x_1 r_1 = x_2 r_2 = \cdots = x_n r_n = 0$, so

$$M = x_1 R \oplus x_2 R \oplus \cdots \oplus x_n R,$$

as asserted.

Finally, note that $\mathrm{ann}(x_i) = (a_i)$ for $i = 1, 2, \ldots, k$ and $\mathrm{ann}(x_i) = 0$ for $i = k + 1, \ldots, n$. Moreover, Proposition 4.3.16 gives $a_i \mid a_{i+1}$ for $i = 1, 2, \ldots, k - 1$, so we have

$$\mathrm{ann}(x_1) \supseteq \mathrm{ann}(x_2) \supseteq \cdots \supseteq \mathrm{ann}(x_n). \qquad \square$$

Recall that if R is an integral domain, then the set $t(M)$ of torsion elements of an R-module M is a submodule of M called the torsion submodule of M.

Proposition 4.3.18. *Let M be a finitely generated module over a principal ideal domain R. Suppose also that M is a direct sum of nonzero cyclic submodules*

$$M = x_1 R \oplus x_2 R \oplus \cdots \oplus x_n R$$

such that $\mathrm{ann}(x_1) \supseteq \mathrm{ann}(x_2) \supseteq \cdots \supseteq \mathrm{ann}(x_n)$. *Then there are nonnegative integers s and k such that $k + s = n$ and such that*

$$M = t(M) \oplus R^{(s)}, \quad \text{where } t(M) = x_1 R \oplus x_2 R \oplus \cdots \oplus x_k R$$

$$\text{and } R^{(s)} = 0 \text{ if } k = n \text{ and } t(M) = 0 \text{ when } s = n.$$

Proof. Suppose that

$$M = x_1 R \oplus x_2 R \oplus \cdots \oplus x_n R \quad \text{and} \quad \mathrm{ann}(x_1) \supseteq \mathrm{ann}(x_2) \supseteq \cdots \supseteq \mathrm{ann}(x_n).$$

If $k \in \{1, 2, \ldots, n\}$ is the largest integer such that $\mathrm{ann}(x_k) \neq 0$, then $\mathrm{ann}(x_{k+1}) = \cdots = \mathrm{ann}(x_n) = 0$, so $x_i R \cong R$ for $i = k + 1, \ldots, n$. Hence, if $s = n - k$, then

$$M = x_1 R \oplus x_2 R \oplus \cdots \oplus x_k R \oplus R^{(s)}.$$

If $x = x_1 r_1 + x_2 r_2 + \cdots + x_k r_k + r_{k+1} + \cdots + r_n$ is a torsion element of M, then there is a nonzero $a \in R$, such that $xa = 0$, But then $r_{k+1}a = \cdots = r_n a = 0$, so $r_{k+1} = \cdots = r_n = 0$. Thus, $t(M) \subseteq x_1 R \oplus x_2 R \oplus \cdots \oplus x_k R$. Conversely, if $\mathrm{ann}(x_k) = (a_k) \neq 0$, then $(x_1 R \oplus x_2 R \oplus \cdots \oplus x_k R)a_k = 0$, so $x_1 R \oplus x_2 R \oplus \cdots \oplus x_k R \subseteq t(M)$. Hence,

$$t(M) = x_1 R \oplus x_2 R \oplus \cdots \oplus x_k R. \qquad \square$$

There is more that can be said about a decomposition of M such as that given in Proposition 4.3.18. Actually, the integers s and k are unique and the summands $x_i R$ are unique up to isomorphism. To address uniqueness, we need the following lemma.

Lemma 4.3.19. *Let M be a finitely generated torsion module over a principal ideal domain R and suppose that p is a prime in R. If $x \in M$ and $p \mid a$, where $(a) = \text{ann}(x)$, then $\text{ann}(xp) = (c)$, where $a = pc$. Furthermore, if $x' \in M$ is such that $\text{ann}(xp) = \text{ann}(x'p)$, then $\text{ann}(x) = \text{ann}(x')$.*

Proof. Suppose that $\text{ann}(x) = (a)$ and let $a = pc$. If $r \in (c)$, then $r = cs$ for some $s \in R$. Hence, $pr = pcs$, so $xpr = xas = 0$. Thus, $r \in \text{ann}(xp)$ and we have $(c) \subseteq \text{ann}(xp)$. Conversely, if $r \in \text{ann}(xp)$, then $xpr = 0$, so $pr \in \text{ann}(x) = (a)$. If $pr = at$, then $pr = pct$, so $r = ct$. Hence, $r \in (c)$, so we have $\text{ann}(xp) \subseteq (c)$ and therefore $\text{ann}(xp) = (c)$.

Finally, let $\text{ann}(xp) = (c)$ and $\text{ann}(x'p) = (c')$, where $a = pc$ and $a' = pc'$. If $\text{ann}(xp) = \text{ann}(x'p)$, then $(c) = (c')$. Hence, there is a unit $u \in R$ such that $c = c'u$. But then $pc = pc'u$, so $a = a'u$. Thus, $(a) = (a')$, so we have $\text{ann}(x) = \text{ann}(x')$. $\qquad\square$

Proposition 4.3.20. *Let M be a finitely generated module over a principal ideal domain. If*

$$M = x_1 R \oplus x_2 R \oplus \cdots \oplus x_k R \oplus R^{(s)}, \qquad (4.13)$$

$$\text{where } \text{ann}(x_1) \supseteq \text{ann}(x_2) \supseteq \cdots \supseteq \text{ann}(x_k) \neq 0$$

and

$$M = x_1' R \oplus x_2' R \oplus \cdots \oplus x_{k'}' \oplus R^{(s')}, \qquad (4.14)$$

$$\text{where } \text{ann}(x_1') \supseteq \text{ann}(x_2') \supseteq \cdots \supseteq \text{ann}(x_{k'}') \neq 0,$$

then $s = s'$, $k = k'$, $\text{ann}(x_i) = \text{ann}(x_i')$ and $x_i R \cong x_i' R$ for $i = 1, 2, \ldots, k$.

Proof. Suppose that M is finitely generated and satisfies the conditions given in (4.13) and (4.14). As in the proof of Proposition 4.3.18, we can show that

$$t(M) = x_1 R \oplus x_2 R \oplus \cdots \oplus x_k R = x_1' R \oplus x_2' R \oplus \cdots \oplus x_{k'}' R.$$

Hence,

$$M = t(M) \oplus R^{(s)} \quad \text{and} \quad M = t(M) \oplus R^{(s')}, \quad \text{so}$$

$$R^{(s)} \cong M/t(M) \cong R^{(s')}.$$

Since R is an IBN-ring, it follows that $s = s'$.

Next, let $\mathfrak{m}$ be a maximal ideal that contains $\text{ann}(x_1)$ Then $\mathfrak{m} = (p)$ and, by Propositions 4.3.5 and 4.3.6, p is a prime. It follows that $p \mid a_i$, where $(a_i) = \text{ann}(x_i)$ for $i = 1, 2, \ldots, k$. Let $T = t(M)$ and consider

$$T/Tp \cong x_1 R/x_1 pR \oplus x_2 R/x_2 pR \oplus \cdots \oplus x_k R/x_k pR \quad \text{and} \qquad (4.15)$$

$$T/Tp \cong x_1' R/x_1' pR \oplus x_2' R/x_2' pR \oplus \cdots \oplus x_{k'}' R/x_{k'}' pR. \qquad (4.16)$$

Now each $x_i R / x_i pR \neq 0$, since if $x_i R = x_i pR$, then $x_i = x_i ps$ for some $s \in R$. But then $(1 - ps) \in \mathrm{ann}(x_i) \subseteq (p)$. If $1 - ps = pt$, then $1 = p(s + t)$ indicates that p is a unit in R which cannot be the case. Similarly, $x_i' R / x_i' pR \neq 0$. Now $R/(p)$ is a field and T/Tp is a vector space over $R/(p)$, so we see from (4.15) that $\dim(T/Tp) = k$. Similarly, from (4.16) we have $\dim(T/Tp) = k'$. But a field is an IBN-ring, so $k = k'$.

Note next that

$$\mathrm{ann}(T) = \mathrm{ann}(x_1) \cap \mathrm{ann}(x_2) \cap \cdots \cap \mathrm{ann}(x_k)$$
$$= \mathrm{ann}(x_k).$$

A similar observation gives $\mathrm{ann}(T) = \mathrm{ann}(x_k')$. Let $\mathrm{ann}(T) = (a)$ and suppose that $a = p_1^{n_1} p_2^{n_2} \cdots p_t^{n_t}$ is a factorization of a. If $n = n_1 + n_2 + \cdots + n_t$, then we set $\#(T) = n$ and proceed by induction on n. If $n = 1$, then $a = p$ and $\mathrm{ann}(x_k) = \mathrm{ann}(x_k') = \mathrm{ann}(T) = (p)$. But (p) is a maximal ideal of R, so we see that

$$\mathrm{ann}(x_1) = \mathrm{ann}(x_2) = \cdots = \mathrm{ann}(x_k) = (p) \quad \text{and}$$
$$\mathrm{ann}(x_1') = \mathrm{ann}(x_2') = \cdots = \mathrm{ann}(x_k') = (p).$$

Hence, the proposition holds if $n = 1$.

Next, suppose that the proposition holds for any torsion module T' satisfying $\#(T') < n$. Now

$$Tp \cong x_1 pR \oplus x_2 pR \oplus \cdots \oplus x_k pR \quad \text{and} \quad Tp \cong x_1' pR \oplus x_2' pR \oplus \cdots \oplus x_k' pR,$$

are torsion R-modules such that

$$\mathrm{ann}(x_1 p) \supseteq \mathrm{ann}(x_2 p) \supseteq \cdots \supseteq \mathrm{ann}(x_k p) \neq 0 \quad \text{and}$$
$$\mathrm{ann}(x_1' p) \supseteq \mathrm{ann}(x_2' p) \supseteq \cdots \supseteq \mathrm{ann}(x_k' p) \neq 0.$$

Since $p \mid a$, where $(a) = \mathrm{ann}(Tp)$, then p is one of the primes in the factorization of a. If $p = p_i$, then $\mathrm{ann}(Tp) = (p_1^{n_1} p_2^{n_2} \cdots p_{i-1}^{n_{i-1}} p_{i+1}^{n_{i+1}} \cdots p_t^{n_t})$, so we see that $\#(Tp) = n - 1$. Hence, the induction hypothesis gives $\mathrm{ann}(x_i p) = \mathrm{ann}(x_i' p)$ for $i = 1, 2, \ldots, k$. Thus, by Lemma 4.3.19, we see that $\mathrm{ann}(x_i) = \mathrm{ann}(x_i')$ for $i = 1, 2, \ldots, k$. It follows that

$$x_i R \cong R / \mathrm{ann}(x_i) = R / \mathrm{ann}(x_i') \cong x_i' R,$$

so the proof is complete.   □

We can now present one form of a decomposition of a finitely generated module over a principal ideal domain.

Proposition 4.3.21 (Invariant Factors). *Let R be a principal ideal domain. Then each finitely generated R-module M has a decomposition*

$$M = x_1 R \oplus x_2 R \oplus \cdots \oplus x_k R \oplus R^{(s)}, \quad where$$

$$\mathrm{ann}(x_1) \supseteq \mathrm{ann}(x_2) \supseteq \cdots \supseteq \mathrm{ann}(x_k) \neq 0.$$

The integers s and k are unique, $x_i R$ is unique up to isomorphism for $i = 1, 2, \ldots, k$ and $t(M) = x_1 R \oplus x_2 R \oplus \cdots \oplus x_k R$. Furthermore, there are $a_1, a_2, \ldots, a_k \in R$ such that

$$M \cong R/(a_1) \oplus R/(a_2) \oplus \cdots \oplus R/(a_k) \oplus R^{(s)} \quad and$$

$$a_i \mid a_{i+1} \quad for\ i = 1, 2, \ldots, k - 1.$$

Proof. Propositions 4.3.17, 4.3.18 and 4.3.20.   □

Corollary 4.3.22. *If R is a principal ideal domain, then a finitely generated R-module is torsion free if and only if it is free.*

Definition 4.3.23. The unique integer s of Proposition 4.3.21 is called the *free-rank* of M. Furthermore, the a_i of Proposition 4.3.21 are unique up to associates and are called the *invariant factors* of M.

Proposition 4.3.21 is not the complete story regarding the decomposition of a finitely generated module over a principal ideal domain. Each cyclic summand $x_i R$ can also be decomposed further as a direct sum of cyclic modules. If $x \in M$ and $\mathrm{ann}(x) = (a)$, then the observation that $xR \cong R/(a)$ leads to the following.

Lemma 4.3.24. *If R is a principal ideal domain and $a \in R$, $a \neq 0$, factors as $a = p_1^{n_1} p_2^{n_2} \cdots p_j^{n_j}$, where the p_i are distinct primes in R and $n_i \geq 1$ for $i = 1, 2, \ldots, j$, then*

$$R/(a) \cong R/(p_1^{n_1}) \oplus R/(p_2^{n_2}) \oplus \cdots \oplus R/(p_j^{n_j}).$$

Proof. Since the p_i are distinct primes, $\gcd(p_h^{n_h}, p_i^{n_i}) = 1$, $1 \leq h, i \leq j$, with $h \neq i$. Thus, Proposition 4.3.3 gives $s_h, s_i \in R$ such that $1 = p_h^{n_h} s_h + p_i^{n_i} s_i$. Hence, if $r \in R$, then $r = p_h^{n_h} s_h r + p_i^{n_i} s_i r$, so $R = (p_h^{n_h}) + (p_i^{n_i})$. Therefore, the ideals $(p_1^{n_1}), (p_2^{n_2}), \ldots, (p_j^{n_j})$ are pairwise comaximal. The ring homomorphism

$$\varphi : R \to R/(p_1^{n_1}) \oplus R/(p_2^{n_2}) \oplus \cdots \oplus R/(p_j^{n_j}) \quad defined\ by$$

$$\varphi(r) = (r + (p_1^{n_1}), r + (p_2^{n_2}), \ldots, r + (p_j^{n_j}))$$

has kernel $(p_1^{n_1}) \cap (p_2^{n_2}) \cap \cdots \cap (p_j^{n_j})$. Since the ideals $(p_1^{n_1}), (p_2^{n_2}), \ldots, (p_j^{n_j})$ are pairwise comaximal, φ is an epimorphism and

$$(a) = (p_1^{n_1})(p_2^{n_2}) \cdots (p_j^{n_j}) = (p_1^{n_1}) \cap (p_2^{n_2}) \cap \cdots \cap (p_j^{n_j}).$$

Thus, the Chinese Remainder Theorem proves the proposition.   □

Proposition 4.3.25 (Elementary Divisors). *If M is a finitely generated module over a principal ideal domain R, then*

$$M \cong R/(p_1^{n_1}) \oplus R/(p_2^{n_2}) \oplus \cdots \oplus R/(p_t^{n_t}) \oplus R^{(s)},$$

where $p_1^{n_1}, p_2^{n_2}, \ldots p_t^{n_t}$ are positive powers of not necessarily distinct primes $p_1, p_2,$ $\ldots, p_t$ in R.

Proof. Proposition 4.3.21 shows that

$$M = x_1 R \oplus x_2 R \oplus \cdots \oplus x_k R \oplus R^{(s)} \quad \text{and}$$

$$\text{ann}(x_1) \supseteq \text{ann}(x_2) \supseteq \cdots \supseteq \text{ann}(x_k) \neq 0,$$

where the integers s and k are unique and each $x_i R$ is unique up to isomorphism. If $a_1, a_2, \ldots, a_k \in R$ are such that $\text{ann}(x_i) = (a_i)$, then $x_i R \cong R/(a_i)$ for $i = 1, 2, \ldots, k$. If $a_i = p_{1_i}^{n_{1_i}} p_{2_i}^{n_{2_i}} \cdots p_{j_i}^{n_{j_i}}$ is a factorization for a_i as a product of distinct primes, for $i = 1, 2, \ldots, k$, then as in Lemma 4.3.24, we see that

$$R/(a_i) \cong R/(p_{1_i}^{n_{1_i}}) \oplus R/(p_{2_i}^{n_{2_i}}) \oplus \cdots \oplus R/(p_{j_i}^{n_{j_i}})$$

for each i. Hence,

$$M \cong \bigoplus_{i=1}^{k} R/(a_i) \oplus R^{(s)}$$

$$\cong \bigoplus_{i=1}^{k} [R/(p_{1_i}^{n_{1_i}}) \oplus R/(p_{2_i}^{n_{2_i}}) \oplus \cdots \oplus R/(p_{j_i}^{n_{j_i}})] \oplus R^{(s)}.$$

If we let $\{p_1^{n_1}, p_2^{n_2}, \ldots, p_t^{n_t}\}$ be the set of primes $\{p_{1_i}^{n_{1_i}}, p_{2_i}^{n_{2_i}}, \ldots, p_{j_i}^{n_{j_i}}\}_{i=1}^{k}$, then the proof is complete. Note that it may be the case that a prime appearing in a factorization of a_i may also appear in a factorization of $a_j, 1 \leq i, j \leq k$, so the primes in $\{p_1^{n_1}, p_2^{n_2}, \ldots, p_t^{n_t}\}$ may not be distinct. $\square$

Definition 4.3.26. The positive powers $p_1^{n_1}, p_2^{n_2}, \ldots, p_t^{n_t}$ of the not necessarily distinct primes $p_1, p_2, \ldots, p_t$, given in Proposition 4.3.25 are called the *elementary divisors* of M.

Example

4. Consider the $\mathbb{Z}$-module $M = \mathbb{Z}_4 \times \mathbb{Z}_4 \times \mathbb{Z}_3 \times \mathbb{Z}_9 \times \mathbb{Z}_5 \times \mathbb{Z}_7 \times \mathbb{Z}_{49}$. The elementary divisors of M are $2^2, 2^2, 3, 3^2, 5, 7$ and 7^2. The largest invariant factor is the least common multiple of the elementary divisors. This gives $2^2 \times 3^2 \times 5 \times 7^2 = 8820$. Eliminating the elementary divisors used in the product $2^2 \times 3^2 \times 5 \times 7^2$ from the elementary divisor list gives $2^2, 3, 7$. The least common multiple of these integers is $2^2 \times 3 \times 7 = 84$. Hence, the invariant factors of M are 84 and 8820, so $M \cong \mathbb{Z}_{84} \times \mathbb{Z}_{8820}$. Note that $84 \mid 8820$.

Problem Set 4.3

1. Verify Examples 1, 2 and 3.

2. Prove parts (1), (2), (3), (5) and (6) of Proposition 4.3.5.

3. Give the elementary divisors of the $\mathbb{Z}$-module

$$M = \mathbb{Z}_2 \times \mathbb{Z}_2 \times \mathbb{Z}_4 \times \mathbb{Z}_8 \times \mathbb{Z}_3 \times \mathbb{Z}_3 \times \mathbb{Z}_3 \times \mathbb{Z}_5 \times \mathbb{Z}_{25}$$

and then compute the invariant factors of M. Write a $\mathbb{Z}$-module isomorphic to M in terms of the invariant factors of M.

4. Let R be a principal ideal domain, let a be a nonzero, nonunit of R such that $M(a) = 0$. If $a = p_1^{n_1} p_2^{n_2} \cdots p_k^{n_k}$ is the factorization of a into distinct primes of R, let $M_i = \operatorname{ann}^M(p_i^{n_i})$ for $i = 1, 2, \ldots, k$. Prove that each M_i is a submodule of M and that $M = M_1 \oplus M_2 \oplus \cdots \oplus M_k$.

5. Let M be a module over a principal ideal domain R and let p be a prime in R. If $M_p = \{x \in M \mid xp^k = 0 \text{ for some } k \geq 1\}$, then M_p is said to be the *p-primary component* of M. If there is a prime p in R such that $M = M_p$, then M is said to be a *p-primary module*.

   (a) Prove that M_p is a submodule of M.

   (b) Show that submodules and factor modules of a p-primary module are p-primary.

6. Let R be a principal ideal domain.

   (a) If $\{M_\alpha\}_\Delta$ is a family of R-modules each of which is p-primary, prove that $\bigoplus_\Delta M_\alpha$ is p-primary.

   (b) If M is finitely generated and torsion, prove that M is a direct sum of its primary submodules, that is, prove that $M = \bigoplus M_p$, where p runs through the primes of R.

Chapter 5

Injective, Projective, and Flat Modules

5.1 Injective Modules

If U is a subspace of a vector space V over a division ring, then U is a direct summand of V. This property follows directly from the fact that Zorn's lemma can be used to extend a basis of U to a basis of V. There are R-modules that possess this summand property even though they may not have a basis. Such modules, called *injective modules*, form an important class of modules. In this section we not only consider injective modules, but we also investigate the effect that such a module M has on the functor $\operatorname{Hom}_R(-, M)$.

Definition 5.1.1. An R-module M is said to be *injective* if every row exact diagram

$$
\begin{array}{ccc}
0 \longrightarrow N_1 & \xrightarrow{\ h\ } & N_2 \\
\quad\ \ \downarrow{\scriptstyle f} & {\scriptstyle g} \\
\quad\ \ M
\end{array}
$$

of R-modules and R-module homomorphisms can be completed commutatively by an R-linear mapping $g : N_2 \to M$.

It is easy to show that if M and N are isomorphic R-modules, then M is injective if and only if N is injective. One can also show that an R-module M is injective if and only if for each R-module N_2 and each submodule N_1 of N_2, every R-linear mapping $f : N_1 \to M$ can be extended to an R-linear mapping $g : N_2 \to M$. So in Definition 5.1.1, we can safely assume that N_1 is a submodule of N_2 and h can be replaced by the canonical injection $i : N_1 \to N_2$.

Our first task is to prove that an injective R-module M has the property that it is a direct summand of every module that contains it as a submodule.

Proposition 5.1.2. *If M is an injective submodule of R-module N, then M is a direct summand of N.*

Proof. Let M be an injective submodule of an R-module N and consider the row exact diagram

$$
\begin{array}{ccc}
0 \longrightarrow M & \overset{i}{\longrightarrow} & N \\
\text{id}_M \downarrow & \swarrow g & \\
M & &
\end{array}
$$

of R-modules, where id_M is the identity mapping on M and i is the canonical injection. If $g : N \to M$ completes the diagram commutatively, then $gi = \text{id}_M$. Hence, g is a splitting map for i. The result follows immediately from Proposition 3.2.3. $\square$

Examples

1. **Not Every R-module Is Injective.**   The $\mathbb{Z}$-module $2\mathbb{Z} \subseteq \mathbb{Z}$ is not injective, since $2\mathbb{Z}$ is not a direct summand of $\mathbb{Z}$.

2. **Vector Spaces.**   If V is a vector space over a division ring D, then V is an injective D-module.

3. **The Rational Numbers.**   The field of rational numbers $\mathbb{Q}$ is an injective $\mathbb{Z}$-module.

The following important and useful proposition is due to R. Baer [50]. It will be referred to as *Baer's criteria* (for injectivity).

Proposition 5.1.3 (Baer's criteria). *The following are equivalent for an R-module M.*

(1) *M is injective.*

(2) *For each right ideal A of R, every R-linear mapping $f : A \to M$ can be extended to an R-linear mapping $g : R \to M$.*

(3) *For each right ideal A of R and each R-linear mapping $f : A \to M$ there is an $x \in M$ such that $f(a) = xa$ for all $a \in A$.*

Proof. (1) $\Rightarrow$ (2) is obvious.

(2) $\Rightarrow$ (3). If g extends f to R, let $g(1) = x$. Then $f(a) = g(a) = g(1)a = xa$ for each $a \in A$.

(3) $\Rightarrow$ (1). If N_1 is a submodule of N_2, $i : N_1 \to N_2$ is the canonical injection and $f : N_1 \to M$ is an R-linear mapping, then we need to show that the diagram

$$
\begin{array}{ccc}
0 \longrightarrow N_1 & \overset{i}{\longrightarrow} & N_2 \\
f \downarrow & \swarrow g & \\
M & &
\end{array}
$$

can be completed commutatively by an R-linear mapping $g : N_2 \to M$. Consider the set $\mathscr{S}$ of ordered pairs (X, g), where X is a submodule of N_2 such that $N_1 \subseteq X$ and g restricted to N_1 gives f. Partial order $\mathscr{S}$ by $(X, g) \leq (X', g')$ if $X \subseteq X'$ and g' restricted to X produces g. Note that $\mathscr{S} \neq \varnothing$ since $(N_1, f) \in \mathscr{S}$. It follows that $\mathscr{S}$ is inductive, so Zorn's lemma produces a maximal element of $\mathscr{S}$, say (X^*, g^*). If $X^* = N_2$, then the proof is complete, so suppose that $X^* \neq N_2$. If $y \in N_2 - X^*$, then we have an R-linear mapping from the right ideal $(X^* : y)$ to M given by $a \mapsto g^*(ya)$. By assumption, this implies that there is a $z \in M$ such that $a \mapsto za$ for all $a \in (X^* : y)$. If $h : X^* + yR \to M$ is defined by $h(x + ya) = g^*(x) + za$, then h extends g^* to $X^* + yR$, which contradicts the maximality of (X^*, g^*) in $\mathscr{S}$. Thus, $X^* = N_2$. $\square$

Baer's criteria shows that the collection of right ideals is a test set for the injectivity of an R-module. Actually, there is an often "smaller" collection of right ideals of R that will perform this task. To show this, we need the following definition.

Definition 5.1.4. A submodule N of an R-module M is said to be an *essential* (or a *large*) *submodule* of M, if $N \cap N' \neq 0$ for each nonzero submodule N' of M. If N is an essential submodule of M, then M is referred to as an *essential extension* of N.

Every nonzero R-module M always has at least one essential submodule, namely M. We now show that for every submodule N of M there is a submodule N_c of M such that the sum $N + N_c$ is direct and such that $N + N_c$ is an essential submodule of M. We will see later that there are R-modules M that have no proper essential submodules. For these modules $N \oplus N_c = M$, so every submodule of M is a direct summand of M. Modules with this property will prove to be of special interest.

Proposition 5.1.5. *If N is a submodule of an R-module M, then there is a submodule N_c of M such that the sum $N + N_c$ is direct and $N + N_c$ is essential in M.*

Proof. Suppose that N is a submodule of M and let $\mathscr{S}$ be the set of submodules N' of M such that $N \cap N' = 0$. Then $\mathscr{S} \neq \varnothing$ since the zero submodule of M is in $\mathscr{S}$. If $\mathscr{S}$ is partially ordered by inclusion, then $\mathscr{S}$ is inductive and Zorn's lemma indicates that $\mathscr{S}$ has a maximal element. If N_c is a maximal element of $\mathscr{S}$, then it is immediate that the sum $N + N_c$ is direct.

We claim that $N + N_c$ is an essential submodule of M. Let X be a nonzero submodule of M and suppose that $(N + N_c) \cap X = 0$. Note that X cannot be contained in N_c, so $X + N_c$ properly contains N_c. Therefore, $N \cap (X + N_c) \neq 0$. Let $z \in N \cap (X + N_c)$, $z \neq 0$, and choose $x \in X$ and $y \in N_c$ to be such that $z = x + y$. Then $z - y = x \in (N + N_c) \cap X = 0$. Hence, $z = y$ and so $z \in N \cap N_c = 0$, a contradiction. Therefore, $(N + N_c) \cap X \neq 0$, which establishes that $N + N_c$ is an essential submodule of M. $\square$

Definition 5.1.6. If N is a submodule of an R-module M, then a submodule N_c of M such that $N \oplus N_c$ is essential in M is said to be a *complement* of N in M.

Proposition 5.1.7. *An R-module M is injective if and only if for every essential A right ideal of R and each R-linear mapping $f : A \to M$, there is an $x \in M$ such that $f(a) = xa$ for all $a \in R$.*

Proof. If M is an injective R-module, then there is nothing to prove, so suppose that the condition holds for each essential right ideal of R. Let A be a right ideal of R and suppose that $f : A \to M$ is an R-linear mapping. If A_c is a complement of A in R, then $A \oplus A_c$ is an essential right ideal of R. If $g : A \oplus A_c \to M$ is such that $g(a + a') = f(a)$, then g is well defined and R-linear, so there is an $x \in M$ such that $g(a + a') = x(a + a')$. Hence, $f(a) = g(a + 0) = xa$ for all $a \in A$, so M is injective by Baer's criteria.   □

Thus, we see that the set of essential right ideals of R will also serve as a test set for the injectivity of an R-module. We now show that every module embeds in an injective module. In order to establish this result, we briefly investigate injective modules over a principal ideal domain. Injective modules over such a domain enjoy a divisibility property that is equivalent to injectivity.

Definition 5.1.8. If R is a principal ideal domain, then an R-module M is said to be *R-divisible* if $Ma = M$ for each nonzero $a \in R$. That is, M is R-divisible if and only if given a nonzero element $a \in R$ and $y \in M$, there is an $x \in M$ such that $xa = y$.

Examples

4. **Homomorphic Images.** Every homomorphic image of an R-divisible R-module is R-divisible.

5. **Direct Sums.** A direct sum (direct product) of R-divisible R-modules is R-divisible and a direct summand of an R-divisible module is R-divisible.

6. **The Rational Numbers.** The field of rational numbers $\mathbb{Q}$ is $\mathbb{Z}$-divisible, so $\mathbb{Q}^{(\Delta)}$ is $\mathbb{Z}$-divisible for any set Δ. Thus, $\mathbb{Q}^{(\Delta)}/N$ is $\mathbb{Z}$-divisible for any subgroup N of $\mathbb{Q}^{(\Delta)}$.

7. If p is a prime and $\mathbb{Z}_{p^\infty} = \{\frac{a}{p^n} + \mathbb{Z} \in \mathbb{Q}/\mathbb{Z} \mid a \in \mathbb{Z} \text{ and } n = 0, 1, 2, \dots\}$, then $\mathbb{Z}_{p^\infty}$ is $\mathbb{Z}$-divisible.

Proposition 5.1.9. *If R is a principal ideal domain, then an R-module is injective if and only if it is R-divisible.*

Proof. Suppose that M is an injective R-module and let $a \in R, a \neq 0$. Then (a) is an ideal of R and since R is an integral domain, if $y \in M$, then $f(ab) = yb$ gives a well-defined R-linear map from (a) to M. But M is injective, so Baer's criteria indicates that there is an $x \in M$ such that $f(c) = xc$ for all $c \in (a)$. In particular, $f(a) = xa$, so $xa = y$. Hence, M is R-divisible.

Conversely, suppose that M is R-divisible. Let (a) be a nonzero ideal of R and suppose that $f : (a) \rightarrow M$ is an R-linear mapping. If $f(a) = y$, then there is an $x \in M$ such that $xa = y$. Consequently, if $ab \in (a)$, then $f(ab) = f(a)b = yb = x(ab)$. Therefore, Baer's criteria is satisfied, so M is an injective R-module.   □

One important aspect of divisible $\mathbb{Z}$-modules is that they can be used to produce injective R-modules.

Proposition 5.1.10. *The following hold for any R-module M.*

(1) *If M is viewed as a $\mathbb{Z}$-module, then M can be embedded in an injective $\mathbb{Z}$-module.*

(2) **Injective Producing Property.**   *If Q is an injective $\mathbb{Z}$-module, then $\mathrm{Hom}_\mathbb{Z}(R, Q)$ is an injective R-module.*

(3) *There is an R-linear embedding of M into an injective R-module.*

Proof. (1) Let M be an R-module. If $\{x_\alpha\}_\Delta$ is a set of generators for M as an additive abelian group, then there is a group epimorphism $\mathbb{Z}^{(\Delta)} \rightarrow M$. If K is the kernel of this map, then $M \cong \mathbb{Z}^{(\Delta)}/K \subsetneq \mathbb{Q}^{(\Delta)}/K$. Example 6 indicates that $\mathbb{Q}^{(\Delta)}/K$ is $\mathbb{Z}$-divisible while Proposition 5.1.9 shows that $\mathbb{Q}^{(\Delta)}/K$ is an injective $\mathbb{Z}$-module.

(2) Since R is an $(R, \mathbb{Z})$-bimodule, $\mathrm{Hom}_\mathbb{Z}(R, Q)$ is an R-module via $(ha)(x) = h(ax)$ for all $a, x \in R$ and $h \in \mathrm{Hom}_\mathbb{Z}(R, Q)$. Now let A be a right ideal of R and suppose that $f : A \rightarrow \mathrm{Hom}_\mathbb{Z}(R, Q)$ is an R-linear mapping. To show that $\mathrm{Hom}_\mathbb{Z}(R, Q)$ is an injective R-module, it suffices, by Baer's criteria, to find an $h \in \mathrm{Hom}_\mathbb{Z}(R, Q)$ such that $f(a) = ha$ for all $a \in A$. If $a \in A$, then $f(a) \in \mathrm{Hom}_\mathbb{Z}(R, Q)$ and for each $x \in R$, $f(a)(x) \in Q$. It follows that $a \mapsto f(a)(1)$ is a $\mathbb{Z}$-linear map $g : A \rightarrow Q$, so if Q is an injective $\mathbb{Z}$-module, then there is a map $h \in \mathrm{Hom}_\mathbb{Z}(R, Q)$ that extends g to R. If $a \in A$ and $b \in R$, then $(ha)(b) = h(ab) = g(ab) = f(ab)(1) = (f(a)b)(1) = f(a)(b)$. Hence, $f(a) = ha$ for all $a \in A$, so $\mathrm{Hom}_\mathbb{Z}(R, Q)$ is an injective R-module.

(3) By (1) there is an embedding $0 \rightarrow M \rightarrow Q$ of M into an injective $\mathbb{Z}$-module Q. Since $\mathrm{Hom}_\mathbb{Z}(R, -)$ is left exact and covariant, we have an exact sequence $0 \rightarrow \mathrm{Hom}_\mathbb{Z}(R, M) \rightarrow \mathrm{Hom}_\mathbb{Z}(R, Q)$. But $M \cong \mathrm{Hom}_R(R, M) \subseteq \mathrm{Hom}_\mathbb{Z}(R, M)$, so it follows that M embeds in $\mathrm{Hom}_\mathbb{Z}(R, Q)$ which is, by (2), an injective R-module. It is not difficult to show that the embedding is an R-linear mapping.   □

Injective Modules and the Functor $\text{Hom}_R(-, M)$

Another property possessed by an injective R-module M is that it renders the contravariant functor $\text{Hom}_R(-, M)$ exact.

Proposition 5.1.11. *An R-module M is injective if and only if for each exact sequence $0 \to N_1 \xrightarrow{g} N_2$ of R-modules and R-module homomorphisms, the sequence $\text{Hom}_R(N_2, M) \xrightarrow{g^*} \text{Hom}_R(N_1, M) \to 0$ is exact in $\mathbf{Ab}$.*

Proof. If M is injective and $f \in \text{Hom}_R(N_1, M)$, then the injectivity of M gives an R-linear mapping $h : N_2 \to M$ such that $f = hg = g^*(h)$. Hence, g^* is an epimorphism, so $\text{Hom}_R(N_2, M) \xrightarrow{g^*} \text{Hom}_R(N_1, M) \to 0$ is exact.

Conversely, if $\text{Hom}_R(N_2, M) \xrightarrow{g^*} \text{Hom}_R(N_1, M) \to 0$ is exact and $f \in \text{Hom}_R(N_1, M)$, then g^* is an epimorphism, so there is an $h \in \text{Hom}_R(N_2, M)$ such that $g^*(h) = f$. But then $hg = f$, so the diagram

is commutative and consequently, M is injective. $\square$

Since $\text{Hom}_R(-, M)$ is left exact for any R-module M, we have the following corollary.

Corollary 5.1.12. *An R-module M is injective if and only if the contravariant functor $\text{Hom}_R(-, M) : \mathbf{Mod}_R \to \mathbf{Ab}$ is exact.*

The characterization of injective modules as precisely those R-modules M for which $\text{Hom}_R(-, M)$ preserves epimorphisms leads to the following proposition.

Proposition 5.1.13. *If $\{M_\alpha\}_\Delta$ is a family of R-modules, then the R-module $\prod_\Delta M_\alpha$ is injective if and only if each M_α is injective.*

Proof. Let $0 \to N_1 \xrightarrow{g} N_2$ be an exact sequence of R-modules and R-module homomorphisms. For each $\alpha \in \Delta$, let

$$g_\alpha^* : \text{Hom}_R(N_2, M_\alpha) \to \text{Hom}_R(N_1, M_\alpha) \quad \text{be such that } g_\alpha^*(f) = fg$$

for each $f \in \text{Hom}_R(N_2, M_\alpha)$. If $\prod_\Delta M_\alpha$ is injective, then

$$\text{Hom}_R\left(N_2, \prod_\Delta M_\alpha\right) \xrightarrow{g^*} \text{Hom}_R\left(N_1, \prod_\Delta M_\alpha\right) \to 0$$

is exact. Part (2) of Proposition 2.1.12 shows that for each $\alpha \in \Delta$ we have a row exact commutative diagram,

$$
\begin{array}{ccccc}
\operatorname{Hom}_R\left(N_2, \prod_{\Delta} M_\alpha\right) & \xrightarrow{\;g^*\;} & \operatorname{Hom}_R\left(N_1, \prod_{\Delta} M_\alpha\right) & \longrightarrow & 0 \\
\Big\downarrow{\scriptstyle\varphi} & & \Big\downarrow{\scriptstyle\varphi'} & & \\
\prod_{\Delta}\operatorname{Hom}_R(N_2, M_\alpha) & \xrightarrow{\;\Pi g_\alpha^*\;} & \prod_{\Delta}\operatorname{Hom}_R(N_1, M_\alpha) & \longrightarrow & 0 \\
\Big\downarrow{\scriptstyle\pi_\alpha} & & \Big\downarrow{\scriptstyle\pi'_\alpha} & & \\
\operatorname{Hom}_R(N_2, M_\alpha) & \xrightarrow{\;g_\alpha^*\;} & \operatorname{Hom}_R(N_1, M_\alpha) & &
\end{array}
$$

where π_α and π'_α are canonical projections, φ and φ' are isomorphisms such that $\varphi(f) = (\pi_\alpha f)$ and $\varphi'(f') = (\pi'_\alpha f')$, respectively, and $\prod g_\alpha^*$ is defined by $\prod g_\alpha^*((f_\alpha)) = (f_\alpha g)$. A simple diagram chase shows that g_α^* is an epimorphism, so each M_α is injective.

Conversely, if M_α is injective for each $\alpha \in \Delta$, then

$$
\prod_{\Delta} g_\alpha^* : \prod_{\Delta}\operatorname{Hom}_R(N_2, M_\alpha) \to \prod_{\Delta}\operatorname{Hom}_R(N_1, M_\alpha)
$$

is an epimorphism. The mappings $\varphi^{-1}((f_\alpha)) = f$, where $f(x) = (f_\alpha(x))$ for $x \in N_2$ and $\varphi'^{-1}((f'_\alpha)) = f'$ with $f'(x) = (f'_\alpha(x))$ for $x \in N_1$ give the commutative diagram

$$
\begin{array}{ccccc}
\prod_{\Delta}\operatorname{Hom}_R(N_2, M_\alpha) & \xrightarrow{\;\Pi g_\alpha^*\;} & \prod_{\Delta}\operatorname{Hom}_R(N_1, M_\alpha) & \longrightarrow & 0 \\
\Big\downarrow{\scriptstyle\varphi^{-1}} & & \Big\downarrow{\scriptstyle\varphi'^{-1}} & & \\
\operatorname{Hom}_R\left(N_2, \prod_{\Delta} M_\alpha\right) & \xrightarrow{\;g^*\;} & \operatorname{Hom}_R\left(N_1, \prod_{\Delta} M_\alpha\right) & &
\end{array}
$$

It is clear that g^* is an epimorphism, so $\prod_{\Delta} M_\alpha$ is injective. $\square$

Corollary 5.1.14. *A direct summand of an injective R-module is injective.*

Proposition 5.1.13 shows that direct products of injective modules are always injective for modules over an arbitrary ring. However, direct sums of injective modules are not always injective unless the ring satisfies certain conditions. The proof of this fact will be delayed until injective envelopes are considered.

Problem Set 5.1

1. (a) If N is a submodule of an R-module M, prove that N is an essential submodule of M if and only if for each nonzero submodule N' of M and each $x \in N'$, $x \neq 0$, there is an $a \in R$ such that $xa \in N$ and $xa \neq 0$.

   (b) Prove that a nonzero submodule N of M is an essential submodule of M if and only if $N' = 0$ whenever N' is a submodule of M such that $N \cap N' = 0$.

2. Revisit Definition 5.1.1 and show that an R-module M is injective if and only if for each R-module N_2 and each submodule N_1 of N_2, every R-linear mapping $f : N_1 \to M$ can be extended to an R-linear mapping $g : N_2 \to M$.

3. Verify Examples 1 through 3.

4. Verify Examples 4 through 7.

5. Let N be an essential submodule of an R-module M. Show that $(N : x) = \{a \in R \mid xa \in N\}$ is an essential right ideal of R for any $x \in N$.

6. If M and N are R-modules and X is an essential submodule of N, prove that $M \oplus X$ is an essential submodule of $M \oplus N$.

7. Show that the embedding map $M \to \operatorname{Hom}_{\mathbb{Z}}(R, Q)$ discussed in the proof of (3) of Proposition 5.1.10 is R-linear.

8. If N is a submodule of an R-module M, prove that complements of N in M are unique up to isomorphism.

9. A monomorphism $f : M \to N$ is said to be an *essential monomorphism* if $f(M)$ is an essential submodule of N.

   (a) Let $f : M \to N$ be an essential monomorphism. If $g : N \to X$ is an R-linear mapping such that gf is an injection, prove that g is an injective mapping.

   (b) Suppose that $f : M \to N$ is a monomorphism such that $g : N \to X$ is an injective R-linear mapping whenever gf is an injective mapping. Prove that f is an essential monomorphism.

10. Suppose that N_1 and N_2 are submodules of an R-module M.

    (a) Prove that if $N_1 \subseteq N_2$, then N_1 is an essential submodule of M if and only if N_1 is an essential submodule of N_2 and N_2 is an essential submodule of M.

    (b) Show that $N_1 \cap N_2$ is an essential submodule of M if and only if N_1 and N_2 are essential submodules of M.

    (c) Prove that a finite intersection of essential submodules of M is an essential submodule of M.

    (d) Show by example that an arbitrary intersection of essential submodules of M need not be an essential submodule of M.

11. (a) **Schanuel's Lemma For Injective Modules.** Suppose that

$$0 \to M \to E_1 \xrightarrow{p_1} C_1 \to 0 \quad \text{and} \quad 0 \to M \to E_2 \xrightarrow{p_2} C_2 \to 0$$

are short exact sequences of R-modules and R-module homomorphisms. If E_1 and E_2 are injective, show that $E_1 \oplus C_2 \cong E_2 \oplus C_1$. [Hint: Consider the row exact diagram

$$
\begin{array}{ccccccccc}
0 & \longrightarrow & M & \longrightarrow & E_1 & \xrightarrow{p_1} & C_1 & \longrightarrow & 0 \\
& & \downarrow{\scriptstyle \mathrm{id}_M} & & \vdots\,\alpha & & \vdots\,\beta & & \\
0 & \longrightarrow & M & \longrightarrow & E_2 & \xrightarrow{p_2} & C_2 & \longrightarrow & 0
\end{array}
$$

Since E_2 is injective, there is a mapping $\alpha : E_1 \to E_2$ making the left-hand square of the diagram commutative. Show that α induces a map $\beta : C_1 \to C_2$ such that the right-hand square is commutative. Now show that the sequence

$0 \to E_1 \xrightarrow{f} E_2 \oplus C_1 \xrightarrow{g} C_2 \to 0$ is split exact, where $f(x) = (\alpha(x), p_1(x))$ and $g((x, y)) = p_2(x) - \beta(y)$.]

(b) Two R-modules M and N are said to be *injectively equivalent* if there are injective R-modules E_1 and E_2 such that $E_1 \oplus M \cong E_2 \oplus N$. Is injective equivalence an equivalence relation on the class of R-modules?

12. Show that a row and column exact diagram

$$
\begin{array}{ccccccccc}
& & 0 & & & & 0 & & \\
& & \downarrow & & & & \downarrow & & \\
0 & \longrightarrow & M_1 & \xrightarrow{f} & M & \xrightarrow{g} & M_2 & \longrightarrow & 0 \\
& & \downarrow{\scriptstyle \alpha} & & & & \downarrow{\scriptstyle \gamma} & & \\
& & E_1 & & & & E_2 & &
\end{array}
$$

of R-modules and R-module homomorphisms, where E_1 and E_2 are injective, can be completed to a row and column exact commutative diagram

$$
\begin{array}{ccccccccc}
& & 0 & & 0 & & 0 & & \\
& & \downarrow & & \downarrow & & \downarrow & & \\
0 & \longrightarrow & M_1 & \xrightarrow{f} & M & \xrightarrow{g} & M_2 & \longrightarrow & 0 \\
& & \downarrow{\scriptstyle \alpha} & & \downarrow{\scriptstyle \beta} & & \downarrow{\scriptstyle \gamma} & & \\
0 & \longrightarrow & E_1 & \xrightarrow{f_1} & E & \xrightarrow{g_1} & E_2 & \longrightarrow & 0
\end{array}
$$

with E injective. [Hint: Set $E = E_1 \oplus E_2$ and let $f_1 : E_1 \to E$ and $g_1 :$ $E \to E_2$ be the canonical injection and the canonical surjection, respectively. If $h : M \to E_1$ is the map given by the injectivity of E_1, let $\beta = f_1 h + \gamma g$.]

13. (a) Let $\{M_\alpha\}_\Delta$ be a family of R-modules. If N_α is an essential submodule of M_α for each $\alpha \in \Delta$, show that $\bigoplus_\Delta N_\alpha$ is an essential submodule of $\bigoplus_\Delta M_\alpha$.

(b) Show that $\mathbb{Z}_\mathbb{Z}$ is an essential submodule of $\mathbb{Q}_\mathbb{Z}$, but that the $\mathbb{Z}$-module $\mathbb{Z}^\mathbb{N}$ is not an essential $\mathbb{Z}$-submodule of $\mathbb{Q}^\mathbb{N}$. Conclude that the property stated in (a) for direct sums does not hold for direct products.

14. (a) A commutative ring R is said to be *self-injective* if R is injective as an R-module. If R is a principal ideal domain and a is a nonzero nonunit in R, prove that $R/(a)$ is a self-injective ring. [Hint: If $(b)/(a)$ is an ideal of $R/(a)$, where $(a) \subseteq (b)$, then it suffices to show that each $R/(a)$-linear map $f : (b)/(a) \to R/(a)$ can be extended to $R/(a)$. If $f(b+(a)) = r+(a)$, then $a = bc$ for some $c \in R$, so $0 = f(a + (a)) = f(bc + (b)) = (r + (a))(c + (a)) = rc + (a)$. Thus, $rc = ad = bcd$ for some $d \in R$.]

(b) Observe that Part (a) does not indicate that the principal ideal domain R itself is self-injective. Give an example of a principal ideal domain that is not self-injective. [Hint: Consider the ring $\mathbb{Z}$.]

15. An element x of an R-module M, where R is a principal ideal domain, is said to have *order* a if $(a) = \operatorname{ann}(x)$. Let M be a finitely generated module over a principal ideal domain R and suppose that p is a prime in R.

(a) Show that M_p, the p-primary component of M, is an $R/(p^n)$-module for some integer $n \geq 1$. Show also that N is an R-submodule of M_p if and only if N is an $R/(p^n)$-submodule of M_p.

(b) If $x \in M_p$, $x \neq 0$, has order p^n, show that $xR \cong R/(p^n)$.

(c) If $x \in M_p$, $x \neq 0$, has order p^n, show that there is an R-submodule N of M_p such that $M_p = xR \oplus N$. [Hint: Consider (a) of Exercise 13.]

5.2   Projective Modules

Projective R-modules are dual to injective modules. We can obtain the definition of a projective R-module simply by reversing the arrows in the diagram given in Definition 5.1.1 of an injective module. More specifically, a projective R-module is defined as follows.

Definition 5.2.1. An R-module M is said to be *projective* if each row exact diagram

$$
\begin{array}{ccc}
& & M \\
& {}^{g}\nearrow & \downarrow {}^{f} \\
N_2 \xrightarrow{\ h\ } & N_1 & \longrightarrow 0
\end{array}
$$

of R-modules and R-module homomorphisms can be completed commutatively by an R-linear mapping $g : M \to N_2$.

Clearly, if M and N are isomorphic R-modules, then M is projective if and only if N is projective. One can also show that an R-module M is projective if and only if for each R-module N and each submodule N' of N, the diagram

$$
\begin{array}{ccc}
 & & M \\
 & {}^{g}\nearrow & \downarrow {}^{f} \\
N & \xrightarrow{\eta} & N/N' \longrightarrow 0
\end{array}
$$

can be completed commutatively by an R-linear mapping $g : M \to N$, where η is the canonical surjection.

Examples

1. **Vector Spaces.** Every vector space over a division ring is projective.

2. **Free Modules.** Every free R-module is projective.

3. **Modules That Are Not Projective.** There are modules that are not projective. For example, $\prod_{\mathbb{N}} \mathbb{Z}_i$, where $\mathbb{Z}_i = \mathbb{Z}$ for $i = 1, 2, 3, \ldots$, is not a projective $\mathbb{Z}$-module. (Details can be found in [26].)

4. The $n \times n$ matrix ring $\mathbb{M}_n(R)$ is projective as an $\mathbb{M}_n(R)$-module and as an R-module.

An injective R-module M is a direct summand of each R-module N that extends M. In fact, if M is injective and $f : M \to N$ is a monomorphism, then there is a submodule X of N such that $M \cong X$ and X is a direct summand of N. Projective modules enjoy a similar property.

Proposition 5.2.2. *If $f : N \to M$ is an epimorphism and M is a projective R-module, then M is isomorphic to a direct summand of N.*

Proof. Since the row exact diagram

$$
\begin{array}{ccc}
 & & M \\
 & {}^{g}\nearrow & \downarrow {}^{1_M} \\
N & \xrightarrow{f} & M \longrightarrow 0
\end{array}
$$

can be completed commutatively by an R-linear mapping $g : M \to N$ such that $fg = \mathrm{id}_M$, g is a splitting map for f and g is a monomorphism. Proposition 3.2.4 shows that $N = \mathrm{Im}\, g \oplus \mathrm{Ker}\, f$, so the result follows since $\mathrm{Im}\, g \cong M$. $\square$

Proposition 5.2.3. *If* $\{M_\alpha\}_\Delta$ *is a family of R-modules, then* $\bigoplus_\Delta M_\alpha$ *is projective if and only if each* M_α *is projective.*

Proof. The proof is dual to that of Proposition 5.1.13.   □

Corollary 5.2.4. *A direct summand of a projective R-module is projective.*

Lemma 5.2.5. *The ring R is a projective R-module.*

Proof. We need to show that any row exact diagram

$$
\begin{array}{ccc}
 & & R \\
 & \overset{g}{\swarrow} & \downarrow f \\
N_2 & \overset{h}{\longrightarrow} & N_1 \longrightarrow 0
\end{array}
$$

can be completed commutatively by an R-linear mapping $g : R \to N_2$. If $f(1) = y$ and $x \in N_2$ is such that $h(x) = y$, let $g : R \to N_2$ be defined by $g(a) = xa$. Then g is well defined, R-linear and $f = hg$.   □

Proposition 5.2.6. *Every free R-module is projective.*

Proof. If F is a free R-module, then there is a set Δ such that $R^{(\Delta)} \cong F$. Using Lemma 5.2.5 we see that R is projective and Proposition 5.2.3 shows that $R^{(\Delta)}$ is projective.   □

Corollary 5.2.7. *Every R-module is a homomorphic image of a projective R-module.*

Proof. Proposition 2.2.6 shows that every R-module is a homomorphic image of a free R-module and a free R-module is, by Proposition 5.2.6, projective.   □

We have seen that every free R-module is projective, but the converse is false. There are projective modules that are not free as shown in the next example.

Example

5. **A Projective Module That Is Not Free.**   The ring $\mathbb{Z}_6$ is a free $\mathbb{Z}_6$-module and so is projective as a $\mathbb{Z}_6$-module. Furthermore, $\mathbb{Z}_6 \cong \mathbb{Z}_2 \oplus \mathbb{Z}_3$, so $\mathbb{Z}_2$ is isomorphic to a direct summand of $\mathbb{Z}_6$ and so is a projective $\mathbb{Z}_6$-module. But if $\mathbb{Z}_2$ is a free $\mathbb{Z}_6$-module, then $\mathbb{Z}_2 \cong \mathbb{Z}_6^{(\Delta)}$ for some set Δ. Hence, if $\mathbb{Z}_2$ is a free $\mathbb{Z}_6$-module, then $\mathbb{Z}_2$ must have at least six elements, which is clearly not the case. Thus, $\mathbb{Z}_2$ is a projective $\mathbb{Z}_6$-module but not a free $\mathbb{Z}_6$-module. Therefore, in general, the class of free R-modules is a proper subclass of the class of projective modules. Later we will see that there are rings for which these two classes coincide.

Proposition 5.2.8. *An R-module M is projective if and only if it is isomorphic to a direct summand of a free R-module.*

Proof. If M is a projective R-module, then Proposition 2.2.6 gives a free R-module $R^{(\Delta)}$ and an epimorphism $f : R^{(\Delta)} \to M$, so apply Proposition 5.2.2. The converse follows from Corollary 5.2.4 and Proposition 5.2.6.   □

Important connections between injective and projective modules are pointed out in the following two propositions.

Proposition 5.2.9. *An R-module M is injective if and only if every row exact diagram of the form*

$$0 \longrightarrow N \longrightarrow P$$
$$\downarrow \qquad \nearrow$$
$$M$$

with P projective can be completed commutatively by an R-linear mapping $P \to M$.

Proof. If M is injective there is nothing to prove, so suppose that every diagram of the form described in the proposition can be completed commutatively. Given the row exact diagram

$$0 \longrightarrow X \xrightarrow{\;i_X\;} Y$$
$$\downarrow f$$
$$M$$

where $X \subseteq Y$ and i_X is the canonical injection, construct the row exact diagram

$$0 \longrightarrow N \xrightarrow{\;i_N\;} P$$
$$\quad \downarrow \alpha \qquad\qquad \downarrow \beta$$
$$0 \longrightarrow X \xrightarrow{\;i_X\;} Y$$
$$\downarrow f$$
$$M$$

as follows: Let P be a projective module such that Y is a homomorphic image of P. If $\beta : P \to Y$ is an epimorphism and $N = \beta^{-1}(X)$, then $N/\operatorname{Ker}\beta \cong X$. If $\alpha : N \to X$ is the obvious epimorphism, then $i_X \alpha = \beta i_N$, where $i_N : N \to P$ is the canonical injection. By assumption there is an R-linear mapping $g : P \to M$ such that $gi_N = f\alpha$. If $x \in \operatorname{Ker}\beta$, then $\alpha(x) = 0$, so $g(x) = 0$. Thus, $g(\operatorname{Ker}\beta) = 0$ and we have an induced map $\bar{g} : Y \to M$ such that $f = \bar{g}i_X$. Hence, M is injective.   □

     The proof of the following proposition is dual to the proof of the preceding proposition and so is left as an exercise.

Proposition 5.2.10. *An R-module M is projective if and only if every row exact diagram of the form*

$$
\begin{array}{ccc}
 & M & \\
 \swarrow & \downarrow & \\
E \longrightarrow N & \longrightarrow & 0
\end{array}
$$

with E injective can be completed commutatively by R-linear mapping $M \to E$.

Projective Modules and the Functor $\operatorname{Hom}_R(M, -)$

We saw earlier that if M is an injective R-module, then the contravariant functor $\operatorname{Hom}_R(-, M)$ is exact. An important property of a projective module R-module M is that it renders the functor $\operatorname{Hom}_R(M, -)$ exact.

Proposition 5.2.11. *An R-module M is projective if and only if for each exact sequence $N_2 \xrightarrow{g} N_1 \to 0$ of R-modules and R-module homomorphisms, the sequence $\operatorname{Hom}_R(M, N_2) \xrightarrow{g_*} \operatorname{Hom}_R(M, N_1)) \to 0$ is exact in $\mathbf{Ab}$.*

Proof. Dual to the proof of Proposition 5.1.11.   □

Corollary 5.2.12. *An R-module M is projective if and only if the functor*

$$\operatorname{Hom}_R(M, -) : \mathbf{Mod}_R \to \mathbf{Ab}$$

is exact.

Hereditary Rings

We now begin a study of rings over which submodules of projective modules are projective and homomorphic images of injective modules are injective. It should come as no surprise that these rings are called hereditary.

Definition 5.2.13. A ring R is said to be *right (left) hereditary* if every right (left) ideal of R is projective. If R is a left and a right hereditary ring, then R is said to be *hereditary*. A hereditary integral domain is referred to as a *Dedekind domain*.

Example

6. If a ring R is without zero divisors and if every right ideal is of the form aR for some $a \in R$, then $aR \cong R$ for each nonzero $a \in R$. Hence, every right ideal of R is projective, so R is right hereditary. In particular, every principal ideal domain is hereditary. Thus, the ring of integers $\mathbb{Z}$ and $K[X]$, K a field, are hereditary.

The following result is due to Kaplansky [63].

Proposition 5.2.14 (Kaplansky). *If R is a right hereditary ring, then a submodule of a free R-module is isomorphic to a direct sum of right ideals of R and is thus projective.*

Proof. Suppose that R is a right hereditary ring, that F is a free R-module and let M be a submodule of F. If $\{x_\alpha\}_\Delta$ is a basis for F, let $\leq$ be a well ordering of Δ. Then Δ can be viewed as a set $\{0, 1, 2, \ldots, \omega, \omega + 1, \ldots\}$ of ordinal numbers with $\alpha < \mathrm{ord}(\Delta)$ for each $\alpha \in \{0, 1, 2, \ldots, \omega, \omega + 1, \ldots\}$. (See Appendix A.) Now let $F_0 = 0$ and set $F_\alpha = \bigoplus_{\beta < \alpha} x_\beta R$ for each $\alpha \in \Delta, \alpha \geq 1$. Then $F_{\alpha+1} = \bigoplus_{\beta \leq \alpha} x_\beta R$ and each $x \in M \cap F_{\alpha+1}$ can be written uniquely as $x = y + x_\alpha a$, where $y \in F_\alpha$ and $a \in R$. Since a is unique for each $x \in M \cap F_{\alpha+1}$, the R-linear mapping $\varphi : M \cap F_{\alpha+1} \to R$ given by $\varphi(x) = a$ determines a right ideal $\mathrm{Im}\, \varphi = A_\alpha$ of R and each A_α is a projective R-module. Note that $\mathrm{Ker}\, \varphi = M \cap F_\alpha$, so since A_α is projective, the sequence

$$0 \to M \cap F_\alpha \to M \cap F_{\alpha+1} \to A_\alpha \to 0$$

splits. Hence, for each $\alpha \in \Delta$, there is a submodule N_α of $M \cap F_{\alpha+1}$ such that $N_\alpha \cong A_\alpha$ and $M \cap F_{\alpha+1} = (M \cap F_\alpha) \oplus N_\alpha$.

The proof will be complete if we can show that $M = \bigoplus_\Delta N_\alpha$. First, we need to show that $M = \sum_\Delta N_\alpha$. Let $N = \sum_\Delta N_\alpha$ and suppose that $N \subsetneq M$. Since Δ is linearly ordered, $F = \bigcup_\Delta F_{\alpha+1}$, so for each $x \in M$ there is an $\alpha(x) \in \Delta$ such that $\alpha(x)$ is the first element in the set $\{\alpha \in \Delta \mid x \in F_{\alpha+1}\}$. Consider the set $\mathscr{S} = \{\alpha(x) \mid x \in M, x \notin N\}$ and suppose that $y \in M$ is such that $\alpha(y)$ is the first element of $\mathscr{S}$. Then $y \in M \cap F_{\alpha(y)+1}$, $y \notin N$ and $y \notin F_{\alpha(y)}$. Since $M \cap F_{\alpha(y)+1} = (M \cap F_{\alpha(y)}) \oplus N_{\alpha(y)}$, let $y = z + w$, where $z \in M \cap F_{\alpha(y)}$ and $w \in N_{\alpha(y)}$. Then $z = y - w \in M$ and $z \notin N$. (Note that $z \notin N$ for if $z \in N$, then $y \in N$.) Now $z \in M \cap F_{\alpha(y)} \subseteq F_{\alpha(y)}$, so $\alpha(z) < \alpha(y)$. But $z \in M$ and $z \notin N$ implies that $\alpha(z) \in \mathscr{S}$, so we have a contradiction. Therefore, it cannot be the case that $N \subsetneq M$, and so $M = \sum_\Delta N_\alpha$.

Finally, we need to show that the sum $\sum_\Delta N_\alpha$ is direct. To this end, suppose that $y_{\alpha_1} + y_{\alpha_2} + \cdots + y_{\alpha_n} = 0$ is a finite sum in $\sum_\Delta N_\alpha$, where $\alpha_1 < \alpha_2 < \cdots < \alpha_n$ is given by the ordering on Δ. We proceed by induction to show that $y_{\alpha_1} + y_{\alpha_2} + \cdots + y_{\alpha_n} = 0$ implies $y_{\alpha_1} = y_{\alpha_2} = \cdots = y_{\alpha_n} = 0$ for each $n \geq 1$. If

$n = 1$, then $y_{\alpha_1} = 0$ and there is nothing to prove. Make the induction hypothesis, that $y_{\alpha_1} + y_{\alpha_2} + \cdots + y_{\alpha_{n-1}} = 0$ implies that $y_{\alpha_1} = y_{\alpha_2} = \cdots = y_{\alpha_{n-1}} = 0$ and suppose that $y_{\alpha_1} + y_{\alpha_2} + \cdots + y_{\alpha_n} = 0$. Then $-y_{\alpha_n} = y_{\alpha_1} + y_{\alpha_2} + \cdots + y_{\alpha_{n-1}} \in (M \cap F_{\alpha_n}) \cap N_{\alpha_n} = 0$ and so it follows that $y_{\alpha_1} = y_{\alpha_2} = \cdots = y_{\alpha_n} = 0$. Hence, the sum $\sum_\Delta N_\alpha$ is direct.   □

Proposition 5.2.15. *The following are equivalent for a ring R.*

(1) *R is a right hereditary ring.*

(2) *Every submodule of a projective R-module is projective.*

(3) *Every factor module of an injective R-module is injective.*

Proof. (1) $\Rightarrow$ (2). If R is right hereditary and M is a projective R-module, then M is, by Proposition 5.2.8, isomorphic to a submodule of a free R-module. Hence, any submodule of M is isomorphic to a submodule of a free R-module and so is projective by Proposition 5.2.14.

(2) $\Rightarrow$ (3). In view of Proposition 5.2.9, an R-module M will be injective if every row exact diagram of the form

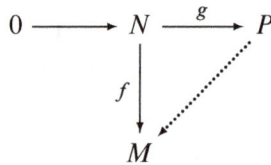

$$0 \longrightarrow N \xrightarrow{\ g\ } P$$
$$f \downarrow \qquad \qquad$$
$$M$$

can be completed commutatively, where P is a projective R-module. So suppose that M is a homomorphic image of an injective module E and consider the row exact diagram

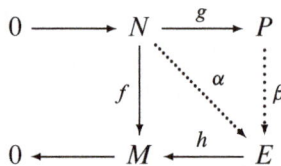

$$0 \longrightarrow N \xrightarrow{\ g\ } P$$
$$f\downarrow \quad {}^{\alpha} \quad \downarrow \beta$$
$$0 \longleftarrow M \xleftarrow{\ h\ } E$$

If (2) holds, then N is projective, so there is an R-linear map $\alpha : N \to E$ such that $h\alpha = f$. But E is injective, so the existence of α gives an R-linear map $\beta : P \to E$ such that $\alpha = \beta g$. Hence, $\gamma = h\beta : P \to M$ is such that $\gamma g = f$, so M is injective.

(3) $\Rightarrow$ (1). Consider the row exact diagram

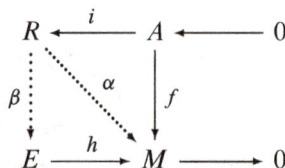

$$R \xleftarrow{\ i\ } A \longleftarrow 0$$
$$\beta \downarrow \quad {}^{\alpha} \quad \downarrow f$$
$$E \xrightarrow{\ h\ } M \longrightarrow 0$$

where A is a right ideal of R, i is the canonical injection and E is injective. If (3) holds, then M is injective, so there is an R-linear map $\alpha : R \to M$ such that $\alpha i = f$. But the fact that R is projective together with the map α gives an R-linear mapping $\beta : R \to E$ such that $h\beta = \alpha$. Since $\gamma = \beta i : A \to E$ is such that $h\gamma = f$, it follows from Proposition 5.2.10 that A is projective. Thus, R is a right hereditary ring. □

Earlier it was pointed out that, in general, the class of free R-modules is "smaller" than the class of projective modules. It was also mentioned that there are rings over which these two classes coincide. Part (2) of the following proposition shows that this is indeed the case. The following proof also fulfills an earlier promise made with regard to Proposition 4.3.11.

Proposition 5.2.16. *If R is a principal ideal domain, then:*

(1) *If M is a submodule of a free R-module F, then M is free and* $\operatorname{rank}(M) \leq \operatorname{rank}(F)$.

(2) *An R-module M is projective if and only if M is a free R-module.*

Proof. We have seen in Example 6 that a principal ideal domain is a hereditary ring.

(1) Because of Proposition 5.2.14, we know that a submodule M of a free R-module F is isomorphic to a direct sum of right ideals of R. But R is a principal ideal domain, so each right ideal of R is of the form aR for some $a \in R$ and $aR \cong R$ when a is nonzero. Hence, M is isomorphic to a direct sum of copies of R and as such is a free R-module. It remains only to show that $\operatorname{rank}(M) \leq \operatorname{rank}(F)$. In the notation of the proof of Proposition 5.2.14, we have $M = \bigoplus_\Delta N_\alpha \subseteq F$, where Δ is an indexing set for a basis of F. Since R is a commutative ring, R is, by Proposition 2.2.11, an IBN-ring and so $\operatorname{rank}(M) \leq \operatorname{card}(\Delta)$. But $\operatorname{card}(\Delta) = \operatorname{rank}(F)$, and so we have the result.

(2) If M is a free R-module, then, by Proposition 5.2.6, M is projective. Conversely, if M is a projective R-module, then M is, by Proposition 5.2.8, isomorphic to a submodule of a free R-module and it follows from (1) that M is a free R-module. □

Corollary 5.2.17. *A principal ideal domain is a Dedekind domain.*

Semihereditary Rings

Definition 5.2.18. A ring R is said to be *right (left) semihereditary* if every finitely generated right (left) ideal of R is projective. A ring that is left and right semihereditary is called a *semihereditary ring*.

A hereditary integral domain was previously called a Dedekind domain. A semihereditary integral domain is referred to as a *Prüfer domain*. Both of these domains are important considerations in number theory and we cite [29] and[40] as references for additional information on these domains.

The following proposition is an analogue of Kaplansky's result for hereditary rings. The proof is left as an exercise.

Proposition 5.2.19. *If R is a right semihereditary ring, then every finitely generated submodule of a free R-module is isomorphic to a direct sum of finitely generated right ideals of R, and is therefore projective.*

Corollary 5.2.20. *A ring R is right semihereditary if and only if finitely generated submodules of projective R-modules are projective.*

Problem Set 5.2

1. Prove that an R-module M is projective if and only if every short exact sequence of the form $0 \to N' \to N \to M \to 0$ splits.

2. (a) Review Definition 5.2.1 and show that an R-module M is projective if and only if for each R-module N and each submodule N' of N, the diagram

$$
\begin{array}{ccc}
 & & M \\
 & \overset{g}{\swarrow} & \downarrow{\scriptstyle f} \\
N & \overset{\eta}{\longrightarrow} N/N' & \longrightarrow 0
\end{array}
$$

can be completed commutatively by an R-linear mapping $g : M \to N$, where η is the canonical surjection.

(b) If R is a commutative ring and F and F' are free R-modules, prove that $F \otimes_R F'$ is a free R-module. [Hint: Use the fact that tensor products commutes with direct sums.]

(c) If R is a commutative ring and if P and P' are projective R-modules, prove that $P \otimes_R P'$ is a projective R-module. [Hint: There are free R-modules F and F' such that $F = P \oplus N$ and $F' = P' \oplus N'$, so consider $F \otimes_R F'$.]

3. Let I be an ideal of R and suppose that $F = R^{(\Delta)}$ is a free R-module.

(a) Show that $FI = I^{(\Delta)}$ and that $F/FI = (R/I)^{(\Delta)}$. Conclude that F/FI is a free R/I-module.

(b) If M is a projective R-module, then $M \oplus N = F$, where F is a free R-module. Show that $F/FI = M/MI \oplus N/NI$. Conclude that M/MI and N/NI are projective R/I-modules.

4. Prove Proposition 5.2.3.

5. Prove Proposition 5.2.10.

6. Prove Proposition 5.2.11.

7. (a) **Schanuel's Lemma for Projective Modules.** Suppose that

$$0 \to K_1 \to P_1 \to M \to 0 \quad \text{and} \quad 0 \to K_2 \to P_2 \to M \to 0$$

are short exact sequences of R-modules and R-module homomorphisms. If P_1 and P_2 are projective, show that $P_1 \oplus K_2 \cong P_2 \oplus K_1$. [Hint: Dualize the proof of (a) of Exercise 11 in Problem Set 5.1.]

(b) Two R-modules M and N are said to be *projectively equivalent* if there exist projective R-modules P_1 and P_2 such that $P_1 \oplus M \cong P_2 \oplus N$. Determine whether or not projective equivalence is an equivalence relation on the class of R-modules.

8. Let

$$
\begin{array}{ccc}
 & & P \\
 & {}^{g}\nearrow & \downarrow {}^{f} \\
M & \xrightarrow{\;h\;} & M'
\end{array}
$$

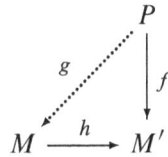

be a diagram of R-modules and R-module homomorphisms. Show that if P is projective, then the following statements are equivalent.

(a) There is an R-linear mapping $g : P \to M$ such that $f = hg$.

(b) $\operatorname{Im} f \subseteq \operatorname{Im} h$.

9. Show that a row and column exact diagram

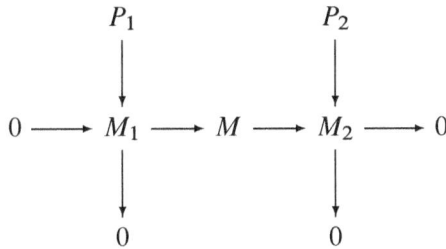

$$
\begin{array}{ccccccc}
 & & P_1 & & P_2 & & \\
 & & \downarrow & & \downarrow & & \\
0 & \longrightarrow & M_1 & \longrightarrow & M \longrightarrow M_2 & \longrightarrow & 0 \\
 & & \downarrow & & \downarrow & & \\
 & & 0 & & 0 & &
\end{array}
$$

of R-modules and R-module homomorphisms, where P_1 and P_2 are projective, can be completed to a row and column exact commutative diagram

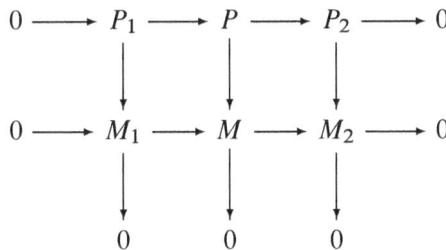

$$
\begin{array}{ccccccccc}
0 & \longrightarrow & P_1 & \longrightarrow & P & \longrightarrow & P_2 & \longrightarrow & 0 \\
 & & \downarrow & & \downarrow & & \downarrow & & \\
0 & \longrightarrow & M_1 & \longrightarrow & M & \longrightarrow & M_2 & \longrightarrow & 0 \\
 & & \downarrow & & \downarrow & & \downarrow & & \\
 & & 0 & & 0 & & 0 & &
\end{array}
$$

where P is projective. [Hint: Dualize the proof of Exercise 12 in Problem Set 5.1.]

10. If M is an R-module, then $M^* = \text{Hom}_R(M, R)$ is said to be the *dual* of M. A pair of indexed sets $\{x_\alpha\}_\Delta \subseteq M$ and $\{f_\alpha\}_\Delta \subseteq M^*$ is said to be a *dual basis* for M if each $x \in M$ can be expressed as $x = \sum_\Delta x_\alpha f_\alpha(x)$, where $f_\alpha(x) = 0$ for almost all $\alpha \in \Delta$.

Dual Basis Lemma. Prove that an R-module M is projective if and only if M has a dual basis. [Hint: If $\{x_\alpha\}_\Delta$ is a set of generators for M, let $\{y_\alpha\}_\Delta$ be a basis for the free R-module $R^{(\Delta)}$ and let $f : R^{(\Delta)} \to M$ be the R-linear mapping defined on basis elements by $f(y_\alpha) = x_\alpha$. Then M is projective if and only if $f : R^{(\Delta)} \to M$ splits.]

11. (a) Prove Proposition 5.2.19.

(b) Prove that the ring direct product of a finite number of right semihereditary rings is right semihereditary. [Hint: Use (a) of Exercise 3 in Problem Set 2.1.]

(c) Prove that the ring direct product of a finite number of right hereditary rings is right hereditary.

12. If P is a projective R-module that generates $\mathbf{Mod}_R$, then P is said to be a *projective generator* for $\mathbf{Mod}_R$. Prove that the following are equivalent for a projective module P.

(a) P is a generator for $\mathbf{Mod}_R$.

(b) $\text{Hom}_R(P, S) \neq 0$ for every simple R-module S.

(c) P generates every simple R-module.

[Hint: For (b) $\Rightarrow$ (a), it suffices to show that M generates R. If $T = \sum_\Delta f(P)$, where $\Delta = \text{Hom}_R(P, R)$, and $T \neq R$, then there is a maximal right ideal $\mathfrak{m}$ of R that contains T and because of (b), $\text{Hom}_R(P, R/\mathfrak{m}) \neq 0$. Use the projectivity of P to produce a contradiction.]

5.3 Flat Modules

Flat Modules and the Functor $M \otimes_R -$

We now consider R-modules M that will turn $M \otimes_R -$ into an exact functor. It was shown in Section 3.3 that $M \otimes_R -$ is right exact for every R-module M, so we need only consider R-modules that preserve monomorphisms when taking tensor products. Because of the symmetry of tensor products, properties of R-modules that render $M \otimes_R -$ exact will also hold for left R-modules M that turn $- \otimes_R M$ into an exact functor.

Definition 5.3.1. An R-module M is said to be *flat* if

$$0 \to M \otimes_R N_1 \xrightarrow{\text{id}_M \otimes f} M \otimes_R N_2$$

is exact in **Ab** whenever $0 \to N_1 \xrightarrow{f} N_2$ is an exact sequence of left R-modules.

Examples

1. **Projective R-modules Are Flat.** Every ring R is flat and in fact every free R-module is flat. The fact that every free R-module is flat implies that every projective R-module is flat.

2. **Not All R-modules Are Flat.** The $\mathbb{Z}$-module $\mathbb{Z}_n$ is not flat. For example, $i : \mathbb{Z} \to \mathbb{Q}$ is an embedding but $\mathrm{id}_{\mathbb{Z}_n} \otimes i : \mathbb{Z}_n \otimes_{\mathbb{Z}} \mathbb{Z} \to \mathbb{Z}_n \otimes_{\mathbb{Z}} \mathbb{Q}$ is not. Note that $\mathbb{Z}_n \otimes_{\mathbb{Z}} \mathbb{Z} \cong \mathbb{Z}_n \neq 0$ and yet $\mathbb{Z}_n \otimes_{\mathbb{Z}} \mathbb{Q} = 0$.

3. **Regular Rings.** A ring R is said to be a *(von Neumann) regular ring* if for each $a \in R$ there is an $r \in R$ such that $a = ara$. Every R-module is flat if and only if R is a regular ring. A proof of this fact will be presented later in this section.

Since $M \otimes_R -$ is always right exact, $M \otimes_R -$ is an exact functor from ${}_R\mathbf{Mod}$ to $\mathbf{Ab}$ when M is flat. This, however, gives us no information about the types of modules that are flat. To develop this information we need the following definition.

Definition 5.3.2. If M is an R-module, then the left R-module $M^+ = \mathrm{Hom}_{\mathbb{Z}}(M, \mathbb{Q}/\mathbb{Z})$ is called the *character module* of M. The left R-module structure on M^+ is given by $(af)(x) = f(xa)$ for all $x \in M$ and $a \in R$.

Proposition 5.3.3. *An R-module M is flat if and only if M^+ is an injective left R-module.*

Proof. Suppose that M is a flat R-module. To show that M^+ is an injective left R-module, it suffices, by Proposition 5.1.11, to show that $\mathrm{Hom}_R(-, M^+)$ is right exact. Let $f : N_1 \to N_2$ be a monomorphism of left R-modules and consider

$$f^* : \mathrm{Hom}_R(N_2, M^+) \to \mathrm{Hom}_R(N_1, M^+).$$

According to Proposition 3.4.6, we have a commutative diagram

$$
\begin{array}{ccc}
\mathrm{Hom}_{\mathbb{Z}}(M \otimes_R N_2, \mathbb{Q}/\mathbb{Z}) & \xrightarrow{\eta_{MN_2}} & \mathrm{Hom}_R(N_2, M^+) \\
\downarrow{\scriptstyle (\mathrm{id}_M \otimes f)^*} & & \downarrow{\scriptstyle f^*} \\
\mathrm{Hom}_{\mathbb{Z}}(M \otimes_R N_1, \mathbb{Q}/\mathbb{Z}) & \xrightarrow{\eta_{MN_1}} & \mathrm{Hom}_R(N_1, M^+)
\end{array}
$$

where η_{MN_1} and η_{MN_2} are isomorphisms. Since M is a flat R-module, the sequence $0 \to M \otimes_R N_1 \xrightarrow{\mathrm{id}_M \otimes f} M \otimes_R N_2$ is exact and since $\mathbb{Q}/\mathbb{Z}$ is an injective $\mathbb{Z}$-module, $(\mathrm{id}_M \otimes f)^*$ is an epimorphism. Hence, f^* is also an epimorphism, so M^+ is injective. The argument easily reverses, so we are done. $\qquad\square$

Corollary 5.3.4. *If $\{M_\alpha\}_\Delta$ is a family of R-modules, then $\bigoplus_\Delta M_\alpha$ is flat if and only if each M_α is flat.*

Proof. The direct sum $\bigoplus_\Delta M_\alpha$ is flat if and only if $(\bigoplus_\Delta M_\alpha)^+$ is injective. Now $(\bigoplus_\Delta M_\alpha)^+ \cong \prod_\Delta M_\alpha^+$, so $\prod_\Delta M_\alpha^+$ is injective if and only if $(\bigoplus_\Delta M_\alpha)^+$ is injective. But according to Proposition 5.1.13 $\prod_\Delta M_\alpha^+$ is injective if and only if each M_α^+ is injective and each M_α^+ is injective if and only if each M_α is flat.   □

If two R-modules are isomorphic, then it is clear that if one of the modules is flat, then the other module is flat as well. Also observe that if

$$0 \to M \xrightarrow{f} N$$

is an exact sequence of left R-modules, then

$$0 \to R \otimes_R M \xrightarrow{\mathrm{id}_R \otimes f} R \otimes_R N$$

is exact due to the fact that $R \otimes_R M \cong M$ and $R \otimes_R N \cong N$. Hence, we immediately see that R is a flat R-module.

Proposition 5.3.5. *Every free R-module is flat.*

Proof. Any free R-module is isomorphic to $R^{(\Delta)}$ for some set Δ. Since $R^{(\Delta)}$ is a direct sum of flat R-modules, $R^{(\Delta)}$ is flat due to Corollary 5.3.4.   □

Corollary 5.3.6. *Every projective R-module is flat.*

Proof. Proposition 5.2.8 shows that every projective module M is isomorphic to a direct summand of a free R-module which is flat by the proposition. Moreover, Corollary 5.3.4 indicates that every direct summand of a flat module is flat.   □

We have seen via Baer's criteria that the right ideals of R form a test set for the injective R-modules. Proposition 5.3.3 can be used in conjunction with Baer's criteria to show that the left ideals of R also form a test set for flat R-modules. The conclusion, as shown in the following proposition, is that in order for an R-module to be flat, it need only preserve canonical injections $A \to R$ for (finitely generated) left ideals A of R when forming tensor products.

Proposition 5.3.7. *The following are equivalent for an R-module M.*

(1) *M is flat.*

(2) *$0 \to M \otimes_R A \to M \otimes_R R \cong M$ is exact for each left ideal A of R.*

(3) *$0 \to M \otimes_R A \to M \otimes_R R \cong M$ is exact for each finitely generated left ideal A of R.*

Proof. The implications (1) $\Rightarrow$ (2) and (2) $\Rightarrow$ (3) require no proof, so suppose that $0 \to M \otimes_R A \to M \otimes_R R$ is exact for each finitely generated left ideal A of R. If B is a left ideal of R and $\sum_{i=1}^{n} (x_i \otimes a_i) \in M \otimes_R B$, then $\{a_1, a_2, \ldots, a_n\}$ is contained in a finitely generated left ideal A of R which in turn is contained in B. Furthermore, by assumption, the composition of the maps

$$M \otimes_R A \to M \otimes_R B \to M \otimes_R R$$

must be a group monomorphism. Hence, if $\sum_{i=1}^{n} (x_i \otimes a_i)$ is zero in $M \otimes_R R$, then $\sum_{i=1}^{n} (x_i \otimes a_i)$ is zero in $M \otimes_R A$ and thus must be zero in $M \otimes_R B$. Therefore, $M \otimes_R B \to M \otimes_R R$ is a group monomorphism for each left ideal B of R whenever $M \otimes_R A \to M \otimes_R R$ is a group monomorphism for each finitely generated left ideal A of R. So (3) $\Rightarrow$ (2).

We complete the proof by showing (2) $\Rightarrow$ (1). If A is a left ideal of R, then we have an exact sequence $0 \to M \otimes_R A \to M \otimes_R R$ in **Ab**. Since $\mathbb{Q}/\mathbb{Z}$ is an injective $\mathbb{Z}$-module, this gives an exact sequence

$$\mathrm{Hom}_{\mathbb{Z}}(M \otimes_R R, \mathbb{Q}/\mathbb{Z}) \to \mathrm{Hom}_{\mathbb{Z}}(M \otimes_R A, \mathbb{Q}/\mathbb{Z}) \to 0.$$

Using Proposition 3.4.6, we see that

$$\mathrm{Hom}_R(R, M^+) \to \mathrm{Hom}_R(A, M^+) \to 0$$

is exact. Therefore, if $f : A \to M^+$ is a left R-linear mapping, then there is a $g \in \mathrm{Hom}_R(R, M^+)$ which extends f to R. Thus, Baer's criteria shows that M^+ is an injective left R-module, so M is, by Proposition 5.3.3, a flat R-module. $\qquad \square$

If M is an R-module and A is a left ideal of R, then MA, the set of all finite sums $\sum_{i=1}^{n} x_i a_i$, where $x_i \in M$ and $a_i \in A$, is a subgroup of M. The following two propositions make additional connections between a flat R-module and the left ideals of R. In preparation for the first proposition, note that since the mapping $M \times A \to MA$ given by $(x, a) \mapsto xa$ is R-balanced, there is a group epimorphism $\varphi : M \otimes_R A \to MA$ such that $\varphi(x \otimes a) = xa$ for each generator $x \otimes a$ in $M \otimes_R A$.

Proposition 5.3.8. *The following are equivalent for an R-module M.*

(1) *M is flat.*

(2) *The group epimorphism $\varphi_A : M \otimes_R A \to MA$ defined by $\varphi_A(\sum_{k=1}^{n} x_k \otimes a_k) = \sum_{k=1}^{n} x_k a_k$ is an isomorphism for each left ideal A of R.*

(3) *The group epimorphism $\varphi_A : M \otimes_R A \to MA$ defined by $\varphi_A(\sum_{k=1}^{n} x_k \otimes a_k) = \sum_{k=1}^{n} x_k a_k$ is an isomorphism for each finitely generated left ideal A of R.*

Proof. (1) $\Rightarrow$ (2). If A is a left ideal of R, let $i : A \to R$ be the canonical injection and consider the commutative diagram

$$
\begin{array}{ccc}
M \otimes_R A & \xrightarrow{\ \mathrm{id}_M \otimes i\ } & M \otimes_R R \\
\varphi_A \downarrow & & \downarrow \varphi_R \\
MA & \xrightarrow{\ j\ } & MR = M
\end{array}
$$

where j is the obvious map. Since φ_R is, by Proposition 2.3.4, an isomorphism, we have $\mathrm{id}_M \otimes i = \varphi_R^{-1} j \varphi_A$, so $\varphi_A(\sum_{k=1}^{n} x_k \otimes a_k) = 0$ leads to $(\mathrm{id}_M \otimes i)(\sum_{k=1}^{n} x_k \otimes a_k) = 0$. But M is flat, so $\mathrm{id}_M \otimes i$ is a monomorphism and this gives $\sum_{k=1}^{n} x_k \otimes a_k = 0$. Hence, φ_A is an injection, so φ_A is an isomorphism.

(2) $\Rightarrow$ (1). Consider the commutative diagram given in the proof that (1) $\Rightarrow$ (2). If φ_A is an isomorphism, then $\mathrm{id}_M \otimes i$ is a monomorphism since j is a monomorphism and φ_R^{-1} is an isomorphism. Proposition 5.3.7 now shows that M is a flat R-module.

No proof is required for (2) $\Rightarrow$ (3), so to complete the proof we need only show that (3) $\Rightarrow$ (2). Let $\sum_{k=1}^{n} x_k \otimes a_k \in M \otimes_R A$, where A is a left ideal of R. If $\sum_{k=1}^{n} x_k \otimes a_k \in \mathrm{Ker}\, \varphi_A$, then $\sum_{k=1}^{n} x_k a_k = 0$. Let $B = \sum_{k=1}^{n} R a_k$, then $\sum_{k=1}^{n} x_k a_k \in MB$. Now B is a finitely generated left ideal of R, so by assumption, $\varphi_B : M \otimes_R A \to MB$ is an isomorphism. Since $\sum_{k=1}^{n} x_k \otimes a_k \in M \otimes_R B$ and since $\varphi_B(\sum_{k=1}^{n} x_k \otimes a_k) = \sum_{k=1}^{n} x_k a_k = 0$, we have $\sum_{k=1}^{n} x_k \otimes a_k = 0$. Therefore, φ_A is an isomorphism. $\square$

Proposition 5.3.9. *If* $0 \to K \to F \xrightarrow{f} M \to 0$ *is exact, where* F *is a flat* R-module, *then the following are equivalent.*

(1) *M is flat.*

(2) *$K \cap FA = KA$ for every left ideal A of R.*

(3) *$K \cap FA = KA$ for every finitely generated left ideal A of R.*

Proof. If A is a (finitely generated) left ideal of R, then since $K \otimes_R -$ is right exact, we have $K \otimes_R A \to F \otimes_R A \to M \otimes_R A \to 0$. Since F is flat, Proposition 5.3.8 gives $F \otimes_R A \cong FA$ and $K \otimes_R A$ corresponds to KA under this isomorphism. Hence, $M \otimes_R A \cong FA/KA$, so using Proposition 5.3.8 again we see that M is flat if and only if $MA \cong FA/KA$ for all (finitely generated) left ideals of R. Now elements of MA can be written as $\sum_{i=1}^{n} f(x_i) a_i = \sum_{i=1}^{n} f(x_i a_i)$, where $x_i \in F$ and $a_i \in A$ for $i = 1, 2, \ldots, n$. Consequently, $MA = f(FA) \cong FA/(K \cap FA)$ for all (finitely generated) left ideals A of R. Therefore, M is flat if and only if $F/A/KA \cong FA/(K \cap FA)$ for all (finitely generated) left ideals of R. Since $KA \subseteq K \cap FA$, it follows that $MA \cong FA/(K \cap FA)$ if and only if $KA = K \cap FA$ for all (finitely generated) left ideals of R. $\square$

The following two propositions provide additional characterizations of flat modules that are often useful. We omit the proof of each and cite [2], [26] and [42] as references.

Proposition 5.3.10. *An R-module M is flat if and only if for each expression $\sum_{j=1}^{n} x_j a_j = 0$, where $x_j \in M$ and $a_j \in R$ for $j = 1, 2, \ldots, n$, there exist $y_i \in M$, $i = 1, 2, \ldots, m$, and a matrix (c_{ij}) in $M_{m \times n}(R)$ such that*

(1) $\sum_{j=1}^{n} c_{ij} a_j = 0$, *for* $i = 1, 2, \ldots, m$, *and*

(2) $\sum_{i=1}^{m} y_i c_{ij} = x_j$, *for* $j = 1, 2, \ldots, n$.

Proposition 5.3.11. *Let $F \to M \to 0$ be an exact sequence in $\mathbf{Mod}_R$, where F is a free R-module, and suppose that K is a submodule of F such that $F/K \cong M$. Then the following are equivalent.*

(1) *M is flat.*

(2) *If $x_1, x_2, \ldots, x_n \in K$, then there is an $f \in \mathrm{Hom}_R(F, K)$ such that $f(x_i) = x_i$ for $i = 1, 2, \ldots, n$.*

(3) *For each $x \in K$, there is an $f \in \mathrm{Hom}_R(F, K)$ such that $f(x) = x$.*

Coherent Rings

We now know that direct sums of flat modules are flat, but a direct product of flat modules need not be flat. In this section we characterize rings over which direct products of flat modules are flat. These rings were first discovered by Chase [54].

Definition 5.3.12. An (A left) *R-module M is said to be* finitely presented *if there is an exact sequence $0 \to K \to F \to M \to 0$ of (left) R-modules, where F is finitely generated and free and K is finitely generated. Such a sequence will be called a finite presentation of M. A finitely presented (left) R-module M is called* (*left*) coherent *if every finitely generated submodule of M is finitely presented. A ring R is said to be* right (*left*) coherent *if it is coherent as an (a left) R-module. A ring that is left and right coherent is a* coherent ring.

The following lemma gives additional information on finitely presented modules.

Lemma 5.3.13. *The following are equivalent for an R-module M.*

(1) *There exists an exact sequence $F_1 \to F_0 \to M \to 0$, where F_1 and F_0 are finitely generated free R-modules.*

(2) *There exist integers m and n such that the sequence $R^{(m)} \to R^{(n)} \to M \to 0$ is exact.*

(3) *M is finitely presented.*

Proof. The equivalence of (1) and (2) is obvious, so suppose that $0 \to K \to F \to M \to 0$ is a finite presentation of M. Since F is finitely generated and free, there is an integer n such that $F \cong R^{(n)}$. If K is generated by m elements, then there is an epimorphism $R^{(m)} \to K$, so we have an exact sequence $R^{(m)} \to R^{(n)} \to M \to 0$. Therefore, (3) $\Rightarrow$ (2). Now suppose that (2) holds and let $K = \operatorname{Ker} \beta$ in the exact sequence $R^{(m)} \xrightarrow{\alpha} R^{(n)} \xrightarrow{\beta} M \to 0$. Then $\operatorname{Im} \alpha = K$, so K is finitely generated. The short exact sequence $0 \to K \to R^{(n)} \xrightarrow{\beta} M \to 0$ shows that M is finitely presented. Thus, (2) $\Rightarrow$ (3). $\square$

It is clear that every finitely presented R-module is finitely generated and the converse holds when R is right noetherian. We have previously seen that if $\{N_\alpha\}_\Delta$ is a family of left R-modules, then $M \otimes_R (\bigoplus_\Delta N_\alpha) \cong \bigoplus_\Delta (M \otimes_R N_\alpha)$ for each R-module M. A similar isomorphism actually holds for every direct product of R-modules if and only if M is finitely presented. The following lemma simplifies the proof of this fact. In the proofs of the following lemma and proposition, the map $\varphi : M \otimes_R (\prod_\Delta N_\alpha) \to \prod_\Delta (M \otimes_R N_\alpha)$ is given by $\varphi(x \otimes (y_\alpha)) = (x \otimes y_\alpha)$. If $N_\alpha = R$ for each $\alpha \in \Delta$, then since $M \otimes_R R \cong M$, if we identify $M \otimes_R R$ with M, we get a map $\varphi : M \otimes_R R^\Delta \to M^\Delta$ such that $x \otimes (a_\alpha) \mapsto (xa_\alpha)$. When more than one module M is involved, this map will be denoted by φ^M.

Lemma 5.3.14. *Let M be an R-module and suppose that φ is the map described in the preceding paragraph. Then the following are equivalent.*

(1) *M is finitely generated.*

(2) *$\varphi : M \otimes_R (\prod_\Delta N_\alpha) \to \prod_\Delta (M \otimes_R N_\alpha)$ is an epimorphism for any family $\{N_\alpha\}_\Delta$ of left R-modules.*

(3) *$\varphi : M \otimes_R R^\Delta \to M^\Delta$ is an epimorphism for any set Δ.*

(4) *$\varphi : M \otimes_R R^M \to M^M$ is an epimorphism.*

Proof. (1) $\Rightarrow$ (2). If M is finitely generated, then there is a short exact sequence $0 \to K \to R^{(n)} \to M \to 0$ for some positive integer n. This leads to a row exact commutative diagram

$$
\begin{array}{ccccccc}
K \otimes_R \left(\prod_\Delta N_\alpha \right) & \longrightarrow & R^{(n)} \otimes_R \left(\prod_\Delta N_\alpha \right) & \longrightarrow & M \otimes_R \left(\prod_\Delta N_\alpha \right) & \longrightarrow & 0 \\
\varphi^K \downarrow & & \varphi^{R^{(n)}} \downarrow & & \varphi^M \downarrow & & \\
\prod_\Delta (K \otimes_R N_\alpha) & \longrightarrow & \Pi_\Delta (R^{(n)} \otimes_R N_\alpha) & \longrightarrow & \prod_\Delta (M \otimes_R N_\alpha) & \longrightarrow & 0
\end{array}
$$

But $\varphi^{R^{(n)}}$ is an isomorphism (See Exercise 12.) and a simple diagram chase shows that φ^M is an epimorphism.

The implications (2) $\Rightarrow$ (3) and (3) $\Rightarrow$ (4) are obvious.

(4) $\Rightarrow$ (1). Suppose that $(x_x) \in M^M$ is such that $x = x_x$ for each $x \in M$. Since φ is an epimorphism, there is an element $\sum_{i=1}^n (x_i \otimes (a_{ix})) \in M \otimes_R R^M$ that is a preimage of (x_x). This gives

$$(x_x) = \varphi\left[\sum_{i=1}^n (x_i \otimes (a_{ix}))\right]$$

$$= \sum_{i=1}^n \varphi[(x_i \otimes (a_{ix}))]$$

$$= \sum_{i=1}^n (x_i a_{ix})$$

$$= \left(\sum_{i=1}^n x_i a_{ix}\right).$$

Thus, for each $x \in M$, we see that $\sum_{i=1}^n x_i a_{ix} = x$, so $x_1, x_2, \ldots, x_n$ is a set of generators for M. $\qquad\square$

Proposition 5.3.15. *The following are equivalent for any R-module M.*

(1) *M is finitely presented.*

(2) *The map $\varphi : M \otimes_R (\prod_\Delta N_\alpha) \to \prod_\Delta (M \otimes_R N_\alpha)$ is an isomorphism for each family $\{N_\alpha\}_\Delta$ of left R-modules.*

(3) *$\varphi : M \otimes_R R^\Delta \to M^\Delta$ is an isomorphism for any set Δ.*

Proof. (1) $\Rightarrow$ (2). Suppose that M is a finitely presented R-module, let $R^{(m)} \to R^{(n)} \to M \to 0$ be as in Lemma 5.3.13 and consider the row exact commutative diagram

$$
\begin{array}{ccccccc}
R^{(m)} \otimes_R \left(\prod N_\alpha\right) & \longrightarrow & R^{(n)} \otimes_R \left(\prod N_\alpha\right) & \longrightarrow & M \otimes_R \left(\prod N_\alpha\right) & \longrightarrow & 0 \\
\quad\Delta & & \quad\Delta & & \quad\Delta & & \\
\varphi^{R^{(m)}}\Big\downarrow & & \varphi^{R^{(n)}}\Big\downarrow & & \varphi^M\Big\downarrow & & \\
\prod (R^{(m)} \otimes_R N_\alpha) & \longrightarrow & \prod (R^{(n)} \otimes_R N_\alpha) & \longrightarrow & \prod (M \otimes_R N_\alpha) & \longrightarrow & 0 \\
\Delta & & \Delta & & \Delta & &
\end{array}
$$

Since $\varphi^{R^{(m)}}$ and $\varphi^{R^{(n)}}$ are isomorphisms, it follows by chasing the diagram that φ^M is an isomorphism.

(2) $\Rightarrow$ (3). Obvious.

(3) $\Rightarrow$ (1). By Lemma 5.3.14, we immediately have that M is finitely generated. So there is an exact sequence $0 \to K \to F \to M \to 0$, where F is a finitely

generated free R-module. Thus, we have a row exact commutative diagram

$$
\begin{array}{ccccccc}
K \otimes_R R^\Delta & \longrightarrow & F \otimes_R R^\Delta & \longrightarrow & M \otimes_R R^\Delta & \longrightarrow & 0 \\
\downarrow{\scriptstyle \varphi^K} & & \downarrow{\scriptstyle \varphi^F} & & \downarrow{\scriptstyle \varphi^M} & & \\
0 \longrightarrow & K^\Delta & \longrightarrow & F^\Delta & \longrightarrow & M^\Delta & \longrightarrow 0
\end{array}
$$

where φ^M and φ^F are isomorphisms. It follows by a diagram chase that φ^K is an epimorphism, so by considering Lemma 5.3.14 again, we have that K is finitely generated. Hence, M is finitely presented. $\qquad\square$

We can now characterize the rings over which direct products of flat modules are always flat.

Proposition 5.3.16 (Chase). *The following are equivalent for a ring R.*

(1) *Every direct product of flat R-modules is flat.*

(2) R^Δ *is a flat R-module for every set Δ.*

(3) *Every finitely presented left R-module is coherent.*

(4) R *is left coherent.*

Proof. (1) $\Rightarrow$ (2). This is clear since R is a flat R-module.

(2) $\Rightarrow$ (3). Let M be a finitely presented left R-module and suppose that N is a finitely generated submodule of M. For any set Δ there is a commutative diagram

$$
\begin{array}{ccc}
R^\Delta \otimes_R N & \longrightarrow & R^\Delta \otimes_R M \\
\downarrow{\scriptstyle \varphi^N} & & \downarrow{\scriptstyle \varphi^M} \\
N^\Delta & \longrightarrow & M^\Delta
\end{array}
$$

where $N^\Delta \to M^\Delta$ is the canonical injection. Since R^Δ is a flat R-module, we also see that $R^\Delta \otimes_R N \to R^\Delta \otimes_R M$ is a monomorphism. Now M is a finitely presented left R-module, so (3) of the left-hand version of Proposition 5.3.15 indicates that φ^M is an isomorphism. It follows that φ^N is also an isomorphism, so another application of Proposition 5.3.15 shows that N is finitely presented.

(3) $\Rightarrow$ (4). This follows easily since $0 \to 0 \to R \to R \to 0$ is a finite presentation of R.

(4) $\Rightarrow$ (1). Let $\{N_\alpha\}_\Delta$ be a family of flat R-modules and suppose that A is a finitely generated left ideal of R. Then A is finitely presented since R is left coherent. So (2) of the left-hand version of Proposition 5.3.15 gives

$$
\left(\prod_\Delta N_\alpha\right) \otimes_R A \cong \prod_\Delta (N_\alpha \otimes_R A) \subseteq \prod_\Delta (N_\alpha \otimes_R R) \cong \prod_\Delta N_\alpha \cong \left(\prod_\Delta N_\alpha\right) \otimes_R R.
$$

Hence, we have an exact sequence $0 \to (\prod_\Delta N_\alpha) \otimes_R A \to (\prod_\Delta N_\alpha) \otimes_R R$, so by (3) of Proposition 5.3.7 we see that $\prod_\Delta N_\alpha$ is a flat R-module. $\qquad\square$

Regular Rings and Flat Modules

In order to characterize rings over which every R-module is flat, we need the following proposition.

Recall that R is a regular ring if for each $a \in R$ there is an $r \in R$ such that $a = ara$.

Proposition 5.3.17. *The following hold for any ring R.*

(1) *R is a regular ring if and only if every principal right (left) ideal of R is generated by an idempotent.*

(2) *If R is a regular ring, then every finitely generated right (left) ideal of R is a principal right (left) ideal of R.*

Proof. (1) Let $a \in R$ and consider the principal right ideal aR. Since R is regular, there is an $r \in R$ such that $ara = a$. Let $e = ar$, then $e^2 = (ar)(ar) = (ara)r = ar = e$, so e is an idempotent of R. Moreover, $e \in aR$ shows that $eR \subseteq aR$. If $c = ab \in aR$, then $ec = arc = arab = ab$, so $ab \in eR$. Hence, $eR = aR$.

Conversely, suppose that the principal right ideal aR is generated by the idempotent e. Then $eR = aR$ and $e = ar$ for some $r \in R$. Hence, $ea = ara$. Since $a \in eR$, $a = eb$ for some $b \in R$, so $ea = e^2b = eb = a$. Therefore, $ara = ea = a$ and we have that R is regular.

(2) The proof is by induction on the number of generators. If the finitely generated right ideal is generated by a single element, there is nothing to prove. Make the induction hypothesis that every right ideal of R generated by k elements, $k < n$, is a principal right ideal of R. If

$$A = a_1 R + a_2 R + \cdots + a_n R$$

is generated by n elements, then, by the induction hypothesis,

$$a_1 R + a_2 R + \cdots + a_{n-1} R$$

is a principal right ideal of R. Consequently, to complete the induction proof, it suffices to prove that a right ideal of the form $aR + bR$ is a principal right ideal of R. From (1), $aR = eR$, where e is an idempotent of R. Since $b = eb + (1 - e)b$, $bR \subseteq ebR + (1 - e)bR$. So it follows that

$$aR + bR = eR + (1 - e)bR = eR + fR,$$

where f is an idempotent element of R such that $ef = 0$. Let $g = f(1 - e)$. Then

$$gf = f(1 - e)f = f(f - ef) = f,$$
$$g^2 = f(1 - e)f(1 - e)$$
$$= f^2(1 - e) = f(1 - e) = g,$$
$$eg = ef(1 - e) = 0 \quad \text{and}$$
$$ge = f(1 - e)e = f(e - e^2) = 0.$$

But $g \in fR$ and $f \in gR$, so $fR = gR$. Therefore, $aR + bR = eR + gR$. We claim that $eR + gR = (e + g)R$. If $er + gs \in eR + gR$, then

$$(e + g)(er + gs) = er + egs + ger + gs$$
$$= er + gs,$$

so $eR + gR \subseteq (e + g)R$. The reverse containment is obvious, so $aR + bR = (e + g)R$ and the proof is complete. (Since the definition of a regular ring is left-right symmetric, the proof for left ideals follows by symmetry.) $\square$

Proposition 5.3.18. *The following are equivalent for any ring R.*

(1) *R is a regular ring.*

(2) *Every R-module is flat.*

(3) *Every cyclic R-module is flat.*

Proof. (1) $\Rightarrow$ (2). If R is a regular ring and A is a finitely generated left ideal of R, then Proposition 5.3.17 shows that $A = Re$ for some idempotent $e \in R$. If B is a left ideal of R such that $R = A \oplus B$ and $i : A \to R$ is the canonical injection, then there is an epimorphism $\pi : R \to A$ such that $\pi i = \mathrm{id}_A$. But then the canonical map

$$0 \to M \otimes_R A \xrightarrow{\ \mathrm{id}_M \otimes i\ } M \otimes_R R$$

has $\mathrm{id}_M \otimes \pi$ as a left inverse. Thus, $\mathrm{id}_M \otimes i$ is an injection and so Proposition 5.3.7 shows that M is a flat R-module.

(2) $\Rightarrow$ (3). Obvious.

(3) $\Rightarrow$ (1). Let $a \in R$ and consider the short exact sequence

$$0 \to aR \to R \to R/aR \to 0.$$

Since R/aR is cyclic and R is flat, in view of (3) of Proposition 5.3.9, we see that $aR \cap Ra = (aR)(Ra) = aRa$. Hence, R is a regular ring, since $a \in aR \cap Ra$. $\square$

Because of the left-right symmetry of regular rings, it is also the case that a ring R is regular if and only if every left R-module is flat. Hence, we have the following corollary to Proposition 5.3.18.

Corollary 5.3.19. *Every regular ring is coherent.*

Problem Set 5.3

1. If $0 \to N_1 \to N \to N_2 \to 0$ is a split short exact sequence of left R-modules and M is an R-module, verify that

$$0 \to M \otimes_R N_1 \to M \otimes_R N \to M \otimes_R N_2 \to 0$$

is split exact in **Ab**. Conclude that $M \otimes_R -$ preserves split short exact sequences regardless of whether or not M is flat.

2. Prove that an R-module M is flat if and only if

$$0 \to M \otimes_R A \to M \otimes_R R \cong M$$

is exact for each essential left ideal A of R. [Hint: If A is a left ideal of R and A_c is a complement of A in R, consider $A \oplus A_c$.]

3. Prove that every flat module over an integral domain is torsion free. [Hint: Consider $f : R \to R$ defined by $f(b) = ba$, $(\mathrm{id}_M \otimes f) : M \otimes_R R \to M \otimes_R R$, where $(\mathrm{id}_M \otimes f)(x \otimes b) = x \otimes ba$, $g : M \otimes_R R \to M$ such that $g(x \otimes b) = xb$ and the diagram

$$
\begin{array}{ccc}
M \otimes_R R & \xrightarrow{\mathrm{id}_M \otimes f} & M \otimes_R R \\
\downarrow{g} & & \downarrow{g} \\
M & \longleftarrow & M
\end{array}
\qquad]
$$

4. Prove Proposition 5.3.9 by using the commutative diagram

$$
\begin{array}{ccccccc}
K \otimes_R A & \xrightarrow{i \otimes \mathrm{id}_A} & F \otimes_R A & \xrightarrow{f \otimes \mathrm{id}_A} & M \otimes_R A & \longrightarrow & 0 \\
& & \downarrow{\varphi_A^F} & & \downarrow{\varphi_A^M} & & \\
0 & \longrightarrow K \cap FA & \longrightarrow & FA & \xrightarrow{\theta} & MA & \longrightarrow 0
\end{array}
$$

where $\theta : FA \to MA$ is such that $\theta(\sum_{i=1}^{n} x_i a_i) = \sum_{i=1}^{n} f(x_i) a_i$, φ_A^F and φ_A^M are the maps defined in Proposition 5.3.8 and $f : F \to M$ is a free module on M.

5. If $0 \to M_1 \xrightarrow{f} M \xrightarrow{g} M_2 \to 0$ is exact sequence of R-modules and M_1 and M_2 are flat, prove that M is flat. [Hint: Let A be a left ideal of R and consider the diagram

$$
\begin{array}{ccccccc}
M_1 \otimes_R A & \xrightarrow{f \otimes \mathrm{id}_A} & M \otimes_R A & \xrightarrow{g \otimes \mathrm{id}_A} & M_2 \otimes_R A & \longrightarrow & 0 \\
\downarrow{\alpha} & & \downarrow{\beta} & & \downarrow{\gamma} & & \\
M_1 A & \longrightarrow & MA & \longrightarrow & M_2 A & \longrightarrow & 0]
\end{array}
$$

6. An exact sequence of R-modules $0 \to M_1 \xrightarrow{f} M \to M_2 \to 0$ is said to be a *pure short exact sequence* if

$$0 \to M_1 \otimes_R N \to M \otimes_R N \to M_2 \otimes_R N \to 0$$

is exact for every left R-module N. In this case we say that $f(M_1)$ is a *pure submodule* of M. Note that Exercise 1 shows that split short exact sequences are pure. Prove that the following are equivalent:

(a) M is a flat R-module.

(b) Every exact sequence of R-modules of the form

$$0 \to M_1 \to M_2 \to M \to 0$$

is pure.

(c) There is a pure exact sequence of the form

$$0 \to M_1 \to M_2 \to M \to 0$$

with M_2 flat.

[Hint: For (a) $\Rightarrow$ (b), let N be a left R-module and suppose that $0 \to K \to F \to N \to 0$ is exact with F a free left R-module. Show that this gives a row and column exact commutative diagram

$$
\begin{array}{ccccccc}
M_1 \otimes_R K & \longrightarrow & M \otimes_R K & \longrightarrow & M_2 \otimes_R K & \longrightarrow & 0 \\
\downarrow & & \downarrow & & \downarrow & & \\
0 \longrightarrow M_1 \otimes_R F & \longrightarrow & M \otimes_R F & \longrightarrow & M_2 \otimes_R F & \longrightarrow & 0 \\
\downarrow & & \downarrow & & \downarrow & & \\
M_1 \otimes_R N & \longrightarrow & M \otimes_R N & \longrightarrow & M_2 \otimes_R N & \longrightarrow & 0 \\
\downarrow & & \downarrow & & \downarrow & & \\
0 & & 0 & & 0 & &
\end{array}
$$

and then show that $M_1 \otimes_R N \to M \otimes_R N$ is an injection. To show that (c) $\Rightarrow$ (a), suppose that $0 \to M_1 \to M_2 \to M \to 0$ is exact, where M_2 a flat R-module, and suppose that $0 \to N_1 \to N \to N_2 \to 0$ is an exact sequence of left R-modules. Show that

$$
\begin{array}{ccccccc}
 & & 0 & & & & \\
 & & \downarrow & & & & \\
M_1 \otimes_R N_1 & \longrightarrow & M_2 \otimes_R N_1 & \longrightarrow & M \otimes_R N_1 & \longrightarrow & 0 \\
\downarrow & & \downarrow & & \downarrow & & \\
M_1 \otimes_R N & \longrightarrow & M_2 \otimes_R N & \longrightarrow & M \otimes_R N & \longrightarrow & 0 \\
\downarrow & & \downarrow & & \downarrow & & \\
M_1 \otimes_R N_2 & \longrightarrow & M_2 \otimes_R N_2 & \longrightarrow & M \otimes_R N_2 & \longrightarrow & 0 \\
\downarrow & & \downarrow & & \downarrow & & \\
0 & & 0 & & 0 & &
\end{array}
$$

is a row and column exact commutative diagram and then show that $M \otimes_R N_1 \to M \otimes_R N$ is an injection.]

7. (a) Is a regular ring left and right semihereditary?

(b) Show that R is a regular ring if and only if for each $a \in R$ there is an element $r' \in R$ such that $a r' a = a$ and $r' a r' = r'$. [Hint: If $r \in R$ is such that $a r a = a$, consider $r' = r a r$.]

(c) Show that R is a regular ring if and only if $A B = A \cap B$ for every right ideal A of R and every left ideal B of R.

(d) Prove that a ring direct product of a finite number of regular rings is a regular ring.

(e) Prove that a regular ring is a division ring if and only if its only idempotents are 0 and 1.

8. If $f : R \to S$ is a ring homomorphism and M is a flat R-module and S is viewed as a left R-module by pullback along f, prove that $M \otimes_R S$ is a flat S-module. [Hint: If N is a left S-module, then N is a left R-module by pullback along f and S is an (R, R)-bimodule by pullback along f. Show that $(M \otimes_R S) \otimes_S N \cong M \otimes_R (S \otimes_S N) \cong M \otimes_R N$.]

9. Clearly, every finitely presented R-module is finitely generated. Prove that the converse holds if R is a right noetherian ring.

10. (a) Prove that every finitely generated projective module is finitely presented. [Hint: If M is a finitely generated projective R-module, consider a short exact sequence $0 \to K \to F \to M \to 0$, where F is a finitely generated free R-module.]

(b) Show that a right semihereditary ring is right coherent.

11. Let $0 \to M_1 \to M \to M_2 \to 0$ be a exact sequence in $\mathbf{Mod}_R$.

    (a) If M_1 and M_2 are finitely presented, show that M is finitely presented. [Hint: Exercise 9 in Problem Set 5.2.]

    (b) Prove that for every positive integer n, a direct sum $M_1 \oplus M_2 \oplus \cdots \oplus M_n$ of R-modules is finitely presented if and only if M_i is finitely presented for $i = 1, 2, \ldots, n$.

12. Prove that the map $\varphi^{R^{(n)}}$ of Proposition 5.3.14 is an isomorphism.

13. Prove that every finitely presented flat R-module M is projective. [Hint: Let $0 \to K \to F \to M \to 0$ be exact where F is finitely generated and free and K is a finitely generated submodule of F. Use Proposition 5.3.11 and show that $0 \to K \to F \to M \to 0$ splits.]

14. An R-module M is said to be *faithfully flat* provided that $0 \to {}_RN_1 \to {}_RN$ is exact if and only if $0 \to M \otimes_R N_1 \to M \otimes_R N$ is exact. Prove that the following are equivalent for an R-module M. Note that a faithfully flat module is clearly flat.

    (a) M is faithfully flat.

    (b) M is flat and $M \otimes_R N = 0$ implies that $N = 0$ for any left R-module N.

    (c) M is flat and $M\mathfrak{m} \neq M$ for every maximal left ideal $\mathfrak{m}$ of R.

    (d) M is flat and $MA \neq M$ for every proper left ideal A of R.

    (e) If $f : N_1 \to N_2$ is R-linear and $\mathrm{id}_M \otimes f : M \otimes_R N_1 \to M \otimes N_2$ is zero, then $f = 0$.

    [(a) $\Leftrightarrow$ (b), Hint: For (a) $\Rightarrow$ (b), consider $0 \to M \otimes_R 0 \to M \otimes_R N \to M \otimes_R 0 \to 0$ and to show that (b) $\Rightarrow$ (a), suppose that $0 \to N_1 \xrightarrow{f} N$ is a sequence of R-modules. If $0 \to M \otimes_R N_1 \xrightarrow{\mathrm{id}_M \otimes f} M \otimes_R N$ is exact, then $M \otimes_R \mathrm{Ker}\, f = 0$.]

    [(b) $\Rightarrow$ (c), Hint: Note that $M/M\mathfrak{m} \cong M \otimes_R (R/\mathfrak{m})$, where $\mathfrak{m}$ is a maximal left ideal of R. See Exercise 2 in Problem Set 2.3.]

    [(c) $\Rightarrow$ (b), Hint: Suppose that $0 \neq x \in N$. Then $Rx \cong R/A$ for some left ideal A of R. If $\mathfrak{m}$ is a maximal left ideal that contains A, then $M \neq M\mathfrak{m} \supseteq MA$, so $M \otimes_R Rx \cong M \otimes_R MA \cong M/MA \neq 0$. Next, consider the map $M \otimes_R Rx \to M \otimes_R N$.]

    [(b) $\Leftrightarrow$ (e), Hint: For (b) $\Rightarrow$ (e), let $N = f(N_1)$ and assume that $\mathrm{id}_M \otimes f : M \otimes_R N_1 \to M \otimes_R N_2$ is zero. Now consider the composition of $M \otimes_R N_1 \xrightarrow{\mathrm{id}_M \otimes f} M \otimes_R N \xrightarrow{\mathrm{id}_M \otimes i} M \otimes_R N_2$, where $i : N \to N_2$ is the canonical injection, and show that $M \otimes_R N = 0$. To show that (e) $\Rightarrow$ (b), consider $\mathrm{id}_M \otimes \mathrm{id}_N : M \otimes_R N \to M \otimes_R N$.]

15. An R-module M is said to be *faithful* provided that $Ma = 0$ implies that $a = 0$. Prove that a faithfully flat R-module is faithful. [Hint: If $Ma = 0$, let $f : R \to R$ be such that $f(b) = ba$ and consider $\mathrm{id}_M \otimes f : M \otimes_R R \to M \otimes_R R$.]

5.4 Quasi-Injective and Quasi-Projective Modules

Injective and projective modules lead to the concepts of quasi-injective and quasi-projective modules. We do little more in this section than give the definitions and examples of these modules. Quasi-injective and quasi-projective modules will be revisited in a subsequent chapter where quasi-injective envelopes and quasi-projective covers will be developed. There we will show that every module has a quasi-projective cover if and only if every module has a projective cover.

Definition 5.4.1. An R-module M is said to be *quasi-injective* if each row exact diagram

$$
\begin{array}{ccccc}
0 & \longrightarrow & N & \xrightarrow{\;f\;} & M \\
 & & \Big\downarrow{\scriptstyle g} & \nearrow & \\
 & & M & &
\end{array}
$$

of R-modules and R-homomorphisms can be completed to a commutative diagram by an endomorphism of M. Dually, M is *quasi-projective* if each row exact diagram

$$
\begin{array}{ccccc}
 & & M & & \\
 & \nearrow & \Big\downarrow{\scriptstyle g} & & \\
M & \xrightarrow{\;f\;} & N & \longrightarrow & 0
\end{array}
$$

of R-modules and R-module homomorphisms can be completed to a commutative diagram by an endomorphism of M.

Examples

1. A simple R-module is quasi-injective and quasi-projective.

2. If an R-module M contains a copy of R_R, then M if is quasi-injective if and only if M is injective. This follows easily from Baer's criteria.

3. If M and N are isomorphic R-modules, then M is quasi-injective (quasi-projective) if and only if N is quasi-injective (quasi-projective).

4. Every injective (projective) R-module is quasi-injective (quasi-projective).

Proofs of the following propositions are left as exercises. Each proof follows directly from the definitions.

Proposition 5.4.2. *An R-module M is quasi-injective if and only if for each pair of submodules N_1 and N_2 of M such that $N_1 \subseteq N_2$ each $f \in \operatorname{Hom}_R(N_1, M)$ can be extended to a $g \in \operatorname{Hom}_R(N_2, M)$.*

Proposition 5.4.3. *An R-module M is quasi-projective if and only if for each pair of submodules N_1 and N_2 of M such that $N_1 \supseteq N_2$, each $f \in \operatorname{Hom}_R(M, M/N_1)$ can be lifted to a $g \in \operatorname{Hom}_R(M, M/N_2)$.*

Proposition 5.4.4. *If $\{M_i\}_{i=1}^{n}$ if a family of R-modules and $\prod_{i=1}^{n} M_i$ is quasi-injective, then each M_i is quasi-injective.*

Remark. There are several concepts that generalize quasi-injective and quasi-projective modules, for instance, continuous modules, quasi-continuous modules, extending modules as well as discrete modules, quasi-discrete modules and lifting modules. These modules constitute an active area of research in ring and module theory. Additional details can be found in [33] and [11].

Problem Set 5.4

1. Verify Example 2.

2. Prove Proposition 5.4.2.

3. Prove Proposition 5.4.3.

4. Prove Proposition 5.4.4. Conclude that a direct summand of a quasi-injective module is quasi-injective.

5. Prove that every R-module is quasi-injective if and only if every R-module is injective.

6. Prove that an R-module M is quasi-injective if and only if for each essential submodule N of M, each R-linear mapping $f : N \to M$ can be extended to M. [Hint: Consider $N \oplus N_c$, where N_c is a complement of N in M.]

7. Prove or find a counterexample to each of the following.
   (a) If $\{M_i\}_{i=1}^{n}$ is a family of R-modules and each M_i is quasi-projective, then $\prod_{i=1}^{n} M_i$ is quasi-projective.
   (b) If $\{M_i\}_{i=1}^{n}$ is a family of R-modules and $\prod_{i=1}^{n} M_i$ is quasi-projective, then each M_i is quasi-projective.

Chapter 6

Classical Ring Theory

If R is a right artinian ring with no nonzero nilpotent ideals (defined below), then R is a finite ring direct product of simple artinian rings. An obstruction to decomposing a right artinian ring in this manner is that it may have nonzero nilpotent ideals. If we could find an ideal of R that contains the nilpotent ideals of R, then the factor ring modulo this ideal will be free of nonzero nilpotent ideals. Fortunately, there is such an ideal rad(R), called the prime radical of R. Furthermore, when R is right artinian, $R/\text{rad}(R)$ is right artinian and rad$(R/\text{rad}(R)) = 0$.

Jacobson [60] discovered another ideal $J(R)$ of R, now called the Jacobson radical, such that rad$(R) \subseteq J(R)$. Moreover, when R is right artinian, $J(R)$ is nilpotent and $J(R) = \text{rad}(R)$. Hence, if R is right artinian, then $R/J(R)$ is a right artinian ring such that $J(R/J(R)) = 0$, so $R/J(R)$ is a finite ring direct product of simple artinian rings. The ideal $J(R)$ is one of the most important ideals in the study of rings and modules.

The main purpose of the chapter is to study right artinian rings R for which $J(R) = 0$. As it turns out, a simple artinian ring is an $n \times n$ matrix ring with entries from a division ring, so a right artinian ring R with $J(R) = 0$ has a very nice description in terms of matrix rings.

The pioneering work on such a decomposition of a ring was done by Wedderburn [71], although in a somewhat different context. Later Artin [47], [48] extended Wedderburn's results to rings that satisfy both the ascending and the descending chain condition. At that time, Artin did not know that the descending chain condition on a ring implies that it satisfies the ascending chain condition (Corollary 6.6.5). This result, and the results of Wedderburn and Artin has led to what is now often referred to as the Wedderburn–Artin theory. To address this theory, we begin with the Jacobson radical.

6.1 The Jacobson Radical

Numerous types of radicals have emerged over the years each with a different foundation. Thus, when the word "radical" is encountered, care must be taken to determine its exact meaning. A discussion of several types of radicals can be found in [9] and [17].

Definition 6.1.1. The *Jacobson radical* [22] of R, denoted by $J(R)$, is the intersection of the maximal right ideals of R. If $J(R) = 0$, then R is said to be a *Jacobson* *semisimple ring.* (Jacobson semisimple rings are also referred to as *J-semisimple rings* and they are sometimes called *semiprimitive rings*.)

At this point, the Jacobson radical of R should probably be called the right Jacobson radical of R since it is formed by taking the intersection of the maximal right ideals of R. However, we will see that $J(R)$ is also the intersection of the maximal left ideals of R, so a designation of left or right is immaterial.

The concept of the Jacobson radical of R carries over to modules. If M is an R-module, then the *radical* of M, denoted by $\mathrm{Rad}(M)$, is the intersection of the maximal submodules of M. If M fails to have maximal submodules, then we set $\mathrm{Rad}(M) = M$. There are modules that fail to have maximal submodules. For example, the $\mathbb{Z}$-module $\mathbb{Q}/\mathbb{Z}$ has no maximal submodules. The following proposition shows that there is a "large" class of modules each of which always has at least one maximal submodule.

Proposition 6.1.2. *If M is a nonzero finitely generated R-module, then M has at least one maximal submodule.*

Proof. If $\{x_1, x_2, \dots, x_n\}$ is a minimal set of generators of M, then

$$x_2 R + x_3 R + \cdots + x_n R = N \subsetneqq M.$$

Let $\mathcal{S}$ be the collection of proper submodules of M that contain N and partial order $\mathcal{S}$ by inclusion. Note that $N' \in \mathcal{S}$ if and only if $N \subseteq N'$ and $x_1 \notin N'$. If $\mathcal{C}$ is a chain of submodules of $\mathcal{S}$, then $x_1 \notin \bigcup_{\mathcal{C}} N'$, so $\bigcup_{\mathcal{C}} N'$ is a proper submodule of M that contains N. Hence, $\mathcal{S}$ is inductive and Zorn's lemma shows that $\mathcal{S}$ has a maximal element, say N^*. If N^* fails to be a maximal submodule of M, then there is a submodule $\bar{N}$ of M such that $N^* \subsetneqq \bar{N} \subsetneqq M$. Now $\bar{N} \notin \mathcal{S}$ for, if so, this would contradict the maximality of N^* in $\mathcal{S}$. But if $\bar{N}$ is not in $\mathcal{S}$, then $x_1 \in \bar{N}$, so $\{x_1, x_2, \dots, x_n\} \subseteq \bar{N}$. Thus, $\bar{N} = M$ and so we have a contradiction. Therefore, N^* is not only maximal in $\mathcal{S}$ but N^* is also a maximal submodule of M. $\square$

Corollary 6.1.3. *If M is a nonzero finitely generated R-module, then $\mathrm{Rad}(M) \neq M$.*

Since R is generated by 1, $J(R) \neq R$. This is also verified by Corollary 1.2.4 which states that R has at least one maximal right ideal. A useful property of the radical of a module is that it is preserved under direct sums.

Proposition 6.1.4. *If $\{M_\alpha\}_\Delta$ is a family of R-modules, then*

$$\mathrm{Rad}\left(\bigoplus_\Delta M_\alpha\right) = \bigoplus_\Delta \mathrm{Rad}(M_\alpha).$$

Proof. We prove the proposition for $\Delta = \{1, 2\}$. The argument for the general case is similar and its proof is left as an exercise. If $i_1 : M_1 \to M_1 \oplus M_2$ and $i_2 : M_2 \to M_1 \oplus M_2$ are the canonical injections, then (a) of Exercise 3 shows that

$$i_k : \mathrm{Rad}(M_k) \to \mathrm{Rad}(M_1 \oplus M_2)$$

for $k = 1, 2$. Consequently, $\mathrm{Rad}(M_1) \oplus \mathrm{Rad}(M_2) \subseteq \mathrm{Rad}(M_1 \oplus M_2)$. Next, note that since $(M_1 \oplus M_2)/(N_1 \oplus M_2) \cong M_1/N_1$, $N_1 \oplus M_2$ is a maximal submodule of $M_1 \oplus M_2$ if and only if N_1 is a maximal submodule of M_1. Therefore, if $(x, y) \in \mathrm{Rad}(M_1 \oplus M_2)$, then $(x, y) \in N_1 \oplus M_2$ for every maximal submodule N_1 of M_1. Hence, $x \in N_1$ for every maximal submodule of M_1, so $x \in \mathrm{Rad}(M_1)$. Similarly, $y \in \mathrm{Rad}(M_2)$ and we have $\mathrm{Rad}(M_1 \oplus M_1) \subseteq \mathrm{Rad}(M_1) \oplus \mathrm{Rad}(M_2)$. Thus, $\mathrm{Rad}(M_1 \oplus M_2) = \mathrm{Rad}(M_1) \oplus \mathrm{Rad}(M_2)$.   □

Corollary 6.1.5. *If F is a free R-module, then* $\mathrm{Rad}(F) = FJ(R)$.

Proof. Since $F \cong R^{(\Delta)}$ for some set Δ, we have

$$\mathrm{Rad}(F) \cong \mathrm{Rad}(R^{(\Delta)}) \cong J(R)^{(\Delta)} \cong R^{(\Delta)}J(R) \cong FJ(R).$$ □

Lemma 6.1.6. *An element a of R is contained in no maximal right ideal of R if and only if a has a right inverse in R.*

Proof. Suppose that a is contained in no maximal right ideal of R. Then aR cannot be a proper right ideal of R since every proper right ideal is, by Proposition 1.2.3, contained in a maximal right ideal of R. Hence, $aR = R$, so there is an $r \in R$ such that $ar = 1$. Conversely, if a has a right inverse and $a \in \mathfrak{m}$, $\mathfrak{m}$ a maximal right ideal of R, then $1 \in \mathfrak{m}$, so $\mathfrak{m} = R$, a contradiction. Hence, $a \notin \mathfrak{m}$, so a is contained in no maximal right ideal of R.   □

Proposition 6.1.7. *The following hold for any ring R.*

(1) *$J(R)$ is an ideal of R that coincides with the intersection of the right annihilator ideals of the simple R-modules.*

(2) *$J(R)$ is the set of all $a \in R$ such that $1 - ar$ has a right inverse for all $r \in R$.*

(3) *$J(R)$ is the largest ideal of R such that for all $a \in J(R)$, $1 - a$ is a unit in R.*

Proof. (1) Let $\mathcal{S}$ be the nonempty class of simple R-modules. We claim that $J(R) = \bigcap_{\mathcal{S}} \mathrm{ann}_r(S)$. Since S is a simple R-module if and only if there is a maximal right ideal $\mathfrak{m}$ of R such that $R/\mathfrak{m} \cong S$, we see that $\mathrm{ann}_r(R/\mathfrak{m}) = \mathrm{ann}_r(S)$. But $a \in \mathrm{ann}_r(R/\mathfrak{m})$ implies that $a + \mathfrak{m} = (1 + \mathfrak{m})a = 0$, so $a \in \mathfrak{m}$. So it follows that $\bigcap_{\mathcal{S}} \mathrm{ann}_r(S) \subseteq J(R)$. Conversely, if $a \in J(R)$, then a is in every maximal right ideal of R. If S is any simple R-module and $x \in S, x \neq 0$, then $xR = S$, so $R/\mathrm{ann}_r(x) \cong S$ and $\mathrm{ann}_r(x)$ is a maximal right ideal of R. Hence, $a \in \mathrm{ann}_r(x)$

for every nonzero $x \in S$. Therefore, $a \in \bigcap_S \mathrm{ann}_r(x) = \mathrm{ann}_r(S)$, so $J(R) \subseteq \bigcap_S \mathrm{ann}_r(S)$.

> Note that since the right annihilator of an R-module is an ideal of R, we
> see that $J(R)$ is an ideal of R.

(2) An element a of R is such that $a \in J(R)$ if and only if a is in every maximal right ideal of R and this in turn is true if and only if $1 - ar$ is in no maximal right ideal of R for any $r \in R$. This observation and Lemma 6.1.6 give the result.

(3) If $a \in J(R)$, then $1 - a$ has a right inverse b in R and $(1 - a)b = 1$ implies that $1 - b = -ab \in J(R)$. Thus, $1 - b$ is in every maximal right ideal of R, so b can be in no maximal right ideal of R. Lemma 6.1.6 now implies that b has a right inverse $c \in R$. But $1 = (1 - a)b$, so $c = (1 - a)bc = 1 - a$ and this shows that $1 = bc = b(1 - a)$. Therefore, b is also a left inverse for $1 - a$. Finally, let I be an ideal of R such that $J(R) \subseteq I$ and such that $1 - a$ is a unit in R for every $a \in I$. If $r \in R$, then $ar \in I$, so $1 - ar$ is a unit in R. But then $1 - ar$ has a right inverse in R, so by (2), $a \in J(R)$. Hence, $I \subseteq J(R)$ and we are done.   □

Part (3) of Proposition 6.1.7 is obviously left-right symmetric and leads to the following proposition.

Proposition 6.1.8. *The following hold for any ring R.*

(1) *$J(R)$ is the intersection of the maximal left ideals of R.*

(2) *$J(R)$ is an ideal of R that coincides with the intersection of the left annihilator ideals of the simple left R-modules.*

(3) *$J(R)$ is the set of all $a \in R$ such that $1 - ra$ has a left inverse for all $r \in R$.*

Because of (1) of Proposition 6.1.7 and (2) of Proposition 6.1.8, we see that $SJ(R) = 0$ for each simple R-module S and $J(R)S = 0$ for each simple left R-module. We now need the following lemma.

Lemma 6.1.9. *If A is a right ideal of R such that $A \subseteq J(R)$, then $MA \subseteq \mathrm{Rad}(M)$ for every R-module M.*

Proof. Suppose that A is a right ideal of R such that $A \subseteq J(R)$. If $\mathrm{Rad}(M) = M$, there is nothing to prove, so suppose that $\mathrm{Rad}(M) \neq M$. Part (1) of Proposition 6.1.7 shows that $SJ(R) = 0$ for every simple R-module S, so $SA = 0$ for all simple R-modules S. If N is a maximal submodule of M, then M/N is a simple R-module, so $(M/N)A = 0$. Thus, $MA \subseteq N$ and we see that MA is contained in every maximal submodule of M. Consequently, $MA \subseteq \mathrm{Rad}(M)$.   □

The notation A^n, not to be confused with $A^{(n)} = A \times A \times \cdots \times A$, denotes the set of all finite sums $\sum a_1 a_2 \cdots a_n$ of products $a_1 a_2 \cdots a_n$ of n elements from a left ideal A, a right ideal A or an ideal A of R. We write $A^n = 0$ if all such finite sums are equal to zero. It follows that $A^n = 0$ if and only if all products $a_1 a_2 \cdots a_n$ of n elements from A are zero. A right ideal (A left ideal, An ideal) A with this property is said to be *nilpotent*. In particular, if A is nilpotent and such that $A^n = 0$, then $a^n = 0$ for each $a \in A$, so every element of A is nilpotent. Nilpotent right ideals of R will be discussed in more detail later in this chapter. The point is that if M is any R-module and A is a nilpotent right ideal of R such that $MA = M$, then $M = 0$. This follows easily since $M = MA = MA^2 = MA^3 = \cdots = MA^n = 0$, where n is a positive integer such that $A^n = 0$. The following important result shows that $MA = M$ implies $M = 0$ for all finitely generated R-modules M and all right ideals A contained in $J(R)$ regardless of whether or not A is nilpotent.

Lemma 6.1.10 (Nakayama's lemma). *If A is a right ideal of R such that $A \subseteq J(R)$, then the following two equivalent conditions hold for every finitely generated R-module M.*

(1) *If N is a submodule of M such that $N + MA = M$, then $N = M$.*

(2) *If $MA = M$, then $M = 0$.*

Proof. Suppose that A is a right ideal of R, that $A \subseteq J(R)$, and that M is finitely generated.

(1) If $M = 0$, the result is obvious, so suppose that $M \neq 0$. If $N + MA = M$ and $x + N \in M/N$, then $x = y + \sum_{k=1}^{n} x_k a_k$, where $y \in N$, $x_k \in M$ and $a_k \in A$ for $k = 1, 2, \ldots, n$. Thus, $x + N = y + \sum_{k=1}^{n} x_k a_k + N = \sum_{k=1}^{n} x_k a_k + N = \sum_{k=1}^{n} (x_k + N) a_k. \in (M/N)A$. This observation and Lemma 6.1.9 show that $M/N \subseteq (M/N)A \subseteq \mathrm{Rad}(M/N)$. Hence, $\mathrm{Rad}(M/N) = M/N$. But if M is finitely generated, then M/N is finitely generated, so if M/N is nonzero, then Corollary 6.1.3 implies that $\mathrm{Rad}(M/N) \neq M/N$. Thus, it must be the case that $M/N = 0$, so $N = M$. Hence, (1) holds for every finitely generated R-module M.

The proof will be completed by showing (1) $\Longleftrightarrow$ (2). Suppose that (1) holds, that $MA = M$ and that $M \neq 0$. Then using Proposition 6.1.2, we see that M has at least one maximal submodule, say N. Since $MA = M$, we have $N + MA = M$ and so, by (1), it must be the case that $N = M$. But maximal submodules of M are proper submodules of M, so $M = 0$ and we have (1) $\Rightarrow$ (2). To see that (2) $\Rightarrow$ (1), suppose that N is a submodule of M such that $N + MA = M$. Then $(M/N)A = (N + MA)/N$, so $(M/N)A = M/N$. But M/N is finitely generated, so (2) shows that $M/N = 0$. Hence, $M = N$. $\square$

Definition 6.1.11. A submodule S of an R-module M is said to be *small* (or *superfluous*) in M if whenever N is a submodule of M such that $S + N = M$, then $N = M$. A right ideal of R is small if it is small when viewed as a submodule of R_R.

Note that every module has at least one small submodule, namely the zero submodule, so the set of small submodules of a given module is always nonempty.

Example

1. **Local Rings.** Recall that a ring is said to be a local ring if it is a commutative ring that has exactly one maximal ideal. Let $\mathfrak{m}$ be the maximal ideal of a local ring R. Then $\mathfrak{m} = J(R)$ and we claim that $\mathfrak{m}$ is small in R. To see this, suppose that I is an ideal of R such that $\mathfrak{m} + I = R$. If I is a proper ideal of R, then because of Proposition 1.2.3, I is contained in a maximal ideal of R which must be $\mathfrak{m}$. But then $\mathfrak{m} + I \subseteq \mathfrak{m} \neq R$, a contradiction. Thus, $I = R$, so $\mathfrak{m}$ is small in R.

Example 1 shows that the Jacobson radical $J(R)$ of a local ring is small in R. The small right ideals of a ring are closely connected to the Jacobson radical of the ring.

Proposition 6.1.12. *A right ideal A is small in R if and only if $A \subseteq J(R)$. Furthermore, if M is a finitely generated R-module, then MA is a small submodule M for each right ideal A of R contained in $J(R)$.*

Proof. Let $A \subseteq J(R)$ be a right ideal of R. Then $A \subseteq RA$, so if B is a right ideal of R such that $B + A = R$, then $B + RA = R$. But R is finitely generated so Nakayama's lemma gives $B = R$. Hence, A is small in R. Conversely, suppose that A is a small right ideal of R. If $A \not\subseteq J(R)$, then there is a maximal right ideal $\mathfrak{m}$ of R such that $A \not\subseteq \mathfrak{m}$, so $A + \mathfrak{m} = R$. This indicates that $\mathfrak{m} = R$ and we have a contradiction. Hence, $A \subseteq J(R)$ when A is a small right ideal of R.

Finally, if M is finitely generated and $N + MA = M$, then $N = M$ by Nakayama's lemma. $\qquad\square$

Corollary 6.1.13. *For any ring R, $J(R)$ is the largest small right (left) ideal of R and as such it is unique.*

The radical of M can also be described in terms of the small submodules of M.

Proposition 6.1.14. *If M is an R-module and $\{S_\alpha\}_\Delta$ is the family of small submodules of M, then $\mathrm{Rad}(M) = \sum_\Delta S_\alpha$.*

Proof. If $\mathrm{Rad}(M) = M$, then it is obvious that $\sum_\Delta S_\alpha \subseteq \mathrm{Rad}(M)$. So suppose that $\mathrm{Rad}(M) \neq M$. If S is a small submodule of M and if there is a maximal submodule N of M such that $S \not\subseteq N$, then $S + N = M$. But this implies that $N = M$ which clearly cannot be the case. Thus, S is contained in every maximal submodule of M, so $S \subseteq \mathrm{Rad}(M)$. Hence, $\sum_\Delta S_\alpha \subseteq \mathrm{Rad}(M)$.

For the reverse containment, let N be a proper submodule of M and suppose that $x \in M - N$. If $M = xR + N$, then we claim that there is a maximal submodule of M that does not contain x. Let $\mathcal{T}$ be the set of submodules N' of M such that $x \notin N'$ and $N' \supseteq N$. Now $\mathcal{T} \neq \varnothing$ since $N \in \mathcal{T}$, so partial order $\mathcal{T}$ by inclusion. If $\mathcal{C}$ is a chain in $\mathcal{T}$, then $\bigcup_\mathcal{C} N'$ is a submodule of M containing N and $x \notin \bigcup_\mathcal{C} N'$.

Hence, $\mathcal{T}$ is inductive, so let N^* be a maximal element of $\mathcal{T}$. We assert that N^* is a maximal submodule of M. If not, there is a submodule X of M such that $N^* \subsetneq X \subsetneq M$ and it must be the case that $x \in X$. Since $xR \subseteq X$ and $N \subseteq N^*$, we see that $M = X + N^* = X$, a contradiction. Hence, N^* is a maximal submodule of M. Therefore, if N is a proper submodule of M and $x \in M - N$ is such that $M = xR + N$, then there is a maximal submodule N^* of M such that $x \notin N^*$. Consequently, $x \notin \operatorname{Rad}(M)$. It follows that if $x \in \operatorname{Rad}(M)$ and $xR + N = M$, then N cannot be a proper submodule of M and this in turn implies that xR is a small submodule of M. Thus, $x \in xR \subseteq \sum_\Delta S_\alpha$ and so $\operatorname{Rad}(M) \subseteq \sum_\Delta S_\alpha$.   □

Corollary 6.1.15. *For any R-module M, $\operatorname{Rad}(M)$ contains every small submodule of M.*

Problem Set 6.1

1. (a) If S is a small submodule of M and S' is a submodule of S, show that S' is a small submodule of M. Conclude that submodules of small submodules are small.

   (b) If $f : M \to N$ is an R-linear mapping and S is a small submodule of M, prove that $f(S)$ is a small submodule of N. [Hint: If N' is a submodule of N such that $f(S) + N' = N$, then $S + f^{-1}(N') = M$.]

   (c) Show that $S_1 \oplus S_2$ is a small submodule of $M_1 \oplus M_2$ if and only if S_1 is a small submodule of M_1 and S_2 is a small submodule of M_2.

2. Let M be an R-module.

   (a) If S_1 and S_2 are submodules of M such that $S_1 \subseteq S_2$, prove that S_2 is small in M if and only if S_1 is small in M and S_2/S_1 is small in M/S_1.

   (b) Show that $S_1 + S_2$ is small in M if and only S_1 and S_2 are small submodules of M.

   (c) Prove that S is a small submodule of an R-module M if and only if $S + N \neq M$ for every proper submodule N of M.

3. Verify each of the following:

   (a) If $f : M \to N$ is an R-linear mapping, then $f(\operatorname{Rad}(M)) \subseteq \operatorname{Rad}(N)$ for all R-modules M and N. In particular, if f is an epimorphism and $\operatorname{Ker} f \subseteq \operatorname{Rad}(M)$, then equality holds. [Hint: If S is a small submodule of M, then $f(S)$ is a small submodule of N. See (b) of Exercise 1. If f is an epimorphism and $\operatorname{Ker} f \subseteq \operatorname{Rad}(M)$, show that there is a one-to-one correspondence among the maximal submodules of N and the maximal submodules of M that contain $\operatorname{Ker} f$.]

   (b) $\operatorname{Rad}(M/\operatorname{Rad}(M)) = 0$ for every R-module M.

   (c) If I is an ideal of R such that $I \subseteq J(R)$, then $J(R/I) = J(R)/I$.

4. (a) If n is a positive integer greater than 1, compute the radical of the $\mathbb{Z}$-module (n). [Hint: Describe $\mathrm{Rad}((n))$ in terms of the prime factorization of n.]

   (b) Compute the Jacobson radical of the rings $\mathbb{Z}$ and $\mathbb{Z}_n$.

   (c) If p is a prime number, what is the Jacobson radical of $\mathbb{Z}_{p^n}$, n a positive integer?

5. Prove that $\mathbb{Z}$-module $\mathbb{Q}/\mathbb{Z}$ has no maximal submodules.   Conclude that $\mathrm{Rad}_{\mathbb{Z}}(\mathbb{Q}/\mathbb{Z}) = \mathbb{Q}/\mathbb{Z}$.

6. Suppose that N is a submodule of an R-module M. Prove that $N = \mathrm{Rad}(M)$ if and only if $N \subseteq \mathrm{Rad}(M)$ and $\mathrm{Rad}(M/N) = 0$.

7. Let M be a nonzero R-module with only one small submodule. Show that M has at least one maximal submodule.

8. (a) Show that $J(R)$ cannot contain a nonzero idempotent element of R. [Hint: If e is an idempotent in $J(R)$, consider $1 - e$ along with (3) of Proposition 6.1.7.]

   (b) Recall that a ring R is said to be a regular ring if for each $a \in R$ there is an $r \in R$ such that $a = ara$. If R is a regular ring, show that R is Jacobson semisimple.

9. If $\prod_{\Delta} R_\alpha$ is the ring direct product of the family $\{R_\alpha\}_{\Delta}$ of rings, show that $J(\prod_{\Delta} R_\alpha) = \prod_{\Delta} J(R_\alpha)$. [Hint: Verify that $(1_\alpha) - (a_\alpha)(b_\alpha)$ has a right inverse in $\prod_{\Delta} R_\alpha$ if and only if $1_\alpha - a_\alpha b_\alpha$ has a right inverse in R_α for each $\alpha \in \Delta$.] Conclude that a ring direct product of Jacobson semisimple rings is Jacobson semisimple.

10. (a) Complete the proof of Proposition 6.1.4 for an arbitrary indexing set Δ.

    (b) Show that if F is a free R-module, then $\mathrm{Rad}(F) = FJ(R)$.

    (c) Prove that $J(\mathbb{M}_n(R)) = \mathbb{M}_n(J(R))$.

11. Prove that $a \in J(R)$ if and only if $1 - ras$ is a unit in R for all $r, s \in R$.

12. (a) If D is a division ring, prove that $J(D[X]) = 0$. [Hint: If $p(X) \in J(D[X])$, then $1 - p(X)$ is a unit in $D[X]$. Show that the only units in $D[X]$ are elements of D, and then show $J(D[X])$ is an ideal of D.]

    (b) Let R be the matrix ring $\left(\begin{smallmatrix} K & K \\ 0 & K \end{smallmatrix}\right)$, where K is a field. Compute $J(R_R)$ and $J(_R R)$.

6.2   The Prime Radical

Recall that if R is a commutative ring, then a proper ideal $\mathfrak{p}$ of R is prime if whenever $a, b \in R$ and $ab \in \mathfrak{p}$, then $a \in \mathfrak{p}$ or $b \in \mathfrak{p}$. If $\mathfrak{p}$ is a prime ideal of a commutative ring R, let A and B be ideals of R such that $AB \subseteq \mathfrak{p}$. If $B \nsubseteq \mathfrak{p}$, let $b \in B$ be such that $b \notin \mathfrak{p}$. Then for any $a \in A$, $ab \in AB \subseteq \mathfrak{p}$, so it must be the case that $a \in \mathfrak{p}$. Hence, $A \subseteq \mathfrak{p}$. So if R is a commutative ring, then an ideal $\mathfrak{p}$ of R is prime if and only if whenever A and B are ideals of R such that $AB \subseteq \mathfrak{p}$, then either $A \subseteq \mathfrak{p}$ or $B \subseteq \mathfrak{p}$. This property of prime ideals in a commutative ring is a property that can be used to extend the concept of a prime ideal to noncommutative rings.

Definition 6.2.1. A proper ideal $\mathfrak{p}$ of a ring R is said to be a *prime ideal* of R if whenever A and B are ideals of R such that $AB \subseteq \mathfrak{p}$, then either $A \subseteq \mathfrak{p}$ or $B \subseteq \mathfrak{p}$.

The following proposition gives several conditions that characterize prime ideals.

Proposition 6.2.2. *If $\mathfrak{p}$ is an ideal of R, then the following are equivalent:*

(1) *$\mathfrak{p}$ is a prime ideal of R.*

(2) *If $a, b \in R$ and $aRb \subseteq \mathfrak{p}$, then $a \in \mathfrak{p}$ or $b \in \mathfrak{p}$.*

(3) *If A and B are right ideals of R such that $AB \subseteq \mathfrak{p}$, then either $A \subseteq \mathfrak{p}$ or $B \subseteq \mathfrak{p}$.*

(4) *If A and B are left ideals of R such that $AB \subseteq \mathfrak{p}$, then either $A \subseteq \mathfrak{p}$ or $B \subseteq \mathfrak{p}$.*

Proof. $(1) \Rightarrow (2)$. If $\mathfrak{p}$ is a prime ideal of R and $a, b \in R$ are such that $aRb \subseteq \mathfrak{p}$, then RaR and RbR are ideals of R and $(RaR)(RbR) = R(aRb)R \subseteq R\mathfrak{p}R \subseteq \mathfrak{p}$. Hence, $a \in RaR \subseteq \mathfrak{p}$ or $b \in RbR \subseteq \mathfrak{p}$.

$(2) \Rightarrow (3)$. Suppose that A and B are right ideals of R such that $AB \subseteq \mathfrak{p}$. If $B \not\subseteq \mathfrak{p}$, then there is a $b \in B$ such that $b \notin \mathfrak{p}$. But then for any $a \in A$, we have $aRb \subseteq AB \subseteq \mathfrak{p}$, so $a \in \mathfrak{p}$. Hence, $A \subseteq \mathfrak{p}$ when $B \not\subseteq \mathfrak{p}$.

$(3) \Rightarrow (1)$. Clear.

A similar proof shows the equivalence of $(1), (2)$ and (4). $\square$

Prime Rings

Definition 6.2.3. The *prime radical* (or the *lower nil radical*) of a ring R, denoted by $\mathrm{rad}(R)$, is the intersection of the prime ideals of R. A ring R is said to be a *prime ring* if zero is a prime ideal of R.

With the following definition, we can give an elementwise characterization of the prime radical of R.

Definition 6.2.4. An element $a \in R$ is said to be *strongly nilpotent* provided that every sequence $a_0, a_1, a_2, \ldots$, where $a = a_0$ and $a_{n+1} \in a_n R a_n$ for $n = 0, 1, 2, \ldots$, is *eventually zero*, that is, if there is an integer $n \geq 0$ such that $a_n = 0$.

Proposition 6.2.5. *The prime radical of a ring R is the set of all strongly nilpotent elements of R.*

Proof. Suppose that $a \in R$ and $a \notin \mathrm{rad}(R)$. Then there is a prime ideal $\mathfrak{p}$ of R such that $a \notin \mathfrak{p}$. Hence, $a_0 R a_0 \not\subseteq \mathfrak{p}$, where $a = a_0$, so there is an $a_1 \in a_0 R a_0$ such that $a_1 \notin \mathfrak{p}$. Thus, $a_1 R a_1 \not\subseteq \mathfrak{p}$. Continuing in this way we obtain a sequence

$a = a_0, a_1, a_2, \cdots$ that is never zero and so a is not strongly nilpotent. Hence, if an element $a \in R$ is strongly nilpotent, then $a \in \text{rad}(R)$.

For the converse, assume that $a \in R$ is not strongly nilpotent. Then there is a sequence $a_0, a_1, a_2, \ldots$, where $a = a_0$ and $a_{n+1} \in a_n R a_n$ for $n = 0, 1, 2, \ldots$, that is never zero. Let $T = \{a_n \mid n \geq 0\}$. Then $0 \notin T$, so suppose that $\mathcal{S}$ is the set of all ideals of R that have empty intersection with T. If $\mathcal{S}$ is partially ordered by inclusion, then $\mathcal{S}$ is inductive, so Zorn's lemma shows there is a maximal element $\mathfrak{p}$ of $\mathcal{S}$. We claim that $\mathfrak{p}$ is a prime ideal of R. Let A and B be ideals of R such that $A \not\subseteq \mathfrak{p}$ and $B \not\subseteq \mathfrak{p}$. Since $\mathfrak{p}$ is maximal in $\mathcal{S}$, both $A + \mathfrak{p}$ and $B + \mathfrak{p}$ have nonempty intersection with T. Let $a_i \in A + \mathfrak{p}$ and $a_j \in B + \mathfrak{p}$. Then $a_{n+1} \in a_n R a_n \subseteq (A + \mathfrak{p})(B + \mathfrak{p}) \subseteq AB + \mathfrak{p}$, where $n = \max(i, j)$. But $a_{n+1} \notin \mathfrak{p}$ since $\mathfrak{p}$ has empty intersection with T, so $AB \not\subseteq \mathfrak{p}$. Therefore, if A and B are ideals of R such that $AB \subseteq \mathfrak{p}$, then $A \subseteq \mathfrak{p}$ or $B \subseteq \mathfrak{p}$. Hence, $\mathfrak{p}$ is a prime ideal of R such that $a \notin \mathfrak{p}$. Consequently, if a is not strongly nilpotent, then $a \notin \text{rad}(R)$. Thus, if $a \in \text{rad}(R)$, then a is strongly nilpotent.   □

Definition 6.2.6. Recall that a right ideal (A left ideal, An ideal) $\mathfrak{n}$ of a ring R is a *nilpotent right ideal* (*nilpotent left ideal, nilpotent ideal*) if $\mathfrak{n}^n = 0$ for some positive integer n. The smallest positive integer n such that $\mathfrak{n}^n = 0$ is the *index of nilpotency* of $\mathfrak{n}$. A right ideal (A left ideal, An ideal) $\mathfrak{n}$ of R is said to be a *nil right ideal* (*nil left ideal, nil ideal*) of R if every element of $\mathfrak{n}$ is nilpotent.

If $\mathfrak{n}^n = 0$, then $a^n = 0$, for each $a \in A$, so every element of $\mathfrak{n}$ is nilpotent. Thus, every nilpotent right (left) ideal is nil. However, the second example of the following two examples shows that it is possible for an ideal in a ring to be nil and yet not be nilpotent.

Examples

1. Consider the ring $\mathbb{Z}_8$. The ideal $[2]\mathbb{Z}_8$ of $\mathbb{Z}_8$ is nilpotent for if

$$[2][a_1], [2][a_2], [2][a_3] \in [2]\mathbb{Z}_8, \quad \text{then}$$

$$([2][a_1])([2][a_2])([2][a_3]) = [8][a_1 a_2 a_2] = [0].$$

   More generally, every proper ideal of $\mathbb{Z}_{p^n}$, p a prime number, is nilpotent.

2. Let p be a prime number and and suppose that $[a_n]$ denotes an element of $\mathbb{Z}_{p^n}$ for $n = 1, 2, \ldots$. Consider the ring direct product $\prod_{\mathbb{N}} \mathbb{Z}_{p^n}$ and suppose that R is the set of all elements of $\prod_{\mathbb{N}} \mathbb{Z}_{p^n}$ of the form

$$([a_1], [a_2], [a_3], \ldots, [a_n], [0], [0], \ldots)$$

   for each integer $n \geq 0$, n not fixed. Then R is a subring of $\prod_{\mathbb{N}} \mathbb{Z}_{p^n}$ that does not have an identity. Next, let I be elements of R of the form

$$a = ([0], [pa_2], \ldots, [pa_n], [0], [0], \ldots).$$

It is routine to show that I is an ideal of R under the coordinatewise operations of addition and multiplication defined on R. Moreover, if $a \in I$ and $a = ([0], [pa_2], \ldots, [pa_n], [0], [0], \ldots)$, then $a^n = 0$. Hence, I is a nil ideal of R. We claim that I is not nilpotent. To show this, it suffices to show that for each $n \geq 1$, there is an $a \in I$ such that $a^n \neq 0$. Let $a = ([0], [p], \ldots, [p], [0], [0], \ldots)$, where the last entry of $[p]$ is in the $(n + 1)$-position. Then $a^n = ([0], \ldots, [0], [p^n], [0], [0], \ldots) \neq 0$ since $[p^n] \neq [0]$ in $\mathbb{Z}_{p^{n+1}}$. Thus, $I^n \neq 0$ for each integer $n \geq 1$, so I is not nilpotent.

Proposition 6.2.7. *If $\mathfrak{n}$ is either a nil left ideal or a nil right ideal of R, then $\mathfrak{n} \subseteq J(R)$.*

Proof. Let $\mathfrak{n}$ be a nil right ideal of R. If $a \in \mathfrak{n}$, then, for any $r \in R$, ar is nilpotent. If $(ar)^n = 0$, then $1 - ar$ has $\sum_{i=0}^{n-1}(ar)^i$ as a right inverse. It follows from (2) of Proposition 6.1.7 that $a \in J(R)$. A similar proof holds if $\mathfrak{n}$ is a nil left ideal of R. $\square$

Corollary 6.2.8. *If $\mathfrak{n}$ is a nilpotent left or right ideal of R, then $\mathfrak{n} \subseteq J(R)$.*

The following proposition shows the connection between strongly nilpotent elements and nilpotent elements of a ring.

Proposition 6.2.9. *If $a \in R$ is strongly nilpotent, then a is nilpotent. Conversely, if R is commutative, then every nilpotent element of R is strongly nilpotent.*

Proof. Suppose that $a \in R$ is strongly nilpotent. Then every sequence $a_0, a_1, a_2, \ldots$, where $a = a_0$ and $a_{n+1} \in a_n R a_n$, for $n = 0, 1, 2, \ldots$, is eventually zero. Choose the sequence defined by $a = a_0$ and $a_{n+1} = a_n^2 = a_n 1 a_n \in a_n R a_n$ for each integer $n \geq 0$. Then $a_1 = a^2, a_2 = a^4, a_3 = a^8, \ldots, a_n = a^{2^n}, \ldots$. But $a_n = 0$ for some integer $n \geq 0$ which shows that a is nilpotent.

Conversely, suppose that R is commutative and that $a \in R$ is nilpotent. We claim that a is strongly nilpotent. Consider a sequence $a_0, a_1, a_2, \ldots$, where $a = a_0$ and $a_{n+1} \in a_n R a_n$, for $n = 0, 1, 2, \ldots$. Then

$$a_1 = a r_0 a = r_0 a^2$$

$$a_2 = a_1 r_1 a_1 = r_1 a_1^2 = r_1 r_0^2 a^4$$

$$a_3 = a_2 r_2 a_2 = r_2 a_2^2 = r_2 r_1^2 r_0^4 a^8$$

$$\vdots$$

$$a_n = r_{n-1} r_{n-2}^2 \cdots r_1^{2^{n-2}} r_0^{2^{n-1}} a^{2^n}$$

$$\vdots$$

Since a is nilpotent, there is an integer m such that $a^m = 0$. If n is chosen so that $2^n \geq m$, then $a_n = 0$ and so a is strongly nilpotent.   □

Corollary 6.2.10. *The prime radical of any ring is a nil ideal of R and* $\mathrm{rad}(R) \subseteq J(R)$.

Proof. In view of Proposition 6.2.5, the fact that $\mathrm{rad}(R)$ is nil follows directly from the proposition and Proposition 6.2.7 shows that $\mathrm{rad}(R) \subseteq J(R)$.   □

Corollary 6.2.11. *If R is a commutative ring, then $\mathrm{rad}(R)$ is the set of all nilpotent elements of R.*

In the opening remarks of this chapter, it was pointed out that $\mathrm{rad}(R)$ contains every nilpotent right ideal of R. The following proposition shows that this is actually the case.

Proposition 6.2.12. *The prime radical of R contains all the nilpotent right ideals and all the nilpotent left ideals of R.*

Proof. If $\mathfrak{n}$ is a nilpotent right ideal of R, then $\mathfrak{n}^n = 0$ for some integer $n \geq 1$. Hence, $\mathfrak{n}^n \subseteq \mathrm{rad}(R)$ which implies that $\mathfrak{n}^n \subseteq \mathfrak{p}$ for every prime ideal $\mathfrak{p}$ of R. If $\mathfrak{n} \nsubseteq \mathfrak{p}$, then $\mathfrak{n}^{n-1} \subseteq \mathfrak{p}$. But $\mathfrak{n}^{n-1} \subseteq \mathfrak{p}$ implies that $\mathfrak{n}^{n-2} \subseteq \mathfrak{p}$ since $\mathfrak{n} \nsubseteq \mathfrak{p}$. Continuing in this way, we eventually come to $\mathfrak{n}^2 \subseteq \mathfrak{p}$, so $\mathfrak{n} \subseteq \mathfrak{p}$, a contradiction. Therefore, it must have been the case that $\mathfrak{n} \subseteq \mathfrak{p}$ to begin with. Hence, $\mathfrak{n}$ is contained in every prime ideal of R and so $\mathfrak{n} \subseteq \mathrm{rad}(R)$. The proof for nilpotent left ideals follows just as easily.   □

Remark. It was pointed out in the opening remarks of Section 6.1 that numerous types of radicals have emerged over the years each with a different foundation. For example, the *upper* and the *Levitzki nil radical* are also useful in the study of the structure of rings. Details on these radicals can be found in [44].

A proper ideal $\mathfrak{p}$ in a commutative ring R is prime if and only if $S = R - \mathfrak{p}$ is a multiplicative system. An analogue holds for noncommutative rings with regard to a generalized multiplicative system called an *m-system* [25].

Definition 6.2.13. A nonempty subset $\mathfrak{M}$ of R is said to be an *m-system* (or a generalized multiplicative system) if $0 \notin \mathfrak{M}$ and if for each $a, b \in \mathfrak{M}$, there is an $r \in R$ such that $arb \in \mathfrak{M}$.

Proposition 6.2.14. *A proper ideal $\mathfrak{p}$ of R is prime if and only if $R - \mathfrak{p}$ is an m-system.*

Proof. Suppose that $\mathfrak{p}$ is a prime ideal of R. If $a, b \in R - \mathfrak{p}$, then $a, b \notin \mathfrak{p}$. By (2) of Proposition 6.2.2, it must be the case that $aRb \nsubseteq \mathfrak{p}$. This gives an $r \in R$ such that $arb \notin \mathfrak{p}$, so there is an $r \in R$ such that $arb \in R - \mathfrak{p}$.

Conversely, suppose that $R - \mathfrak{p}$ is an m-system and let $a, b \in R$. If $aRb \subseteq \mathfrak{p}$, then we must show that either $a \in \mathfrak{p}$ or $b \in \mathfrak{p}$. If neither a nor b is in $\mathfrak{p}$, then $a, b \in R - \mathfrak{p}$. But $R - \mathfrak{p}$ is an m-system, so there is an $r \in R$ such that $arb \in R - \mathfrak{p}$. Hence, $aRb \not\subseteq \mathfrak{p}$, a contradiction. Therefore, if $aRb \subseteq \mathfrak{p}$, then either $a \in \mathfrak{p}$ or $b \in \mathfrak{p}$, so $\mathfrak{p}$ is a prime ideal of R. $\qquad\qquad\qquad\qquad\qquad\qquad\qquad\qquad\qquad\qquad\qquad\qquad\qquad\quad$ □

Clearly, a multiplicative system in a commutative R is an m-system. However, an m-system need not be multiplicatively closed. For example, if a is a nonzero nonunit of an integral domain R, then $\mathfrak{M} = \{a, a^2, a^4, a^8, \ldots\}$ is an m-system that is not multiplicatively closed.

Proposition 6.2.15. *If $\mathfrak{M}$ is an m-system in R, then an ideal that is maximal with respect to having empty intersection with $\mathfrak{M}$ is a prime ideal of R.*

Proof. First, we show that ideals that satisfy this property do indeed exist. Let $\mathcal{S}$ be the set of all ideals of R that are disjoint from $\mathfrak{M}$. The set $\mathcal{S}$ is nonempty since $0 \in \mathcal{S}$. Moreover, if $\mathcal{S}$ is partially ordered by inclusion, then it is easy to verify that $\mathcal{S}$ is inductive. Hence, Zorn's lemma shows that $\mathcal{S}$ has a maximal element, say $\mathfrak{p}$. Thus, $\mathfrak{p}$ is an ideal of R that is maximal with respect to having empty intersection with $\mathfrak{M}$.

Finally, we claim that $\mathfrak{p}$ is a prime ideal of R. Suppose that $a \notin \mathfrak{p}$, $b \notin \mathfrak{p}$ and $aRb \subseteq \mathfrak{p}$. Then RaR and RbR are ideals of R, so by the maximality of $\mathfrak{p}$, $RaR + \mathfrak{p}$ and $RbR + \mathfrak{p}$ have nonempty intersection with $\mathfrak{M}$. If

$$a' \in (RaR + \mathfrak{p}) \cap \mathfrak{M} \quad \text{and}$$
$$b' \in (RbR + \mathfrak{p}) \cap \mathfrak{M},$$

choose $r \in R$ to be such that $a'rb' \in \mathfrak{M}$. Then

$$a'rb' \in (RaR)(RbR) + \mathfrak{p} \subseteq \mathfrak{p}$$

and so $\mathfrak{p} \cap \mathfrak{M} \neq \varnothing$, a contradiction. Hence, if $aRb \subseteq \mathfrak{p}$, then $a \in \mathfrak{p}$ or $b \in \mathfrak{p}$ and so $\mathfrak{p}$ is prime. $\qquad\qquad\qquad\qquad\qquad\qquad\qquad\qquad\qquad\qquad\qquad\qquad\qquad\qquad\qquad\quad$ □

Definition 6.2.16. If I is an ideal of R, then the *radical* of I is the set $\sqrt{I} = \{a \in R \mid$ every m-system containing a has nonempty intersection with $I\}$. If $\sqrt{I} = I$, then I is said to be a *radical ideal*.

Proposition 6.2.17. *If I is an ideal of R, then $\sqrt{I}$ is the intersection of the prime ideals of R that contain I.*

Proof. Let $\mathcal{P}$ be the set of prime ideals of R that contain I. If $a \in \sqrt{I}$ and $\mathfrak{p}$ is a prime ideal of R that contains I, then $R - \mathfrak{p} \subseteq R - I$. (Such a prime ideal of R exists, for if $\mathfrak{m}$ is a maximal ideal of R that contains I, then, due to Exercise 3, $\mathfrak{m}$ is a prime ideal of R.) Since $R - \mathfrak{p}$ is an m-system, this system cannot contain a for if it does, then it has nonempty intersection with I. Consequently, $R - I$ would have a nonempty intersection with I, a contradiction. Thus, $a \in \mathfrak{p}$. Hence, each $a \in \sqrt{I}$ is in every prime ideal containing I, so $\sqrt{I} \subseteq \bigcap_{\mathfrak{p}} \mathfrak{p}$.

Conversely, suppose that $a \notin \sqrt{I}$. Then there is an m-system $\mathfrak{M}$ containing a that has empty intersection with I. If $\mathfrak{p}$ is an ideal of R containing I that is maximal with respect to the property of having empty intersection with $\mathfrak{M}$, then, by Proposition 6.2.15, $\mathfrak{p}$ is prime. Consequently, $a \notin \mathfrak{p}$, so $a \notin \bigcap_{\mathfrak{p}} \mathfrak{p}$. Hence, $a \in \bigcap_{\mathfrak{p}} \mathfrak{p}$ gives $a \in \sqrt{I}$ and so we have $\bigcap_{\mathfrak{p}} \mathfrak{p} \subseteq \sqrt{I}$. □

Corollary 6.2.18. *For any ideal I of a ring R, $\sqrt{I} = \mathrm{rad}(R/I)$, so $\sqrt{0} = \mathrm{rad}(R)$.*

Semiprime Rings

If $\mathfrak{p}$ is a prime ideal of R and I is an ideal of R such that $I^2 \subseteq \mathfrak{p}$, then $I \subseteq \mathfrak{p}$. However, the condition $I^2 \subseteq \mathfrak{p}$ implies $I \subseteq \mathfrak{p}$ does not mean that $\mathfrak{p}$ is a prime ideal. We use this "weaker" condition to define the concept of a semiprime ideal.

Definition 6.2.19. A proper ideal $\mathfrak{s}$ of R is said to be a *semiprime ideal* of R if whenever I is an ideal of R such that $I^2 \subseteq \mathfrak{s}$, then $I \subseteq \mathfrak{s}$. A ring R is said to be a *semiprime ring* if the zero ideal is a semiprime ideal of R.

Clearly any prime ideal of R is semiprime. The proof of the following proposition is similar to the proof of Proposition 6.2.2, so the proof is omitted. The proposition gives several conditions that can be used to show that an ideal of R is semiprime.

Proposition 6.2.20. *The following are equivalent for a proper ideal $\mathfrak{s}$ of R.*

(1) $\mathfrak{s}$ *is semiprime.*

(2) *If $a \in R$ and $aRa \subseteq \mathfrak{s}$, then $a \in \mathfrak{s}$.*

(3) *If A is a right ideal of R and $A^2 \subseteq \mathfrak{s}$, then $A \subseteq \mathfrak{s}$.*

(4) *If A is a left ideal of R and $A^2 \subseteq \mathfrak{s}$, then $A \subseteq \mathfrak{s}$.*

An m-system was used to characterize prime ideals of R. Likewise, an *n-system* can be defined and used to characterize semiprime ideals.

Definition 6.2.21. A nonempty subset $\mathfrak{N}$ of R is said to be an *n-system* if $0 \notin \mathfrak{N}$ and if for each $a \in \mathfrak{N}$ there is an $r \in R$ such that $ara \in \mathfrak{N}$.

The proof of the following proposition is similar to that of Proposition 6.2.14 and so is left as an exercise.

Proposition 6.2.22. *A proper ideal $\mathfrak{s}$ of R is semiprime if and only if $R - \mathfrak{s}$ is an n-system.*

Lemma 6.2.23. *Let $\mathfrak{N}$ be an n-system in a ring R and suppose that $a \in \mathfrak{N}$. Then there is an m-system $\mathfrak{M} \subseteq \mathfrak{N}$ such that $a \in \mathfrak{M}$.*

Proof. Suppose that $\mathfrak{N}$ is an n-system and let $a \in \mathfrak{N}$. Next, let $a_1 = a$ and for each integer $n \geq 1$, suppose that $r_n \in R$ is such that $a_{n+1} = a_n r_n a_n \in \mathfrak{N}$. Let $\mathfrak{M} = \{a_1, a_2, a_3, \ldots\}$. By construction, $\mathfrak{M} \subseteq \mathfrak{N}$ and $a \in \mathfrak{M}$, so the proof will be complete if we can show that $\mathfrak{M}$ is an m-system. To show this, it suffices to show that for any $a_i, a_j \in \mathfrak{M}$, there is an $r \in R$ such that $a_i r a_j \in \mathfrak{M}$. If $i \leq j$, then $a_{i+1} = a_i r_i a_i, a_{i+2} = a_{i+1} r_{i+1} a_{i+1}, \ldots, a_{j+1} = a_j r_j a_j$. Hence,

$$a_{j+1} = a_j r_j a_j = (a_{j-1} r_{j-1} a_{j-1}) r_j a_j = \cdots$$
$$= a_i (r_i a_i \cdots r_{j-1} a_{j-1} r_j) a_j \in a_i R a_j,$$

so there is an $r \in R$ such that $a_i r a_j = a_{j+1} \in \mathfrak{M}$. Thus, $\mathfrak{M}$ is an m-system. $\square$

Proposition 6.2.24. *The following are equivalent for a proper ideal $\mathfrak{s}$ of R.*

(1) $\mathfrak{s}$ is semiprime.

(2) $\mathfrak{s}$ is a radical ideal.

(3) $\mathfrak{s}$ is an intersection of prime ideals.

Proof. (2) $\Rightarrow$ (3) is Proposition 6.2.17 and (3) $\Rightarrow$ (1) is clear since it is easy to show that the intersection of any collection of prime ideals is a semiprime ideal. Hence, we need to show that (1) $\Rightarrow$ (2). Since $\sqrt{\mathfrak{s}}$ is the intersection of the prime ideals of R that contain $\mathfrak{s}$, $\mathfrak{s} \subseteq \sqrt{\mathfrak{s}}$, so we are only required to show that $\sqrt{\mathfrak{s}} \subseteq \mathfrak{s}$. If $a \notin \mathfrak{s}$, then $a \in R - \mathfrak{s}$. Proposition 6.2.22 indicates that $R - \mathfrak{s}$ is an n-system and, by Lemma 6.2.23, there is an m-system $\mathfrak{M} \subseteq R - \mathfrak{s}$ such that $a \in \mathfrak{M}$. But $\mathfrak{M}$ has an empty intersection with $\mathfrak{s}$, so Definition 6.2.16 gives $a \notin \sqrt{\mathfrak{s}}$. Hence, $\sqrt{\mathfrak{s}} \subseteq \mathfrak{s}$ and so $\mathfrak{s}$ is a radical ideal of R. $\square$

Proposition 6.2.25. *A ring R is semiprime if and only if $\mathrm{rad}(R) = 0$.*

Proof. Suppose that $\mathrm{rad}(R) = 0$, then zero is a semiprime ideal of R, so R is a semiprime ring. Conversely, if R is a semiprime ring, then, by Proposition 6.2.24, zero is a radical ideal of R. Hence, by Corollary 6.2.18, $0 = \sqrt{0} = \mathrm{rad}(R)$. $\square$

Corollary 6.2.26. *For any ring R, $R/\mathrm{rad}(R)$ is a semiprime ring. In particular, $\mathrm{rad}(R)$ is the smallest ideal I of R such that R/I is semiprime.*

Proof. The fact that $R/\mathrm{rad}(R)$ is semiprime is immediate since $\mathrm{rad}(R/\mathrm{rad}(R)) = 0$. (See Exercise 2.) If I is an ideal of R such that R/I is semiprime, then $\mathrm{rad}(R/I) = 0$ and so the intersection of the prime ideals of R/I is zero. But there is a one-to-one correspondence between the prime ideals of R containing I and the prime ideals of R/I. It follows that R/I is semiprime if and only if I is the intersection of the prime ideals of R that contain I. Clearly $\mathrm{rad}(R)$ is the smallest such ideal of R. $\square$

We conclude our discussion of radicals with the following proposition.

Proposition 6.2.27. *The following are equivalent for a ring R.*

(1) *R is semiprime.*

(2) *The zero ideal is the only nilpotent ideal of R.*

(3) *If A and B are right (left) ideals of R such that $AB = 0$, then $A \cap B = 0$.*

Proof. (1) $\Rightarrow$ (3). If A and B are right ideals of R such that $AB = 0$, then $AB \subseteq \mathfrak{p}$ for every prime ideal $\mathfrak{p}$ of R. Hence, $A \subseteq \mathfrak{p}$ or $B \subseteq \mathfrak{p}$ and so $A \cap B \subseteq \mathfrak{p}$ for every prime $\mathfrak{p}$. Thus, $A \cap B \subseteq \mathrm{rad}(R) = 0$.

(3) $\Rightarrow$ (2). Let $\mathfrak{n}$ be a nilpotent ideal of R. If $\mathfrak{n}^n = 0$, then it follows from (3) that $\mathfrak{n} = \mathfrak{n}_1 \cap \mathfrak{n}_2 \cap \cdots \cap \mathfrak{n}_n = 0$, where $\mathfrak{n}_i = \mathfrak{n}$ for $i = 1, 2, \ldots, n$.

(2) $\Rightarrow$ (1). If $0 \neq a \in R$, let $a = a_0$. Then $Ra_0R \neq 0$ and so the ideal Ra_0R is not nilpotent. Hence, we can pick $a_1 \in Ra_0R$, $a_1 \neq 0$. For the same reasons, we can select a nonzero $a_2 \in Ra_1R$ and so on. Thus, a is not strongly nilpotent, so $a \notin \mathrm{rad}(R)$. Hence, $\mathrm{rad}(R) = 0$ and R is therefore semiprime. $\square$

Problem Set 6.2

1. (a) If A and B are nilpotent ideals of R, prove that $A + B$ is also a nilpotent ideal of R. [Hint: If $A^m = B^n = 0$ and $a_1 + b_1, a_2 + b_2 \in A + B$, then $(a_1 + b_1)(a_2 + b_2) = a_1a_2 + b_1a_2 + a_1b_1 + b_1b_2 = a_1a_2 + b$, where $b = b_1a_2 + a_1b_1 + b_1b_2 \in B$. Show that if $a_1 + b_1, a_2 + b_2, \ldots, a_m + b_m \in A + B$, then $(a_1 + b_1)(a_2 + b_2) \cdots (a_m + b_m) = a_1a_2 \cdots a_m + b$ for some $b \in B$. But $A^m = 0$ gives $a_1a_2 \cdots a_m = 0$, so $(a_1 + b_1)(a_2 + b_2) \cdots (a_m + b_m) \in B$. Hence, $(A + B)^m \subseteq B$ indicates that $(A + B)^{mn} = 0$.]

   (b) Show that if $A_1, A_2, \ldots, A_n$ are nilpotent ideals of R, then $A_1 + A_2 + \cdots + A_n$ is nilpotent.

2. Verify that rad as defined in Definition 6.2.3 satisfies the following conditions:

   (a) If $f : R \to R'$ is a ring homomorphism, then $f(\mathrm{rad}(R)) \subseteq \mathrm{rad}(R')$. [Hint: If $\mathfrak{p}'$ is a prime ideal of R, show that $f^{-1}(\mathfrak{p}')$ is a prime ideal of R.]

   (b) $\mathrm{rad}(R/\mathrm{rad}(R)) = 0$ for every ring R.

3. (a) Prove that a maximal ideal $\mathfrak{m}$ in a ring R is a prime ideal of R. [Hint: Let A and B be ideals of R not contained in $\mathfrak{m}$ and consider $(A + \mathfrak{m})(B + \mathfrak{m}) = AB + \mathfrak{m}$.]

   (b) Prove that a ring R is prime if and only if for all nonzero $a, b \in R$, there is an $r \in R$ such that $arb \neq 0$.

4. (a) Let $\mathfrak{p}$ be an ideal of a ring R. Prove that $\mathfrak{p}$ is a prime ideal of R if and only if $\mathbb{M}_n(\mathfrak{p})$ is a prime ideal of $\mathbb{M}_n(R)$. [Hint: Exercise 12 in Problem Set 1.2.] Conclude that R is a prime ring if and only if $\mathbb{M}_n(R)$ is prime.

   (b) Show that $\mathbb{M}_n(\mathrm{rad}(R)) = \mathrm{rad}(\mathbb{M}_n(R))$. Conclude that R is semiprime if and only if $\mathbb{M}_n(R)$ is semiprime.

5. Prove Proposition 6.2.20.

6. Prove Proposition 6.2.22.

7. Fill in the details of the proof of $(3) \Rightarrow (2)$ in Proposition 6.2.27.

8. Prove for any ideal I of R that $\sqrt{I}$ is the smallest semiprime ideal of R that contains I.

9. (a) If R is a commutative ring and I is an ideal of R, prove that $\sqrt{I} = \{a \in R \mid a^n \in I \text{ for some positive integer } n\}$.

   (b) If p is a prime number, show that $\sqrt{(p^m)} = (p)$ in the ring $\mathbb{Z}$.

10. A ring R is a *subdirect product* of a family $\{R_\alpha\}_\Delta$ of rings if there is an injective ring homomorphism $\varphi : R \to \prod_\Delta R_\alpha$ such that $\pi_\alpha \varphi$ is a surjective ring homomorphism for each $\alpha \in \Delta$, where $\pi_\alpha : \prod_\Delta R_\alpha \to R_\alpha$ is the canonical projection.

    (a) Show that R is a subdirect product of a family $\{R_\alpha\}_\Delta$ of rings if and only if there is a family $\{I_\alpha\}_\Delta$ of ideals of R such that $R_\alpha \cong R/I_\alpha$ for each $\alpha \in \Delta$ and $\bigcap_\Delta I_\alpha = 0$.

    (b) Prove that a ring R is a subdirect product of prime rings if and only if $\mathrm{rad}(R) = 0$. Conclude that a ring is a semiprime ring if and only if it is a subdirect product of prime rings.

11. Prove each of the following for a commutative ring R.

    (a) R is a subdirect product of fields if and only if R is Jacobson semisimple.

    (b) R is semiprime if and only if R is a subdirect product of integral domains.

12. A ring R is said to be *subdirectly irreducible* if the intersection of the nonzero ideals of R is not zero.

    (a) Prove that R is subdirectly irreducible if and only if whenever R is the subdirect product of a family $\{R_\alpha\}_\Delta$ of rings, one of the rings R_α is isomorphic to R.

    (b) Show that every ring R is a subdirect product of subdirectly irreducible rings. [Hint: For each $a \in R$, $a \neq 0$, show that there is an ideal A_a which is maximal among the ideals contained in $R - \{a\}$. Then $\bigcap_{a \in R, a \neq 0} A_a = 0$, so consider the family of rings $\{R/A_a\}_{a \in R, a \neq 0}$.]

6.3   Radicals and Chain Conditions

Proposition 6.3.1. *If R is a right artinian ring, then $J(R)$ is nilpotent.*

Proof. Let $J = J(R)$. Since R is right artinian, the decreasing chain $J \supseteq J^2 \supseteq J^3 \supseteq \cdots$ must terminate. Suppose that $n \geq 2$ is the smallest positive integer such that $J^n = J^{n+1} = \cdots$. Then $J^{2n} = J^n$. Assume that $J^n \neq 0$ and let A be minimal among the right ideals of R contained in J^n such that $AJ^n \neq 0$. Next, let $a \in A$ be such that $aJ^n \neq 0$. Now $aJ^n \subseteq A \subseteq J^n$ and $(aJ^n)J^n = aJ^{2n} = aJ^n$, so $(aJ^n)J^n \neq 0$. Hence, by the minimality of A, $aJ^n = A$. Finally, let $b \in J^n \subseteq J$ be such that $ab = a$. Then (2) of Proposition 6.1.7 shows there is a $c \in R$ such that $1 = (1-b)c$. This gives $a = a(1-b)c = (a-ab)c = 0$, an obvious contradiction. Hence, $J^n = 0$.   □

Corollary 6.3.2. *If R is a right artinian ring, then $J(R)$ is the largest nilpotent left ideal of R as well as the largest nilpotent right ideal of R.*

Proof. Corollary 6.2.8 shows that every nilpotent left ideal and every nilpotent right ideal of R is contained in $J(R)$.   □

Corollary 6.3.3. *If R is a right artinian ring, then any nil left or nil right ideal of R is nilpotent.*

Proof. By referring to Proposition 6.2.7, we see that every nil right (left) ideal $\mathfrak{n}$ of R is contained in $J(R)$. So $\mathfrak{n}^n \subseteq J(R)^n = 0$ for some positive integer n.   □

Corollary 6.3.4. *If R is a right artinian ring, then $\mathrm{rad}(R) = J(R)$.*

Proof. Corollary 6.2.10 shows that $\mathrm{rad}(R) \subseteq J(R)$ and Proposition 6.2.12 indicates that $\mathrm{rad}(R)$ contains all the nilpotent ideals of R.   □

We now need the following lemma.

Lemma 6.3.5. *The following hold for any ring R.*

(1) *If A is a minimal right ideal of R, then either $A^2 = 0$ or $A = eR$ for some idempotent e of R.*

(2) *If R is semiprime, then every minimal right ideal of R is generated by an idempotent.*

Proof. (1) Let A be a minimal right ideal of R and assume that $A^2 \neq 0$. Then there is an $a \in A$ such that $aA \neq 0$. Since aA is a nonzero right ideal of R contained in A, we must have $A = aA$. Let $e \in A$ be such that $a = ae$. If $B = \mathrm{ann}_r(a)$, then B is a right ideal of R and $A \cap B \neq A$. Hence, $A \cap B = 0$. But $ae = ae^2$, so $a(e - e^2) = 0$.

Therefore, $e - e^2 \in A \cap B$, so $e = e^2$. Hence, e is an idempotent of R and $e \neq 0$, since $a \neq 0$. Thus, $0 \neq eR \subseteq A$ gives $eR = A$.

(2) Suppose that R is semiprime and let A be a minimal right ideal of R. Then $A^2 = 0$ would imply that $A = 0$ since zero is a semiprime ideal of R. This shows that $A^2 \neq 0$, so (1) indicates that there is an idempotent $e \in R$ such that $A = eR$. □

When R is right artinian, we have seen in Corollaries 6.3.2 and 6.3.4 that $\mathrm{rad}(R)$ is the largest nilpotent left ideal as well as the largest nilpotent right ideal of R. We can now show that if R is right noetherian, then the same result also holds for $\mathrm{rad}(R)$.

Proposition 6.3.6. *If R is a right noetherian ring, then $\mathrm{rad}(R)$ is nilpotent and, in fact, when R is right noetherian, $\mathrm{rad}(R)$ is the largest nilpotent ideal of R.*

Proof. If R is right noetherian, then because of Proposition 4.2.3 any nonempty collection of right ideals of R, when ordered by inclusion, has a maximal element. So among all the nilpotent ideals of R, there is at least one, say $\mathfrak{n}$, that is maximal. If $\mathfrak{n}'$ is another nilpotent ideal of R, then Exercise 1 in Problem Set 6.2 shows that $\mathfrak{n} + \mathfrak{n}'$ is also a nilpotent ideal of R. But $\mathfrak{n} \subseteq \mathfrak{n} + \mathfrak{n}'$, so the maximality of $\mathfrak{n}$ implies that $\mathfrak{n} = \mathfrak{n} + \mathfrak{n}'$. Therefore, $\mathfrak{n}' \subseteq \mathfrak{n}$ and we see that $\mathfrak{n}$ is the largest nilpotent ideal of R. Moreover, since $\mathfrak{n}$ is nilpotent, it follows from Proposition 6.2.12 that $\mathfrak{n} \subseteq \mathrm{rad}(R)$. Hence, $\mathrm{rad}(R)$ contains every nilpotent ideal of R. The proof will be complete if we can show that $\mathrm{rad}(R)$ is nilpotent. We show first that the ring $R/\mathfrak{n}$ is semiprime. Let $A + \mathfrak{n}$ be a nilpotent ideal of $R/\mathfrak{n}$. Then $A^n + \mathfrak{n} = (A + \mathfrak{n})^n = 0$ for some integer n. Hence, $A^n \subseteq \mathfrak{n}$ and $A^{nm} \subseteq \mathfrak{n}^m = 0$. Thus, A is nilpotent and we have $A \subseteq \mathfrak{n}$. Therefore, $R/\mathfrak{n}$ has no nonzero nilpotent ideals and Proposition 6.2.27 indicates that $R/\mathfrak{n}$ is semiprime. Using Corollary 6.2.26, we see that $\mathrm{rad}(R)$ is the smallest ideal of R such that $R/\mathrm{rad}(R)$ is semiprime, so $\mathrm{rad}(R) \subseteq \mathfrak{n}$. Hence, we have $\mathrm{rad}(R) = \mathfrak{n}$ which shows that $\mathrm{rad}(R)$ is the largest nilpotent ideal of R. □

We have seen in Corollary 6.3.3 that if R is a right artinian ring, then every nil left ideal as well as every nil right ideal of R is nilpotent. This property is also shared by right noetherian rings.

Proposition 6.3.7. *If R is a right noetherian ring, then every nil left ideal and every nil right ideal of R is nilpotent.*

Proof. We begin by assuming that R is semiprime. Let $\mathfrak{n}$ be a nil left ideal of R and assume that R is right noetherian. Then the set of right ideals of R of the form $\mathrm{ann}_r(c)$, where $c \in \mathfrak{n}, c \neq 0$, has a maximal element, say $\mathrm{ann}_r(a)$. Choose any $b \in R$ such that $0 \neq ba \in \mathfrak{n}$ and let k be the smallest positive integer such that $(ba)^k = 0$. Then $(ba)^{k-1} \neq 0$ and we clearly have $\mathrm{ann}_r(a) \subseteq \mathrm{ann}_r(ba)^{k-1}$. Thus, $\mathrm{ann}_r(a) = \mathrm{ann}_r(ba)^{k-1}$ by the maximality of $\mathrm{ann}_r(a)$. It follows that $ba \in \mathrm{ann}_r(a)$, so $aba = 0$. This shows that $aRa = 0$ and we have $a = 0$ since R is semiprime.

Therefore, $\mathfrak{n} = 0$ and so a semiprime right noetherian ring has no nonzero nil left ideals. Next, suppose that $\mathfrak{n}$ is a nil right ideal of R. If $a \in \mathfrak{n}$, then aR is a nil right ideal for R. Note that Ra is a nil left ideal of R, for if $(ab)^n = 0$, then $(ba)^{n+1} = 0$. Hence, it is also the case that a semiprime right noetherian ring can have no nonzero nil right ideals. Finally, drop the assumption that R is semiprime and suppose that $\mathfrak{n}$ is a nil left (or right) ideal of R. Then $(\mathfrak{n} + \mathrm{rad}(R))/\mathrm{rad}(R)$ is a nil left (or right) ideal of the semiprime right noetherian ring $R/\mathrm{rad}(R)$, so $(\mathfrak{n} + \mathrm{rad}(R))/\mathrm{rad}(R)$ must be zero. Thus, $\mathfrak{n} \subseteq \mathrm{rad}(R)$ and we have by Proposition 6.3.6 that there is an integer m such that $\mathfrak{n}^m \subseteq \mathrm{rad}(R)^m = 0$.   □

Corollary 6.3.8. *If R is a right noetherian ring, then $\mathrm{rad}(R)$ is the largest nil left ideal as well as the largest nil right ideal of R.*

Proof. Corollary 6.2.10 shows that $\mathrm{rad}(R)$ is a nil ideal of R and we saw in the proof of the proposition that if $\mathfrak{n}$ is a nil left or right ideal of R, then $\mathfrak{n} \subseteq \mathrm{rad}(R)$.   □

Problem Set 6.3

1. Let $f : R \to S$ be a surjective ring homomorphism and suppose that R is a right artinian (right noetherian) ring. Prove that S is right artinian (right noetherian).

2. (a) Prove that every finite ring direct product of right noetherian (right artinian) rings is right noetherian (right artinian). [Hint: Exercise 3 in Problem Set 2.1.]
   (b) Let $\{E_{ij}\}_{i,j=1}^{n}$ be the matrix units of $M_n(R)$. If A is a right ideal of $M_n(R) = \bigoplus_{i,j=1}^{n} E_{ij}R$, prove that $E_{11}A$ is an R-submodule of the free R-module $E_{11}M_n(R) = \bigoplus_{j=1}^{n} E_{1j}R$. Moreover, show that if $\overline{A}$ is an R-submodule of $E_{11}M_n(R)$, then $\sum_{i=1}^{n} E_{i1}\overline{A}$ is a right ideal of $M_n(R)$. Show also that this establishes an order preserving one-to-one correspondence among the right ideals of $M_n(R)$ and the R-submodules of $E_{11}M_n(R)$.
   (c) Show that a ring R is right noetherian (artinian) if and only if the matrix ring $M_n(R)$ is right noetherian (artinian). [Hint: Use (a) and (b).]

3. Suppose that R contains a subring D that is a division ring and that R_D is a finite dimensional vector space over D. Show that R is right artinian. [Hint: Note that D is an IBN-ring and if A and B are right ideals of R such that $A \supsetneq B$, then A_D and B_D are subspaces of R_D and $A_D \supsetneq B_D$.]

4. Prove that a right artinian ring without zero divisors is a division ring. [Hint: If $a \in R, a \neq 0$, consider $aR \supseteq a^2 R \supseteq a^3 R \supseteq \cdots$.]

5. When R is a commutative ring, deduce that the nilpotent elements of R form the ideal $\mathrm{rad}(R)$. Note that this result does not hold for noncommutative rings. To see this, consider the matrix ring $M_n(D)$, where D is a division ring and $n \geq 2$. Part (c) of Exercise 2 shows that $M_n(D)$ is a right noetherian ring and Proposition 6.3.6 shows that $\mathrm{rad}(M_n(D))$ is nilpotent. Furthermore, Part (b)

of Exercise 4 in Problem Set 6.2 indicates that $\mathbb{M}_n(D)$ is semiprime, so $\text{rad}(\mathbb{M}_n(D)) = 0$. Consequently, if the nilpotent elements of $\mathbb{M}_n(D)$ were to form $\text{rad}(\mathbb{M}_n(D))$, then $\mathbb{M}_n(D)$ would have no nonzero nilpotent elements. However, Exercise 5 in Problem Set 1.1 shows that $\mathbb{M}_n(D)$ has an abundance nonzero nilpotent elements.

6.4 Wedderburn–Artin Theory

If V is a vector space over a division ring D, then V can be decomposed as a direct sum of its 1-dimensional subspaces. Furthermore, every subspace is a direct summand of V and V is an injective and a projective D-module. Our goal is to classify rings over which every module will exhibit similar properties. Since an arbitrary R-module may fail to have a basis, we will see that, under certain conditions, the simple submodules of a given module will play a role in its decomposition similar to that played by the 1-dimensional subspaces in the decomposition of a vector space. As mentioned earlier, this theory is often referred to as the Wedderburn–Artin theory due to the work of Wedderburn [71] and Artin [47], [48].

Definition 6.4.1. The *socle* of an R-module M, denoted by $\text{Soc}(M)$, is the sum of the simple submodules of M. If M fails to have a simple submodule, then $\text{Soc}(M) = 0$. An R-module M is said to be *semisimple* (or *completely reducible*) if $M = \text{Soc}(M)$. *Semisimple left R-modules* are defined analogously. If every module in $\mathbf{Mod}_R$ ($_R\mathbf{Mod}$) is semisimple, then we will refer to $\mathbf{Mod}_R$ ($_R\mathbf{Mod}$) as a *semisimple category*. A ring R is said to be *right (left) semisimple* if R is semisimple as an (a left) R-module. A ring that is left and right semisimple will be referred to as a *semisimple ring*.

Examples

1. If D is a division ring, then $M = D \times D$ is a D-module under componentwise addition and the D-action on M given by $(x, y)a = (xa, ya)$. The submodules $D \times 0$ and $0 \times D$ are the simple submodules of M and $M = (D \times 0) \oplus (0 \times D)$. Hence, M is a semisimple D-module. In general, the direct product $\prod_{i=1}^n D_i$, with $D_i = D$ for each i, is a semisimple D-module under componentwise addition and a similar D-action as that defined on M.

2. If D is a division ring, then $\mathbb{M}_n(D)$ is a simple ring. To see this, observe that if $\bar{I}$ is an ideal of $M_n(D)$ and I is the set of all elements that are entries in the first row and first column of a matrix in $\bar{I}$, then I is a uniquely determined ideal of D such that $\bar{I} = M_n(I)$. Thus, $\bar{I} = 0$ or $\bar{I} = M_n(D)$ since 0 and D are the only ideals of D. It is also the case that $\mathbb{M}_n(D)$ is a left and a right semisimple ring with $\mathbb{M}_n(D) = \bigoplus_{i=1}^n c_i(D) = \bigoplus_{i=1}^n r_i(D)$, where each minimal left

ideal $c_i(D)$ is the set of all column matrices with entries from D in the ith column and zeroes everywhere else and each minimal right ideal $r_i(D)$ is the set of all row matrices in $\mathbb{M}_n(D)$ with entries from D in the ith row and zeroes elsewhere. Thus, $\mathbb{M}_n(D)$ is a semisimple ring. This property of $\mathbb{M}_n(D)$ is not out of the ordinary. We will see later that if a ring is left or right semisimple, then it is semisimple.

3. Since $\mathrm{Soc}(0) = 0$, the zero module is, by definition, semisimple.

Note that if the submodules of M are ordered by inclusion, then the minimal nonzero submodules of M under this ordering are just the simple submodules of M. The sum of the minimal submodules of M is dual to the intersection of the maximal submodules of M, that is, $\mathrm{Soc}(M)$ is dual to $\mathrm{Rad}(M)$. Note also that if two R-modules are isomorphic, then they have isomorphic socles. Thus, if one of the modules is semisimple, then the other module will also be semisimple.

We saw in Example 1 that the semisimple D-module M is not only the sum of its simple submodules but that the sum is actually direct. The following proposition shows that a similar case holds for a nonzero semisimple module M, although if $M = \bigoplus_\Delta S_\alpha$, then the family $\{S_\alpha\}_\Delta$ of simple submodules of M may be a proper subset of the collection of all simple submodules of M.

Proposition 6.4.2. *If the socle of an R-module M is nonzero, then $\mathrm{Soc}(M)$ is a direct sum of a subfamily of the family $\{S_\alpha\}_\Delta$ of the simple submodules of M. Moreover, if $f : M \to M$ is an R-linear mapping then $f(\mathrm{Soc}(M)) \subseteq \mathrm{Soc}(M)$; that is, $\mathrm{Soc}(M)$ is stable under every endomorphism of M.*

Proof. Suppose that $\mathrm{Soc}(M) \neq 0$ and let $\{S_\alpha\}_\Delta$ be the family of simple submodules of M. Then $\mathrm{Soc}(M) = \sum_\Delta S_\alpha$. Suppose next that $\mathcal{S}$ is the collection of subsets Γ of Δ such that the sum $\sum_\Gamma S_\alpha$ is direct. The set $\mathcal{S}$ is nonempty since singleton subsets of Δ are in $\mathcal{S}$. Partial order $\mathcal{S}$ by inclusion. If $\mathcal{C}$ is a chain in $\mathcal{S}$, then we claim that $\Lambda = \bigcup_\mathcal{C} \Gamma$ is in $\mathcal{S}$. Let $\sum_\Lambda x_\alpha \in \sum_\Lambda S_\alpha$. If $\sum_\Lambda x_\alpha = 0$, then the $x_\alpha \neq 0$ in this sum are at most finite in number. The fact that $\mathcal{C}$ is a chain, implies that there is a $\Gamma \in \mathcal{C}$ such that the subscripts of the possibly nonzero x_α are in Γ. But $\sum_\Gamma S_\alpha$ is direct, so these x_α must also be zero. Therefore, the sum $\sum_\Lambda S_\alpha$ is direct, so $\mathcal{S}$ is inductive. Apply Zorn's lemma and choose Γ^* to be a maximal element of $\mathcal{S}$. If $\sum_{\Gamma^*} S_\alpha \neq \mathrm{Soc}(M)$, then there is a simple submodule S_β of M such that S_β is not contained in the sum $\sum_{\Gamma^*} S_\alpha$. It follows that $S_\beta \cap \sum_{\Gamma^*} S_\alpha = 0$, so the sum $S_\beta + \sum_{\Gamma^*} S_\alpha$ is direct. But then the set $\Gamma^* \cup \{\beta\}$ contradicts the maximality of Γ^* and thus it must be the case that $\mathrm{Soc}(M) = \bigoplus_{\Gamma^*} S_\alpha$.

Finally, if $f : M \to M$ is an R-linear mapping and S is a simple submodule of M, then either $f(S) = 0$ or $f(S)$ is a simple submodule of M. It follows from this observation that $f(\mathrm{Soc}(M)) \subseteq \mathrm{Soc}(M)$.   □

Corollary 6.4.3. *For any R-module M, Soc(M) is the unique, largest semi-simple submodule of M.*

When is $\mathbf{Mod}_R$ a semisimple category? As it turns out, R is a right semisimple ring if and only if $\mathbf{Mod}_R$ is a semisimple category. We need the following two lemmas to establish this fact.

Lemma 6.4.4. *Let M be an R-module with the property that every submodule of M is a direct summand of M. Then every submodule of M also has this property.*

Proof. Let M be an R-module with the property that every submodule of M is a direct summand of M. Let N be a submodule of M and suppose that N_1 is a submodule of N. We claim that N_1 is a direct summand of N. Since N_1 is a direct summand of M, there is a submodule N_2 of M such that $M = N_1 \oplus N_2$. From this we have $N = N \cap M = N \cap (N_1 \oplus N_2)$. By the modularity property of modules, given in Example 10 of Section 1.4, we see that $N = N_1 \oplus (N \cap N_2)$, so N_1 is a direct summand of N.   □

Lemma 6.4.5. *An R-module M is semisimple if and only if every submodule of M is a direct summand of M.*

Proof. Suppose that M is a semisimple R-module and that $M = \bigoplus_\Lambda S_\alpha$, where S_α is a simple submodule of M for each $\alpha \in \Delta$. Let N be a submodule of M. If $N = 0$ or if $N = M$, then there is nothing to prove, so suppose N is proper and nonzero. Let $\mathscr{S}$ be the set of all subsets Γ of Δ such that $\bigoplus_\Gamma S_\alpha \cap N = 0$. The set $\mathscr{S}$ is nonempty since if $\mathscr{S} = \varnothing$, then $S_\alpha \subseteq N$ for each $\alpha \in \Delta$ which implies that $N = M$. Partial order $\mathscr{S}$ by inclusion and apply Zorn's lemma to obtain a subset Λ of Δ that is maximal in $\mathscr{S}$. If $\beta \in \Delta - \Lambda$, then $(S_\beta + \bigoplus_\Lambda S_\alpha) \cap N \neq 0$. So if $z \in (S_\beta + \bigoplus_\Lambda S_\alpha) \cap N$ is nonzero, let $x \in S_\beta$ and $y \in \bigoplus_\Lambda S_\alpha$ be such that $z = x + y$. Then $x = z - y \in (N + \bigoplus_\Lambda S_\alpha) \cap S_\beta$ and $x \neq 0$. Thus, $(N + \bigoplus_\Lambda S_\alpha) \cap S_\beta \neq 0$ and, since S_β is simple, $(N + \bigoplus_\Lambda S_\alpha) \cap S_\beta = S_\beta$. Consequently, $S_\beta \subseteq N + \bigoplus_\Lambda S_\alpha$. But this means that $M = \bigoplus_\Lambda S_\alpha \subseteq N + \bigoplus_\Lambda S_\alpha$, so $M = N + \bigoplus_\Lambda S_\alpha$ and $N \cap \bigoplus_\Lambda S_\alpha = 0$. Therefore, N is a direct summand of M.

To prove the converse, the first step is to show that every nonzero submodule of M contains a simple submodule. Let N be a nonzero submodule of M and suppose that $x \in N$, $x \neq 0$. Apply Zorn's lemma and let $\overline{N}$ be maximal among the submodules of N that do not contain x. Then by Lemma 6.4.4, there is a submodule S of N such that $N = \overline{N} \oplus S$. We claim that S is simple. If S is not simple, let S_1 be a proper nonzero submodule of S. Then by applying Lemma 6.4.4 again, we see that there is a nonzero submodule S_2 of S such that $S = S_1 \oplus S_2$. Hence, $N = \overline{N} \oplus S_1 \oplus S_2$. Since $\overline{N} \subsetneq \overline{N} \oplus S_1$ and $\overline{N} \subsetneq \overline{N} \oplus S_2$, it must be the case that $x \in \overline{N} \oplus S_1$ and $x \in \overline{N} \oplus S_2$. Therefore, there exist $x_1, x_2 \in \overline{N}$, $y_1 \in S_1$ and $y_2 \in S_2$ such that $x = x_1 + y_1$ and $x = x_2 + y_2$. Thus, $0 = (x_1 - x_2) + y_1 - y_2 \in \overline{N} \oplus S_1 \oplus S_2$ which gives $x_1 = x_2$ and $y_1 = y_2 = 0$. Consequently, $x = x_1 = x_2 \in \bar{N}$, contradicting the fact that $x \notin \bar{N}$, so S must be a simple R-module.

Finally, let $\{S_\alpha\}_\Delta$ be the family of simple submodules of M. Then as we saw in the proof of Proposition 6.4.2, there is a subset Γ^* of Δ that is maximal among the subsets Γ of Δ such that the sum $\sum_\Gamma S_\alpha$ is direct. If N is a nonzero submodule of M such that $M = N \oplus (\bigoplus_{\Gamma^*} S_\alpha)$, then from what was proved in the previous paragraph, we see that N contains a simple submodule S_β, $\beta \notin \Gamma^*$. Since the sum $S_\beta + \bigoplus_{\Gamma^*} S_\alpha$ is direct, the set $\Gamma^* \cup \{\beta\}$ contradicts the maximality of Γ^*. Hence, it must be the case that $N = 0$. Therefore, $M = \bigoplus_{\Gamma^*} S_\alpha$, so M is semisimple.   □

Corollary 6.4.6. *An R-module M is semisimple if and only if M has no proper essential submodules.*

Proof. If M is a semisimple R-module, then it is clear that M has no proper essential submodules. Conversely, if N is a submodule of M and N_c is a complement of N in M, then $N \oplus N_c$ is an essential submodule of M. Thus, $N \oplus N_c = M$, so N is a direct summand of M.   □

We are now in a position to give several conditions that will ensure that the ring is right semisimple.

Proposition 6.4.7. *The following are equivalent for a ring R.*

(1) *R is right semisimple.*

(2) *Every right ideal of R is a direct summand of R.*

(3) *$\mathbf{Mod}_R$ is a semisimple category.*

(4) *For every R-module M, each submodule of M is a direct summand of M.*

(5) *Every short exact sequence $0 \to M_1 \to M \to M_2 \to 0$ in $\mathbf{Mod}_R$ splits.*

(6) *Every R-module is injective.*

(7) *Every R-module is projective.*

Proof. The fact that (1) and (2) are equivalent follows from Lemma 6.4.5 as does the equivalence of (3) and (4). The equivalence of (4) and (5) is clear and the implication (5) $\Rightarrow$ (6) is obvious, since we have previously seen (Proposition 5.1.10) that there is an injective R-module E and an embedding $M \to E$. This gives a short exact sequence

$$0 \to M \to E \to E/M \to 0$$

that splits. The implication (6) $\Rightarrow$ (5) follows from Proposition 5.1.2, so (5) and (6) are equivalent. Baer's criteria shows that (2) $\Rightarrow$ (6) and (6) $\Rightarrow$ (2) follows from Proposition 5.1.2. Thus, we have that (1) through (6) are equivalent. To show that

(6) $\Rightarrow$ (7), let M be an R-module and consider the short exact sequence $0 \to K \to F \to M \to 0$, where F is a free R-module. Since K is injective, we have $F \cong K \oplus M$, so M is projective since F is. To see that (7) $\Rightarrow$ (6), consider the exact sequence

$$0 \to M \to E \to E/M \to 0,$$

where E is an injective R-module. Since E/M is projective, the sequence splits, so $E \cong M \oplus E/M$. Hence, M is injective, so (7) $\Rightarrow$ (6). $\square$

Corollary 6.4.8. *Every submodule and every homomorphic image of a semisimple module is semisimple.*

Proof. Lemmas 6.4.4 and 6.4.5 show that every submodule of a semisimple module is semisimple. If $f : M \to N$ is an epimorphism and M is a semisimple R-module, then $0 \to \operatorname{Ker} f \to M \to N \to 0$ splits since (4) of the proposition indicates that $\operatorname{Ker} f$ is a direct summand of M. Hence, N is isomorphic to a semisimple submodule of M and is thus semisimple. $\square$

Remark. Proposition 6.4.7 points out that a ring R is semisimple if and only if every R-module is projective, which in turn holds if and only if every R-module is injective. It also follows that R is semisimple if and only if every cyclic R-module is projective. (See Exercise 9.) A similar result holds for cyclic injective modules. Osofsky proved in [39] that a ring R is semisimple if and only if every cyclic R-module is injective.

We have previously seen that the radical of an R-module M is the sum of the small submodules of M. Dually, the socle of M can be described in terms of the essential (or large) submodules of M.

Proposition 6.4.9. *If $\{N_\alpha\}_\Delta$ is the family of essential submodules of an R-module M, then $\operatorname{Soc}(M) = \bigcap_\Delta N_\alpha$.*

Proof. If S is a simple submodule of M and N is an essential submodule of M, then $S \cap N$ is a nonzero submodule of S. Hence, $S = S \cap N \subseteq N$, so $\operatorname{Soc}(M) \subseteq N$. Thus, $\operatorname{Soc}(M) \subseteq \bigcap_\Delta N_\alpha$. Conversely, let $N = \bigcap_\Delta N_\alpha$ and suppose that X is a submodule of N. If X_c is a complement in M of X, then $X \oplus X_c$ is, by Proposition 5.1.5, an essential submodule of M. Since $X \subseteq N \subseteq X \oplus X_c$, we see, by modularity, that $N = N \cap (X \oplus X_c) = X \oplus (N \cap X_c)$. Hence, each submodule of N is a direct summand of N, so N is, by Lemma 6.4.5, a semisimple submodule of M. Thus, it follows from Corollary 6.4.3 that $N \subseteq \operatorname{Soc}(M)$. Therefore, $\operatorname{Soc}(M) = \bigcap_\Delta N_\alpha$. $\square$

Actually, a great deal more can be said about right semisimple rings. We will see that these rings have a nice structure in terms of matrix rings and this structure enables us to conclude that right semisimple rings are also left semisimple and conversely.

Definition 6.4.10. A set $\{e_1, e_2, \ldots, e_n\}$ of idempotents of a ring R is said to be an *orthogonal set of idempotents* of R, if $e_i e_j = 0$ for all i and j, $i \neq j$. If $1 = e_1 + e_2 + \cdots + e_n$, then $\{e_1, e_2, \ldots, e_n\}$ is a *complete set of idempotents* of R. An idempotent e of a ring R is said to be a *central idempotent* of R if e is in the center of R, that is, if $ea = ae$ for all $a \in R$.

Proposition 6.4.11. *The following hold for each right semisimple ring R.*

(1) *There exist minimal right ideals $A_1, A_2, \ldots, A_n$ of R such that*

$$R = A_1 \oplus A_2 \oplus \cdots \oplus A_n.$$

(2) *If $A_1, A_2, \ldots, A_n$ and $B_1, B_2, \cdots, B_m$ are minimal right ideals of R such that*

$$R = A_1 \oplus A_2 \oplus \cdots \oplus A_n \quad and \quad R = B_1 \oplus B_2 \oplus \cdots \oplus B_m,$$

then $n = m$ and there is a permutation $\sigma : \{1, 2, \ldots, n\} \to \{1, 2, \ldots, n\}$ such that $A_i \cong B_{\sigma(i)}$ for $i = 1, 2, \ldots, n$.

(3) *If $A_1, A_2, \ldots, A_n$ is a set of minimal right ideals of R such that*

$$R = A_1 \oplus A_2 \oplus \cdots \oplus A_n,$$

then there is a complete set $\{e_1, e_2, \ldots, e_n\}$ of orthogonal idempotents of R such that

$$R = e_1 R \oplus e_2 R \oplus \cdots \oplus e_n R$$

and $A_i = e_i R$ for $i = 1, 2, \ldots, n$. Furthermore, the idempotents in $\{e_1, e_2, \ldots, e_n\}$ are unique.

Proof. (1) Let R be a right semisimple ring and suppose that $R = \bigoplus_\Gamma A_\alpha$, where A_α is a minimal right ideal of R for each $\alpha \in \Gamma$. Since $1 \in R$, there is a finite subset $\Delta \subseteq \Gamma$ such that $1 \in \bigoplus_\Delta A_\alpha$. This clearly implies that $R = \bigoplus_\Delta A_\alpha$. After renumbering, we can let Δ be the set $\{1, 2, \ldots, n\}$ for some integer $n \geq 1$.

(2) Suppose that $R = A_1 \oplus A_2 \oplus \cdots \oplus A_n$, where the A_i are minimal right ideals of R. If we let $\bar{A}_0 = R$, $\bar{A}_n = 0$ and $\bar{A}_i = A_{i+1} \oplus \cdots \oplus A_n$ for $i = 1, 2, \ldots, n-1$, then $\bar{A}_{i-1}/\bar{A}_i \cong A_i$ for each i and

$$R = \bar{A}_0 \supseteq \bar{A}_1 \supseteq \bar{A}_2 \supseteq \cdots \supseteq \bar{A}_n = 0$$

is a composition series of R of length n. Similarly, we can construct a composition series from $R = B_1 \oplus B_2 \oplus \cdots \oplus B_m$ of R of length m. Proposition 4.2.16, the Jordan–Hölder theorem, shows that these two composition series of R are equivalent and this gives (2).

(3) If $R = A_1 \oplus A_2 \oplus \cdots \oplus A_n$, let $e_1 \in A_1, e_2 \in A_2, \ldots, e_n \in A_n$ be such that $1 = e_1 + e_2 + \cdots + e_n$. If j is such that $1 \leq j \leq n$, then we have

$$e_j = e_1 e_j + e_2 e_j + \cdots + e_j^2 + \cdots + e_n e_j, \quad \text{so}$$

$$e_j - e_j^2 = e_1 e_j + e_2 e_j + \cdots + e_{j-1} e_j + e_{j+1} e_j + \cdots + e_n e_j \in A_j \cap \bigoplus_{i \neq j} A_i = 0.$$

Therefore, $e_j = e_j^2$. Also from

$$e_1 e_j + e_2 e_j + \cdots + e_{j-1} e_j + e_{j+1} e_j + \cdots + e_n e_j = 0,$$

we see that $e_i e_j = 0$ for all i and j, $i \neq j$, since the sum $\sum_{i=1}^n A_i$ is direct. Hence, $\{e_1, e_2, \ldots, e_n\}$ is a complete set of orthogonal idempotents of R.

Next, we need to show that $A_i = e_i R$ for $i = 1, 2, \ldots, n$. Let $a \in A_i$. Since $1 = e_1 + e_2 + \cdots + e_n$, we have $a = e_1 a + e_2 a + \cdots + e_i a + \cdots + e_n a$. Thus,

$$a - e_i a = e_1 a + e_2 a + \cdots + e_{i-1} a + e_{i+1} a + \cdots + e_n a \in A_i \cap \bigoplus_{j \neq i} A_j = 0.$$

Therefore, $a = e_i a \in e_i R$, so $A_i \subseteq e_i R$. But $e_i \in A_i$ implies that $e_i R \subseteq A_i$, so $A_i = e_i R$.

Finally, let $\{f_1, f_2, \ldots, f_n\}$ be a complete set of orthogonal idempotents of R such that $A_i = f_i R$ for $i = 1, 2, \ldots, n$. Then $e_1 + e_2 + \cdots + e_n = f_1 + f_2 + \cdots + f_n$, so for each i, $1 \leq i \leq n$,

$$e_i - f_i = \sum_{j \neq i} (f_j - e_j) \in A_i \cap \bigoplus_{j \neq i} A_j = 0.$$

Therefore, we have uniqueness.   □

Note that (2) of the proposition shows that the number of summands in any decomposition of R as a direct sum of minimal right ideals is unique. Moreover, the minimal right ideals of R appearing in any such decomposition of R are unique up to isomorphism, although repetitions of isomorphic minimal right ideals may occur. The following is also a consequence of the proposition.

Corollary 6.4.12. *If R is a right semisimple ring, then R is right artinian, right noetherian and Jacobson semisimple.*

Proof. A minimal right ideal of R is clearly right artinian and right noetherian. If R is a right semisimple ring, then $R = A_1 \oplus A_2 \oplus \cdots \oplus A_n$, where the A_i are minimal right ideals of R. By applying Proposition 4.2.7, we see that R is right artinian and right noetherian. Finally, let $\bar{A}_j = \bigoplus_{i \neq j} A_i$ for $j = 1, 2, \ldots, n$. Then $R/\bar{A}_j \cong A_j$, so each $\bar{A}_j$ is a maximal right ideal of R. Since $J(R) \subseteq \bigcap_{j=1}^n \bar{A}_j = 0$, R is Jacobson semisimple.   □

Although the following lemma is quite easy to prove, it plays an important role in the development of the structure of semisimple rings.

Lemma 6.4.13 (Schur's lemma). *Let M and S be R-modules and suppose that S is simple.*

(1) *If $f : S \to M$ is a nonzero R-linear mapping, then f is a monomorphism.*

(2) *If $f : M \to S$ is a nonzero R-linear mapping, then f is an epimorphism.*

(3) $\operatorname{End}_R(S)$ *is a division ring.*

Proof. If $f : S \to M$ is nonzero and R-linear, then $\operatorname{Ker} f \neq S$. Hence, $\operatorname{Ker} f = 0$, so f is a monomorphism and therefore (1) holds. If $f : M \to S$ is R-linear and nonzero, then $f(M)$ is a nonzero submodule of S, so $f(M) = S$ and we have (2). Part (3) follows immediately from (1) and (2).   □

If $\mathscr{S}$ is the class of simple R-modules and the relation $\sim$ is defined on $\mathscr{S}$ by $S \sim S'$ if and only if S and S' are isomorphic, then $\sim$ is an equivalence relation on $\mathscr{S}$. An equivalence class $[S]$ in $\mathscr{S}$ determined by $\sim$ is said to be an *isomorphism class of simple R-modules.*

Proposition 6.4.14. *If R is a right semisimple ring, then there are only a finite number of isomorphism classes of simple R-modules.*

Proof. Let $[S]$ be an isomorphism class of simple R-modules and suppose that $R = A_1 \oplus A_2 \oplus \cdots \oplus A_n$, where each A_i is a minimal right ideal of R. We claim that S is isomorphic to one of the A_i. If not, then Schur's lemma shows that $\operatorname{Hom}_R(A_i, S) = 0$ for each i. Thus, $\prod_{i=1}^{n} \operatorname{Hom}_R(A_i, S) = 0$. But

$$S \cong \operatorname{Hom}_R(R, S) = \operatorname{Hom}_R(A_1 \oplus A_2 \oplus \cdots \oplus A_n, S) \cong \prod_{i=1}^{n} \operatorname{Hom}_R(A_i, S)$$

and $S \neq 0$, so we have a contradiction. Therefore, each representative of every isomorphism class of simple R-modules is isomorphic to one of the minimal right ideals $A_1, A_2, \ldots, A_n$. Consequently, the isomorphism classes of simple R-modules must be finite in number.   □

From the proof of the preceding proposition we see that the number of distinct isomorphism classes of simple R-modules is less than or equal to the number of summands in the decomposition of R as a direct sum of its minimal right ideals. Moreover, the number of distinct isomorphism classes of simple R-modules is equal to the number of summands if no two of the minimal right ideals in the decomposition of R are isomorphic.

Proposition 6.4.15. *If S is a simple R-module, then for any positive integer n, $\mathrm{End}_R(S^{(n)})$ is isomorphic to $\mathbb{M}_n(D)$, where D is the division ring $\mathrm{End}_R(S)$.*

Proof. Proposition 2.1.12 gives

$$\mathrm{End}_R(S^{(n)}) = \mathrm{Hom}_R(S^{(n)}, S^{(n)}) \cong \prod_{i=1}^{n}\prod_{j=1}^{n} \mathrm{Hom}_R(S_i, S_j),$$

where $S_i = S_j = S$ for each i and j. If we let $D_{ij} = \mathrm{Hom}_R(S_i, S_j) = \mathrm{End}_R(S)$ and agree to write $\prod_{i=1}^{n}\prod_{j=1}^{n} D_{ij}$ as the matrix

$$\mathbb{M}_n(D) = \begin{pmatrix} D_{11} & D_{12} & \cdots & D_{1n} \\ D_{21} & D_{22} & \cdots & D_{2n} \\ \vdots & \vdots & \ddots & \vdots \\ D_{n1} & D_{n2} & \cdots & D_{nn} \end{pmatrix},$$

where $D = D_{ij}$, $1 \leq i, j \leq n$, then $\mathrm{End}_R(S^{(n)}) \cong \mathbb{M}_n(D)$ and Schur's lemma indicates that D is a division ring. $\square$

Definition 6.4.16. Let $R = A_1 \oplus A_2 \oplus \cdots \oplus A_n$ be a decomposition of R, where the A_i are minimal right ideals of R. Arrange the minimal right ideals A_i into isomorphism classes and renumber with double subscripts so that

$$R = (A_{11} \oplus A_{12} \oplus \cdots \oplus A_{1n_1})$$
$$\oplus (A_{21} \oplus A_{22} \oplus \cdots \oplus A_{2n_2})$$
$$\vdots$$
$$\oplus (A_{m1} \oplus A_{m2} \oplus \cdots \oplus A_{mn_m}).$$

If $H_i = A_{i1} \oplus A_{i2} \oplus \cdots \oplus A_{in_i}$ for $i = 1, 2, \ldots, m$, then $n = n_1 + n_2 + \cdots + n_m$ and

$$R = H_1 \oplus H_2 \oplus \cdots \oplus H_m.$$

The H_i are said to be the *homogeneous components* of R.

If $R = A_1 \oplus A_2 \oplus \cdots \oplus A_n$ is a decomposition of R by minimal right ideals and if $[S]$ is an isomorphism class of simple R-modules, then we have seen that S is isomorphic to one of $A_1, A_2, \cdots, A_n$. So if m is the number of homogeneous components of R, then it follows that there are exactly m isomorphism classes of simple R-modules.

Proposition 6.4.17. *The following hold for any right semisimple ring R with decomposition $R = A_1 \oplus A_2 \oplus \cdots \oplus A_n$ as a direct sum of minimal right ideals.*

(1) *The homogeneous components $\{H_i\}_{i=1}^m$ are ideals of R and there is a complete orthogonal set $\{\bar{e}_1, \bar{e}_2, \ldots, \bar{e}_m\}$ of central idempotents of R such that $R = \bar{e}_1 R \oplus \bar{e}_2 R \oplus \cdots \oplus \bar{e}_m R$ and $H_i = \bar{e}_i R$ for $i = 1, 2, \ldots, m$.*

(2) $\text{End}_R(H_i)$ *is isomorphic to an $n_i \times n_i$ matrix ring with entries from a division ring D_i for $i = 1, 2, \ldots, m$.*

Proof. (1) Let $\{e_i\}_{i=1}^n$ be the idempotents of R determined by the decomposition $R = A_1 \oplus A_2 \oplus \cdots \oplus A_n$. If $H_i = A_{i1} \oplus A_{i2} \oplus \cdots \oplus A_{in_i}$, $i = 1, 2, \ldots, m$, are the homogeneous components of R, renumber the e_i using double subscripts so that $e_{ij} \in A_{ij}$, for $i = 1, 2, \ldots, m$ and $j = 1, 2, \ldots, n_i$, and let $\bar{e}_i = e_{i1} + e_{i2} + \cdots + e_{in_i}$. Then $\{\bar{e}_i\}_{i=1}^m$ is a complete set of orthogonal idempotents of R and $\bar{e}_i \in H_i$ for each i. It is easy to show that $R = \bar{e}_1 R \oplus \bar{e}_2 R \oplus \cdots \oplus \bar{e}_m R$ and the fact that $\bar{e}_i R = H_i$, for $i = 1, 2, \ldots, m$, follows as in the proof of $A_i = e_i R$ in (3) of Proposition 6.4.11. Next, we claim that each H_i is an ideal of R. Let

$$A \in \{A_1, A_2, \ldots, A_n\} - \{A_{i1}, A_{i2}, \cdots, A_{in_i}\}.$$

If $a \in A$ and A_{ij} is any one of the minimal right ideals $A_{i1}, A_{i2}, \cdots, A_{in_i}$, then $f : A_{ij} \to R$ defined by $f(r) = ar$ is an R-linear mapping. Since A_{ij} is a minimal right ideal of R, either $f = 0$ or f is a monomorphism. If f is a monomorphism, then $A_{ij} \cong f(A_{ij})$, so $f(A_{ij})$ is a minimal right ideal of R. Now every simple R-module is isomorphic to one of $A_1, A_2, \ldots, A_n$, so this must be true of $f(A_{ij})$ as well. But every minimal right ideal in $\{A_1, A_2, \ldots, A_n\}$ that is isomorphic to A_{ij} is in the set $\{A_{i1}, A_{i2}, \cdots, A_{in_i}\}$. Hence, $a A_{ij} = f(A_{ij}) \subseteq H_i$, so we have $0 \neq a A_{ij} \subseteq A \cap H_i = 0$, an obvious contradiction. Consequently, it must be the case that $f = 0$, so $a A_{ij} = 0$ for all $a \in A$. Hence, $A A_{ij} = 0$ and it follows that $A H_i = 0$, so let $a \in R$ and suppose that $a = a_1 + a_2 + \cdots + a_m \in H_1 \oplus H_2 \oplus \cdots \oplus H_m$. From what we just demonstrated, $a_j H_i = 0$ whenever $j \neq i$, so $a H_i = a_i H_i \subseteq H_i$. Therefore, H_i is an ideal of R. Finally, we need to show that each of the idempotents $\bar{e}_i$ is in the center of R. If $a \in R$, then $a = a \bar{e}_1 + a \bar{e}_2 + \cdots + a \bar{e}_m = \bar{e}_1 a + \bar{e}_2 a + \cdots + \bar{e}_m a$, so since the sum $H_1 + H_2 + \cdots + H_m$ is direct, we have $a \bar{e}_i = \bar{e}_i a$ for $i = 1, 2, \ldots, m$. It is clear that H_i is a ring with identity $\bar{e}_i$.

(2) Since $H_i = A_{i1} \oplus A_{i2} \oplus \cdots \oplus A_{in_i}$ and $A_{i1} \cong A_{i2} \cong \cdots \cong A_{in_i}$, if we let A be any one of the A_{ij}, then $H_i \cong A^{(n_i)}$ and $\text{End}_R(H_i) \cong \text{End}_R(A^{(n_i)})$. Proposition 6.4.15 shows that $\text{End}_R(H_i) \cong M_{n_i}(D_i)$, where D_i is the division ring $\text{End}_R(A)$. $\square$

It follows from Proposition 6.4.17 that if R is a right semisimple ring, then each homogeneous component H_i of R is a ring with identity $\bar{e}_i$. In general, $\bar{e}_i \neq 1_R$, but as we will see in the last proposition of this section, there are rings with only one homogeneous component H_1 in which case $\bar{e}_1 = 1_R$.

Proposition 6.4.18 (Wedderburn–Artin). *A ring R is right semisimple if and only if there exist division rings $D_1, D_2, \ldots, D_m$ such that*

$$R \cong \mathbb{M}_{n_1}(D_1) \times \mathbb{M}_{n_2}(D_2) \times \ldots \times \mathbb{M}_{n_m}(D_m).$$

Proof. If R is right semisimple, then $R = H_1 \oplus H_2 \oplus \cdots \oplus H_m$, where $H_1, H_2, \ldots, H_m$ are the homogeneous components of R. Hence,

$$R \cong \mathrm{End}_R(R) = \mathrm{Hom}_R(H_1 \oplus H_2 \oplus \cdots \oplus H_m, \; H_1 \oplus H_2 \oplus \cdots \oplus H_m)$$

$$\cong \prod_{i=1}^{m} \prod_{j=1}^{m} \mathrm{Hom}_R(H_i, H_j).$$

But $\mathrm{Hom}_R(H_i, H_j) = 0$ if $i \neq j$, so we have

$$R \cong \prod_{i=1}^{m} \mathrm{End}_R(H_i).$$

Schur's lemma shows that if A_i is a representative of the isomorphic minimal right ideals of R that form H_i, then $D_i = \mathrm{End}_R(A_i)$ is a division ring for $i = 1, 2, \ldots, m$. We also have, by the proof of (2) in Proposition 6.4.17, that $\mathrm{End}_R(H_i) \cong \mathbb{M}_{n_i}(D_i)$ for each i, so

$$R \cong \mathbb{M}_{n_1}(D_1) \times \mathbb{M}_{n_2}(D_2) \times \cdots \times \mathbb{M}_{n_m}(D_m).$$

Conversely, suppose that there are division rings $D_1, D_2, \ldots, D_m$ such that

$$R \cong \mathbb{M}_{n_1}(D_1) \times \mathbb{M}_{n_2}(D_2) \times \cdots \times \mathbb{M}_{n_m}(D_m).$$

Note first that $\mathbb{M}_{n_i}(D_i) = \bigoplus_{k=1}^{n_i} r_k(D_i)$, where each $r_k(D_i)$ is the minimal right ideal of $\mathbb{M}_{n_i}(D_i)$ composed of matrices with arbitrary entries from D_i in the kth row and zeros elsewhere. Therefore, each $\mathbb{M}_{n_i}(D_i)$ is a right semisimple ring. But Exercise 7 indicates that a finite ring direct product of right semisimple rings is a right semisimple ring, so we are done. $\qquad\square$

Remark. Due to the Wedderburn–Artin proposition given above, we see that R is right semisimple if and only if

$$R \cong \mathbb{M}_{n_1}(D_1) \times \mathbb{M}_{n_2}(D_2) \times \cdots \times \mathbb{M}_{n_m}(D_m),$$

where $D_1, D_2, \ldots, D_m$ are division rings. This is clearly left-right symmetric since, as observed in Example 2, a matrix ring over a division ring is left and right semisimple.

Thus, right semisimple rings are left semisimple and conversely, so we may now refer to such a ring simply as being semisimple.

Consequently, Proposition 6.4.7 will hold when "right semisimple" is replaced by "left semisimple" in condition (1) of that proposition. Moreover, Proposition 6.4.7 will hold if the condition "right" is, at random, replaced by "left" in conditions (1) through (7).

There is still more to be said about semisimple rings. Recall that a ring R is a simple ring if zero is the only proper ideal of R. Simple right artinian rings are also semisimple rings, but their structure is less complex than that of general semisimple rings.

Proposition 6.4.19 (Wedderburn–Artin). *A ring R is a simple right artinian ring if and only if there is a division ring D such that $R \cong M_n(D)$ for some integer $n \geq 1$.*

Proof. If R is a simple right artinian ring, then R has a minimal right ideal, say A. Note that $\operatorname{Hom}_R(A, R) \neq 0$ since the canonical embedding of A into R is in $\operatorname{Hom}_R(A, R)$. If $I = \sum f(A)$, where f varies throughout $\operatorname{Hom}_R(A, R)$, then we claim that I is an ideal of R. If $a \in R$, then $g : R \to R$ such that $g(r) = ar$ for all $r \in R$ is an R-linear mapping. Moreover, if $f \in \operatorname{Hom}_R(A, R)$, then gf in $\operatorname{Hom}_R(A, R)$. From this we see that $af(A) = (gf)(A) \subseteq I$, so it follows that $aI \subseteq I$. Now $A \subseteq I$, so $I \neq 0$ which means that $I = R$. Consequently, $R = \sum f(A)$ and each nonzero summand $f(A)$ is isomorphic to A. This means that R is a semisimple R-module, so there are $f_1, f_2, \ldots, f_n \in \operatorname{Hom}_R(A, R)$ such that $R = \bigoplus_{i=1}^{n} f_i(A) \cong A^{(n)}$. So R is a semisimple ring with one homogeneous component $H \cong A^{(n)}$. Hence, $R \cong \operatorname{End}_R(R) \cong \operatorname{End}_R(A^{(n)}) \cong M_n(D)$ and $D = \operatorname{End}_R(A)$ is a division ring.

Conversely, suppose that $R \cong M_n(D)$ for some integer $n \geq 1$. If $n = 1$, then $R \cong D$ and D is clearly simple and right artinian, so suppose that $n \geq 2$. If I is a nonzero ideal of $M_n(D)$, let m be a nonzero matrix in I and suppose that m has the nonzero element $a \in D$ in the (s, t)th position. Then $\sum_{i=1}^{n} E_{is} m E_{ti} = \sum_{i=1}^{n} a E_{ii} = a \sum_{i=1}^{n} E_{ii}$ is in I, where $\{E_{ij}\}, 1 \leq i, j \leq n$, are the matrix units of $M_n(D)$. But $\sum_{i=1}^{n} E_{ii}$ is the identity matrix, so $a \sum_{i=1}^{n} E_{ii}$ is an invertible matrix with inverse $a^{-1} \sum_{i=1}^{n} E_{ii}$. Thus, the identity matrix is in I, so $I = M_n(D)$. Hence, $M_n(D)$ is a simple ring. Finally, observe that $M_n(D) = \bigoplus_{i=1}^{n} r_i(D)$ and each set $r_i(D)$ of ith row matrices is a minimal right ideal of $M_n(D)$. Since simple $M_n(D)$-modules are right artinian and since finite direct sums of right artinian $M_n(D)$-modules are right artinian, we see that $M_n(D)$ is a right artinian ring. □

If R is a simple right artinian ring, then $R \cong M_n(D)$ for some division ring D. Thus, R is also a simple left artinian ring, so we can refer to these rings as *simple artinian rings*.

Corollary 6.4.20. *A ring R is semisimple if and only if R is a ring direct product of a finite number of simple artinian rings.*

Problem Set 6.4

1. Suppose that M is a semisimple R-module.

   (a) If $M = \bigoplus_{i=1}^{n} S_i$ and $M = \bigoplus_{i=1}^{m} S_i'$ are decompositions of M as direct sums of simple submodules, show that $m = n$ and that there is a permutation $\sigma : \{1, 2, \ldots, n\} \to \{1, 2, \ldots, n\}$ such that $S_i \cong S_{\sigma(i)}'$ for $i = 1, 2, \ldots, n$. [Hint: Jordan–Hölder.]

   (b) Under the conditions given in (a), show that M is artinian and noetherian and that $\text{Rad}(M) = 0$.

2. Prove that an R-module M is a direct sum of a finite number of simple submodules if and only if M is finitely generated semisimple R-module. [Hint: If $M = \bigoplus_\Delta S_\alpha$, where S_α is a simple R-module for each $\alpha \in \Delta$, and if $\{x_i\}_{i=1}^{n}$ is a set of generators of M, show that for each i there is a finite set $\Delta_i \subseteq \Delta$ such that $x_i R \subseteq \bigoplus_{\Delta_i} S_\alpha$.]

3. Prove that the following hold for an idempotent $e \in R$.

   (a) $\text{End}_R(eR) \cong eRe$ and if R is semiprime, then eR is a minimal right ideal of R if and only if eRe is a division ring. [Hint: Define $\varphi : eRe \to \text{End}_R(eR)$ by $\varphi(eae) = f_{ea}$, where $f_{ea} : eR \to eR$ is such that $f_{ea}(eb) = eaeb$ for any $b \in R$ and show that φ is a well defined ring isomorphism. If eR is a minimal right ideal of R, use Shur's lemma to conclude that $\text{End}_R(eR)$ is a division ring. For the converse, if eRe is a division ring, let A be a right ideal of R such that $A \subseteq eR$. If $a \in A$, $a \neq 0$, then $a = eb$ for some $b \in R$, $b \neq 0$. Hence, $ebReb \neq 0$, so $ebce \neq 0$ for some $c \in R$. If ede is the inverse of $ebce$, then $ebcede = e$, so that $ebR = eR$.]

   (b) If R is a semiprime ring, then a minimal right ideal A of R must be for the form $A = eR$ for some idempotent e of R. Moreover, eR is a minimal right ideal of R if and only if Re is a minimal left ideal of R. [Hint: $A^2 \neq 0$, so $aA \neq 0$ for some $a \in A$. Hence, $aA = A$. Also $\text{ann}_r(a) \cap A \neq A$, so $\text{ann}_R(a) \cap A = 0$. Now let $e \in A$ be such that $ae = a$.]

   (c) If R is a semiprime ring, then $\text{Soc}(R_R) = \text{Soc}(_R R)$. [Hint: Show that $\text{Soc}(R_R) = \sum eR$, where e varies over the idempotents $e \in R$ such that eRe is a division ring. Similarly, $\text{Soc}(_R R) = \sum Re$. Show also that $\text{Soc}(R_R)$ and $\text{Soc}(_R R)$ are ideals of R and that $\text{Soc}(_R R) \subseteq \text{Soc}(R_R)$ and $\text{Soc}(R_R) \subseteq \text{Soc}(_R R)$.]

4. Let M be a semisimple R-module and suppose that $M = \bigoplus_\Delta S_\alpha$ is a decomposition of M as a direct sum of simple submodules. Arrange the simple submodules S_α into isomorphism classes and call each isomorphism class a *homogeneous component* H of M.

(a) If M is semisimple and finitely generated, show that M has only a finite number of homogeneous components H_i, $i = 1, 2, \ldots, m$, and that each homogeneous component is a finite sum of isomorphic simple submodules of M. [Hint: Exercise 2.]

(b) If M is semisimple and finitely generated, prove that

$$\operatorname{End}_R(M) \cong \mathbb{M}_{n_1}(D_1) \oplus \mathbb{M}_{n_2}(D_2) \oplus \cdots \oplus \mathbb{M}_{n_m}(D_m),$$

where, for each i, $D_i = \operatorname{End}_R(S_i)$ and S_i is a representative of the isomorphism class of the n_i simple submodules of M whose sum gives H_i.

5. Let $D_1, D_2, \ldots, D_n$ be division rings.

(a) What are the minimal left and minimal right ideals of the ring direct product $D_1 \times D_2 \times \cdots \times D_n$?

(b) Show that the minimal right (left) ideals of (a) can be used to decompose $D_1 \times D_2 \times \cdots \times D_n$ as a direct sum of minimal right (left) ideals. Conclude that $D_1 \times D_2 \times \cdots \times D_n$ is a semisimple ring.

(c) Let $\{D_\alpha\}_\Delta$ be a family of division rings. Deduce that the ring direct product $\prod_\Delta D_\alpha$ is a semisimple ring if and only if Δ is a finite set.

6. Show that a finite ring direct product of semisimple rings is a semisimple ring.

7. (a) Let I be an ideal of R and suppose that M is an R-module such that $MI = 0$. Prove that M is semisimple as an R-module if and only if M is semisimple as an R/I-module.

(b) If R is a semisimple ring and I is a proper ideal of R, prove that R/I is a semisimple ring. Conclude that if $f : R \to S$ is a surjective ring homomorphism and R is semisimple, then so is S.

(c) Show that a subring of a semisimple ring need not be semisimple.

8. Prove that the following are equivalent for a ring R.

(a) R has a simple generator.

(b) R is a simple ring and R_R is semisimple.

9. Show that the following are equivalent for a ring R.

(a) R is a semisimple ring.

(b) Every finitely generated R-module is projective.

(c) Every cyclic R-module is projective. [Hint: If A is a right ideal of R, consider R/A.]

(d) Show that (a), (b) and (c) are equivalent if R-module is replaced with left R-module in (b) and (c).

10. Let $\mathcal{S}$ be a *minimal set of simple generators* for $\mathbf{Mod}_R$, in the sense that no set of simple R-modules with fewer elements can generate $\mathbf{Mod}_R$.

(a) Prove that if $\text{card}(\mathcal{S}) = n$, then there are exactly n isomorphism classes of simple R-modules. In this situation, we say that $\mathbf{Mod}_R$ has a *set of n simple generators*.

(b) Prove that a ring R is semisimple if and only if $\mathbf{Mod}_R$ has a finite set of simple generators.

(c) Prove that R is a simple artinian ring if and only if $\mathbf{Mod}_R$ is generated by a simple R-module. In this case, we say that $\mathbf{Mod}_R$ has a *simple generator*. Conclude that for a simple artinian ring there is exactly one isomorphism class of simple R-modules.

6.5 Primitive Rings and Density

We now investigate a class of rings that generalizes the class of simple artinian rings. To develop these rings we need the following bimodule structures. If M is an R-module and $H = \text{End}_R(M)$, then M is a left H-module under the addition already present on M and under the H-action on M given by $hx = h(x)$ for $h \in H$ and $x \in M$. In fact, it's easy to see that M is an (H, R)-bimodule. If we now form the ring $\text{End}_H(M)$ and write $f \in \text{End}_H(M)$ on the right of the argument $x \in M$ opposite that of the ring action of H on M, then M is an $(H, \text{End}_H(M))$-bimodule. With these bimodule structures in mind, the following properties hold.

1. Let $\varphi : R \to \text{End}_H(M)$ be such that $\varphi(a) = f_a$, where $f_a : {}_HM \to {}_HM$ is defined by $(x)f_a = xa$ for all $x \in M$. Then

$$(x + y)f_a = (x + y)a = xa + ya = (x)f_a + (y)f_a \quad \text{and}$$
$$(hx)f_a = (hx)a = h(x)a = h(x)f_a$$

for all $x, y \in M$ and $h \in H$. Hence f_a is an H-linear mapping and we also have

$$(x)f_{a+b} = x(a + b) = xa + xb = (x)f_a + (x)f_b,$$
$$(x)f_{ab} = x(ab) = (xa)b = (xa)f_b = (x)f_a f_b \quad \text{and}$$
$$(x)f_1 = x$$

for all $x \in M$ and $a, b \in R$. Thus,

$$\varphi(a + b) = \varphi(a) + \varphi(b),$$
$$\varphi(ab) = \varphi(a)\varphi(b) \quad \text{and}$$
$$\varphi(1) = \text{id}_{\text{End}_H(M)}.$$

Therefore, φ is an identity preserving ring homomorphism.

2. The map φ is a monomorphism if and only if $Ma = 0$ implies $a = 0$ for all $a \in R$. Recall that an R-module M is said to be faithful if $Ma = 0$ gives $a = 0$ for all $a \in R$. When M is faithful, the embedding $\varphi : R \to \mathrm{End}_H(M)$ is said to be a *faithful representation* of R. In this case, we will refer to φ simply as the *canonical embedding* of R into $\mathrm{End}_H(M)$.

3. If S is a simple R-module, then $R/\mathrm{ann}_r(x) \cong S$, where $x \in S$ and $x \neq 0$. Note that $\mathrm{ann}_r(S) \subseteq \mathrm{ann}_r(x)$ and that $\mathrm{ann}_r(S)$ is an ideal contained in the maximal right ideal $\mathrm{ann}_r(x)$ of R. If R has the property that zero is the largest ideal contained in $\mathrm{ann}_r(x)$, then $\mathrm{ann}_r(S) = 0$ and S is a faithful simple R-module. When this is the case, R embeds in $\mathrm{End}_D(S)$ and $D = \mathrm{End}_R(S)$ is a division ring. So if R admits a faithful simple module, then R embeds in the ring of linear transformations of the left D-vector space S.

Definition 6.5.1. If M is an R-module and $H = \mathrm{End}_R(M)$, then the ring $\mathrm{End}_H(M)$, often denoted by $\mathrm{BiEnd}_R(M)$, is said to be the *biendomorphism ring* of M and elements of $\mathrm{End}_H(M)$ are said to be *biendomorphisms* of M. An ideal $\mathfrak{p}$ of R is said to be a *right primitive ideal* of R if $\mathfrak{p}$ is the largest ideal contained in some maximal right ideal of R, and R is said to be a *right primitive ring* if zero is a right primitive ideal. *Left primitive ideals* and *left primitive rings* are similarly defined. A ring that is left and right primitive is said to be a *primitive ring*.

There are rings that are right primitive but not left primitive and, conversely, there are left primitive rings that are not right primitive. (See [52] for the details.)

Examples

1. Every simple ring is right primitive, so if D is a division ring, then $\mathbb{M}_n(D)$ is a right primitive ring.

2. Every maximal ideal is a right primitive ideal.

3. A field is a primitive ring.

If $\mathfrak{p}$ is an ideal of R, then $\mathfrak{p}$ is the largest ideal contained in the maximal right ideal $\mathfrak{m}$ of R if and only if zero is the largest ideal contained in the maximal right ideal $\mathfrak{m}/\mathfrak{p}$ of $R/\mathfrak{p}$. Hence, we see that an ideal $\mathfrak{p}$ of R is right primitive if and only if $R/\mathfrak{p}$ is a right primitive ring.

A right primitive ring is Jacobson semisimple. This follows, for if R is a right primitive ring, then there is a maximal right ideal $\mathfrak{m}$ of R such that zero is the largest ideal contained in $\mathfrak{m}$. But $J(R)$ is an ideal of R and $J(R) \subseteq \mathfrak{m}$, so $J(R) = 0$. The converse is false. There are Jacobson semisimple rings that are not right primitive. For example, the ring of integers $\mathbb{Z}$ is such that $J(\mathbb{Z}) = 0$, but the zero ideal is not the largest ideal contained in any maximal ideal of $\mathbb{Z}$. Indeed, every maximal ideal of $\mathbb{Z}$ looks like (p), where p is a prime number. But then we have $(0) \subsetneqq (p^2) \subsetneqq (p)$, so

$\mathbb{Z}$ cannot be a primitive ring. In fact, (p) is the largest ideal contained in the maximal ideal (p).

Proposition 6.5.2. *A ring R is right primitive if and only if R admits a faithful simple R-module.*

Proof. If R is a right primitive ring, then zero is the largest ideal contained in some maximal right ideal $\mathfrak{m}$ of R. But then $R/\mathfrak{m}$ is a simple R-module and the right annihilator ideal A of $R/\mathfrak{m}$ is contained in $\mathfrak{m}$. Hence, $A = 0$, so $R/\mathfrak{m}$ is a faithful simple R-module.

Conversely, suppose that R admits a faithful simple R-module S. Then $R/\operatorname{ann}_r(x) \cong xR = S$ for any $x \in S$, $x \neq 0$ and $\operatorname{ann}_r(x)$ is a maximal right ideal of R. If I is an ideal of R such that $I \subseteq \operatorname{ann}_r(x)$, then $SI = 0$. Hence, $I = 0$ since S is faithful. Thus, zero is the largest ideal contained in $\operatorname{ann}_r(x)$, so R is a right primitive ring.   □

As indicated earlier, right primitive rings provide a generalization of simple artinian rings. To continue with the development of this generalization, we need the following definition.

Definition 6.5.3. Let R be a subring of $\operatorname{End}_D(M)$, where $_DM$ is a left vector space over a division ring D, and suppose that $\operatorname{End}_D(M)|_N$ denotes the D-endomorphisms of $_DM$ restricted to a D-subspace $_DN$ of $_DM$. Then R is said to be a *dense subring* of $\operatorname{End}_D(M)$ provided that $\operatorname{End}_D(M)|_N \subseteq R|_N$ for each finite dimensional subspace $_DN$ of $_DM$, that is, for each $f \in \operatorname{End}_D(M)$ there is an $a \in R$ such that $f|_N = a|_N$.

In view of Definition 6.5.3, if $\varphi : R \to \operatorname{End}_D(S)$ is the canonical embedding, where S is a faithful simple R-module and $D = \operatorname{End}_R(S)$, then $\varphi(R)$ will be dense in $\operatorname{End}_D(S)$ if for each finite dimensional subspace $_DN$ of $_DS$ and every $f \in \operatorname{End}_D(S)$, there is an $a \in R$ such that $f|_N = f_a|_N$. Thus, we can say, loosely speaking, that locally (= on finite dimensional subspaces) every linear transformation $_DS \to_D S$ is an element of $\varphi(R)$.

A simple artinian ring R is isomorphic to a dense subring of the endomorphism ring of a vector space since R is isomorphic to the ring of linear transformations of a finite dimensional vector space over a division ring. Indeed, when R is a simple artinian ring, we have seen in the proof of Proposition 6.4.19 that there is a minimal right ideal A of R such that if $D = \operatorname{End}_R(A)$, then $R \cong \mathbb{M}_n(D)$ for some integer $n \geq 1$ and D is a division ring. Now suppose that M is a left D-vector space of dimension n and that $\{x_1, x_2, \dots, x_n\}$ is a basis for $_DM$. If $x = \sum_{i=1}^n k_i x_i$ is an element of $_DM$, then a matrix $(a_{ij}) \in \mathbb{M}_n(D)$

$$(x)f_{(a_{ij})} = \left(\sum_{i=1}^n k_i a_{i1}\right)x_1 + \left(\sum_{i=1}^n k_i a_{i2}\right)x_2 + \cdots + \left(\sum_{i=1}^n k_i a_{in}\right)x_n$$

defines a linear transformation $f_{(a_{ij})} \in \mathrm{End}_D(M)$. Conversely, if $f \in \mathrm{End}_D(M)$
and

$$(x_1)f = a_{11}x_1 + a_{12}x_2 + \cdots + a_{1n}x_n$$
$$(x_2)f = a_{21}x_1 + a_{22}x_2 + \cdots + a_{2n}x_n$$
$$\vdots$$
$$(x_n)f = a_{n1}x_1 + a_{n2}x_2 + \cdots + a_{nn}x_n$$

then the array gives a matrix $(a_{ij})_f \in \mathbb{M}_n(D)$ determined by f. It is left to the reader
to show that the map

$$\mathbb{M}_n(D) \to \mathrm{End}_D(M) \quad \text{such that } (a_{ij}) \mapsto f_{(a_{ij})}$$

is a ring isomorphism with inverse function

$$\mathrm{End}_D(M) \to \mathbb{M}_n(D) \quad \text{given by } f \mapsto (a_{ij})_f.$$

Hence, $R \cong \mathbb{M}_n(D) \cong \mathrm{End}_D(M)$, so it's trivial that R is isomorphism to a dense
subring of $\mathrm{End}_D(M)$.

We now show that a right primitive ring provides a "local form", in the sense described above, of this property of simple artinian rings.

Proposition 6.5.4 (Jacobson's density theorem). *A ring R is isomorphic to a dense
subring of the biendomorphism ring of a faithful simple R-module if and only if R is
a right primitive ring.*

Proof. If R is isomorphic to a dense subring of the biendomorphism ring of a faithful
simple R-module, then, by assumption, R admits a faithful simple R-module, so
Proposition 6.5.2 shows that R is a right primitive ring.

Conversely, if R is a right primitive ring, then R admits a faithful simple R-module
S. If $D = \mathrm{End}_R(S)$, then D is a division ring and we have a canonical embedding
$\varphi : R \to \mathrm{End}_D(S)$. Now suppose that $_D N$ is a finite dimensional subspace of $_D S$
and let $\{x_i\}_{i=1}^n$ be a basis for N. Since $\sum_{i=1}^n x_i \neq 0$ and since S is a simple R-mod-
ule, $(\sum_{i=1}^n x_i)R = S$. If $f \in \mathrm{End}_D(S)$, then $\sum_{i=1}^n (x_i)f \in S$, so there is an $a \in R$
such that $\sum_{i=1}^n x_i a = (\sum_{i=1}^n x_i)a = \sum_{i=1}^n (x_i)f$. Hence, $(x_i)f = x_i a = (x_i)f_a$
for $i = 1, 2, \ldots, n$. Since f and f_a agree on basis elements, f and f_a agree on N.
Thus, $\varphi(R)$ is dense in $\mathrm{End}_D(S)$.   □

Remark. There is a topological version of Proposition 6.5.4 in which R is isomorphic
to a topologically dense subring of the biendomorphism ring of a faithful simple R-
module if and only if R is a right primitive ring. This accounts for the term "dense"
in Definition 6.5.3. (Details of the topological version of Jacobson's density theorem
can be found in [22].)

Corollary 6.5.5. *The following hold if R is a right primitive ring with faithful simple R-module S.*

(1) *For each D-linearly independent subset $\{x_i\}_{i=1}^n$ of the vector space $_D S$ over the division ring $D = \operatorname{End}_R(S)$ and for each arbitrary set $\{y_i\}_{i=1}^n$ of elements of $_D S$, there is an $a \in R$ such that $x_i a = y_i$ for $i = 1, 2, \ldots, n$.*

(2) *If $\{x_i\}_{i=1}^{n+1}$ is a D-linearly independent set of elements of $_D S$, then there is an $a \in R$ such that $x_i a = 0$ for $i = 1, 2, \ldots, n$ and $x_{n+1} a \neq 0$.*

Proof. (1) Let $R, S, \{x_i\}_{i=1}^n$ and $\{y_i\}_{i=1}^n$ be as described. Then, by (2) of Proposition 2.2.9, there is a basis $\mathcal{B}$ of $_D S$ that contains $\{x_i\}_{i=1}^n$. Define $f : \mathcal{B} \to _D S$ by $(x_i) f = y_i$ for $i = 1, 2, \ldots, n$ and $(x) f = 0$ if $x \in \mathcal{B} - \{x_i\}_{i=1}^n$. If f is extended D-linearly to $_D S$ and if this extension is also denoted by f, then Jacobson's density theorem shows that there is an $a \in R$ such that $x_i a = (x_i) f = y_i$ for $i = 1, 2, \ldots n$.

(2) Choose $\{y_i\}_{i=1}^{n+1}$ to be such that $y_i = 0$ for $i = 1, 2, \ldots, n$ and $y_{n+1} \neq 0$ and apply (1). $\qquad\square$

A simple ring is clearly right primitive and, in fact, when R is right artinian, R is a simple ring if and only if it is right primitive.

Proposition 6.5.6. *The following are equivalent for a ring R.*

(1) *R is right artinian and right primitive.*

(2) *R is right artinian and simple.*

Proof. Since a simple ring is right primitive, we are only required to show that $(1) \Rightarrow (2)$. Let R be a right artinian right primitive ring. Then R admits a faithful simple R-module S, and if $D = \operatorname{End}_R(S)$, then S is a left vector space over D. Let $\{x_\alpha\}_\Delta$ be a basis for $_D S$ and suppose that $\dim_D(S) = \infty$. Select a subset $\{x_1, x_2, x_3, \ldots\}$ of $\{x_\alpha\}_\Delta$ and, for $n = 1, 2, \ldots$, let A_n denote the annihilator right ideal of the set $\{x_1, x_2, \ldots, x_n\}$. Then $A_1 \supseteq A_2 \supseteq A_3 \supseteq \cdots$ is a descending chain of right ideals of R. Using (2) of Corollary 6.5.5, we see that for each n there is an $a \in R$ such that $x_i a = 0$ for $i = 1, 2, \ldots, n$ and $x_{n+1} a \neq 0$. Thus, $A_n \supsetneq A_{n+1}$ for each n which contradicts the fact that R is right artinian. Therefore, $\dim_D(S)$ is finite, so let $\{x_i\}_{i=1}^n$ be a basis for $_D S$. Also let $\varphi : R \to \operatorname{End}_D(S)$ be the canonical embedding and note that

$$\operatorname{End}_D(S) = \operatorname{Hom}_D \left(\bigoplus_{i=1}^n D x_i, \bigoplus_{i=1}^n D x_i \right) \cong \prod_{i=1}^n \prod_{j=1}^n \operatorname{Hom}_D(D x_i, D x_j).$$

Now $D x_i$ and $D x_j$ are simple left D-modules such that $D x_i \cong D x_j \cong D$ for each i and j and it follows that $\operatorname{End}_D(S) \cong \mathbb{M}_n(D)$. If $f \in \operatorname{End}_D(S)$, then f is completely determined by its action on the basis elements $x_1, x_2, \ldots, x_n$. But since R is right primitive, $\varphi(R)$ is dense in $\operatorname{End}_D(R)$, so there is an $a \in R$ such that

$(x_i)f = (x_i)f_a$. Thus, $f \in \varphi(R)$ which shows that φ is a ring isomorphism. Hence, $R \cong \mathbb{M}_n(D)$ and this establishes, by Proposition 6.4.19, that R is a simple artinian ring. $\square$

Problem Set 6.5

1. Prove that a commutative ring is primitive if and only if it is a field. [$\mathfrak{m} \subseteq$ ann$(R/\mathfrak{m})$ for every maximal ideal of R.]

2. Let M be an R-module.

   (a) Prove M is a faithful $R/\text{ann}_r(M)$-module.

   (b) Prove that M is faithful if and only if the canonical ring homomorphism $\varphi : R \to \text{BiEnd}_R(M)$ is an embedding.

3. (a) Show that if R is a simple ring, then every simple R-module is faithful.

   (b) Prove that a simple ring embeds in the biendomorphism ring of each of its simple R-modules.

4. (a) Prove that an ideal $\mathfrak{p}$ of a ring R is right primitive if and only if it is the right annihilator ideal of a faithful simple R-module.

   (b) Let S be a simple R-module and suppose that A is the right annihilator ideal of S. Show that R/A is a right primitive ring.

5. Prove that $J(R)$ is the intersection of the right (left) primitive ideals of R. [Hint: Proposition 6.1.7.]

6. Is a right primitive right artinian ring also left primitive and left artinian?

7. We have seen in the introduction to this section that if M is an R-module, $H = \text{End}_R(M)$ and biendomorphisms $f \in \text{End}_H(M)$ are written on the right of elements of M, then the canonical embedding $\varphi : R \to \text{End}_H(M)$ given by $\varphi(a) = f_a$ will be a ring homomorphism. Show that if biendomorphisms $f \in \text{End}_H(M)$ are written on the left of elements of M and composition of functions is applied in the usual fashion, then φ is an anti-ring homomorphism.

8. Suppose that e is a nonzero idempotent element of R. If R is right primitive, show that eRe is also a right primitive ring. [Hint: If R is a right primitive ring, then R admits a faithful simple R-module S. If $D = \text{End}_R(S)$, then D is a division ring and we have a canonical embedding $\varphi : R \to \text{End}_D(S)$. Observe that eS is a vector space over D and show that there is a canonical embedding $\varphi_e : eRe \to \text{End}_D(eS)$.]

9. Deduce that if the category Mod_R has a simple generator S, then S is a faithful R-module. [Hint: Exercise 1 in Problem Set 4.1.]

10. (a) In the discussion immediately preceding Jacobson's density theorem, it was
    stated that a matrix $(a_{ij}) \in \mathbb{M}_n(D)$ defines a linear transformation

$$(x)f_{(a_{ij})} = \left(\sum_{i=1}^{n} k_i a_{i1}\right)x_1 + \left(\sum_{i=1}^{n} k_i a_{i2}\right)x_2 + \cdots + \left(\sum_{i=1}^{n} k_i a_{in}\right)x_n$$

in $\text{End}_D(M)$. Show that this is actually the case.

(b) Show that the map

$$\mathbb{M}_n(D) \to \text{End}_D(M) \quad \text{such that } (a_{ij}) \mapsto f_{(a_{ij})}$$

is a ring isomorphism with inverse function

$$\text{End}_D(M) \to \mathbb{M}_n(D) \quad \text{given by } f \mapsto (a_{ij})_f.$$

6.6 Rings that Are Semisimple

Previously we defined and investigated right primitive, Jacobson semisimple, prime,
semiprime, and semisimple rings. These classes of rings are not necessarily disjoint.
Most of the implications in the following table have already been pointed out.

$$
\begin{array}{ccccc}
\text{simple} & \Rightarrow & \text{right primitive} & \Rightarrow & \text{prime} \\
\Downarrow +\text{artinian} & & \Downarrow & & \Downarrow \\
\text{semisimple} & \Rightarrow & \text{Jacobson semisimple} & \Rightarrow & \text{semiprime} \\
& & \Uparrow & & \\
& & \text{regular} & &
\end{array}
\qquad (6.1)
$$

It is trivial that a simple ring is right primitive and that a prime ring is semiprime.
A right primitive ring R is certainly Jacobson semisimple, since in a right primitive
ring the zero ideal is the largest ideal contained in a maximal right ideal of R. We
have also seen in Corollary 6.4.12 that a semisimple ring is Jacobson semisimple.
Moreover, it follows that a Jacobson semisimple ring is semiprime since, by Corollary 6.2.10, $\text{rad}(R) \subseteq J(R)$ and, by Proposition 6.2.25, a ring R is semiprime if and
only if $\text{rad}(R) = 0$. Finally, if R is a regular ring and $a \in J(R)$, then there is an $r \in R$
such that $a = ara$. But then $ra \in J(R)$ and ra is idempotent, so $a = ara = 0$ since
$J(R)$ contains no nonzero idempotents. Hence, a regular ring is Jacobson semisimple.
Thus, to complete the verification of table (6.1) we need to prove that a right primitive
ring is prime.

Proposition 6.6.1. *If R is right primitive, then R is prime.*

Proof. Let R be a right primitive ring and suppose that S is a faithful simple R-module. Let A and B be nonzero ideals of R and suppose that $AB = 0$ with $A \neq 0$ and $B \neq 0$. Then $S(AB) = 0$. But $S(AB) = (SA)B = SB = S \neq 0$ and we have a contradiction. Hence, if $AB = 0$, then either $A = 0$ or $B = 0$, so zero is a prime ideal of R. Consequently, R is prime.   ☐

Our intention now is to show that several of the implications in table (6.1) reverse when R is right artinian. We also show that semisimple rings are the artinian Jacobson semisimple rings mentioned in the opening remarks of this chapter.

Proposition 6.6.2. *The following are equivalent for a ring* R.

(1) R *is semisimple.*

(2) R *is right artinian and regular.*

(3) R *is right artinian and Jacobson semisimple.*

(4) R *is right artinian and semiprime.*

(5) R *is right noetherian and regular.*

Proof. If R is semisimple, then, by Corollary 6.4.12, R is artinian and noetherian. Furthermore, every right ideal of R is a direct summand of R. In particular, every principal right ideal of R is a direct summand, so it follows that R is regular. Hence, (1) implies (2) and (5). Table (6.1) shows that $(2) \Rightarrow (3) \Rightarrow (4)$. If (4) holds, then by Corollary 6.3.4, $J(R) = \mathrm{rad}(R)$ and Proposition 6.2.25 indicates that $\mathrm{rad}(R) = 0$. Therefore, the intersection of the maximal right ideals of R is zero. Since R is right artinian, there is a finite set $\{\mathfrak{m}_i\}_{i=1}^n$ of maximal right ideals of R such that $\mathfrak{m}_1 \cap \mathfrak{m}_2 \cap \cdots \cap \mathfrak{m}_n = 0$. We may assume that $A_i = \bigcap_{j \neq i} \mathfrak{m}_j \not\subseteq \mathfrak{m}_i$ for each i, for if $A_i \subseteq \mathfrak{m}_i$, then $\mathfrak{m}_i$ can be eliminated from the set $\{\mathfrak{m}_i\}_{i=1}^n$ and we still have a set of maximal right ideals of R with zero intersection. Hence, $A_i + \mathfrak{m}_i = R$ for $i = 1, 2, \ldots, n$. But $A_i \cap \mathfrak{m}_i = 0$, so $A_i \oplus \mathfrak{m}_i = R$ and $A_i \cong R/\mathfrak{m}_i$ is a minimal right ideal of R. Therefore, since R is semiprime, by considering (2) of Lemma 6.3.5, we see that $A_i = e_i R$ for each i, where e_i is an idempotent of R. Hence, $\mathfrak{m}_i = (1 - e_i)R$. Let $e = \sum_{i=1}^n e_i$. Then $1 - e = (1 - e_i) - \sum_{j \neq i} e_j \in \mathfrak{m}_i$, for each $i \neq j$, since $i \neq j$ implies that $e_j \in A_j \subseteq \mathfrak{m}_i$. Therefore, $1 - e \in \bigcap_{i=1}^n \mathfrak{m}_i = 0$ and so $1 = e = \sum_{i=1}^n e_i$. It follows that $R = \bigoplus_{i=1}^n A_i$, so R is semisimple and we have $(4) \Rightarrow (1)$. We complete the proof by showing $(5) \Rightarrow (1)$. If R is right noetherian and A is a right ideal of R, then A is finitely generated. But R is regular and so, by Proposition 5.3.17, we see that A is a principal right ideal of R generated by an idempotent. Hence, A is a direct summand of R, so Proposition 6.4.7 indicates that R is semisimple.   ☐

We now have the following augmentation of table (6.1). The abbreviations are the obvious ones and the equivalence rt. art. + simple $\Leftrightarrow$ rt. art. + rt. primitive is Proposition 6.5.6.

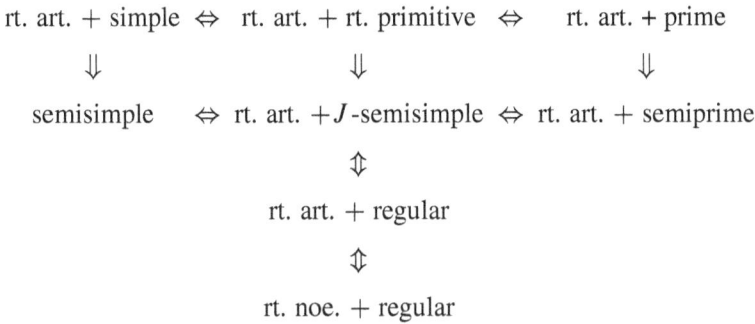

$$\text{rt. art. + simple} \Leftrightarrow \text{rt. art. + rt. primitive} \Leftrightarrow \text{rt. art. + prime}$$

$$\Downarrow \qquad\qquad\qquad \Downarrow \qquad\qquad\qquad \Downarrow$$

$$\text{semisimple} \quad \Leftrightarrow \text{rt. art.} + J\text{-semisimple} \Leftrightarrow \text{rt. art. + semiprime}$$

$$\Updownarrow$$

$$\text{rt. art. + regular}$$

$$\Updownarrow$$

$$\text{rt. noe. + regular}$$

It was pointed out in Example 5 in Section 4.2 that the ring $\mathbb{Z}$ is noetherian but not artinian. Hence, there are right noetherian rings that are not right artinian. However, as we will now show, every right artinian ring is right noetherian. For this we need the following two lemmas.

But first recall that if $\{M_\alpha\}_\Delta$ is a family of submodules of an R-module M, where Δ is well ordered, then $\{M_\alpha\}_\Delta$ is said to be a increasing chain (decreasing chain) of submodules of M if $M_\alpha \subseteq M_\beta (M_\alpha \supseteq M_\beta)$ whenever $\alpha \leq \beta$.

Lemma 6.6.3. *An R-module M is noetherian (artinian) if and only if every increasing (decreasing) chain of submodules of M indexed over a well-ordered set terminates.*

Proof. We prove the lemma for the artinian case and leave the noetherian case to the reader. If every decreasing chain of submodules of M indexed over a well ordered set terminates, then every decreasing chain $M_1 \supseteq M_2 \supseteq M_3 \supseteq \cdots$ of submodules of M terminates since $\mathbb{N}$ is well ordered. Thus, M is artinian. Conversely, suppose that M is artinian and let $\{M_\alpha\}_\Delta$ be a chain of submodules of M, where Δ is a well ordered set. Since Δ is well ordered, then we can assume that Δ is the set of ordinal numbers $\{0, 1, 2, \ldots, \omega, \omega + 1, \ldots\}$ with $\alpha < \mathrm{ord}(\Delta)$ for each $\alpha \in \{0, 1, 2, \ldots, \omega, \omega + 1, \ldots\}$. (See Appendix A.) Since M is artinian Proposition 4.2.4 shows that the collection of submodules $\{M_\alpha\}_\Delta$ has a minimal element. Hence, the chain $M_0 \supseteq M_1 \supseteq M_2 \supseteq \cdots \supseteq M_\omega \supseteq M_{\omega+1} \supseteq \cdots$ must terminate. $\qquad\square$

Lemma 6.6.4. *Let M be an R-module such that $M = \bigoplus_\Delta S_\alpha$, where each S_α is a simple submodule of M. If M is either artinian or noetherian, then Δ is finite.*

Proof. Suppose that M is artinian and that Δ is an infinite set. Now Δ can be well ordered, so, as in the proof of the previous lemma, we can assume that Δ is the set of ordinal numbers $\{0, 1, 2, \ldots, \omega, \omega + 1, \ldots\}$ with $\alpha < \text{ord}(\Delta)$ for each $\alpha \in \{0, 1, 2, \ldots, \omega, \omega + 1, \ldots\}$. If for each ordinal $\alpha \in \{0, 1, 2, \ldots, \omega, \omega + 1, \ldots\}$, we set $N_\alpha = 0 \oplus 0 \oplus \cdots \oplus S_\alpha \oplus S_{\alpha+1} \oplus \cdots$, then

$$M = N_0 \supseteq N_1 \supseteq N_2 \supseteq \cdots \supseteq N_\omega \supseteq N_{\omega+1} \supseteq \cdots$$

is a decreasing chain of submodules of M indexed over a well ordered set that fails to terminate. Hence, the lemma above indicates that M is not artinian and so we have a contradiction. Thus, Δ must be a finite set. A similar proof holds if M is noetherian. $\square$

The following proposition is due to Hopkins [59] who proved that a right artinian ring must be right noetherian.

Proposition 6.6.5 (Hopkins). *If R is a right artinian ring, then an R-module is noetherian if and only if it is artinian.*

Proof. Let J denote the Jacobson radical of R. If R is right artinian, then J is nilpotent, so $J^n = 0$ for some integer $n \geq 1$. Hence, we have a descending chain of submodules

$$M \supseteq MJ \supseteq MJ^2 \supseteq \cdots \supseteq MJ^n = 0$$

of M. Consider the factor modules MJ^{i-1}/MJ^i for $i = 1, 2, \ldots, n$, where we let $J^0 = R$. Since $(MJ^{i-1}/MJ^i)J = 0$, each MJ^{i-1}/MJ^i is an R/J-module. Since there is a one-to-one correspondence between the right ideals of R that contain J and the right ideals of R/J, it follows that R/J is right artinian. But R/J is also Jacobson semisimple, so by Proposition 6.6.2, R/J is a semisimple ring. Thus, each MJ^{i-1}/MJ^i is a semisimple R/J-module. Consequently, each MJ^{i-1}/MJ^i is semisimple as an R-module. Therefore, MJ^{i-1}/MJ^i is a direct sum of simple R-modules and if M is artinian, then MJ^{i-1}/MJ^i is artinian and we see from Lemma 6.6.3 that this sum of simple R-modules is finite. Hence, MJ^{i-1}/MJ^i has an R-composition series, so by Proposition 4.2.14, we have that MJ^{i-1}/MJ^i is a noetherian R-module for $i = 1, 2, \ldots, n$. Now consider the short exact sequences

$$0 \to MJ^{n-1} \to MJ^{n-2} \to MJ^{n-2}/MJ^{n-1} \to 0 \qquad (6.2)$$

$$0 \to MJ^{n-2} \to MJ^{n-3} \to MJ^{n-3}/MJ^{n-2} \to 0 \qquad (6.3)$$

$$\vdots$$

$$0 \to MJ \to M \to M/MJ \to 0. \qquad (6.4)$$

Since $(MJ^{n-1})J = 0$, MJ^{n-1} is an R/J-module and so MJ^{n-1} is noetherian. Hence, MJ^{n-1} and MJ^{n-2}/MJ^{n-1} are noetherian, so Corollary 4.2.6 together with the short exact sequence (6.2) shows that MJ^{n-2} is noetherian. Likewise, (6.3) indicates that MJ^{n-3} is noetherian and so on until we finally arrive at (6.4) which shows that M is noetherian. Thus, over a right artinian ring, every artinian R-module is noetherian. This proof can be easily adapted to show that when the ring is right artinian, every noetherian R-module is artinian.   □

Corollary 6.6.6. *If R is a right artinian ring, then R is right noetherian.*

Remark. Researchers have also investigated the case as to when a right noetherian ring is right artinian. For example, Akizuki proved that a commutative noetherian ring is artinian if and only its prime ideals are maximal. (See [34].) For the noncommutative case, it is known that a right noetherian ring R is right artinian if and only if for every prime ideal $\mathfrak{p}$ of R, $R/\mathfrak{p}$ is right artinian. Moreover, Kertesz has shown in [64] that a right noetherian ring R is right artinian if and only if for every prime ideal $\mathfrak{p}$ of R, $R/\mathfrak{p}$ is right artinian.

Problem Set 6.6

1. Prove Lemma 6.6.3 and Lemma 6.6.4 for the noetherian case.

2. Prove Proposition 6.6.5 for the noetherian case.

3. If M is a semisimple left R-module and $E = \operatorname{End}_R(M)$, prove that M_E is a semisimple E-module.

4. If R is an integral domain, prove that R is semisimple if and only if R is a field.

5. (a) Show that the center of a simple ring is a field.

   (b) Prove that the center of a semisimple ring is a finite direct product of fields.

6. Give necessary and sufficient conditions on n for $\mathbb{Z}_n$ to be semisimple. [Hint: $\mathbb{Z}_n$ is artinian, so $\mathbb{Z}_n$ will be semisimple if and only if $\operatorname{Rad}(\mathbb{Z}_n) = 0$.]

Chapter 7

Envelopes and Covers

There are several types of envelopes and covers. There are injective envelopes, projective envelopes, flat envelopes, injective covers, flat covers, projective covers, etc. An excellent bibliography as well as a novel approach to envelopes and covers can be found in [12] and [46]. As a way of introducing this subject, we restrict our attention to developing injective envelopes and projective covers of modules whenever the latter can be shown to exist. Injective envelopes were first discovered by Eckmann and Schöpf [55] and rings over which every module has a projective cover were characterized by Bass [51]. We will also develop quasi-injective envelopes and quasi-projective covers, developed by Johnson and Wong [61] and by Wu and Jans [72], respectively.

7.1 Injective Envelopes

We have seen that injective modules exist, though not every module is injective. Even though there exist modules that are not injective, it is the case that for every R-module M there is an embedding $M \to E$, where E is an injective module. Among the injective modules that contain a submodule isomorphic to M, there is one, called the *injective envelope* of M, that is unique up to isomorphism. This envelope is, in some sense, the "*best approximation*" of M by an injective module. If the class of injective modules that contain a copy of M is ordered by inclusion, we will show that there exists at least one injective module that is a minimal element of this class.

Definition 7.1.1. An *injective envelope* of an R-module M is an injective module $E(M)$ together with a monomorphism $\varphi : M \to E(M)$ such that $E(M)$ is an essential extension of $\varphi(M)$. An injective envelope $\varphi : M \to E(M)$ of M is said to be *unique up to isomorphism* if whenever $\varphi' : M \to E(M)'$ is another injective envelope of M, there exists an isomorphism $f : E(M)' \to E(M)$ such that $f\varphi' = \varphi$.

Lemma 7.1.2. *If M is an essential submodule of an R-module M' and $f : M \to N$ is a monomorphism of M into an R-module N, then any extension $g : M' \to N$ of f to M' is also a monomorphism.*

Proof. Let $g : M' \to N$ be an extension of a monomorphism $f : M \to N$. If $x \in M \cap \operatorname{Ker} g$, then $f(x) = g(x) = 0$, so $x = 0$ since f is an injection. Hence, $M \cap \operatorname{Ker} g = 0$, so $\operatorname{Ker} g = 0$ since M is an essential submodule of M'. $\square$

The following proposition is due to Eckmann and Schöpf [55].

Proposition 7.1.3 (Eckmann–Schöpf). *Every R-module has an injective envelope.*

Proof. If M is any R-module, then, by Proposition 5.1.10, there is a monomorphism $\varphi : M \to E$, where E is an injective R-module. Let $\mathcal{S}$ be the set of all submodules N of E such that N is an essential extension of $\varphi(M)$. Then $\mathcal{S} \neq \varnothing$ since $\varphi(M) \in \mathcal{S}$. If $\{N_\alpha\}_\Delta$ is a chain in $\mathcal{S}$, then $\bigcup_\Delta N_\alpha \in \mathcal{S}$ serves as an upper bound for the chain. Hence, $\mathcal{S}$ is inductive, so it follows from Zorn's lemma that $\mathcal{S}$ has a maximal element, say $E(M)$.

We claim that $\varphi : M \to E(M)$ is an injective envelope of M. Observe first that if E_c is a complement of $E(M)$ in E, then $E(M) \cong (E(M) \oplus E_c)/E_c$ is an essential submodule of E/E_c. This follows since if N/E_c is a submodule of E/E_c such that $((E(M) \oplus E_c)/E_c) \cap (N/E_c) = 0$, then $(E(M) \oplus E_c) \cap N \subseteq E_c$. Modularity (Example 10 in Section 1.4) gives $(E(M) \cap N) \oplus E_c = (E(M) \oplus E_c) \cap N$, so we have $E_c \subseteq (E(M) \cap N) \oplus E_c \subseteq E_c$. Therefore, $E(M) \cap N = 0$, so by the maximality of E_c, $E_c = N$. Thus, $N/E_c = 0$ and so $(E(M) \oplus E_c)/E_c$ is essential in E/E_c.

Next, we claim that $E = E(M) \oplus E_c$. To see this, consider the commutative diagram

$$
\begin{array}{ccc}
0 \longrightarrow (E(M) \oplus E_c)/E_c & \overset{i}{\longrightarrow} & E/E_c \\
\Big\downarrow{\scriptstyle\psi} & & \Big\downarrow{\scriptstyle g} \\
E(M) & \overset{i'}{\longrightarrow} & E
\end{array}
$$

where ψ is the isomorphism $E(M) \cong (E(M) \oplus E_c)/E_c$, i and i' are canonical injections and g is the map given by the injectivity of E. Note that, $i'\psi$ is an injection, so Lemma 7.1.2 shows that g is an injection as well. From the diagram we have $E(M) = \mathrm{Im}(i'\psi) = g((E(M) \oplus E_c)/E_c) \subseteq g(E/E_c)$. But $\varphi(M)$ is essential in $E(M)$ and it follows that $E(M) = g((E(M) \oplus E_c)/E_c)$ is essential in $g(E/E_c)$, since $(E(M) \oplus E_c)/E_c$ is essential in E/E_c. Hence, $\varphi(M)$ is essential in $g(E/E_c)$. Now $E(M)$ is a maximal essential extension of $\varphi(M)$ in E, so by the maximality of $E(M)$, $g((E(M) \oplus E_c)/E_c) = g(E/E_c)$. Hence, we see that $E(M) \oplus E_c = E$. Since Corollary 5.1.14 shows that a direct summand of an injective module is injective, we have that $E(M)$ is an injective R-module. Thus, $\varphi : M \to E(M)$ is an injective envelope of M. $\qquad\square$

Proposition 7.1.4. *Injective envelopes are unique up to isomorphism.*

Proof. Let M be an R-module and suppose that $\varphi_1 : M \to E_1$ and $\varphi_2 : M \to E_2$ are injective envelopes of M. Then Lemma 7.1.2 shows that the diagram

$$
\begin{array}{ccc}
0 \longrightarrow M & \xrightarrow{\;\varphi_2\;} & E_2 \\
\varphi_1 \downarrow & \swarrow f & \\
E_1 & &
\end{array}
$$

can be completed commutatively by a monomorphism $f : E_2 \to E_1$. But this indicates that $f(E_2)$ is an injective submodule of E_1 and as such is, by Proposition 5.1.2, a direct summand of E_1. If $f(E_2) \oplus N = E_1$, then $\varphi_1(M) \cap N = 0$ since $\varphi_1(M) \subseteq f(E_2)$. But $\varphi_1(M)$ is essential in E_1, so $N = 0$. Hence, f is also an epimorphism. $\qquad\qquad\qquad\qquad\qquad\qquad\qquad\qquad\qquad\qquad\qquad\square$

Since an injective envelope $\varphi : M \to E(M)$ of an R-module M is unique up to isomorphism, we can speak of the injective envelope of M. There is no loss of generality in identifying M with $\varphi(M)$ and considering M to be a submodule of $E(M)$. The map φ can now be replaced by the canonical injection $i : M \to E(M)$.

The proof of the following proposition is left as an exercise.

Proposition 7.1.5. *The following properties hold for injective envelopes.*

(1) *If M is a submodule of an injective R-module E, then $E \cong E(M) \oplus E'$ for some (necessarily injective) submodule E' of E.*

(2) *If M is an essential submodule of an R-module N, then $E(M) \cong E(N)$.*

(3) *If $\{M_\alpha\}_\Delta$ is a family of R-modules, then $\bigoplus_\Delta E(M_\alpha)$ embeds in $E(\bigoplus_\Delta M_\alpha)$ and the embedding is an isomorphism if $\bigoplus_\Delta E(M_\alpha)$ is injective.*

It follows immediately from (3) of the preceding proposition, that if $\{M_\alpha\}_\Delta$ is a family of R-modules and if the indexing set Δ is finite, then $\bigoplus_\Delta E(M_\alpha) \cong E(\bigoplus_\Delta M_\alpha)$ since finite direct sums of injective modules are injective. We are now in a position to show that arbitrary direct sums of injective modules are injective if and only if the ring is right noetherian, a result due to Matlis [68] and Papp [70].

Proposition 7.1.6 (Matlis–Papp). *The following are equivalent for a ring R.*

(1) *Every direct sum of injective R-modules is injective.*

(2) *If $\{M_\alpha\}_\Delta$ is a family of R-modules, then $\bigoplus_\Delta E(M_\alpha) \cong E(\bigoplus_\Delta M_\alpha)$.*

(3) *R is a right noetherian ring.*

Proof. (1) $\Rightarrow$ (2). This follows immediately from (3) of Proposition 7.1.5.

(2) $\Rightarrow$ (3). Let $A_1 \subseteq A_2 \subseteq A_3 \subseteq \cdots$ be an increasing chain of right ideals of R. If $A = \bigcup_{\mathbb{N}} A_i$, then A is a right ideal of R and if $a \in A$, then $a \in A_i$ for almost

all $i \in \mathbb{N}$. Consequently, $a + A_i = 0$ in R/A_i for almost all $i \in \mathbb{N}$. Thus, we have R-linear mappings

$$0 \to A \xrightarrow{f} \bigoplus_{\mathbb{N}} R/A_i \subseteq \bigoplus_{\mathbb{N}} E(R/A_i) \xrightarrow{\pi_i} E(R/A_i) \to 0,$$

where $f : A \to \bigoplus_{\mathbb{N}} R/A_i$ is defined by $f(a) = (a + A_i)$ and π_i is the canonical projection for each i. Note that $\pi_i f(A) = A/A_i$ for $i = 1, 2, 3, \ldots$. By assumption $\bigoplus_{\mathbb{N}} E(R/A_i) \cong E(\bigoplus_{\mathbb{N}} R/A_i)$, so $\bigoplus_{\mathbb{N}} E(R/A_i)$ is injective. Hence, Baer's criteria gives an $x \in \bigoplus_{\mathbb{N}} E(R/A_i)$ such that $f(a) = xa$ for all $a \in A$. Since $x \in \bigoplus_{\mathbb{N}} E(R/A_i)$, there is a positive integer n such that $\pi_i(x) = 0$ for all $i \geq n$. If $i \geq n$, then $A/A_i = \pi_i f(A) = \pi_i(xA) = \pi_i(x)A = 0$. Therefore, $A_i = A$ for all $i \geq n$, so R is right noetherian.

(3) $\Rightarrow$ (1). Let A be a right ideal of R and suppose that $\{E_\alpha\}_\Delta$ is a family of injective R-modules. If $f : A \to \bigoplus_\Delta E_\alpha$ is an R-linear mapping, then, since A is finitely generated, there is a finite subset Γ of Δ such that $f(A) \subseteq \bigoplus_\Gamma E_\alpha$. But $\bigoplus_\Gamma E_\alpha$ is injective, so there is an $x \in \bigoplus_\Gamma E_\alpha$ such that $f(a) = xa$ for all $a \in A$. Since $\bigoplus_\Gamma E_\alpha$ embeds in $\bigoplus_\Delta E_\alpha$, we can assume that $x \in \bigoplus_\Delta E_\alpha$. Thus, Baer's criteria for injectivity holds, so $\bigoplus_\Delta E_\alpha$ is injective.   □

Problem Set 7.1

1. (a) If M and N are R-modules, then we say that N *extends* M if there is a monomorphism $f : M \to N$. Prove that every injective module E that extends M also extends $E(M)$ and that $E(M)$ is isomorphic to a direct summand of E. Conclude that every injective R-module E that contains a copy of M also contains a copy of $E(M)$.

   (b) Prove that every short exact sequence of the form

$$0 \to M_1 \to M \to M_2 \to 0$$

   splits if and only if M_1 is injective.

2. (a) Let $\{M_\alpha\}_\Delta$ be a family of R-modules. If N_α is an essential submodule of M_α for each $\alpha \in \Delta$, show that $\bigoplus_\Delta N_\alpha$ is an essential submodule of $\bigoplus_\Delta M_\alpha$.

   (b) Show that $\mathbb{Z}_\mathbb{Z}$ is an essential submodule of $\mathbb{Q}_\mathbb{Z}$, but that the $\mathbb{Z}$-module $\mathbb{Z}^\mathbb{N}$ is not an essential $\mathbb{Z}$-submodule of $\mathbb{Q}^\mathbb{N}$. [Hint: Consider Exercise 1 in Problem Set 5.1 and $p/q\mathbb{Z} \times p/q^2\mathbb{Z} \times \cdots \times p/q^n\mathbb{Z} \times \cdots$.] Conclude that the property stated in (a) for direct sums does not always hold for direct products.

3. Prove Proposition 7.1.5.

4. Prove that an R-module M is injective if and only if M has no proper essential extensions. Conclude that an injective module E containing M is an injective envelope of M if and only if E is a *maximal essential extension* of M.

5. Suppose that M is a submodule of an injective R-module E. Show that E is an injective envelope of M if and only if E is minimal among the injective R-modules that contain M. Conclude that an injective envelope of M is a *minimal injective extension* of M.

6. If M is a submodule of an R-module E, prove that E is an injective envelope of M if and only if E is an injective essential extension of M. Conclude that an injective envelope of M is an *injective essential extension* of M.

Remark. Exercises 4, 5 and 6 show that the following are equivalent.

(1) E is a maximal essential extension of M.

(2) E is a minimal injective extension of M.

(3) E is an injective essential extension of M.

(4) E is an injective envelope of M.

7. Let $\mathcal{X}$ be a class of R-modules that satisfies the following three conditions.

(i) $\mathcal{X}$ is *closed under isomorphisms*: If $X \in \mathcal{X}$ and $X' \cong X$, then $X' \in \mathcal{X}$.

(ii) $\mathcal{X}$ is *closed under finite direct sums*: If $\{X_i\}_{i=1}^{n}$ is a family of modules in $\mathcal{X}$, then $\bigoplus_{i=1}^{n} X_i \in \mathcal{X}$.

(iii) $\mathcal{X}$ is *closed under direct summands*: If $X \in \mathcal{X}$ and X' is a direct summand of X, then $X' \in \mathcal{X}$.

For an R-module M, an $\mathcal{X}$-*envelope* of M is an $X \in \mathcal{X}$ together with an R-linear mapping $\varphi : M \to X$ such that the following two conditions are satisfied.

(1) For any R-linear mapping $\varphi' : M \to X'$ with $X' \in \mathcal{X}$, the diagram

$$
\begin{array}{ccc}
M & \xrightarrow{\ \varphi\ } & X \\
{\scriptstyle \varphi'}\downarrow & \swarrow{\scriptstyle f} & \\
X' & &
\end{array}
$$

can be completed commutatively by an R-linear mapping $f : X \to X'$.

(2) The diagram

$$
\begin{array}{ccc}
M & \xrightarrow{\ \varphi\ } & X \\
{\scriptstyle \varphi}\downarrow & \swarrow & \\
X & &
\end{array}
$$

can be completed commutatively only by automorphisms of X.

If $\varphi : M \to X$ satisfies (1) but perhaps not (2), then $\varphi : M \to X$ is said to be an $\mathcal{X}$-*preenvelope* of M. Additional information on $\mathcal{X}$-(pre)envelopes can be found in [12] and [46].

(a) If an R-module M has an $\mathcal{X}$-envelope $\varphi : M \to X$, prove that X is unique up to isomorphism. [Hint: Let $\varphi : M \to X$ and $\varphi' : M \to X'$ be $\mathcal{X}$-envelopes of M and consider the diagrams

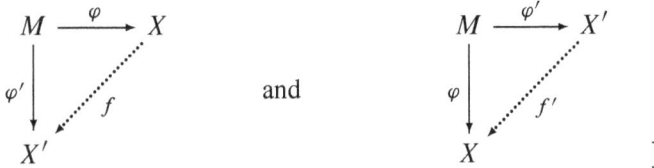

$$
\begin{array}{ccc}
M & \xrightarrow{\ \varphi\ } & X \\
{\scriptstyle \varphi'}\downarrow & \nearrow{\scriptstyle f} & \\
X' & &
\end{array}
\qquad \text{and} \qquad
\begin{array}{ccc}
M & \xrightarrow{\ \varphi'\ } & X' \\
{\scriptstyle \varphi}\downarrow & \nwarrow{\scriptstyle f'} & \\
X & &
\end{array}
\qquad]
$$

(b) If $\mathcal{X}$ contains all the injective modules, prove that for an $\mathcal{X}$-preenvelope $\varphi : M \to X$ of M, the R-linear mapping φ is an injection. [Hint: By Proposition 5.1.10, there is a monomorphism $\varphi' : M \to E$, where E is an injective R-module, so consider (1).]

(c) Suppose that M has an $\mathcal{X}$-envelope $\varphi : M \to X$. Prove that if $\varphi' : M \to X'$ is an $\mathcal{X}$-preenvelope of M, then X is isomorphic to a direct summand of X'. [Hint: Consider the diagram

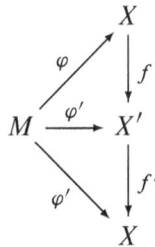

$$
\begin{array}{ccc}
 & & X \\
 & {\scriptstyle \varphi}\nearrow & \downarrow{\scriptstyle f} \\
M & \xrightarrow{\ \varphi'\ } & X' \\
 & {\scriptstyle \varphi}\searrow & \downarrow{\scriptstyle f'} \\
 & & X
\end{array}
$$

and note that $f'f$ is an automorphism.]

8. Let $\mathcal{E}$ be the class of injective R-modules and note that $\mathcal{E}$ satisfies conditions (i), (ii) and (iii) given in Exercise 7 for a class of R-modules. Prove that the following are equivalent.

(a) M has an $\mathcal{E}$-envelope $\varphi : M \to E$, where $\mathcal{E}$-envelope is defined as in Exercise 7.

(b) M has an injective envelope $\varphi : M \to E(M)$ in the sense of Definition 7.1.1.

7.2 Projective Covers

We know that every R-module M is the homomorphic image of a projective module. Among the projective modules that cover M, there may be one that is, in some sense, minimal. Such a cover of M, if it exists, can be viewed as a *"best approximation"* of M by a projective module. As we will see, there are modules that fail to have a projective cover.

Definition 7.2.1. A *projective cover* of an R-module M is a projective R-module $P(M)$ together with an epimorphism $\varphi : P(M) \rightarrow M$ such that $\operatorname{Ker} \varphi$ is small in $P(M)$. A projective cover $\varphi : P(M) \rightarrow M$ of M is said to be *unique up to isomorphism* if whenever $\varphi' : P(M)' \rightarrow M$ is another projective cover of M, there is an isomorphism $f : P(M)' \rightarrow P(M)$ such that $\varphi f = \varphi'$. The projective module $P(M)$ will often be denoted simply by P, when M is understood.

We will see later in this section that a projective cover of a module may fail to exist.

Proposition 7.2.2. *A projective cover of an R-module M is unique up to isomorphism whenever it can be shown to exist.*

Proof. Suppose that $\varphi_1 : P_1 \rightarrow M$ and $\varphi_2 : P_2 \rightarrow M$ are projective covers of M. Then the projectivity of P_2 gives an R-linear mapping $f : P_2 \rightarrow P_1$ such that $\varphi_1 f = \varphi_2$. If $x \in P_1$, then $\varphi_1(x) \in M$, so since φ_2 is an epimorphism, there is a $y \in P_2$ such that $\varphi_2(y) = \varphi_1(x)$. If $z = x - f(y)$, then $z \in \operatorname{Ker} \varphi_1$, so $x = z + f(y) \in \operatorname{Ker} \varphi_1 + \operatorname{Im} f$. Thus, $P_1 = \operatorname{Ker} \varphi_1 + \operatorname{Im} f$ and so f is an epimorphism since $\operatorname{Ker} \varphi_1$ is small in P_1. But P_1 is projective, so the epimorphism $f : P_2 \rightarrow P_1$ splits. If $f' : P_1 \rightarrow P_2$ is a splitting map for f, then f' is a monomorphism and $P_2 = \operatorname{Im} f' \oplus \operatorname{Ker} f$. But $\operatorname{Ker} f \subseteq \operatorname{Ker} \varphi_2$, so $\operatorname{Ker} f$ is small in P_2 since submodules of small submodules are small. Hence, $P_2 = \operatorname{Im} f' \cong P_1$. $\square$

Examples

1. **Projective Modules.** Every projective module has a projective cover, namely, itself.

2. **Local Rings and Projective Covers.** Let R be a commutative ring that has a unique maximal ideal $\mathfrak{m}$. Then R together with the natural mapping $R \rightarrow R/\mathfrak{m}$ is a projective cover of $R/\mathfrak{m}$. Actually, much more can be said about projective covers when R is a local ring. If M is a finitely generated R-module, then M has a projective cover. Indeed, if M is finitely generated, then $M/M\mathfrak{m}$ is a finite dimensional vector space over the field $R/\mathfrak{m}$. So there is a positive integer n such that $(R/\mathfrak{m})^{(n)} \cong M/M\mathfrak{m}$. In addition, the canonical mapping $f : R^{(n)} \rightarrow M/M\mathfrak{m}$ has small kernel. Since $R^{(n)}$ is a projective R-module, there is an R-module homomorphism $\varphi : R^{(n)} \rightarrow M$ such that the diagram

$$R^{(n)}$$

$$M \xrightarrow{\ \eta\ } M/M\mathfrak{m} \longrightarrow 0$$

is commutative, where $\eta : M \rightarrow M/M\mathfrak{m}$ is the natural mapping. Since $\operatorname{Ker} \varphi \subseteq \operatorname{Ker} f$, $\operatorname{Ker} \varphi$ is a small submodule of $R^{(n)}$ and it follows that $\varphi : R^{(n)} \rightarrow M$ is a projective cover of M. Later in the text we will extend the concept of a commutative local ring to noncommutative rings.

The next proposition will subsequently prove to be useful. It shows that a finite direct sum of projective covers is a projective cover.

Proposition 7.2.3. *Let $\{M_k\}_{k=1}^{n}$ be a finite family of R-modules.*

(1) *If S_k is a small submodule of M_k, for $k = 1, 2, \ldots, n$, then $\bigoplus_{k=1}^{n} S_k$ is a small submodule of $\bigoplus_{k=1}^{n} M_k$.*

(2) *If each M_k has a projective cover $\varphi_k : P_k \rightarrow M_k$, then $\bigoplus_{k=1}^{n} M_k$ has a projective cover $\bigoplus_{k=1}^{n} \varphi_k : \bigoplus_{k=1}^{n} P_k \rightarrow \bigoplus_{k=1}^{n} M_k$ and if $\varphi : P \rightarrow \bigoplus_{k=1}^{n} M_k$ is a projective cover of $\bigoplus_{k=1}^{n} M_k$, then there is a family $\{\bar{P}_k\}_{k=1}^{n}$ of submodules of P such that $\bar{P}_k \cong P_k$ for each k.*

Proof. (1) Let $i_j : M_j \rightarrow \bigoplus_{k=1}^{n} M_k$ and $\pi_j : \bigoplus_{k=1}^{n} M_k \rightarrow M_j$ be the jth canonical injection and the jth canonical projection, respectively, for $j = 1, 2, \ldots, n$. Let $S = \bigoplus_{k=1}^{n} S_k$ and suppose that N is a submodule of $\bigoplus_{k=1}^{n} M_k$ such that $S + N = \bigoplus_{k=1}^{n} M_k$. Then $S_j + \pi_j(N) = M_j$, so $\pi_j(N) = M_j$ for $j = 1, 2, \ldots, n$. But $\sum_{k=1}^{n}(i_k \pi_k) = \mathrm{id}_{\bigoplus_{k=1}^{n} M_k}$, so we have $N = (\sum_{k=1}^{n} i_k \pi_k)(N) = \sum_{k=1}^{n} i_k \pi_k(N) = \sum_{k=1}^{n} i_k(M_k) = \bigoplus_{k=1}^{n} M_k$. Hence, $\bigoplus_{k=1}^{n} S_k$ is a small submodule of $\bigoplus_{k=1}^{n} M_k$.

(2) Let $\varphi_k : P_k \rightarrow M_k$ be a projective cover of M_k for $k = 1, 2, \ldots, n$. Then (1) and the fact that direct sums of projective R-modules are projective shows that $\bigoplus_{k=1}^{n} \varphi_k : \bigoplus_{k=1}^{n} P_k \rightarrow \bigoplus_{k=1}^{n} M_k$ is a projective cover of $\bigoplus_{k=1}^{n} M_k$. But projective covers are unique up to isomorphism, so if $\varphi : P \rightarrow \bigoplus_{k=1}^{n} M_k$ is a projective cover of $\bigoplus_{k=1}^{n} M_k$, then $P \cong \bigoplus_{k=1}^{n} P_k$. The existence of the family $\{\bar{P}_k\}_{k=1}^{n}$ of submodules of P such that $\bar{P}_k \cong P_k$ for each k follows easily. $\square$

The Radical of a Projective Module

Corollary 6.1.5 indicates that every free R-module F has the property that $\operatorname{Rad}(F) = FJ(R)$. In fact, every projective R-module has this property.

Proposition 7.2.4. *If M is a projective R-module, then $\operatorname{Rad}(M) = MJ(R)$.*

Proof. If M is a projective R-module, then Proposition 5.2.8 shows that there is an R-module N such that $R^{(\Delta)} \cong M \oplus N$ for some set Δ. Hence, by Proposition 6.1.4

and Corollary 6.1.5, we see that

$$\mathrm{Rad}(M) \oplus \mathrm{Rad}(N) = \mathrm{Rad}(M \oplus N)$$

$$\cong \mathrm{Rad}(R^{(\Delta)}), \quad \text{so}$$

$$\mathrm{Rad}(M) \oplus \mathrm{Rad}(N) \cong R^{(\Delta)} J(R).$$

Therefore,

$$\mathrm{Rad}(M) \oplus \mathrm{Rad}(N) \cong (M \oplus N) J(R)$$

$$\subseteq M J(R) \oplus N J(R).$$

Thus, if $x \in \mathrm{Rad}(M)$, then $x = y + z$, where $y \in MJ(R)$ and $z \in NJ(R)$. But by Lemma 6.1.9, $MJ(R) \subseteq \mathrm{Rad}(M)$ and $NJ(R) \subseteq \mathrm{Rad}(N)$, so we see that $z = x - y \in \mathrm{Rad}(M) \cap \mathrm{Rad}(N) = 0$. It follows that $\mathrm{Rad}(M) \subseteq MJ(R)$ and so $\mathrm{Rad}(M) = MJ(R)$.   □

The fact that $\mathrm{Rad}(M) = MJ(R)$ when M is a projective module can be used to show that a projective cover may fail to exist.

Example

3. **A Module that does not have a Projective Cover.** Suppose that R is a Jacobson semisimple ring. If $\varphi : P \to M$ is a projective cover of M, then $\mathrm{Ker}\,\varphi$ is a small submodule of P and $\mathrm{Ker}\,\varphi \subseteq \mathrm{Rad}(P) = PJ(R) = 0$. Consequently, φ is an isomorphism and so over a Jacobson semisimple ring an R-module M has a projective cover if and only if it is projective. The ring $\mathbb{Z}$ is Jacobson semisimple, so the only $\mathbb{Z}$-modules with projective covers are the free $\mathbb{Z}$-modules. Thus, $\mathbb{Z}_n$ does not have a projective cover, since $\mathbb{Z}_n$ is not a free $\mathbb{Z}$-module for any integer $n \geq 2$.

Since there are modules that fail to have a projective cover, this brings up the question "Are there rings over which every module has a projective cover?" Such rings do indeed exist and our goal now is to characterize these rings. In the process, rings over which every finitely generated module has a projective cover will also be described. Before we can begin, we need several additional results concerning the radical of a projective module.

Proposition 7.2.5. *If M is a projective R-module, then $f \in J(\mathrm{End}_R(M))$ if and only if $\mathrm{Im}\,f$ is a small submodule of M.*

Proof. Let $H = \text{End}_R(M)$, suppose that $f \in J(H)$ and let N be a submodule of M such that $\text{Im } f + N = M$. If $\eta : M \to M/N$ is the natural mapping and $x + N \in M/N$, then $x = f(y) + z$ for some $y \in M$ and $z \in N$. Hence, we see that $x + N = \eta f(y)$ and so $\eta f : M \to M/N$ is an epimorphism. If $g : M \to M$ is the completing map for the diagram

$$
\begin{array}{ccc}
 & & M \\
 & \nearrow^{g} & \downarrow{\eta} \\
M & \xrightarrow{\eta f} & M/N \longrightarrow 0
\end{array}
$$

given by the projectivity of M, then $\eta(\text{id}_H - fg) = 0$. But Proposition 6.1.7 shows that $\text{id}_H - fg$ has a right inverse in H. Thus, $\eta = 0$, so $N = M$. Therefore, $\text{Im } f$ is a small submodule of M.

Conversely, suppose that $\text{Im } f$ is small in M and let $fH = \{fg \mid g \in H\}$. In view of Corollary 6.1.15, we will have $f \in fH \subseteq J(H)$ if we can show that fH is small in H. If A is a right ideal of H such that $fH + A = H$, then $\text{id}_M = fg + h$ for some $g \in H$ and $h \in A$. This gives $M = \text{id}_M(M) = fg(M) + h(M) \subseteq \text{Im } f + h(M) \subseteq M$, so $h(M) = M$ since $\text{Im } f$ is small in H. Hence, h is an epimorphism and M is projective, so the short exact sequence $0 \to \text{Ker } h \to M \xrightarrow{h} M \to 0$ splits. Consequently, there is a monomorphism $h' : M \to M$ such that $\text{id}_M = hh' \in A$ and so $A = H$. Thus, fH is small in H and this completes the proof. $\square$

Corollary 7.2.6. *If M is a projective R-module and $MJ(R)$ is a small submodule of M, then*

(1) $J(\text{End}_R(M)) = \text{Hom}_R(M, MJ(R))$ *and*

(2) $\text{End}_R(M)/J(\text{End}_R(M)) \cong \text{End}_R(M/MJ(R))$.

Proof. Let $H = \text{End}_R(M)$. By Proposition 7.2.4, $\text{Rad}(M) = MJ(R)$, so if $MJ(R)$ is small in M, then using Corollary 6.1.15 we can conclude that a submodule of M is small if and only if it is contained in $MJ(R)$. Thus, if $f \in J(H)$, then the proposition indicates that $\text{Im } f \subseteq MJ(R)$, so $f \in \text{Hom}_R(M, MJ(R))$. Conversely, if $f \in \text{Hom}_R(M, MJ(R))$, then $\text{Im } f \subseteq MJ(R)$, so the proposition gives $f \in J(H)$ and we have (1).

Next, consider the ring homomorphism $\varphi : H \to \text{End}_R(M/MJ(R))$ defined by $\varphi(f) = \bar{f}$, where $\bar{f} : M/MJ(R) \to M/MJ(R)$ is such that $\bar{f}(x + MJ(R)) = f(x) + MJ(R)$. Since M is a projective R-module, it is easy to show that φ is an epimorphism. Moreover, $\text{Ker } \varphi = \text{Hom}_R(M, MJ(R)) = J(H)$, so the isomorphism asserted in (2) is immediate. $\square$

Corollary 7.2.7. $J(\mathbb{M}_n(R)) = \mathbb{M}_n(J(R))$.

Proof. Using Proposition 2.1.12, we see that

$$\mathrm{End}_R(R^{(n)}) = \mathrm{Hom}_R(R^{(n)}, R^{(n)}) \cong \prod_{i=1}^{n} \prod_{j=1}^{n} \mathrm{Hom}_R(R_i, R_j),$$

where $R_i = R_j = R$ for each i and j. But $\mathrm{Hom}_R(R, R) \cong R$, so

$$\mathrm{End}_R(R^{(n)}) \cong \prod_{i=1}^{n} \prod_{j=1}^{n} R_{ij} \cong \mathbb{M}_n(R),$$

with $R_{ij} = R$ for each i and j. If $\varphi : \mathbb{M}_n(R) \to \mathrm{End}_R(R^{(n)})$ is an isomorphism, then $(a_{ij}) \in J(\mathbb{M}_n(R))$ if and only if $\varphi((a_{ij})) \in J(\mathrm{End}_R(R^{(n)}))$. Since $R^{(n)}$ is a projective R-module, the preceding corollary shows that

$$J(\mathbb{M}_n(R)) \cong J(\mathrm{End}_R(R^{(n)})) = \mathrm{Hom}_R(R^{(n)}, R^{(n)}J(R)).$$

But $R^{(n)}J(R) = J(R)^{(n)}$, so

$$J(\mathbb{M}_n(R)) = \mathrm{Hom}_R(R^{(n)}, J(R)^{(n)}).$$

Hence, $(a_{ij}) \in J(\mathbb{M}_n(R))$ if and only if $\varphi((a_{ij})) \in \mathrm{Hom}_R(R^{(n)}, J(R)^{(n)})$ and

$$\mathrm{Hom}_R(R^{(n)}, J(R)^{(n)}) \cong \prod_{i=1}^{n} \prod_{j=1}^{n} \mathrm{Hom}_R(R_i, J(R)_j)$$

$$\cong \prod_{i=1}^{n} \prod_{j=1}^{n} J(R)_{ij}$$

$$\cong \mathbb{M}_n(J(R)).$$

It follows that $(a_{ij}) \in J(\mathbb{M}_n(R))$ if and only if $(a_{ij}) \in \mathbb{M}_n(J(R))$ and so $J(\mathbb{M}_n(R)) = \mathbb{M}_n(J(R))$. $\square$

We have seen (Proposition 6.1.2) that every nonzero finitely generated module contains a maximal submodule. The same is true of projective R-modules.

Proposition 7.2.8. *If M is a nonzero projective R-module, then M contains a maximal submodule.*

Proof. Let M be a nonzero projective R-module and suppose that M does not have a maximal submodule. Then $\mathrm{Rad}(M) = M$. But Proposition 7.2.4 shows that $\mathrm{Rad}(M) = MJ(R)$, so we have $MJ(R) = M$. By Proposition 5.2.8, there is a free R-module F such that $F \cong M \oplus N$ for some R-module N. We can, without loss of generality, identify M and N with their images in F and assume that

$F = M \oplus N$. Let $\pi : F \to M$ be the canonical projection and assume that $\{x_\alpha\}_\Delta$ is a basis for F. If $x \in M$, then there is a finite subset Γ of Δ such that $x = \sum_\Gamma x_\alpha a_\alpha$. Now $M = MJ(R)$, so for each $\alpha \in \Gamma$, there is a finite subset $\Gamma_\alpha \subseteq \Delta$ such that $\pi(x_\alpha) = \sum_{\Gamma_\alpha} x_\beta a_{\alpha\beta}$ with each $a_{\alpha\beta}$ in $J(R)$. This shows that there is finite subset $\Lambda \subseteq \Delta$ such that

$$0 = x - \pi(x) = \sum_{\alpha \in \Lambda} x_\alpha a_\alpha - \sum_{\alpha \in \Lambda} \pi(x_\alpha) a_\alpha$$

$$= \left(\sum_{\alpha \in \Lambda} \left(\sum_{\beta \in \Lambda} x_\beta \delta_{\alpha\beta} \right) a_\alpha \right) - \left(\sum_{\alpha \in \Lambda} \left(\sum_{\beta \in \Lambda} x_\beta a_{\alpha\beta} \right) a_\alpha \right)$$

$$= \sum_{\beta \in \Lambda} x_\beta \left(\sum_{\alpha \in \Lambda} (\delta_{\alpha\beta} - a_{\alpha\beta}) a_\alpha \right),$$

where $\delta_{\alpha\beta}$ is such that $\delta_{\alpha\beta} = 0$ if $\alpha \neq \beta$ and $\delta_{\alpha\beta} = 1$ when $\alpha = \beta$. If $\operatorname{card}(\Lambda) = n$, then, since the x_β are linearly independent, this gives n equations

$$\sum_{\alpha \in \Lambda} (\delta_{\alpha\beta} - a_{\alpha\beta}) a_\alpha = 0 \quad \text{for each } \beta \in \Lambda.$$

This in turn leads to a matrix equation

$$(I_n - (a_{\alpha\beta}))(a_\alpha) = 0,$$

where I_n is the $n \times n$ identity matrix of $\mathbb{M}_n(R)$ and $(a_{\alpha\beta})$ is a matrix in $\mathbb{M}_n(J(R))$. Hence, we see from Corollary 7.2.7 that $(a_{\alpha\beta}) \in J(\mathbb{M}_n(R))$, so Proposition 6.1.8 indicates that $I_n - (a_{\alpha\beta})$ has a left inverse in $\mathbb{M}_n(R)$. Thus, $(a_\alpha) = 0$ and so $x = 0$. But x was chosen arbitrarily in M, so $M = 0$, contradicting the assumption that $M \neq 0$. Consequently, a nonzero projective R-module must have a maximal submodule. □

Corollary 7.2.9. *If M is a projective R-module, then* $\operatorname{Rad}(M) \subsetneqq M$.

We now consider rings over which every finitely generated module has a projective cover.

Semiperfect Rings

Definition 7.2.10. A ring R is said to be a *semiperfect ring* if every finitely generated R-module has a projective cover.

Remark. Later we will see that semiperfect rings are left-right symmetric, so that the omission of the modifier "right" is justified in the definition of a semiperfect ring.

Previously we called a commutative ring R local if R had a unique maximal ideal. It was shown in Example 2 that local rings are semiperfect. We now extend the concept of a local ring to noncommutative rings. These noncommutative local rings are also semiperfect.

Proposition 7.2.11. *The following are equivalent for a ring R.*

(1) *R has a unique maximal right ideal.*

(2) *$J(R)$ is a unique maximal right ideal of R.*

(3) *$R/J(R)$ is a division ring.*

(4) *$J(R) = \{a \in R \mid a$ is not a unit in $R\}$.*

(5) *If $a \in R$, then either a or $1 - a$ is a unit.*

(6) *R has a unique maximal left ideal.*

(7) *$J(R)$ is a unique maximal left ideal of R.*

(8) *If $U(R)$ is the the group of units of R, then $a + b \in U(R)$ implies that $a \in U(R)$ or $b \in U(R)$.*

Proof. (1) $\Rightarrow$ (2). If R has a unique maximal right ideal $\mathfrak{m}$, then $\mathfrak{m} = J(R)$, so (2) is immediate. Now assume (2). If $a + J(R) \in R/J(R)$ and $a \notin J(R)$, then $aR + J(R) = R$ and so $ab + c = 1$ for some $b \in R$ and $c \in J(R)$. Hence, $ab - 1 \in J(R)$ which implies that $(a + J(R))(b + J(R)) = 1 + J(R)$. Thus, every nonzero $a + J(R)$ has a right inverse in $R/J(R)$. Therefore, $R/J(R)$ is a division ring, so we have that (2) $\Rightarrow$ (3). To see that (3) $\Rightarrow$ (1), let $\mathfrak{m}$ be a maximal right ideal of R. Then $J(R) \subseteq \mathfrak{m}$ and $\mathfrak{m}/J(R)$ is a maximal right ideal of $R/J(R)$. But $R/J(R)$ is a division ring and as such has no proper nonzero right ideals. Consequently, $\mathfrak{m}/J(R) = 0$ and we have $\mathfrak{m} = J(R)$. Thus, every maximal right ideal of R coincides with $J(R)$ and so R has a unique maximal right ideal. Therefore, (1) $\Leftrightarrow$ (2) $\Leftrightarrow$ (3) and in a similar fashion we can show (3) $\Leftrightarrow$ (6) $\Leftrightarrow$ (7). Next, assume that (2) holds. We claim that $J(R)$ is the set of nonunits of R. If $a \in R - J(R)$ and if aR is a proper right ideal, then we know that aR is contained in a maximal right ideal of R which must be $J(R)$. Hence, $a \in aR \subseteq J(R)$, a contradiction. Thus, $aR = R$ and so there is a $b \in R$ such that $ab = 1$. Now $J(R)$ is a proper ideal of R, so $b \notin J(R)$. If the same argument is now applied to b, then there is a $c \in R$ such that $bc = 1$. Consequently, b has a left and a right inverse, so it follows that b is a unit of R with $b^{-1} = a = c$. Therefore, a is a unit with inverse b and so every $a \in R - J(R)$ is a unit of R. Since $J(R)$ can contain only nonunits, we see that the nonunits of R form the ideal $J(R)$. Thus, (2) $\Rightarrow$ (4). Now suppose that (4) holds. If $a \in R$ and $a \notin J(R)$, then a is a unit since every nonunit of R is in $J(R)$. If $a \in R$ and $a \in J(R)$, then Proposition 6.1.7 shows that $1 - a$ is a unit. Hence, (4) $\Rightarrow$ (5). Finally, we claim that (5) $\Rightarrow$ (2). If $\mathfrak{m}$ is a maximal right ideal of R, then $J(R) \subseteq \mathfrak{m}$. If $a \in \mathfrak{m}$, then a is not a unit of R.

But if (5) holds, then $1 - a$ is a unit in R and it follows from Proposition 6.1.7 that $a \in J(R)$. Therefore, $J(R) = \mathfrak{m}$, so $J(R)$ is a unique maximal right ideal of R.

Finally, we show (3) $\Leftrightarrow$ (8). Suppose that (3) holds and let $a + b \in U(R)$. If $a \notin U(R)$, then (4) indicates that $a \in J(R)$. If $c \in R$ is such that $c(a + b) = (a + b)c = 1$, then $bc = 1 - ac$ has a right inverse in R. Thus, there is a $d \in R$ such that $bcd = (1 - ac)d = 1$. Thus, b has a right inverse in R. Similarly, b has a left inverse in R, so $b \in U(R)$. Conversely, suppose that (8) holds and let $a + J(R)$ be a nonzero element of $R/J(R)$. Then $a \notin J(R)$, so there is a maximal right ideal $\mathfrak{m}$ of R such that $a \notin \mathfrak{m}$. This gives $aR + \mathfrak{m} = R$ and so $ab + c = 1$ for some $b \in R$ and $c \in \mathfrak{m}$. It follows that $(a + \mathfrak{m})(b + \mathfrak{m}) = 1 + \mathfrak{m}$, so every nonzero element of $R/J(R)$ has a right inverse in $R/J(R)$ and this suffices to show that $R/J(R)$ is a division ring. $\square$

Definition 7.2.12. A ring R is said to be a *local ring* if R satisfies one of the equivalent conditions of Proposition 7.2.11.

From this point forward, the term "local ring" will mean noncommutative local ring unless stated otherwise.

Examples

4. If D is a division ring, then D is a local ring.

5. If R is a local ring, then $R[[X]]$, the ring of formal power series over R, is a local ring. This follows since $J(R[[X]])$ consists of all power series each of which has its constant term in $J(R)$. Hence, we see that $R/J(R) \cong R[[X]]/J(R[[X]])$ is a division ring.

Remark. A local ring contains no idempotents other that 0 and 1. To see this, suppose that e is an idempotent of a local ring R. Due to Proposition 7.2.11 either e or $1 - e$ is a unit. If e is a unit, then there is an $a \in R$ such that $ea = ae = 1$. Hence, $ea = e$ gives $e(a - 1) = 0$ which in turn gives $a = 1$. But if $a = 1$, then $ea = 1$ produces $e = 1$. Likewise, if $1 - e$ is a unit, then $1 - e = 1$, so $e = 0$.

Local rings are also semiperfect. The proof given in Example 2 for the commutative case carries over with only minor changes required. Because of Proposition 7.2.11, local rings are left-right symmetric. It follows that if R is a local ring, then every finitely generated R-module has a projective cover as does every finitely generated left R-module. Right artinian rings are also semiperfect. To show this, we need the concept of *lifting idempotents*.

Definition 7.2.13. If I is an ideal of R and $\bar{f}$ is an idempotent in the ring R/I, then we say that $\bar{f}$ *can be lifted* to R if there is an idempotent $e \in R$ such that $e + I = \bar{f}$.

If e is an idempotent of R such that $e + I = \bar{f}$, then we also say that e lifts $\bar{f}$ modulo I or that $\bar{f}$ can be lifted to R modulo I.

There is no assurance that idempotents in R/I can be lifted to R. For example, the ring of integers $\mathbb{Z}$ has only 0 and 1 as idempotents while $[0], [1], [3]$ and $[4]$ are idempotents in $\mathbb{Z}_6$. Clearly, the idempotents $[3]$ and $[4]$ of $\mathbb{Z}_6$ do not lift to $\mathbb{Z}$. The ability to lift idempotents from R/I to R is often determined by properties of the ideal I.

Proposition 7.2.14. *If $\mathfrak{n}$ is a nil ideal of R, then idempotents of $R/\mathfrak{n}$ can be lifted to R.*

Proof. Let $\mathfrak{n}$ be a nil ideal of R, suppose that $\bar{f}$ is an idempotent of $R/\mathfrak{n}$ and let $u \in R$ be such that $u + \mathfrak{n} = \bar{f}$. Then $u^2 + \mathfrak{n} = \bar{f}^2 = \bar{f} = u + \mathfrak{n}$, so $u^n + \mathfrak{n} = \bar{f}$ for any integer $n \geq 1$. Furthermore, $u - u^2 \in \mathfrak{n}$, so there is a positive integer n such that $(u - u^2)^n = 0$. Thus, we see, via a binomial expansion, that

$$u^n - u^{n+1}g = 0, \tag{7.1}$$

where $g = g(u)$ is a polynomial in u. Furthermore, $ug = gu$ and (7.1) gives $u^n = u^{n+1}g$. So if $e = u^n g^n$, then

$$e^2 = u^{2n}g^{2n} = u^{n-1}(u^{n+1}g)g^{2n-1} = u^{n-1}u^n g^{2n-1} = u^{2n-1}g^{2n-1}$$
$$= u^{n-2}(u^{n+1}g)g^{2n-2} = u^{n-2}u^n g^{2n-2} = u^{2n-2}g^{2n-2} = \cdots = u^n g^n = e.$$

Hence, e is an idempotent element of R. We also have

$$e + \mathfrak{n} = u^n g^n + \mathfrak{n} = (u^n + \mathfrak{n})(g^n + \mathfrak{n}) = \bar{f}(g^n + \mathfrak{n})$$

and $u^n = u^{n+1}g$ implies that

$$\bar{f} = u^n + \mathfrak{n} = u^{n+1}g + \mathfrak{n} = u^n ug + \mathfrak{n} = u^{n+1}gug + \mathfrak{n}$$
$$= u^{n+2}g^2 + \mathfrak{n} = \cdots = u^{2n}g^n + \mathfrak{n}$$
$$= (u^{2n} + \mathfrak{n})(g^n + \mathfrak{n}) = \bar{f}^2(g^n + \mathfrak{n}) = \bar{f}(g^n + \mathfrak{n}).$$
$$= e + \mathfrak{n}.$$

Therefore, e lifts $\bar{f}$ to R.   □

If $\bar{f}_1$ and $\bar{f}_2$ are orthogonal idempotents of R/I that lift to the idempotents e_1 and e_2 of R, then there is no assurance that e_1 and e_2 are orthogonal. However if $I \subseteq J(R)$, then orthogonality can be preserved.

Proposition 7.2.15. *If idempotents of R/I can be lifted to R modulo an ideal I contained in $J(R)$, then any countable set of orthogonal idempotents of R/I can be lifted to a countable set of orthogonal idempotents of R. Furthermore, a complete set $\{\bar{f}_1, \bar{f}_2, \ldots, \bar{f}_n\}$ of orthogonal idempotents of R/I can be lifted to a complete set $\{e_1, e_2, \ldots, e_n\}$ of orthogonal idempotents of R.*

Proof. The proof is by induction. Let I be an ideal of R contained in $J(R)$ and suppose that idempotents in R/I can be lifted to R. Suppose also that $\{\bar{f}_1, \bar{f}_2, \ldots, \bar{f}_n, \ldots\}$ is a countable set of orthogonal idempotents in R/I. By assumption, there is an idempotent e_1 in R such that $e_1 + I = \bar{f}_1$, so make the induction hypothesis that there are orthogonal idempotents $e_1, e_2, \ldots, e_j$ of R such that $e_i + I = \bar{f}_i$ for $i = 1, 2, \ldots, j$. If $e = e_1 + e_2 + \cdots + e_j$, then e is an idempotent of R and

$$\bar{f}_{j+1}(e + I) = \bar{f}_{j+1}(e_1 + I + e_2 + I + \cdots + e_j + I)$$
$$= \bar{f}_{j+1}(\bar{f}_1 + \bar{f}_2 + \cdots + \bar{f}_j)$$
$$= \bar{f}_{j+1}\bar{f}_1 + \bar{f}_{j+1}\bar{f}_2 + \cdots + \bar{f}_{j+1}\bar{f}_j$$
$$= 0.$$

Similarly, $(e + I)\bar{f}_{j+1} = 0$. Next, let g be an idempotent of R such that $g + I = \bar{f}_{j+1}$. Then $ge + I = (g + I)(e + I) = \bar{f}_{j+1}(e + I) = 0$, so $ge \in I \subseteq J(R)$. Likewise, $eg + I = 0$, so we have $ge + I = eg + I$. Since $ge \in J(R)$, $1 - ge$ is, by Proposition 6.1.7, a unit in R. Let $e_{j+1} = (1-e)(1-ge)^{-1}g(1-e)$. We claim that $e_{j+1} + I = \bar{f}_{j+1}$ and that e_{j+1} is orthogonal to e_i for $i = 1, 2, \ldots, j$. First, note that

$$e_{j+1} + I = (1-e)(1-ge)^{-1}g(1-e) + I$$
$$= (1-e)(1-ge)^{-1}(g - g^2e) + I$$
$$= (1-e)(1-ge)^{-1}(1-ge)g + I$$
$$= (1-e)g + I$$
$$= g + I - eg + I$$
$$= g + I$$
$$= \bar{f}_{j+1}.$$

Next, let i be such that $1 \le i \le j$. Then

$$e_{j+1}e_i = (1-e)(1-ge)^{-1}g(1-e)e_i$$
$$= (1-e)(1-ge)^{-1}g(e_i - ee_i)$$
$$= (1-e)(1-ge)^{-1}g(e_i - e_i)$$
$$= 0.$$

Similarly, $e_i e_{j+1} = 0$, so we have by induction that a countable set $\{\bar{f}_1, \bar{f}_2, \ldots,$
$\bar{f}_n, \ldots\}$ of orthogonal idempotents of R/I can be lifted to a countable set $\{e_1, e_2, \ldots,$
$e_n, \ldots\}$ of orthogonal idempotents of R.

Finally, suppose that $\{\bar{f}_1, \bar{f}_2, \ldots, \bar{f}_n\}$ is a complete set of orthogonal idempotents
of R/I. Then $\{\bar{f}_1, \bar{f}_2, \ldots, \bar{f}_n\}$ can be lifted to a set $\{e_1, e_2, \ldots, e_n\}$ of orthogonal
idempotents of R. We claim that the set $\{e_1, e_2, \ldots, e_n\}$ is complete. Let $e = e_1 +$
$e_2 + \cdots + e_n$. Then $e + I = \bar{f}_1 + \bar{f}_2 + \cdots + \bar{f}_n = 1 + I$. Thus, $1 - e \in I \subseteq J(R)$
and this implies that $e = 1$, since $J(R)$ cannot contain nonzero idempotents. $\square$

Corollary 7.2.16. *If R is a right (or left) artinian ring, then any countable set of or-*
thogonal idempotents of $R/J(R)$ can be lifted to a countable set of orthogonal idem-
potents of R. Furthermore, any complete set of orthogonal idempotents of $R/J(R)$
can be lifted to a complete set of orthogonal idempotents of R.

Proof. If R is right (or left) artinian, then, by Proposition 6.3.1 we have that $J(R)$ is
a nil ideal of R, so Propositions 7.2.14 and 7.2.15 give the result. $\square$

Lemma 7.2.17. *A cyclic R-module M has a projective cover if and only if there is an*
idempotent e in R and a right ideal A of R contained in $J(R)$ such that $M \cong eR/eA$.
Under these conditions, the natural mapping $\eta : eR \to eR/eA$ composed with the
isomorphism $eR/eA \cong M$ produces a projective cover of M.

Proof. Suppose that $M \cong eR/eA$, where e is an idempotent of R and A is a right
ideal of R such that $A \subseteq J(R)$. Now $A \subseteq J(R)$ implies that $eA \subseteq eJ(R)$ and since
$J(R)$ is small in R, $eJ(R)$ is small in eR. It follows that eA is small in eR. Note also
that eR is a projective R-module since eR is a direct summand of R. The fact that
$\operatorname{Ker} \eta = eA$ shows that $\eta : eR \to eR/eA \cong M$ is an R-projective cover of M.

Conversely, suppose that $\varphi : P \to M$ is a projective cover of the cyclic R-mod-
ule M. Since M is cyclic, there is an epimorphism $f : R \to M$. If $g : R \to P$ is the
completing map for the diagram

$$
\begin{array}{ccc}
 & & R \\
 & {}^{g}\swarrow & \big\downarrow {}^{f} \\
P & \xrightarrow{\ \varphi\ } & M \longrightarrow 0
\end{array}
$$

given by the projectivity of R, then $P = g(R) + \operatorname{Ker} \varphi$. But $\operatorname{Ker} \varphi$ is small in P, so
g is an epimorphism. Hence, the short exact sequence

$$0 \to \operatorname{Ker} g \to R \xrightarrow{\ g\ } P \to 0$$

splits and we have $R \cong P \oplus \operatorname{Ker} g$. Thus, there is an idempotent e of R such that
$eR \cong P$. If we identify P with eR under this isomorphism and let $B = \{b \in R \mid$

$\varphi(eb) = 0\}$, then B is a right ideal of R such that $eB = \text{Ker}\,\varphi$. Therefore, eB is small in eR and $eB \subseteq eJ(R) \subseteq J(R)$. If we let $A = eB$, then $A \subseteq J(R)$ and $eR/eA \cong M$. The fact that $\eta : eR \to eR/eA$ composed with the isomorphism $eR/eA \cong M$ produces a projective cover of M is now immediate. □

Corollary 7.2.18. *If R is a right artinian ring and $\bar{f}$ is an idempotent of $R/J(R)$, then $\bar{f}(R/J(R))$ has a projective cover as an R-module.*

Proof. If R is right artinian, then $J(R)$ is a nil ideal of R. So if $\bar{f}$ is an idempotent of $R/J(R)$, then $\bar{f}$ can be lifted to an idempotent e of R. The proposition and the fact that $eR/eJ(R)$ and $\bar{f}(R/J(R))$ are isomorphic R-modules establishes the corollary. □

We can now show that right (left) artinian rings are semiperfect. We prove the case for right artinian rings.

Proposition 7.2.19. *If R is a right artinian ring, then R is semiperfect.*

Proof. If R is right artinian, then $R/J(R)$ is right artinian and Jacobson semisimple and thus $R/J(R)$ is, by Proposition 6.6.2, a semisimple ring. Hence, there is a complete set $\{\bar{f}_i\}_{i=1}^{n}$ of orthogonal idempotents of $R/J(R)$ such that $R/J(R) = \bigoplus_{i=1}^{n} \bar{f}_i(R/J(R))$ and each $\bar{f}_i(R/J(R))$ is a minimal right ideal of $R/J(R)$. Now suppose that M is a finitely generated R-module. Then $M/MJ(R)$ is an $R/J(R)$-module and so $M/MJ(R)$ is a direct sum $\bigoplus_{\Delta} S_{\alpha}$ of simple $R/J(R)$-submodules of $M/MJ(R)$. Since $M/MJ(R)$ is finitely generated, the S_{α} are finite in number and each S_{α} is isomorphic to $\bar{f}_i(R/J(R))$ for some $i, 1 \leq i \leq n$. If these isomorphisms are used to reindex the S_{α} with the corresponding $i, 1 \leq i \leq n$, then Corollary 7.2.18 shows that there is an R-projective cover $\varphi_i : P_i \to S_i$ and so, by Proposition 7.2.3, we have an R-projective cover $\varphi : \bigoplus_{i=1}^{n} P_i \to M/MJ(R)$ of $M/MJ(R)$, where $\varphi = \bigoplus_{i=1}^{n} \varphi_i$. Next, consider the commutative diagram

$$
\begin{array}{ccc}
 & \bigoplus_{i=1}^{n} P_i & \\
 & \swarrow^{\bar{\varphi}} \qquad \downarrow^{\varphi} & \\
M \xrightarrow{\;\eta\;} & M/MJ(R) \longrightarrow 0 &
\end{array}
$$

where $\eta : M \to M/MJ(R)$ is the natural mapping and $\bar{\varphi}$ is the completing map given by the R-projectivity of $\bigoplus_{i=1}^{n} P_i$. Now φ is an epimorphism, so $M = \text{Im}\,\bar{\varphi} + MJ(R)$. But Proposition 6.1.12 indicates that $MJ(R)$ is small in M, so $\bar{\varphi}$ is an epimorphism. Since $\text{Ker}\,\bar{\varphi} \subseteq \text{Ker}\,\varphi$, $\text{Ker}\,\bar{\varphi}$ is a small submodule of $\bigoplus_{i=1}^{n} P_i$ and we have that $\bar{\varphi} : \bigoplus_{i=1}^{n} P_i \to M$ is a projective cover of M. □

We have seen that local rings and right (left) artinian rings are semiperfect. Our goal now is to give conditions that will characterize semiperfect rings. We begin with the following proposition.

Proposition 7.2.20. *The following statements about a nonzero projective R-module P are equivalent.*

(1) *P is a projective cover of a simple R-module.*

(2) *$PJ(R)$ is a small, maximal submodule of P.*

(3) $\text{End}_R(P)$ *is a local ring.*

Moreover, if $\varphi : P \to S$ is a projective cover of a simple R-module S, then $P \cong eR$ for some idempotent e of R.

Proof. (1) $\Rightarrow$ (2). Suppose that $\varphi : P \to S$ is a projective cover of the simple R-module S. Then $\text{Ker}\,\varphi$ is a small, maximal submodule of P. Now by Proposition 7.2.4, $\text{Rad}(P) = PJ(R)$, so $PJ(R)$ is the intersection of the maximal submodules of P. Hence, $PJ(R) \subseteq \text{Ker}\,\varphi$. But Corollary 6.1.15 shows that $PJ(R)$ contains every small submodule of P, so $\text{Ker}\,\varphi \subseteq PJ(R)$. Thus, $\text{Ker}\,\varphi = PJ(R)$.

(2) $\Rightarrow$ (3). If $PJ(R)$ is a small submodule of P, then by Corollary 7.2.6 we have $\text{End}_R(P)/J(\text{End}_R(P)) \cong \text{End}_R(P/PJ(R))$. But since $PJ(R)$ is a maximal submodule of P, $P/PJ(R)$ is a simple R-module, so Schur's lemma indicates that $\text{End}_R(P/PJ(R))$ is a division ring. Thus, $\text{End}_R(P)/J(\text{End}_R(P))$ is a division ring and Proposition 7.2.11 shows that $\text{End}_R(P)$ is a local ring.

(3) $\Rightarrow$ (1). Assume that $\text{End}_R(P)$ is a local ring. Proposition 7.2.8 indicates that P has a maximal submodule, say N. We claim that P together with the natural map $\eta : P \to P/N$ is a projective cover of the simple R-module P/N. This requires that we show that N is a small submodule of P. Suppose that N' is a submodule of P such that $N + N' = P$. Then $P/N = (N + N')/N \cong N'/(N \cap N')$. If $\eta : N' \to N'/(N \cap N')$ is the natural map, then, since P is projective,

$$\text{Hom}_R(P, N') \xrightarrow{\eta_*} \text{Hom}_R(P, N'/(N \cap N')) \to 0$$

is exact. Thus, if $f \in \text{Hom}_R(P, N'/(N \cap N'))$, then there is a $g \in \text{Hom}_R(P, N')$ such that $f = \eta_*(g) = \eta g$. Therefore, if $f \neq 0$, then $\text{Im}\,g \nsubseteq N$, so it follows from Corollary 6.1.15 that $\text{Im}\,g$ cannot be a small submodule of P. Proposition 7.2.5 now shows that $g \notin J(\text{End}_R(P))$ and, by Proposition 7.2.11, we see that g is a unit in $\text{End}_R(P)$. Hence, $P = g(P) \subseteq N'$, so $N' = P$. Thus, N is a small submodule of P and so we have that (1), (2) and (3) are equivalent.

Finally, if S is a simple R-module and $x \in S$, $x \neq 0$, then $f : R \to S$ defined by $a \mapsto xa$ is an epimorphism. If $\varphi : P \to S$ is a projective cover of S, then the

diagram

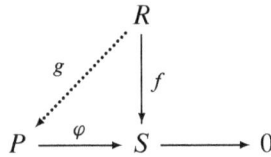

can be completed commutatively by an R-linear mapping $g : R \to P$. But $P = \operatorname{Im} g + \operatorname{Ker} \varphi$ and $\operatorname{Ker} \varphi$ is small in P, so g is an epimorphism. Thus, the sequence $0 \to \operatorname{Ker} g \to R \xrightarrow{g} P \to 0$ splits and we have $R \cong \operatorname{Ker} g \oplus P$. Hence, there is an idempotent $e \in R$ such that $P \cong eR$. $\qquad\qquad\qquad\qquad\Box$

Corollary 7.2.21. *The following are equivalent for a ring R and an idempotent e of R.*

(1) *$eR/eJ(R)$ is a simple R-module.*

(2) *$eJ(R)$ is a unique maximal submodule of eR.*

(3) *eRe is a local ring.*

Proof. Left to the reader. Note that $\operatorname{End}_R(eR)$ and eRe are isomorphic rings. $\qquad\Box$

Clearly, if we replace eR and $eJ(R)$ by Re and $J(R)e$, respectively, in (1) and (2) of Corollary 7.2.21, then the "new" (1) and (2) are each equivalent to (3).

Proposition 7.2.22. *If I is an ideal of R such that $I \subseteq J(R)$, then the following are equivalent.*

(1) *Idempotents of R/I lift to R.*

(2) *Every direct summand of the R-module R/I has an R-projective cover.*

(3) *Every complete set of orthogonal idempotents of R/I lifts to a complete set of orthogonal idempotents of R.*

Proof. (1) $\Rightarrow$ (2). Let A be a right ideal of R containing I such that A/I is a direct summand of the R-module R/I. Then A/I is a direct summand of R/I as an R/I-module. Such an R/I-summand of R/I is generated by an idempotent $\bar{f} \in R/I$ which, by (1), can be lifted to an idempotent e of R. Since $eR/eI \cong \bar{f}(R/I) = A/I$, we can apply Lemma 7.2.17 to obtain the result.

(2) $\Rightarrow$ (3). Let $\{\bar{f}_1, \bar{f}_2, \ldots, \bar{f}_n\}$ be a complete set of orthogonal idempotents of R/I. Then

$$R/I = \bar{f}_1(R/I) \oplus \bar{f}_2(R/I) \oplus \cdots \oplus \bar{f}_n(R/I)$$

and since $I \subseteq J(R)$, Proposition 6.1.12 shows that I is small in R. Hence, the natural map $\eta : R \to R/I$ is an R-projective cover of R/I. By assumption each cyclic R-module $\bar{f}_i(R/I)$ has an R-projective cover $\varphi_i : P_i \to \bar{f}_i(R/I)$. It follows from

Proposition 7.2.15 and Lemma 7.2.17 that there is a complete set $\{e_1, e_2, \ldots, e_n\}$ of orthogonal idempotents of R such that $R = e_1 R \oplus e_2 R \oplus + \cdots + \oplus e_n R$ with $e_i R \cong P_i$ for $i = 1, 2, \ldots, n$. Furthermore, for each i, $(e_i + I)R/I = \eta(e_i R) = \bar{f}_i(R/I)$, so it follows that e_i lifts $\bar{f}_i$ to R for $i = 1, 2, \ldots, n$.

(3) $\Rightarrow$ (1). If $\bar{f}$ is an idempotent of R/I, then $\{\bar{f}, 1_{R/I} - \bar{f}\}$ is a complete set of orthogonal idempotents of R/I which lifts to a complete set $\{e, 1_R - e\}$ of R. Thus, e lifts $\bar{f}$ to R modulo I. $\square$

Remark. If P is a projective module such that $\mathrm{Hom}_R(P, S) \neq 0$ for every simple R-module S, then P generates $\mathbf{Mod}_R$. To show that P generates $\mathbf{Mod}_R$, it suffices to show that P generates R. Let $T = \sum_\Delta f(P)$, where $\Delta = \mathrm{Hom}_R(P, R)$. If $T \neq R$, then there is a maximal right ideal $\mathfrak{m}$ of R that contains T and $\mathrm{Hom}_R(P, R/\mathfrak{m}) \neq 0$. If $f \in \mathrm{Hom}_R(P, R/\mathfrak{m})$ is nonzero, then the projectivity of P produces a nonzero R-linear map $g : P \to R$ such that $f = \eta g$, where $\eta : R \to R/\mathfrak{m}$ is the canonical map. But $\mathrm{Im}\, g \subseteq T \subseteq \mathfrak{m}$ and so $f = \eta g = 0$, a contradiction. Thus, $T = R$, so P generates R.

Finally, we are in a position to prove several equivalent conditions, due to Bass [51], that are necessary and sufficient for a ring to be semiperfect.

Proposition 7.2.23 (Bass). *The following are equivalent for a ring R.*

(1) *R is semiperfect.*

(2) *$R/J(R)$ is semisimple and idempotents of $R/J(R)$ can be lifted to R.*

(3) *R has a complete set $\{e_1, e_2, \ldots, e_n\}$ of orthogonal idempotents such that $e_i R e_i$ is a local ring for $i = 1, 2, \ldots, n$.*

(4) *Every simple R-module has a projective cover.*

Proof. (1) $\Rightarrow$ (2). If (1) holds, then every finitely generated R-module has a projective cover. In particular, every direct summand of $R/J(R)$ has an R-projective cover, so by Proposition 7.2.22, idempotents of $R/J(R)$ lift to R. We claim that $R/J(R)$ is a semisimple ring. Suppose that A is a right ideal of R such that $J(R) \subseteq A$. Since the cyclic R-module R/A has a projective cover, we see by Lemma 7.2.17 that there is a right ideal B of R such that $B \subseteq J(R)$ and an idempotent e of R such that $R/A \cong eR/eB$. But then $eB \subseteq eJ(R)$, so $(eR/eB)J(R) \cong (R/A)J(R) = 0$. Hence, $eJ(R) = (eR)J(R) \subseteq eB$ and so $eB = eJ(R)$. Therefore, $R/A \cong (e + J(R))R/J(R)$ is a projective $R/J(R)$-module. Consequently, the short exact sequence

$$0 \to A/J(R) \to R/J(R) \to R/A \to 0$$

splits which shows that $A/J(R)$ is a direct summand of $R/J(R)$. Thus, $R/J(R)$ is a semisimple ring.

(2) $\Rightarrow$ (3). If $R/J(R)$ is a semisimple ring, then $R/J(R) = \bigoplus_{i=1}^{n} \bar{f}_i(R/J(R))$, where $\{\bar{f}_1, \bar{f}_2, \ldots, \bar{f}_n\}$ is a complete set of orthogonal idempotents of $R/J(R)$, and each $\bar{f}_i(R/J(R))$ is a simple $R/J(R)$-module. Since idempotents of $R/J(R)$ can be lifted to R, it follows from Proposition 7.2.15 that the set of idempotents $\{\bar{f}_1, \bar{f}_2, \ldots, \bar{f}_n\}$ can be lifted to a complete set $\{e_1, e_2, \ldots, e_n\}$ of orthogonal idempotents of R. Also each $\bar{f}_i(R/J(R))$ is a simple $R/J(R)$-module, so each $\bar{f}_i(R/J(R))$ is a simple R-module. But for each i, $e_i R/e_i J(R) \cong \bar{f}_i(R/J(R))$, so the fact that each $e_i Re_i$ is a local ring is a consequence of Corollary 7.2.21.

(3) $\Rightarrow$ (4). If (3) holds, then Corollary 7.2.21 indicates that each $e_i R/e_i J(R)$ is a simple R-module. Furthermore, since

$$R/J(R) = e_1 R/e_1 J(R) \oplus e_2 R/e_2 J(R) \oplus \cdots \oplus e_n R/e_n J(R),$$

each simple R-module is isomorphic to one of the R-simple summands of $R/J(R)$. Lemma 7.2.17 now shows that each simple R-module has an R-projective cover.

(4) $\Rightarrow$ (1). Since the isomorphism classes of simple R-modules form a set, we can choose a set $\mathcal{S}$ of representatives of simple R-modules, exactly one from each isomorphism class. Next, for each $S \in \mathcal{S}$, choose one and only one projective cover $P(S) \to S$ of S. If $\mathcal{P}$ is this set of projective covers, then $\mathcal{P}$ generates every simple R-module S', so $P = \bigoplus_{\mathcal{P}} P(S)$ is such that $\mathrm{Hom}_R(P, S') \neq 0$ for every simple R-module S'. The preceding Remark shows that P is a projective generator for $\mathbf{Mod}_R$ and since $\mathcal{P}$ generates P, $\mathcal{P}$ generates $\mathbf{Mod}_R$. Thus, if M is a finitely generated R-module, then there is a finite set $\{P_1, P_2, \ldots, P_n\}$ of modules in $\mathcal{P}$ and an epimorphism

$$\varphi : Q = P_1 \oplus P_2 \oplus \cdots \oplus P_n \to M.$$

But $\varphi(QJ(R)) = MJ(R)$ and it follows that we have an induced epimorphism

$$P_1/P_1 J(R) \oplus P_2/P_2 J(R) \oplus \cdots \oplus P_n/P_n J(R) \to M/MJ(R).$$

Using Proposition 7.2.20, we see that each $P_i/P_i J(R)$ is a simple R-module, so $M/MJ(R)$ is a direct sum of simple R-modules. Hence, by Proposition 7.2.3, $M/MJ(R)$ has a projective cover $\varphi : P \to M/MJ(R)$. If $\eta : M \to M/MJ(R)$ is the natural mapping and $f : P \to M$ is such that $\eta f = \varphi$, then $M = \mathrm{Im}\, f + \mathrm{Ker}\, \eta$. But Proposition 6.1.12 indicates that $\mathrm{Ker}\, \eta$ is small in M and so f is an epimorphism. Moreover, $\mathrm{Ker}\, f \subseteq \mathrm{Ker}\, \varphi$, so $f : P \to M$ is a projective cover of M and we have that R is semiperfect. $\square$

Perfect Rings

We have seen that local and right artinian rings are semiperfect and Proposition 7.2.23 shows that a ring R is semiperfect if and only if $R/J(R)$ is semisimple and idempotents of $R/J(R)$ can be lifted to R. Thus, a semiperfect ring is left-right symmetric, so every finitely generated R-module has a projective cover if and only if every finitely generated left R-module has a projective cover. We will now characterize the right perfect rings of Bass [51], the rings over which every R-module has a projective cover. The description of right perfect rings begins with the following lemma.

Lemma 7.2.24. *Let $a_1, a_2, \ldots$ be a sequence of elements of R. If F is a free R-module with basis $\{x_i\}_\mathbb{N}$, let $y_i = x_i - x_{i+1}a_i$ for each $i \in \mathbb{N}$. If M is the submodule of F generated by $\{y_i\}_\mathbb{N}$, then*

(1) M *is a free R-module with basis $\{y_n\}_\mathbb{N}$ and*

(2) $M = F$ *if and only if for each $k \in \mathbb{N}$, there is an integer $n \geq k$ such that $a_n a_{n-1} \cdots a_k = 0$.*

Proof. (1) Since $\{y_i\}_\mathbb{N}$ generates M, we need only show that $\{y_i\}_\mathbb{N}$ is a set of linearly independent elements of M. Let $y_k, y_{k+1}, \ldots, y_n$ be a finite set of the y_i and suppose that $b_k, b_{k+1}, \ldots, b_n \in R$ are such that

$$y_k b_k + y_{k+1} b_{k+1} + \cdots + y_n b_n = 0.$$

Then

$$
\begin{aligned}
y_k b_k + y_{k+1} b_{k+1} + \cdots + y_n b_n &= (x_k - x_{k+1}a_k)b_k + (x_{k+1} - x_{k+2}a_{k+1})b_{k+1} \\
&\quad + \cdots + (x_n - x_{n+1}a_n)b_n \\
&= x_k b_k + x_{k+1}(b_{k+1} - a_k b_k) + \cdots \\
&\quad + x_n(b_n - a_{n-1}b_{n-1}) - x_{n+1}a_n b_n
\end{aligned}
$$

gives $b_k = b_{k+1} - a_k b_k = \cdots = b_n - a_{n-1}b_{n-1} = a_n b_n = 0$. But this implies that $b_k = b_{k+1} = \cdots = b_n = 0$ and so the y_i are linearly independent.

(2) If $M = F$, then $x_k \in M$ for each $k \in \mathbb{N}$, so we can write $x_k = y_1 b_1 + y_2 b_2 + \cdots + y_n b_n$ for some integer $n \geq 1$. But then, as in the proof of (1),

$$
\begin{aligned}
x_k &= x_1 b_1 + x_2(b_2 - a_1 b_1) + \cdots + x_{k-1}(b_{k-1} - a_{k-2}b_{k-2}) \\
&\quad + x_k(b_k - a_{k-1}b_{k-1}) + x_{k+1}(b_{k+1} - a_k b_k) \\
&\quad + \cdots + x_n(b_n - a_{n-1}b_{n-1}) - x_{n+1}a_n b_n.
\end{aligned}
$$

Hence, $b_1 = b_2 = \cdots = b_{k-1} = 0$, $b_k = 1$ and $b_{k+1} - a_k b_k = \cdots = b_n - a_{n-1}b_{n-1} = a_n b_n = 0$. But $b_k = 1$ and $b_{k+1} - a_k b_k = \cdots = b_n - a_{n-1}b_{n-1} = a_n b_n = 0$ show that $a_n a_{n-1} \cdots a_k = 0$ and $n \geq k$.

Conversely, suppose that for each $k \in \mathbb{N}$, there is an integer $n \geq k$ such that $a_n a_{n-1} \cdots a_k = 0$. Let $k \in \mathbb{N}$ and suppose that $n \geq k$ is such an integer. Now

$y_i = x_i - x_{i+1}a_i$, so $x_i = y_i + x_{i+1}a_i$, for each $i \in \mathbb{N}$. Hence, we see that

$$
\begin{aligned}
x_k &= y_k + x_{k+1}a_k \\
&= y_k + (y_{k+1} + x_{k+2}a_{k+1})a_k \\
&= y_k + y_{k+1}a_k + x_{k+2}(a_{k+1}a_k) \\
&= y_k + y_{k+1}a_k + (y_{k+2} + x_{k+3}a_{k+2})a_{k+1}a_k \\
&= y_k + y_{k+1}a_k + y_{k+2}(a_{k+1}a_k) + x_{k+3}(a_{k+2}a_{k+1}a_k) \\
&\quad\vdots \\
&= y_k + y_{k+1}a_k + \cdots + y_n(a_{n-1}a_{n-2}\cdots a_k) + x_{n+1}(a_n a_{n-1}\cdots a_k).
\end{aligned}
$$

But $a_n a_{n-1} \cdots a_k = 0$, so $x_k \in M$. Therefore, $M = F$.   □

The necessary and sufficient condition for $M = F$ in (2) of Lemma 7.2.24 leads to the following definition.

Definition 7.2.25. A subset K of R is said to be *right T-nilpotent* (T for transfinite) if for each sequence $a_1, a_2, a_3, \ldots$ of elements of K, $a_n \cdots a_2 a_1 = 0$ for some integer $n \in \mathbb{N}$ and *left T-nilpotent* if $a_1 a_2 \cdots a_n = 0$ for some $n \in \mathbb{N}$.

Note that if a right ideal A of R is left or right T-nilpotent, then it is nil. Indeed, if $a \in A$, then $a, a, a, \ldots$ is a sequence in A, so $a^n = 0$ for some $n \in \mathbb{N}$.

Lemma 7.2.26. *If A is a right ideal of R, then the following are equivalent.*

(1) *A is right T-nilpotent.*

(2) *$MA \subsetneqq M$ for every nonzero R-module M.*

(3) *MA is a small submodule of M for every nonzero R-module M.*

(4) *FA is a small submodule of every countably generated free R-module F.*

Proof. (1) $\Rightarrow$ (2). If $MA = M \neq 0$, then $A \nsubseteq \operatorname{ann}_r(M)$, so there is at least one $a_1 \in A$ such that $Ma_1 \neq 0$. Next, let $n \in \mathbb{N}$ and assume that we can find $a_1, a_2, \ldots, a_n \in A$ such that $Ma_n a_{n-1} \cdots a_1 \neq 0$. If $a_1, a_2, \ldots, a_n \in A$ is such a sequence, then $MAa_n a_{n-1} \cdots a_1 = Ma_n a_{n-1} \cdots a_1 \neq 0$. Hence, there are $a_1, a_2, \ldots,$ $a_n, a_{n+1} \in A$ such that $Ma_{n+1}a_n \cdots a_1 \neq 0$. Consequently, induction shows that there is a sequence $a_1, a_2, \ldots$ of elements of A such that $Ma_n a_{n-1} \cdots a_1 \neq 0$ for each $n \in \mathbb{N}$. So for this sequence we clearly have $a_n a_{n-1} \cdots a_1 \neq 0$ for each $n \in \mathbb{N}$, so A cannot be right T-nilpotent. Hence, if A is right T-nilpotent, then $MA \subsetneqq M$.

(2) $\Rightarrow$ (3). If N is a proper submodule of $M \neq 0$, then $M/N \neq 0$. From (2) we have $(M/N)A \neq M/N$. But $(M/N)A = (MA+N)/N$, so $(MA+N)/N \neq M/N$. Hence, $MA + N \neq M$. Therefore, for every proper submodule N of M, we have $MA + N \neq M$. Thus, MA is a small submodule of M.

(3) $\Rightarrow$ (4). Obvious.

(4) $\Rightarrow$ (1). Let F be a free R-module with basis $\{x_i\}_\mathbb{N}$ and suppose that $a_1, a_2, \ldots$ is a sequence in A. Then by (1) of Lemma 7.2.24 we have that $\{x_i - x_{i+1}a_i\}_\mathbb{N}$ is a basis for $M = \sum_\mathbb{N}(x_i - x_{i+1}a_i)R$. If $x \in F$, then $x = \sum_\mathbb{N} x_i b_i$, where each b_i is in R and $b_i = 0$ for almost all i. Thus, we have $x = \sum_\mathbb{N}(x_i - x_{i+1}a_i)b_i + \sum_\mathbb{N} x_{i+1}(a_i b_i) \in M + FA$, so $M + FA = F$. But we are assuming that FA is a small submodule of F, so it must be the case that $M = F$. Lemma 7.2.24 now shows that A is right T-nilpotent. $\square$

Definition 7.2.27. A ring R is said to be a *right perfect ring*, if every R-module has a projective cover. Left perfect rings are defined similarly. A ring that is left and right perfect is referred to as a *perfect ring*.

Bass [51] has given the following characterization of rings over which every module has a projective cover.

Proposition 7.2.28 (Bass). *The following are equivalent for a ring R.*

(1) *R is a right perfect ring.*

(2) *$R/J(R)$ is semisimple and every nonzero R-module contains a maximal submodule.*

(3) *$R/J(R)$ is semisimple and $J(R)$ is right T-nilpotent.*

Proof. (1) $\Rightarrow$ (2). If R is right perfect, then R is semiperfect, so $R/J(R)$ is a semi-simple ring. If M is a nonzero R-module, then M has a projective cover $\varphi : P \to M$. Since P is projective, it follows from Proposition 7.2.8 that P has a maximal submodule, say N. Also $\operatorname{Ker}\varphi$ is small in P, so it must be the case that $\operatorname{Ker}\varphi \subseteq \operatorname{Rad}(P) \subseteq N$. Hence, $N/\operatorname{Ker}\varphi$ is a maximal submodule of $P/\operatorname{Ker}\varphi \cong M$.

(2) $\Rightarrow$ (3). If M is a nonzero R-module and N is a maximal submodule of M, then M/N is a simple R-module. Since $(M/N)J(R) = 0$, we have $MJ(R) \subseteq N \subsetneq M$. Hence, Lemma 7.2.26 indicates that $J(R)$ is right T-nilpotent.

(3) $\Rightarrow$ (1). Since $J(R)$ is right T-nilpotent, $J(R)$ is a nil ideal of R, so Proposition 7.2.14 shows that idempotents of $R/J(R)$ can be lifted to R. Thus, (2) of Proposition 7.2.23 shows that R is a semiperfect ring and (4) of Proposition 7.2.23 indicates that every simple R-module has a projective cover. Next, suppose that M is a nonzero R-module. Since $R/J(R)$ is semisimple, $M/MJ(R)$ is a semisimple $R/J(R)$-module and, as a consequence, is a direct sum of simple $R/J(R)$-modules. Let $M/MJ(R) = \bigoplus_\Delta S_\alpha$, where each S_α is a simple $R/J(R)$-module. Since each S_α is also a simple R-module, it follows from Lemma 7.2.17 that, for each $\alpha \in \Delta$, there is an idempotent e_α of R such that $e_\alpha R/e_\alpha J(R) \cong S_\alpha$. If $P = \bigoplus_\Delta e_\alpha R$, then $P/PJ(R) \cong \bigoplus_\Delta e_\alpha R/e_\alpha J(R) \cong M/MJ(R)$ and we have an epimorphism $f : P \to M/MJ(R)$ with kernel $PJ(R)$. Now (3) of Proposition 7.2.26 shows that

$PJ(R)$ is small in P and that $MJ(R)$ is small in M. Thus, if $\varphi : P \to M$ completes the diagram

$$
\begin{array}{ccc}
 & & P \\
 & \varphi\nearrow & \downarrow f \\
M & \xrightarrow{\eta} & M/MJ(R) \longrightarrow 0
\end{array}
$$

commutatively, then $\operatorname{Im}\varphi + MJ(R) = M$ which shows that φ is an epimorphism. Since $\operatorname{Ker}\varphi \subseteq \operatorname{Ker} f = PJ(R)$, we see that $\varphi : P \to M$ is a projective cover of M. Hence, every R-module has a projective cover, so R is a right perfect ring.   □

There are additional conditions that describe right perfect rings. The proof of the following proposition can be found in [51].

Proposition 7.2.29 (Bass). *The following are equivalent for a ring R.*

(1) *R is a right perfect ring.*

(2) *R satisfies the descending chain condition on principal left ideals.*

(3) *Every flat R-module is projective.*

(4) *R contains no infinite set of orthogonal idempotents and every nonzero left R-module contains a simple submodule.*

We saw in Examples 1 and 2 of Section 5.3 that every projective module is flat but that there are rings over which a flat module need not be projective. When R is right perfect the preceding proposition points out that the class of projective R-modules and the class of flat R-modules coincide.

Remark. Bonah has shown in [62] that if R satisfies the descending chain condition on principal left ideals, that is, if R is right perfect, then R satisfies the ascending chain condition on principal right ideals. However, the ring $\mathbb{Z}$ shows that the converse fails. Thus, these two conditions are not equivalent. Furthermore, Björk proved in [53] that a ring R is right perfect if and only if R satisfies the descending chain condition on finitely generated left ideals.

Problem Set 7.2

1. Prove that a (not necessarily commutative) local ring is semiperfect. [Hint: Similar to Example 2.]

2. Let R be any ring.
   (a) Prove that $J(R[[X]])$ consists of all power series with constant term in $J(R)$. [Hint: If $A = \bigoplus_{k=1}^{\infty} X^k R$, then A is a right ideal of R and $R[[X]]/A \cong R$ via $\bigoplus_{k=0}^{\infty} X^k a_k + A \mapsto a_0$.]

(b) If R is a local ring, prove that $R[[X]]/J(R[[X]]) \cong R/J(R)$. [Hint: Consider $\sum_{k=0}^{\infty} X^k a_k + J(R[[X]]) \mapsto a_0 + J(R)$.] Conclude that $R[[X]]$ is a local ring.

3. Establish the isomorphism $eR/eJ(R) \cong \bar{f}(R/J(R))$ of the proof of Corollary 7.2.18.

4. Prove Corollary 7.2.21.

5. A nonzero idempotent e of R is said to be a *primitive idempotent* if e cannot be written as $e = f + g$, where f and g are nonzero orthogonal idempotents of R. An idempotent e of R is a *local idempotent* if eRe is a local ring.

   (a) Show that a nonzero idempotent e in R is primitive if and only if the only idempotents in the ring eRe are 0 and e.

   (b) Suppose that idempotents in R/I can be lifted to R, where $I \subseteq J(R)$. If e is a primitive idempotent of R, prove that $e + I$ is primitive in R/I.

   (c) If e is a local idempotent of R, prove that e is primitive. [Hint: Suppose that $e = f + g$, where f and g are orthogonal idempotents of R and show that f and g are in the local ring eRe.]

   (d) Prove that an idempotent e of R is primitive if and only if eR is an indecomposable right ideal of R. [Hint: If $e = f + g$, where f and g are orthogonal idempotents of R, then $eR = fR \oplus gR$.] Conclude by symmetry that e is primitive if and only if Re is an indecomposable left ideal of R.

   (e) Show by example that a primitive idempotent need not be local.

   (f) If e is a primitive idempotent in a regular ring, prove that eRe is a division ring.

6. (a) Let I be an ideal of R and suppose that M is an R/I-module. If $\varphi : P \to M$ is a projective cover of M as an R-module, show that $\bar{\varphi} : P/PI \to M$ is a projective cover of M as an R/I-module, where $\bar{\varphi}$ is the induced map. [Hint: Exercise 1 in Problem Set 6.1.]

   (b) If R is a right perfect ring and I is an ideal of R, prove that the ring R/I is also right perfect.

7. If P is a projective R-module, prove that $P/PJ(R)$ has a projective cover if and only if $PJ(R)$ is a small submodule of P.

8. If M and N are R-modules, then we will say that N *covers* M if there is an epimorphism $f : N \to M$. Let M be an R-module that has a projective cover $\varphi : P \to M$. Prove that every projective R-module Q that covers M also covers P and that P is isomorphic to a direct summand of Q. Note that this is the dual of (a) of Exercise 1 in Problem Set 7.1. [Hint: If $f : Q \to M$ is a cover of M and $g : Q \to P$ is such that $\varphi g = f$, show that $P = \text{Im } g + \text{Ker } \varphi$.]

9. Prove that a finite ring direct product of semiperfect rings is a semiperfect ring.

10. Show that the following are equivalent for a right ideal A of R.

(a) A is right T-nilpotent.

(b) For any left R-module N, $\text{ann}_r^N(A) = 0$ implies that $N = 0$.

(c) For any R-module M, if $MA = M$, then $M = 0$.

(d) If N is a submodule of an R-module M and $MA + N = M$, then $N = M$.

[$(a) \Rightarrow (b)$, Hint: Assume that $\text{ann}_r^N(A) = 0$ and $N \neq 0$. If $x \in N$, $x \neq 0$, then there is an $a_1 \in A$ be such that $a_1 x \neq 0$. Next, let $a_2 \in A$ be such that $a_2 a_1 x \neq 0$ and so forth.]

[$(b) \Rightarrow (c)$, Hint: Assume that $M \neq 0$ and let $B = \text{ann}_r(M)$, then $B \subsetneq R$ is an ideal. If $N = R/B$, then $\text{ann}_\ell^N(A) = (B : A)/B$. Now show that $MA \neq M$.]

[$(c) \Rightarrow (a)$, Hint: Let $a_1, a_2, a_3, \ldots$ be a sequence of element of A and suppose that $F = \bigoplus_{i=1}^\infty R_i$ is a free R-module with basis $\{x_i\}_{i=0}^\infty$, where $R_i = R$ for each i. Let S be the submodule of F generated by elements of the form $x_n - x_{n+1} a_{n+1}$ for $n = 0, 1, 2, \ldots$. If $M = F/S$, then $\overline{x_n} = \overline{x_{n+1} a_{n+1}}$ in M gives $MA = M$ and so, by assumption, $M = 0$. Hence, $F = S$. This means that we can write $x_0 = (x_0 - x_1 a_1) b_1 + (x_1 - x_2 a_2) b_2 + \cdots + (x_{n-1} - x_n a_n) b_n$. Compare coefficients of $x_0, x_1, \ldots, x_n$ on the right of this equation with the coefficients of $x_0, x_1, \ldots, x_n$ on the left and show that $a_n a_{n-1} \cdots a_2 a_1 = 0$.]

Conclude that the concept of right T-nilpotent extends the ideas presented in Nakayama's lemma to arbitrary modules.

11. Let $f : M \to N$ be an epimorphism with small kernel. Prove that $\varphi : P \to M$ is a projective cover of M if and only if $\varphi f : P \to N$ is a projective cover of N.

12. Prove that a ring R is semiperfect if and only if R has a complete set $\{e_1, e_2, \ldots, e_n\}$ of orthogonal idempotents such that $e_i R e_i$ is a local ring for $i = 1, 2, \ldots, n$.

13. Let $\mathcal{X}$ be a class of R-modules that satisfies the following three conditions.

(i) $\mathcal{X}$ is closed under isomorphisms: If $X \in \mathcal{X}$ and $X' \cong X$, then $X' \in \mathcal{X}$.

(ii) $\mathcal{X}$ is closed under finite direct sums: If $\{X_i\}_{i=1}^n$ is a family of modules in $\mathcal{X}$, then $\bigoplus_{i=1}^n X_i \in \mathcal{X}$.

(iii) $\mathcal{X}$ is closed under direct summands: If $X \in \mathcal{X}$ and X' is a direct summand of X, then $X' \in \mathcal{X}$.

For an R-module M, an $\mathcal{X}$-cover of M is an $X \in \mathcal{X}$ together with an R-linear mapping $\varphi : X \to M$ such that the following two conditions are satisfied.

(1) For any R-linear mapping $\varphi' : X' \to M$ with $X' \in \mathcal{X}$, the diagram

can be completed commutatively by an R-linear mapping $f : X' \to X$.

(2) The diagram

can be completed commutatively only by automorphisms of X.

If $\varphi : X \to M$ satisfies (1) but maybe not (2), then $\varphi : X \to M$ is said to be an $\mathcal{X}$-*precover* of M. Additional information on $\mathcal{X}$-(pre)covers can be found in [12] and [46].

(a) If an R-module M has an $\mathcal{X}$-cover $\varphi : X \to M$, prove that X is unique up to isomorphism.

(b) Suppose that M has an $\mathcal{X}$-cover $\varphi : X \to M$. Prove that if $\varphi' : X' \to X$ is any $\mathcal{X}$-precover of M, then X is isomorphic to a direct summand of X'.

(c) If every R-module M has an $\mathcal{X}$-cover $\varphi : X \to M$, show that for every such $\mathcal{X}$-cover the mapping φ is a surjection if and only if $\mathcal{X}$ contains all the projective R-modules.

[Hint: The definition of an $\mathcal{X}$-(pre)cover and (a), (b) and (c) are dual to the definition of an $\mathcal{X}$-(pre)envelope and the exercises of Exercise 7 in Problem Set 7.1.]

14. Let $\mathcal{P}$ be the class of projective R-modules and note that $\mathcal{P}$ satisfies conditions (i), (ii) and (iii) given in Exercise 13 for a class of R-modules. Prove that the following are equivalent.

   (a) M has a $\mathcal{P}$-cover $\varphi : P \to M$, where $\mathcal{P}$-cover is defined as in Exercise 13.

   (b) M has a projective cover $\varphi : P(M) \to M$ in the sense of Definition 7.2.1.

   [Hint: This is the dual of Exercise 8 in Problem Set 7.1.]

7.3 Quasi-Injective Envelopes and Quasi-Projective Covers

In this section we investigate quasi-injective envelopes and quasi-injective covers. We will see that every module has a quasi-injective envelope that is unique up to isomorphism. As in the case of projective covers, a module may fail to have a quasi-projective cover. However, we will show that every module will have a quasi-projective cover if and only if every module has a projective cover, that is, if and only if the ring is right perfect.

Quasi-Injective Envelopes

Definition 7.3.1. If M is an R-module and $f : M \to N$ is a monomorphism, where N is a quasi-injective R-module, then $f : M \to N$ is said to be a *quasi-injective extension* of M. A quasi-injective extension $\varphi : M \to E_q(M)$ of M is a *quasi-injective envelope* of M if whenever $f : M \to N$ is a quasi-injective extension of M, there is a monomorphism $g : E_q(M) \to N$ such that the diagram

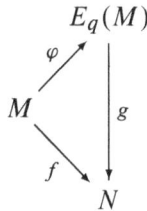

$$
\begin{array}{ccc}
 & E_q(M) & \\
{\scriptstyle \varphi}\nearrow & & \Big\downarrow {\scriptstyle g} \\
M & & \\
 & {\scriptstyle f}\searrow & \\
 & & N
\end{array}
$$

is commutative. To simplify notation, a quasi-injective envelope will often be denoted by $\varphi : M \to E_q$. We say that a quasi-injective envelope $\varphi : M \to E_q$ of M is *unique up to isomorphism* if whenever $\varphi' : M \to E_q'$ is another quasi-injective envelope of M, there is an isomorphism $g : E_q \to E_q'$ such that $g\varphi = \varphi'$.

Since $M \subseteq E(M)$, every module is contained in a quasi-injective module. However, Johnson and Wong were able to show in [61] that there is a smallest quasi-injective submodule of $E(M)$ that contains M.

Proposition 7.3.2. *If M is an R-module and $H = \operatorname{End}_R(E(M))$, then:*

(1) *HM is a submodule of $E(M)$ containing M and HM is quasi-injective.*

(2) *HM is the intersection of the quasi-injective submodules of $E(M)$ that contain M.*

(3) *$M = HM$ if and only if M is quasi-injective.*

Proof. (1) It follows easily that HM is a submodule of $E(M)$ and $M \subseteq HM$ since $M \subseteq E(M)$ and $M = \operatorname{id}_{E(M)}(M) \subseteq HM$. If N is a submodule of HM and $f : N \to HM$ is R-linear, then f extends to a $g \in H$. But then $g(HM) \subseteq HM$, so if $\bar{g} = g|_{HM}$, then $\bar{g} : HM \to HM$ and $\bar{g}|_N = f$. Thus, HM is quasi-injective.

(2) Let Q be a quasi-injective submodule of $E(M)$ that contains M. We claim that $HM \subseteq Q$. Let $f \in H$. Then $Q_f = \{x \in Q \mid f(x) \in Q\}$ is a submodule of Q, so $g : Q_f \to Q$ defined by $g(x) = f(x)$ is an R-linear map. But Q is quasi-injective, so g extends to an R-linear mapping $h : Q \to Q$. Using the injectivity of $E(M)$, h can be extended to an $\overline{h} \in H$. If $Q_f \neq Q$, then $(\overline{h} - f)(Q) \neq 0$ and, since $M \subseteq Q \subseteq E(M)$ implies that Q is an essential submodule of $E(M)$, $Q \cap (\overline{h} - f)(Q) \neq 0$. Let $y \in Q \cap (\overline{h} - f)(Q)$, $y \neq 0$, and suppose that $x \in Q$ is such that $(\overline{h} - f)(x) = y$. Then $\overline{h}(x) - f(x) \in Q$. But $\overline{h}$ agrees with h on Q and $h(x) \in Q$, so it follows that $f(x) \in Q$. Therefore, $\overline{h}(x) = f(x)$, so $y = 0$, a contradiction that gives $Q_f = Q$. Hence, $f(Q) \subseteq Q$. Also $M \subseteq Q$, so $HM \subseteq HQ \subseteq Q$. Hence, if $\{Q_\alpha\}_\Delta$ is the family of quasi-injective submodules of $E(M)$ each of which contains M, we see that $HM \subseteq \bigcap_\Delta Q_\alpha$. Moreover, (1) shows that HM is one of the Q_α, so we have $\bigcap_\Delta Q_\alpha \subseteq HM$. Thus, $HM = \bigcap_\Delta Q_\alpha$.

Parts (1) and (2) clearly give (3).   □

Corollary 7.3.3. *HM is the smallest quasi-injective submodule of $E(M)$ that contains M. Furthermore, HM is an essential extension of M.*

Because of the minimality of HM in $E(M)$, HM together with the canonical embedding $i : M \to HM$ is a candidate for a quasi-injective envelope of M. To establish that $i : M \to HM$ actually is a quasi-injective envelope, we need to show that $i : M \to HM$ satisfies the requirements of Definition 7.3.1.

If N is a submodule of M, then we call N *closed* in M if N has no proper essential extensions in M.

Lemma 7.3.4. *Let M be a quasi-injective R-module and suppose that N is a closed submodule of M. If L is a submodule of M, then any R-linear mapping $f : L \to N$ can be extended to an R-linear mapping $g : M \to N$.*

Proof. If $\mathcal{S}$ is the family $\{f' : L' \to N\}$ of all R-linear mappings, where $M \supseteq L' \supseteq L$ and $f'|_L = f$, then an application of Zorn's lemma shows that $\mathcal{S}$ has a maximal element. Let $\overline{f} : \overline{L} \to N$ be a maximal element of $\mathcal{S}$. Since M is quasi-injective, $\overline{f}$ can be extended to a map $g \in \mathrm{End}_R(M)$. Suppose that $g(M) \nsubseteq N$ and let N_c be a complement of N in M. Then since N is closed in M, N is a complement of N_c. Since $g(M) + N \nsupseteq N$, we have $(g(M) + N) \cap N_c \neq 0$. Let

$$x = y + z \in (g(M) + N) \cap N_c, \quad \text{where } x \neq 0, \ y \in g(M) \text{ and } z \in N.$$

If $y \in N$, then $x \in N \cap N_c = 0$, a contradiction. Therefore, $y \notin N$ and $y = x - z \in N_c \oplus N$. If $X = \{w \in M \mid g(w) \in N_c \oplus N\}$, then X is a submodule of M that contains $\overline{L}$. If $w \in M$ is such that $g(w) = y$, then $w \in X$, but $w \notin \overline{L}$ since $y \notin N$.

Consequently, if $\pi : N_c \oplus N \rightarrow N$ is the canonical projection, then $\pi g : X \rightarrow N$ is a proper extension of $f : L \rightarrow N$. Therefore, the assumption that $g(M) \nsubseteq N$ gives a contradiction, so $g(M) \subseteq N$ and the proof is complete. $\square$

Proposition 7.3.5. *If M is a quasi-injective R-module and N is a submodule of M, then M contains a maximal essential extension of N that is quasi-injective and a direct summand of M.*

Proof. Let $\mathcal{S}$ be the collection of submodules of M that are essential extensions of N. Then $\mathcal{S} \neq \emptyset$ and an easy application of Zorn's lemma shows that $\mathcal{S}$ has a maximal element, say $\overline{N}$. Clearly $\overline{N}$ is closed in M and, due to Lemma 7.3.4, the identity map $\mathrm{id}_{\overline{N}} : \overline{N} \rightarrow \overline{N}$ can be extended to a mapping $f : M \rightarrow \overline{N}$ such that $fi = \mathrm{id}_{\overline{N}}$, where $i : \overline{N} \rightarrow M$ is the canonical injection. Thus, i is a splitting map for f, so it follows that $M = \overline{N} \oplus \mathrm{Ker}\, f$. Hence, by Proposition 5.4.4, $\overline{N}$ is quasi-injective. $\square$

Corollary 7.3.6. *If $\varphi : M \rightarrow E_q$ is a quasi-injective envelope of M, then $\varphi(M)$ is an essential submodule of E_q.*

Proof. Suppose that $\varphi : M \rightarrow E_q$ is a quasi-injective envelope of M. Then by the preceding proposition, there is a quasi-injective extension $\overline{M}$ of $\varphi(M)$ contained in E_q that is a maximal essential extension of $\varphi(M)$. Since $\varphi : M \rightarrow E_q$ is a quasi-injective envelope, it follows from Lemma 7.3.4 that there is a monomorphism $g : E_q \rightarrow \overline{M}$ such that $g\varphi = f$, where $f : M \rightarrow \overline{M}$ is such that $f(x) = \varphi(x)$. If X is a submodule of E_q and $\varphi(M) \cap X = 0$, then $g\varphi(M) \cap g(X) = g(\varphi(M) \cap X) = 0$. But $\varphi(M) = f(M) = g\varphi(M)$, so $\varphi(M) \cap g(X) = 0$ in $\overline{M}$. Thus, $g(X) = 0$ and this gives $X = 0$. $\square$

Proposition 7.3.7. *Every R-module has a quasi-injective envelope $\varphi : M \rightarrow E_q$ that is unique up to isomorphism.*

Proof. Suppose that $f : M \rightarrow N$ is a quasi-injective extension of M and let $H^* = \mathrm{End}_R(E(N))$. Then $H^*N = N$ and if $H = \mathrm{End}_R(E(M))$, then $i : M \rightarrow HM$ is a quasi-injective essential extension of M. Thus, f can be extended to a monomorphism $g : HM \rightarrow E(N)$ such that the diagram

$$
\begin{array}{ccc}
0 \longrightarrow M & \overset{i}{\longrightarrow} & HM \\
 \downarrow f & \swarrow g & \\
E(N) & &
\end{array}
$$

is commutative. Now $g(HM)$ is quasi-injective, so it follows that $H^*g(HM) \subseteq g(HM)$ and so $H^*X \subseteq X$, where $X = N \cap g(HM)$. Therefore, X is quasi-injective and $g^{-1}(X)$ is a quasi-injective extension of M contained in HM. But HM is the smallest quasi-injective extension of M contained in $E(M)$, so $g^{-1}(X) = HM$. Thus, $g(HM) = X \subseteq N$, so the diagram

$$
\begin{array}{ccc}
0 \longrightarrow & M & \xrightarrow{\ i\ } HM \\
 & \downarrow{\scriptstyle f} & \swarrow{\scriptstyle g} \\
 & N &
\end{array}
$$

is commutative and g is an injection. Therefore, $i : M \to HM$ is a quasi-injective envelope of M. Uniqueness up to isomorphism of quasi-injective envelopes now follows easily. $\square$

Quasi-Projective Covers

Definition 7.3.8. A quasi-projective module $P_q(M)$ together with an epimorphism $\varphi : P_q(M) \to M$ is said to be a *quasi-projective cover* of M if $\operatorname{Ker}\varphi$ is small in $P_q(M)$ and if K is a nonzero submodule of $\operatorname{Ker}\varphi$, then $P_q(M)/K$ is not quasi-projective. A quasi-projective cover of M will often be denoted simply by $\varphi : P_q \to M$. A quasi-projective cover $\varphi : P_q \to M$ of M, is said to be *unique up to isomorphism* if whenever $\varphi' : P'_q \to M$ is another quasi-projective cover of M, there is an isomorphism $g : P_q \to P'_q$ such that $\varphi'g = \varphi$. A submodule N of an R-module M is said to be *stable under endomorphisms* of M if $f(N) \subseteq N$ for each $f \in \operatorname{End}_R(M)$.

The proof of the following proposition is left as an exercise.

Proposition 7.3.9. *A quasi-projective cover of an R-module M, if it exists, is unique up to isomorphism.*

Wu and Jans proved in [72] that if a module has a projective cover, then it has a quasi-projective cover. To establish this result, we need the following two lemmas.

Lemma 7.3.10. *If $\varphi : P \to M$ is a projective cover of M and $\operatorname{Ker}\varphi$ is stable under endomorphisms of P, then M is quasi-projective.*

Proof. Consider the diagram

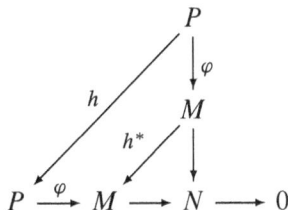

$$
\begin{array}{ccc}
 & & P \\
 & \overset{h}{\nearrow} & \downarrow{\varphi} \\
 & & M \\
 & \overset{h^*}{\nearrow} & \downarrow \\
P \overset{\varphi}{\longrightarrow} M & \longrightarrow & N \longrightarrow 0
\end{array}
$$

where h is given by the projectivity of P. Since $\operatorname{Ker}\varphi$ is stable under endomorphisms of P, $h(\operatorname{Ker}\varphi) \subseteq \operatorname{Ker}\varphi$, so there is an induced map $h^* : P/\operatorname{Ker}\varphi \to P/\operatorname{Ker}\varphi$. But $P/\operatorname{Ker}\varphi \cong M$, so we have a mapping $h^* : M \to M$ that makes the inner triangle commute. Thus, M is quasi-projective.   □

Lemma 7.3.11. *If $\varphi : P \to P/K$ is a projective cover of the quasi-projective module P/K, where φ is the natural surjection, then $K = \operatorname{Ker}\varphi$ is stable under endomorphisms of P.*

Proof. If $f \in \operatorname{End}_R(P)$, then f induces an R-linear mapping $f^* : P/K \to P/(K + f(K))$ given by $f^*(x + K) = f(x) + K + f(K)$, so consider the diagram

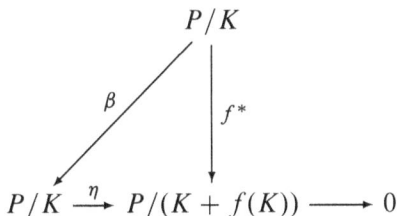

$$
\begin{array}{ccc}
 & P/K & \\
\overset{\beta}{\swarrow} & & \downarrow{f^*} \\
P/K \overset{\eta}{\longrightarrow} & P/(K + f(K)) & \longrightarrow 0
\end{array}
$$

where η is the natural mapping and β is the map given by the quasi-projectivity of P/K. This gives a commutative diagram

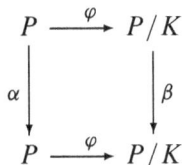

$$
\begin{array}{ccc}
P & \overset{\varphi}{\longrightarrow} & P/K \\
\downarrow{\alpha} & & \downarrow{\beta} \\
P & \overset{\varphi}{\longrightarrow} & P/K
\end{array}
$$

where α is given by the projectivity of P. Now let

$$
X = \{x \in P \mid f(x) - \alpha(x) \in K\}.
$$

We claim that $X = P$. Since $\varphi\alpha(K) = \beta\varphi(K) = 0$, we have $\alpha(K) \subseteq K$ and α gives an induced map

$$
\alpha^* : P/K \to P/(K + f(K)) \text{ such that } \alpha^*(x + K) = \alpha(x) + K + f(K).
$$

Hence,

$$(f^* - \alpha^*)(x + K) = f^*(x + K) - \alpha^*(x + K)$$
$$= \eta\beta(x + K) - (\alpha(x) + K + f(K))$$
$$= \eta\beta(x + K) - \eta\varphi\alpha(x)$$
$$= \eta\beta(x + K) - \eta\beta\varphi(x)$$
$$= \eta\beta(x + K) - \eta\beta(x + K)$$
$$= 0.$$

Therefore, $f(x) + K + f(K) - (\alpha(x) + K + f(K)) = 0$ and this gives $f(x) - \alpha(x) \in K + f(K)$. Now let $f(x) - \alpha(x) = k_1 + f(k_2)$, where $k_1, k_2 \in K$. Since $\alpha(k_2) \in \alpha(K) \subseteq K$, we see that $f(x - k_2) - \alpha(x - k_2) = k_1 + \alpha(k_2) \in K$. Hence, $x - k_2 \in X$ and so $P = X + K$. But K is small in P so $X = P$. Hence, if $x \in K$, then $x \in X$, so $f(x) - \alpha(x) \in K$. Thus, $f(x) \in K$, since $\alpha(x) \in K$. Therefore, $f(K) \subseteq K$ and K is stable under endomorphisms of P. □

Proposition 7.3.12. *If an R-module has a projective cover, then it has a quasi-projective cover.*

Proof. Let $\varphi' : P \to M$ be a projective cover of M and via Zorn's lemma choose X maximal in $K = \operatorname{Ker}\varphi'$ such that X is stable under endomorphisms of P. If $\varphi : P/X \to M$ is such that $\varphi(x + X) = \varphi'(x)$, then we claim that $\varphi : P/X \to M$ is a quasi-projective cover of M. First, note that $\operatorname{Ker}\varphi = K/X$ and since K is small in P, it follows that $\operatorname{Ker}\varphi$ is small in P/X. Since $X \subseteq K$, the natural map $P \to P/X$ is a projective cover of P/X, so since X is stable under endomorphisms of P, Lemma 7.3.10 shows that P/X is quasi-projective. Finally, if Y is such that $X \subseteq Y \subseteq K$, then $Y/X \subseteq K/X$ and $(P/X)/(Y/X) \cong P/Y$. It follows that the natural map $P \to P/Y$ gives a projective cover of P/Y. But in this setting, Lemma 7.3.11 shows that Y is stable under endomorphisms of P. Thus, the maximality of X gives $X = Y$. Hence, $Y/X = 0$ and we have that P/X is a quasi-projective cover of M. □

In the previous section, rings were characterized over which every module has a projective cover. It is natural to ask the question, What are the characteristics of a ring over which every module has a quasi-projective cover? The following proposition is due to Koehler [65].

Proposition 7.3.13. *The following are equivalent for a ring R.*

1. *R is right perfect.*

2. *Every R-module has a quasi-projective cover.*

Proof. (1) $\Rightarrow$ (2) is the result of Proposition 7.3.12, so we need only prove (2) $\Rightarrow$ (1). Assume that every R-module has a quasi-projective cover, let M be an R-module and let $\theta : R^{(M)} \to M$ be a free module on M. If $\varphi : P_q \to R^{(M)} \oplus M$ is a quasi-projective cover of $R^{(M)} \oplus M$, then we have a commutative diagram

$$
\begin{array}{ccc}
 & R^{(M)} & \\
 f \nearrow & \downarrow \mathrm{id}_{R^{(M)}} & \\
P_q \xrightarrow{\ \varphi\ } R^{(M)} \oplus M \xrightarrow{\ \pi_1\ } R^{(M)} \longrightarrow 0
\end{array}
$$

where π_1 is the canonical projection and f is the completing map given by the projectivity of $R^{(M)}$. Since $\pi_1 \varphi f = \mathrm{id}_{R^{(M)}}$, if $M^* = \mathrm{Ker}(\pi_1\varphi)$, then, since f is a monomorphism, we can assume that $P_q = R^{(M)} \oplus M^*$. If $\varphi^* = \varphi|_{M^*}$, then $\varphi^* : M^* \to M$ is an epimorphism and we claim that $\varphi^* : M^* \to M$ is a projective cover of M. To show this, let N be a submodule of M^* such that $\mathrm{Ker}\,\varphi^* + N = M^*$. Then $R^{(M)} \oplus (\mathrm{Ker}\,\varphi^* + N) = R^{(M)} \oplus M^*$ and $\mathrm{Ker}\,\varphi^* \subseteq \mathrm{Ker}\,\varphi$ give $R^{(M)} \oplus N = R^{(M)} \oplus M^*$. Thus, $N = M^*$ and so $\mathrm{Ker}\,\varphi^*$ is small in M^*. Next, consider the commutative diagram

$$
\begin{array}{ccc}
 & R^{(M)} & \\
 \theta^* \nearrow & \downarrow \theta & \\
M^* \xrightarrow{\ \varphi^*\ } M \longrightarrow 0
\end{array}
$$

where θ^* is given by the projectivity of $R^{(M)}$ and note that θ^* is an epimorphism since $\mathrm{Ker}\,\varphi^*$ is small in M^*. Hence, we have a commutative diagram

$$
\begin{array}{ccc}
 & R^{(M)} \oplus M^* & \\
 g \nearrow & i_2 \uparrow\ \downarrow \pi_2 & \\
 & M^* & \\
 g^* \nearrow & \downarrow \mathrm{id}_{M^*} & \\
R^{(M)} \oplus M^* \xrightarrow{\ \pi_1\ } R^{(M)} \xrightarrow{\ \theta^*\ } M^* \longrightarrow 0
\end{array}
$$

with the map g being given by the quasi-projectivity of $R^{(M)} \oplus M^*$. If $g^* = \pi_1 g i_2$, then the inner triangle is commutative and g^* is a monomorphism. Therefore, M^* is isomorphic to a direct summand of $R^{(M)}$ and so M^* is a projective R-module. Hence, every R-module has a projective cover, so R is a right perfect ring.   □

Problem Set 7.3

1. Let N be a submodule of M and suppose that N_c is a complement of N in M. Show that there is a complement $N_{cc} = (N_c)_c$ in M of N_c such that $N_{cc} \supseteq N$. Show that N_{cc} is a maximal essential extension of N in M. Conclude that if N is a closed submodule of M and if N_c is a complement of N in M, then N is a complement in M of N_c.

2. Prove that an R-module M is quasi-injective if and only if every $f \in \mathrm{Hom}_R(N, M)$ can be extended to an endomorphism of M for each essential submodule N of M.

3. If M is an R-module, let A be an ideal of R and set $N = \mathrm{ann}_\ell^M(A)$.

   (a) Prove that M is a quasi-injective R-module if and only if N is a quasi-injective R/A-module.

   (b) Prove that if $MA = 0$, then M is a quasi-injective R-module if and only if M is a quasi-injective R/A-module.

4. Complete the proof of Proposition 7.3.7 by showing that quasi-injective envelopes are unique up to isomorphism.

5. Prove Proposition 7.3.9.

6. (a) If $\{M_\alpha\}_\Delta$ is a family of R-modules such that $\bigoplus_\Delta M_\alpha$ is quasi-injective, prove that each M_α is quasi-injective.

   (b) Show that the $\mathbb{Z}$-modules $\mathbb{Q}$ and $\mathbb{Z}_p$ are quasi-injective, where p is a prime number and consider the $\mathbb{Z}$-module $M = \mathbb{Q} \oplus \mathbb{Z}_p$. Show that the canonical epimorphism $\eta : \mathbb{Z} \to \mathbb{Z}_p$ cannot be extended to a $\mathbb{Z}$-linear mapping $\mathbb{Q} \to \mathbb{Z}_p$. Conclude that the map $\mathbb{Z} \oplus 0 \to \mathbb{Q} \oplus \mathbb{Z}_p$ such that $(n, 0) \mapsto (0, [n])$ can not be extended to a $\mathbb{Z}$-linear map in $\mathrm{End}_\mathbb{Z}(M)$ and therefore that M cannot be quasi-injective. Thus, a direct sum of quasi-injectives need not be quasi-injective and so the converse of (a) may not hold.

7. (a) If a quasi-injective R-module M contains a copy of R_R, show that M is injective.

   (b) Prove that if the direct sum of every pair of quasi-injective modules is quasi-injective, then every quasi-injective module is injective. [Hint: Consider $E(R) \oplus M$.]

8. Show that if direct sums of quasi-injective R-modules are quasi-injective, then R is right noetherian. Note the converse fails since (b) of Exercise 6 gives

a direct sum of a pair quasi-injective modules over the noetherian ring $\mathbb{Z}$ that is not quasi-injective.

Note that this Exercise shows that noetherian rings are not characterized by direct sums of quasi-injective modules being quasi-injective, whereas Proposition 7.1.6 indicates that this is the case if direct sums of injective modules are injective.

9. Prove that the following are equivalent.

(a) R is a semisimple ring.

(b) Every R-module is quai-injective.

Chapter 8

Rings and Modules of Quotients

It is well known that if R is an integral domain and R^* is the set of nonzero elements of R, then the relation $\sim$ defined on $R \times R^*$ by $(a, b) \sim (c, d)$ if and only if $ad = bc$ is an equivalence relation. If a/b denotes the equivalence class determined by $(a, b) \in R \times R^*$ and Q is the set of all such equivalence classes, then Q is a field, called the *field of fractions* of R, if addition and multiplication are defined on Q by

$$\frac{a}{b} + \frac{c}{d} = \frac{ad + bc}{bd} \quad \text{and} \quad \frac{a}{b}\frac{c}{d} = \frac{ac}{bd}$$

for all $a/b, c/d \in Q$. The additive identity of Q is $0/1$ and the multiplicative identity is $1/1$. The mapping $\varphi : R \rightarrow Q$ defined by $a \mapsto a/1$ is an injective ring homomorphism, so R embeds in Q. In particular, if $R = \mathbb{Z}$, then $Q = \mathbb{Q}$, the *field of rational numbers*. This procedure for constructing the field of fractions of an integral domain provides a model for construction of a ring of quotients for a suitable ring which may not be an integral domain. In fact, there is no need for such a ring to be commutative.

8.1 Rings of Quotients

The Noncommutative Case

If $\varphi : R \rightarrow Q$ is the field of fractions of an integral domain, then there are three properties that are evident from the construction of Q.

1. R^*, the set of nonzero elements of R, is closed under multiplication, $1 \in R^*$ and $0 \notin R^*$.

2. If $b \in R^*$, then $\varphi(b)$ has a multiplicative inverse in Q.

3. Every element of Q can be written as $\varphi(a)\varphi(b)^{-1}$ for some $a \in R$ and $b \in R^*$.

These properties motivate the following definition. The ring R is not assumed to be commutative and it can have zero divisors.

Definition 8.1.1. A nonempty subset S of R is said to be *multiplicatively closed* if $st \in S$ whenever $s, t \in S$. If S is a multiplicatively closed subset of R, then S is a *multiplicative system* in R if $1 \in S$ and $0 \notin S$.

Examples

1. If $a \in R$ is not nilpotent, then $S = \{a^n \mid n = 0, 1, 2, \ldots\}$ is a multiplicative system in R.

2. If $\mathfrak{p}$ is a prime ideal of a commutative ring R and $S = R - \mathfrak{p}$, then S is a multiplicative system. This follows since if $a, b \in S$ and $ab \notin S$, then $ab \in \mathfrak{p}$, so either $a \in \mathfrak{p}$ or $b \in \mathfrak{p}$, a contradiction. Hence, if $a, b \in S$, then $ab \in S$.

3. A nonzero element of R is said to be a *regular element* of R if it is not a zero divisor in R. If $\mathcal{R}$ is the set of regular elements of R, then $\mathcal{R}$ is a multiplicative system in R.

Definition 8.1.2. Let S be a multiplicative system in R and suppose that $\varphi : R \to R'$ is a ring homomorphism. Then φ is said to be an *S-inverting homomorphism* if $\varphi(s)$ has a multiplicative inverse in R' for each $s \in S$. A ring R_S together with a ring homomorphism $\varphi : R \to R_S$ is said to be a *right ring of quotients* of R at S provided that the following three conditions hold.

(1) φ is S-inverting.

(2) Every element of R_S is of the form $\varphi(a)\varphi(s)^{-1}$ for some $(a, s) \in R \times S$.

(3) $\operatorname{Ker}\varphi = \{a \in R \mid as = 0 \text{ for some } s \in S\}$.

The modifier "right" will now be suppressed so that *ring of quotients of R* will mean right ring of quotients of R. A *left ring of quotients* of R at S is similarly defined.

By assuming that R has a ring of quotients at S, conditions can be found that will allow us to construct a ring of quotients of R at S.

Proposition 8.1.3. *Let S be a multiplicative system in R. If R has a ring of quotients $\varphi : R \to R_S$ at S, then:*

(1) *For each $(a, s) \in R \times S$, there is a $(b, t) \in R \times S$ such that $at = sb$.*

(2) *If $sa = 0$ for $(a, s) \in R \times S$, then there is a $t \in S$ such that $at = 0$.*

Proof. (1) If $(a, s) \in R \times S$, consider the element $\varphi(s)^{-1}\varphi(a)$ of R_S and let $(b, t) \in R \times S$ be such that $\varphi(s)^{-1}\varphi(a) = \varphi(b)\varphi(t)^{-1}$. Then $at - sb \in \operatorname{Ker}\varphi$, so there is an $s' \in S$ such that $(at - sb)s' = 0$. This gives $a(ts') = s(bs')$ and $(bs', ts') \in R \times S$.

(2) If $sa = 0$, then $\varphi(s)\varphi(a) = 0$, so $\varphi(a) = 0$ since $\varphi(s)$ has an inverse in R_S. Therefore, $a \in \operatorname{Ker}\varphi$, so there is a $t \in S$ such that $at = 0$.   □

Definition 8.1.4. If S is a multiplicative system in a ring R, then S is a *right permutable set* if (1) of Proposition 8.1.3 holds. If S satisfies condition (2), then S is said to be *right reversible*. If S is right permutable and right reversible, then S is a *right denominator set* in R. *Left permutable, left reversible* and *left denominator sets* are defined in the obvious way.

Remark. One method that can be used to distinguish right permutable sets from those that are left permutable is given by the following.

(1) **A right permutable set.** If $(a, s) \in R \times S$, then there is a $(b, t) \in R \times S$ such that $at = sb$: Write (b, t) on the right of (a, s) as in $(a, s)(b, t)$. Then the product of the outside members of the ordered pairs is equal to the product of the inside members with order maintained in multiplication.

(2) **A left permutable set.** If $(s, a) \in S \times R$, then there is a $(t, b) \in S \times R$ such that $ta = bs$. If we write $(t, b)(s, a)$ and then apply the same procedure as in (1) we get $ta = bs$.

Lemma 8.1.5. *Let S be a right denominator set in R and suppose that the relation $\sim$ is defined on $R \times S$ by $(a, s) \sim (a', s')$ if and only if there are $b, b' \in R$ such that $sb = s'b' \in S$ and $ab = a'b'$. Then $\sim$ is an equivalence relation on $R \times S$.*

Proof. It is easy to show that $\sim$ is reflexive and symmetric, so we will only verify that $\sim$ is transitive. Suppose that $(a, s) \sim (a', s')$ and $(a', s') \sim (a'', s'')$. Then there are $b, b', c, c' \in R$ such that $sb = s'b' \in S$, $ab = a'b'$, $s'c = s''c' \in S$ and $a'c = a''c'$. Now $(s'c, s'b') \in R \times S$, so since S is right permutable, there is an $(r, t) \in R \times S$ such that $s'ct = s'b'r$. Thus, $s'(ct - b'r) = 0$ and so since S is right reversible, there is a $t' \in S$ such that $ctt' = b'rt'$. Hence,

$$sbr = s'b'r = s'ct = s''c't \in S \quad \text{implies that } s(brt') = s''(c'tt') \in S.$$

We also have

$$a(brt') = a'b'rt' = a'ctt' = a''(c'tt'),$$

so $(a, s) \sim (a'', s'')$.   □

Notation. If S is a right denominator set in R and $\sim$ is the equivalence relation defined on $R \times S$ as in Lemma 8.1.5, then RS^{-1} will denote the set of all equivalence classes a/s determined by the ordered pairs $(a, s) \in R \times S$.

If $\varphi : R \to RS^{-1}$ is such that $a \mapsto a/1$ for all $a \in R$, then we claim that $\varphi : R \to RS^{-1}$ is a ring of quotients of R. Showing that RS^{-1} is a ring in the following proposition is technical and often quite tedious. For this reason the proof is sketched. Proofs of the ring properties of RS^{-1} are left to the interested reader.

Proposition 8.1.6. *Let S be a multiplicative system in R. Then $\varphi : R \to RS^{-1}$ is a ring of quotients of R at S if and only if S is a right denominator set in R.*

Proof. If $\varphi : R \to RS^{-1}$ is a ring of quotients at S, then Proposition 8.1.3 shows that S is a right denominator set in R. Conversely, assume that S is a right denominator set in R and define addition and multiplication on RS^{-1} by

$$a/s + b/t = (au + bc)/su, \quad \text{where } (c, u) \in R \times S \text{ is such that } su = tc,$$

and

$$(a/s)(b/t) = (ac)/(tu), \quad \text{where } (c, u) \in R \times S \text{ is such that } bu = sc.$$

These well-defined operations turn RS^{-1} into a ring with additive identity $0/1$ and multiplicative identity $1/1$. It is easy to verify that $ar/sr = a/s$ for any $r \in R$ such that $sr \in S$. Now let $\varphi : R \to RS^{-1}$ be such that $\varphi(a) = a/1$. If $a, b \in R$, then $\varphi(a)\varphi(b) = (a/1)(b/1) = (ac)/u$, where $(c, u) \in R \times S$ is such that $bu = 1c$. Hence, $(ac)/u = (abu)/u = ab/1 = \varphi(ab)$. Similarly, $\varphi(a + b) = \varphi(a) + \varphi(b)$, so φ is a ring homomorphism which is clearly S-inverting. Moreover, every element of RS^{-1} is of the form $\varphi(a)\varphi(s)^{-1}$. Finally, $a \in \operatorname{Ker}\varphi$ if and only if $a/1 = 0/1$, which is true if and only if $(a, 1) \sim (0, 1)$. That is, if and only if there are $b, b' \in R$ such that $1b = 1b' \in S$ and $ab = 0b' = 0$. Hence, $a \in \operatorname{Ker}\varphi$ if and only if there is an $s \in S$ such that $as = 0$. Thus, $\varphi : R \to RS^{-1}$ is a ring of quotients of R at S. $\quad\square$

Definition 8.1.7. If S is a right denominator set in R and $\varphi : R \to RS^{-1}$ is a ring of quotients of R at S, then elements a/s of RS^{-1} will be called *fractions* or *quotients* and the ring homomorphism φ will be referred to as the *canonical ring homomorphism* from R to RS^{-1}. If $\varphi : R \to RS^{-1}$ is a ring of quotients of R at S, we will often refer to RS^{-1} as a ring of quotients of R at S with the map φ understood.

Remark. The ring of quotients RS^{-1} is called the *localization of R at S*, even though RS^{-1} may not be a local ring. This is probably due to the fact that the construction of RS^{-1} is a generalization of the localization of R at a prime ideal in a commutative ring which is a local ring. The localization of a commutative ring at a prime ideal will be discussed later.

Proposition 8.1.8. *If S is a right denominator set in R and $\varphi : R \to RS^{-1}$ is a ring of quotients of R at S, then $\varphi : R \to RS^{-1}$ has the universal mapping property in the sense that for every S-inverting ring homomorphism $f : R \to R'$, there is a unique ring homomorphism $g : RS^{-1} \to R'$ such that the diagram*

$$
\begin{array}{ccc}
 & & RS^{-1} \\
 & \nearrow^{\varphi} & \big\downarrow{\scriptstyle g} \\
R & & \\
 & \searrow_{f} & \\
 & & R'
\end{array}
$$

is commutative.

Proof. If we define $g : RS^{-1} \to R'$ by $g(a/s) = f(a)f(s)^{-1}$, then $g\varphi(a) = g(a/1) = f(a)f(1)^{-1} = f(a)$, so the diagram is commutative. The fact that g is a ring homomorphism follows directly from the fact that f is a ring homomorphism. If $g' : RS^{-1} \to R'$ is also a ring homomorphism which makes the diagram commutative, then $(g - g')(\varphi(a)) = 0$ for each $a \in R$. In particular, $(g - g')(1/1) = (g - g')(\varphi(1)) = 0$, so $g = g'$.   □

Corollary 8.1.9. *If* $\varphi : R \to RS^{-1}$ *is a ring of quotients of R at S, then* RS^{-1} *is unique up to ring isomorphism.*

Proof. If we form the category $\mathcal{C}$ whose objects are S-inverting ring homomorphisms $f : R \to R'$ and whose morphism sets

$$\mathrm{Mor}(f : R \to R', g : R \to R'')$$

are composed of ring homomorphisms $h : R' \to R''$ such that $hf = g$, then the proposition shows that $\varphi : R \to RS^{-1}$ is an initial object in $\mathcal{C}$. As a consequence, RS^{-1} is unique up to ring isomorphism.   □

If S is a left denominator set in R, then $S^{-1}R$ will denote a left ring of quotients of R at S with the canonical map $\varphi : R \to S^{-1}R$ understood. If S is a left and a right denominator set in R, then it follows that $S^{-1}R \cong RS^{-1}$.

Examples

4. **Ore's Condition.** If $\mathcal{R}$ is the set of regular elements of R, then $\mathcal{R}$ is clearly a multiplicatively closed set that is right reversible. If $\mathcal{R}$ is right permutable, then R is said to satisfy the *right Ore condition* and the ring of quotients $R\mathcal{R}^{-1}$ of R at $\mathcal{R}$, denoted by $Q_{\mathrm{cl}}^r(R)$, is the *classical (right) ring of quotients* of R. In this case, $\varphi : R \to Q_{\mathrm{cl}}^r(R)$ is an injective ring homomorphism and we may consider R to be a subring of $Q_{\mathrm{cl}}^r(R)$ by identifying $a \in R$ with $a/1$ in $Q_{\mathrm{cl}}^r(R)$. If R satisfies the right Ore condition, then R is said to be a *right Ore ring*. *Left Ore rings* are defined similarly. If R is a left and right Ore ring, then R is simply referred to as an *Ore ring*. If R is an Ore ring, then one can show that $Q_{\mathrm{cl}}^\ell(R) \cong Q_{\mathrm{cl}}^r(R)$. Clearly, any commutative ring is an Ore ring. If R is a right Ore ring without zero divisors, then $Q_{\mathrm{cl}}^r(R)$ is a division ring. In particular, if R is an integral domain, then $Q_{\mathrm{cl}}^r(R)$ is the field of fractions of R. Right (and left) Ore rings are named in honor of Oystein Ore who discovered the right (left) permutable condition on $\mathcal{R}$. Details can be found in [69].

5. If R is a right noetherian ring without zero divisors, then R is a right Ore ring. Since $\mathcal{R}$ is a right reversible, to prove that R is a right Ore ring we need only show that if $(a, s) \in R \times \mathcal{R}$, then there is a $(b, t) \in R \times \mathcal{R}$ such that $at = sb$.

If $(0, s) \in R \times \mathcal{R}$, then $(0, t) \in R \times \mathcal{R}$ is such that $0t = s0$, for any $t \in \mathcal{R}$. Thus, we can assume that $a \neq 0$. Consider the ascending chain

$$sR \subseteq sR + asR \subseteq \cdots \subseteq sR + asR + a^2 sR + \cdots + a^n sR \subseteq \cdots$$

of right ideals of R. Since R is right noetherian, there is a smallest integer $n \geq 0$ such that

$$sR + asR + a^2 sR + \cdots + a^n sR = sR + asR + a^2 sR + \cdots + a^n sR + a^{n+1} sR.$$

Let $r_0, r_1, \ldots, r_n \in R$ be such that

$$a^{n+1} s = s r_0 + a s r_1 + a^2 s r_2 + \cdots + a^n s r_n.$$

This gives

$$a(s r_1 + a s r_2 + \cdots + a^{n-1} s r_n - a^n s) = s(-r_0),$$

so if we can show that

$$s r_1 + a s r_2 + \cdots + a^{n-1} s r_n - a^n s \neq 0,$$

we will be finished for then we can let

$$b = -r_0 \quad \text{and} \quad t = s r_1 + a s r_2 + \cdots + a^{n-1} s r_n - a^n s.$$

If

$$s r_1 + a s r_2 + \cdots + a^{n-1} s r_n - a^n s = 0,$$

then

$$a^n sR = (s r_1 + a s r_2 + \cdots + a^{n-1} s r_n) R \subseteq sR + asR + \cdots + a^{n-1} sR$$

and this leads to a contradiction of the minimality of n.

One property that holds in RS^{-1} that is often useful for proving properties of fractions is the following common denominator property.

Lemma 8.1.10 (Common Denominator Property). *Let S be a right denominator set in R. For any integer $n \geq 1$, if $s_1, s_2, \ldots, s_n \in S$, then there exist $a_1, a_2, \ldots, a_n \in R$ and $s \in S$ such that $1/s_i = a_i/s$ for $i = 1, 2, \ldots, n$.*

Proof. If $n = 1$, let $a_1 = s_1$ and $s = s_1^2$. Then $1/s_1 = s_1/s_1^2 = a_1/s$, so the lemma holds when $n = 1$. Next, make the hypothesis that if $n = k$ and if $s_1, s_2, \ldots, s_k \in S$, then there exist $a_1, a_2, \ldots, a_k \in R$ and $s \in S$ such that $1/s_i = a_i/s$ for $i = 1, 2, \ldots, k$. If $s_1, s_2, \ldots, s_{k+1} \in S$, then, by hypothesis, there are $a_1, a_2, \ldots, a_k \in R$

and $s \in S$ such that $1/s_i = a_i/s$ for $i = 1, 2, \ldots, k$. Since $(s_{k+1}, s) \in R \times S$, there is an $(r, a_{k+1}^*) \in R \times S$ such that $s_{k+1} a_{k+1}^* = sr$. If we let $s^* = s_{k+1} a_{k+1}^* = sr$, then $s^* \in S$ and $1/s_{k+1} = a_{k+1}^*/s^*$. Now let $a_i^* = a_i r$ for $i = 1, 2, \ldots, k$. Then we also have $1/s_i = a_i/s = a_i r/sr = a_i^*/s^*$, for $i = 1, 2, \ldots, k$, and the lemma follows by induction.   □

Corollary 8.1.11. *If S is a right denominator set in R and if $a_1/s_1, a_2/s_2, \ldots, a_n/s_n \in RS^{-1}$, then there is an $s \in S$ such that*

$$(a_1/s_1)s, (a_2/s_2)s, \ldots, (a_n/s_n)s \in R$$

and there are $b_1, b_2, \ldots, b_n \in R$ such that

$$a_1/s_1 + a_2/s_2 + \cdots + a_n/s_n = (b_1 + b_2 + \cdots + b_n)/s.$$

If S is a right denominator set in R, then we need to establish a connection between the right ideals of R and the right ideals of RS^{-1}. If $\varphi : R \to RS^{-1}$ is the canonical ring homomorphism and A is a right ideal of R, then $\varphi(A)(RS^{-1})$ is a right ideal of RS^{-1} that will be denoted by A^e and referred to as the *extension of A to RS^{-1}*. Moreover, if A is a right ideal of RS^{-1}, then $A^c = \varphi^{-1}(A)$ is a right ideal of R called the *contraction of A to R*. (Note the distinction between A^c and A_c, the complement of a right ideal A in R.)

Proposition 8.1.12. *The following hold for any right denominator set S in R.*

(1) *If A is a right ideal of R, then $A^e = \{a/s \mid a \in A \text{ and } s \in S\}$.*

(2) *If A is a right ideal of R, then $A^{ec} = \{a \in R \mid as \in A \text{ for some } s \in S\}$.*

(3) *If A is a right ideal of RS^{-1}, then $A^{ce} = A$.*

(4) *Suppose that $\bigoplus_\Delta A_\alpha$ is a direct sum of right ideals of RS^{-1} and suppose that $\varphi|_{A_\alpha^c}$ is an injection for each $\alpha \in \Delta$, where $\varphi : R \to RS^{-1}$ is the canonical ring homomorphism. Then $\bigoplus_\Delta A_\alpha^c$ is a direct sum of right ideals of R.*

(5) *If $\bigoplus_\Delta A_\alpha$ is a direct sum of right ideals of R, then $\bigoplus_\Delta A_\alpha^e$ is a direct sum of right ideals of RS^{-1}.*

(6) *If R is right artinian (right noetherian), then so is RS^{-1}.*

Proof. We prove (5) and leave the proofs of (1) through (4) and (6) as exercises. Suppose that $\bigoplus_\Delta A_\alpha$ is a direct sum of right ideals of R, then by (1) each finite sum in $\sum_\Delta A_\alpha^e$ can be expressed as $a_1/s_1 + a_2/s_2 + \cdots + a_n/s_n$, where $a_i \in A_{\alpha_i}$ and $s_i \in S$ for $i = 1, 2, \ldots, n$. By Lemma 8.1.10, there are $b_1, b_2, \ldots, b_n \in R$ and $s \in R$ such that $1/s_i = b_i/s$ for $i = 1, 2, \ldots, n$. Hence, $a_i/s_i = a_i b_i/s$ for $i = 1, 2, \ldots, n$. From these observations, we have

$$a_1/s_1 + a_2/s_2 + \cdots + a_n/s_n = (a_1 b_1 + a_1 b_2 + \cdots + a_n b_n)/s.$$

Therefore, if $a_1/s_1 + a_2/s_2 + \cdots + a_n/s_n = 0/1$ in RS^{-1}, then it follows that $(a_1b_1 + a_2b_2 + \cdots + a_nb_n)/1 = 0/1$ in RS^{-1}. Hence, $a_1b_1 + a_2b_2 + \cdots + a_nb_n$ is in the kernel of the canonical map $\varphi : R \rightarrow RS^{-1}$, so there is a $t \in S$ such that $a_1b_1t + a_2b_2t + \cdots + a_nb_nt = (a_1b_1 + a_2b_2 + \cdots + a_nb_n)t = 0$ in R. Moreover, $a_ib_it \in A_{\alpha_i}$ for each i, so $a_1b_1t = a_2b_2t = \cdots = a_nb_nt = 0$, since the sum $\bigoplus_\Delta A_\alpha$ is direct. But this indicates that $a_ib_i \in \operatorname{Ker} \varphi$ for each i and so $a_ib_i/1 = 0/1$ in RS^{-1} for $i = 1, 2, \ldots, n$. Thus, for each i, we have $a_i/s_i = a_ib_i/s = 0/1$ in RS^{-1} which shows that the sum $\bigoplus_\Delta A_\alpha^e$ is direct.   □

The proof of the following proposition is left to the reader.

Proposition 8.1.13. *The following hold for a right denominator set S of regular elements of R.*

(1) *If A and B are right ideals of R such that $A \subseteq B$, then A is essential in B if and only if A^e is essential in B^e.*

(2) *If A and B are right ideals of RS^{-1} such that $A \subseteq B$, then A^c is essential in B^c if and only if A essential in B.*

The Commutative Case

When R is a commutative ring, every multiplicative system in R is a denominator set and the equivalence relation established in Lemma 8.1.5 takes the form given in (2) of the following proposition.

Proposition 8.1.14. *The following are equivalent for a commutative ring R and a multiplicative system S in R.*

(1) *If $(a, s), (b, t) \in R \times S$, then there are $c, c' \in R$ such that $sc = tc' \in S$ and $ac = bc'$.*

(2) *If $(a, s), (b, t) \in R \times S$, then there is a $u \in S$ such that $(at - sb)u = 0$.*

Proof. (1) $\Rightarrow$ (2). Let $(a, s), (b, t) \in R \times S$ and suppose that there are $c, c' \in R$ such that $sc = tc' \in S$ and $ac = bc'$. Multiplying the first equation by b and the second by t gives $atc = sbc$, so $(at - sb)c = 0$. If we let $u = sc \in S$, then $(at - sb)u = 0$.
   (2) $\Rightarrow$ (1). Suppose $(a, s), (b, t) \in R \times S$ and let $u \in S$ be such that $atu = bsu$. If $c = tu$ and $c' = su$, then $sc = tc'$ and $ac = bc'$, so we are finished.   □

Corollary 8.1.15. *If R is a commutative ring and S is a multiplicative system in R, then the relation $\sim$ defined on $R \times S$ by $(a, s) \sim (b, t)$ if and only if there is a $u \in S$ such that $(at - sb)u = 0$ is an equivalence relation on $R \times S$. Furthermore, $\sim$ yields the same equivalence classes as the equivalence relation of Lemma 8.1.5 and the* *operations of addition and multiplication defined on RS^{-1} in the proof of Proposition 8.1.6 become*

$$a/s + b/t = (at + sb)/st \quad and \quad (a/s)(b/t) = ab/st.$$

It follows that the lemmas, propositions, corollaries and examples that were given in the section on noncommutative localization also hold for commutative rings.

Examples

6. If R is a commutative ring and $\mathfrak{p}$ is a prime ideal of R, then we saw in Example 2 that $S = R - \mathfrak{p}$ is a multiplicative system in R. In this case, RS^{-1}, denoted by $R_{\mathfrak{p}}$, is a local ring with unique maximal ideal $\mathfrak{p}S^{-1} = \{a/s \mid a \in \mathfrak{p}, s \in S\}$.

7. If K is a field, then $K[X]$ is an integral domain and $K[X]\mathcal{R}^{-1}$ is the *field of rational functions*, where $\mathcal{R}$ is the multiplicative system of nonzero polynomials in $K[X]$. If R is a commutative ring, let $R[[X]]$ be the ring of power series in X over R. If $S = \{X^n \mid n = 0, 1, 2, \ldots\}$, then S is a multiplicative system in $R[[X]]$ and $R[[X]]S^{-1}$ is the ring of *Laurent power series* $R[[X]]S^{-1}$. Every element of $R[[X]]S^{-1}$ can be written as

$$\frac{a_0 + Xa_1 + X^2a_2 + X^3a_3 + \cdots}{X^n}$$

which, with a change of notation, can be written as

$$X^{-n}a_{-n} + X^{-n+1}a_{-n+1} + \cdots + Xa_{-1} + a_0 + Xa_1 + X^2a_2 + \cdots,$$

where $a_k \in R$ for each integer $k \geq -n$. The ring $R[[X]]S^{-1}$ is usually denoted by $R[[X, X^{-1}]]$. Note that

$$\frac{a_0 + Xa_1 + X^2a_2 + X^3a_3 + \cdots}{1} = a_0 + Xa_1 + X^2a_2 + X^3a_3 + \cdots,$$

so it follows that $R[[X]]$ is a subring of $R[[X, X^{-1}]]$.

Problem Set 8.1

In the following exercises, S is a right denominator set in R.

1. Let $\varphi : R \to RS^{-1}$ be a ring of quotients of R at S. Then an element $a \in R$ is said to be S-*torsion* if there is an $s \in R$ such that $as = 0$.

   (a) If $t(R)$ is the set of all S-torsion elements of R, prove that $t(R)$ is a right ideal of R and note that $\text{Ker}\,\varphi = t(R)$. Conclude that $t(R)$ is an ideal of R.

   (b) Prove that the induced ring homomorphism $\bar{\varphi} : R/t(R) \to RS^{-1}$ is an injection.

   (c) If $\eta : R \to \bar{R} = R/t(R)$ is the canonical surjection, prove that $\bar{S} = \eta(S)$ is a right denominator set in $\bar{R}$ and that $RS^{-1} \cong \bar{R}\bar{S}^{-1}$ as rings.

2. Prove Corollary 8.1.11.

3. Prove properties (1) through (4) and (6) of Proposition 8.1.12. [(1), Hint: Write the sum $a_1/s_1 + a_2/s_2 + \cdots + a_n/s_n$ in $\{a/s \in R \mid a \in A$ and $s \in S\}$ with a common denominator.] Note that (3) gives (6).

4. Prove Proposition 8.1.13. [(1), Hint: If A is essential in B and a/s is a nonzero element of B^e, let ar be nonzero in A. For the converse, assume that A^e is essential in B^e and let a be a nonzero element of B. Then $a(rs^{-1}) = bt^{-1}$ is nonzero for some $r \in R$, $b \in A$ and $s, t \in S$. Now write rs^{-1} and bt^{-1} with a common denominator.]

5. Prove that the canonical map $\varphi : R \to RS^{-1}$ is an isomorphism if and only if every element of S is a unit in R.

In the following exercises, R is a commutative ring and S is a multiplicative system in R. Additional information on localization of commutative rings at a multiplicative system can be found in [6].

6. Prove Corollary 8.1.15.

7. (a) Suppose 0 is permitted to be an element of S in the construction of RS^{-1}. Show that if $0 \in S$, then $RS^{-1} = 0$.

   (b) Suppose that R is a ring without an identity and define a multiplicative system S in R to be a multiplicatively closed subset of R such that $0 \notin S$. Prove that RS^{-1} can be constructed in the same manner as when R has an identity and $1 \in S$. Show that under this "new" construction of RS^{-1}, s/s is an identity for RS^{-1} for any $s \in S$. Observe that this construction of RS^{-1} works even if R has an identity and $1 \notin S$.

   (c) If R has an identity and $1 \notin S$, let $S' = S \cup \{1\}$. Prove that $RS^{-1} \cong RS'^{-1}$ as rings, where RS^{-1} is constructed as in (b) and RS'^{-1} is constructed in the usual fashion.

   (d) Prove that any ring without zero divisors can be embedded in a field even if the ring fails to have an identity.

8. (a) Let R and R' be isomorphic rings, both without identities, that are free of zero divisors. If S and S' denote the set of nonzero elements of R and R', respectively, construct RS^{-1} and $R'S'^{-1}$ as in (b) of Exercise 9. Prove that RS^{-1} is isomorphic to $R'S'^{-1}$. Conclude that isomorphic rings that are free of zero divisors have isomorphic fields of fractions. Show also that $RS^{-1} \cong R'S'^{-1}$ if R and R' have identities.

   (b) The converse of (a) is false: If S is the set of nonzero elements of $\mathbb{Z}$, let $2\mathbb{Z}$ denote the ring of even integers and let $2S$ denote the set of nonzero even integers. Prove that $\mathbb{Z}S^{-1} \cong (2\mathbb{Z})(2S)^{-1}$ as rings, where $(2\mathbb{Z})(2S)^{-1}$ is constructed by the method suggested in (b) of Exercise 9. Note that $\mathbb{Z}S^{-1} = \mathbb{Q}$, so we have $(2\mathbb{Z})(2S)^{-1} \cong \mathbb{Q}$. What is the isomorphism $(2\mathbb{Z})(2S)^{-1} \to \mathbb{Q}$?

Conclude that it is possible for the fields of fractions of two rings that are free of zero divisors to be isomorphic and yet the two rings are not isomorphic.

9. Let $R = \{a + b\sqrt{n} \mid a, b \in \mathbb{Z}\}$, where n is a square free integer, and define addition and multiplication on R in the usual way. If S is the set of nonzero elements of R, form RS^{-1} and prove that RS^{-1} and the quadratic field $Q(n) = \{\frac{a}{b} + \frac{c}{d}\sqrt{n} \mid \frac{a}{b}, \frac{c}{d} \in \mathbb{Q}\}$ are isomorphic rings. [Hint: Example 10 in Section 1.1.]

8.2 Modules of Quotients

Let S be a right denominator set in a ring R. If M is an R-module, then a construction similar to that developed for the construction of RS^{-1} can be carried out to establish a module of quotients of M which, as it turns out, is an RS^{-1}-module. Of course, any RS^{-1}-module can be viewed as an R-module by pullback along the canonical map from R to RS^{-1} which we now denote by $\varphi_R : R \to RS^{-1}$. Furthermore, RS^{-1} is an (R, R)-bimodule by pullback along φ_R.

Definition 8.2.1. Let S be a right denominator set in R and suppose that M is an R-module. An RS^{-1}-module MS^{-1} is said to be a *module of quotients* of M at S if there is an R-linear mapping $\varphi : M \to MS^{-1}$ such that

(1) Every element of MS^{-1} can be written as $\varphi(x)\varphi_R(s)^{-1}$, where $(x, s) \in M \times S$, and

(2) $\operatorname{Ker}\varphi = \{x \in M \mid xs = 0 \text{ for some } s \in S\}$.

The R-linear mapping φ is called the *canonical map* from M to MS^{-1}.

Lemma 8.2.2. *Let S be a right denominator set in R and suppose that M is an R-module. If the relation $\sim$ is defined on $M \times S$ by $(x, s) \sim (y, t)$ if and only if there are $a, a' \in R$ such that $sa = ta' \in S$ and $xa = ya'$, then $\sim$ is an equivalence relation on $M \times S$.*

Proof. It is straightforward to show that $\sim$ is reflexive and symmetric, so let us show that $\sim$ is transitive. Suppose that $(x, s) \sim (y, t)$ and $(y, t) \sim (z, u)$. Then there are $a, a', b, b' \in R$ such that $sa = ta' \in S$, $xa = ya'$, $tb = ub' \in S$ and $yb = zb'$. Now $(tb, ta') \in R \times S$, so since S is right permutable, there is a $(r, w) \in R \times S$ such that $tbw = ta'r$. Thus, $t(bw - a'r) = 0$ and so since S is right reversible, there is a $w' \in S$ such that $bww' = a'rw'$. Hence,

$$sar = ta'r = tbw = ub'w \in S \quad \text{implies that } s(arw') = u(b'ww') \in S.$$

We also have

$$x(arw') = ya'rw' = ybww' = z(b'ww'),$$

so $(x, s) \sim (z, u)$. $\square$

Proposition 8.2.3. *If S is a right denominator set in R, then every R-module has a module of quotients at S.*

Proof (sketched). Let $\sim$ be the equivalence relation on $M \times S$ of Lemma 8.2.2 and let MS^{-1} denote the set of all equivalence classes x/s determined by the ordered pairs $(x, s) \in M \times S$. Define addition and the RS^{-1}-action on MS^{-1} by

$$x/s + y/t = (xu + yc)/su, \quad \text{where } (c, u) \in R \times S \text{ is such that } su = tc,$$

and

$$(x/s)(a/t) = xc/tu, \quad \text{where } (c, u) \in R \times S \text{ is such that } au = sc.$$

These well-defined operations turn MS^{-1} into an RS^{-1}-module with additive identity $0/1$. If $\varphi : M \to MS^{-1}$ is such that $\varphi(x) = x/1$, then φ is an R-linear mapping and every element of MS^{-1} is of the form $\varphi(x)\varphi(s)^{-1} = x/s$. Finally, note that $x \in \operatorname{Ker}\varphi$ if and only if $x/1 = 0/1$ and this holds if and only if $(x, 1) \sim (0, 1)$. That is, if and only if there are $b, b' \in R$ such that $1b = 1b' \in S$ and $xb = 0b' = 0$. Hence, $x \in \operatorname{Ker}\varphi$ if and only if there is an $s \in S$ such that $xs = 0$. Thus, φ is the canonical map from M to MS^{-1} and so MS^{-1} together with φ is a module of quotients of M.   □

We now need the following lemma.

Lemma 8.2.4. *If S is a right denominator set in R, then for any R-module M, $M \otimes_R RS^{-1} \cong MS^{-1}$ as RS^{-1}-modules.*

Proof. Consider the diagram

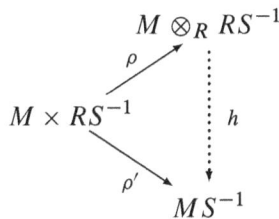

$$
\begin{array}{ccc}
 & & M \otimes_R RS^{-1} \\
 & \nearrow^{\rho} & \vdots \\
M \times RS^{-1} & & \vdots \ h \\
 & \searrow_{\rho'} & \vdots \\
 & & MS^{-1}
\end{array}
$$

where $\rho((x, a/s)) = x \otimes a/s$ and $\rho'((x, a/s)) = xa/s$. Then ρ and ρ' are R-balanced maps, so the definition of a tensor product ensures the existence of a unique group homomorphism $h : M \otimes_R RS^{-1} \to MS^{-1}$ such that $h(x \otimes a/s) = xa/s$ for all $x \otimes a/s$ in $M \otimes_R RS^{-1}$. Moreover, it follows easily that h is an RS^{-1}-epimorphism. Note that h is also injective for if $h(x \otimes a/s) = 0$, then $xa/s = 0/1$, so $(xa, s) \sim (0, 1)$. Hence, there are $b, b' \in R$ such that $xab = 0b' = 0$ and $sb = 1b' \in S$. Thus, $x \otimes a/s = x \otimes ab/sb = xab \otimes 1/sb = 0$, so h is injective. Therefore, h is an RS^{-1}-isomorphism.   □

Proposition 8.2.5. *If S is a right denominator set in R and $\varphi : M \to MS^{-1}$ is a module of quotients of M at S, then for every RS^{-1}-module N and every R-linear mapping $f : M \to N$, there is a unique RS^{-1}-linear mapping $g : MS^{-1} \to N$ such that the diagram*

$$
\begin{array}{ccc}
 & & MS^{-1} \\
 & {}^{\varphi}\nearrow & \Big\downarrow {}^{g} \\
M & & \\
 & {}_{f}\searrow & \\
 & & N
\end{array}
$$

is commutative.

Proof. Using Proposition 3.4.5, adjoint associativity, we see that

$$\operatorname{Hom}_{RS^{-1}}(M \otimes_R RS^{-1}, N) \cong \operatorname{Hom}_R(M, \operatorname{Hom}_{RS^{-1}}(RS^{-1}, N)) \cong \operatorname{Hom}_R(M, N).$$

Hence, using the lemma immediately above, we have

$$\operatorname{Hom}_{RS^{-1}}(MS^{-1}, N) \cong \operatorname{Hom}_R(M, N).$$

Therefore, given an R-linear map $f : M \to N$, there is a corresponding RS^{-1}-linear map $g : MS^{-1} \to N$ such that if $x \in M$, then $g(x/1) = f(x)$. It follows that $g(xa/s) = g(x/1)a/s = f(x)a/s$ for all $x/1 \in MS^{-1}$ and $a/s \in RS^{-1}$. Hence, if $x \in M$ and $\varphi : M \to MS^{-1}$ is the canonical map, then $g\varphi(x) = g(x/1) = f(x)$, so the diagram

$$
\begin{array}{ccc}
 & & MS^{-1} \\
 & {}^{\varphi}\nearrow & \Big\downarrow {}^{g} \\
M & & \\
 & {}_{f}\searrow & \\
 & & N
\end{array}
$$

is commutative.

If $g' : MS^{-1} \to N$ is also an RS^{-1}-linear mapping that makes the diagram commutative, then $g\varphi(x) = g'\varphi(x)$ for each $x \in M$. Thus, $g(x/1) = g'(x/1)$ for all $x \in M$, so for each $s \in S$ we have $g(x/1)(1/s) = g'(x/1)(1/s)$. But g and g' are RS^{-1}-linear, so $g(x/s) = g'(x/s)$. Therefore, $g = g'$ and we have that g is unique. $\qquad\square$

Corollary 8.2.6. *The module MS^{-1} is unique up to isomorphism in $\mathbf{Mod}_{RS^{-1}}$.*

Proof. The proposition shows that $\varphi : M \to MS^{-1}$ has the universal mapping property and a category $\mathcal{C}$ can be constructed that has $\varphi : M \to MS^{-1}$ as an initial object. Furthermore, morphisms in $\mathcal{C}$ are RS^{-1}-linear mappings, so MS^{-1} is unique up to RS^{-1}-isomorphism.   □

Note that if R is a commutative ring and S is a multiplicative system in R, then the equivalence relation $\sim$ defined on $M \times S$ in Lemma 8.2.2 becomes $(x, s) \sim (y, t)$ if and only if there is a $u \in S$ such that $(xt - ys)u = 0$. If MS^{-1} is the set of all equivalence classes of $M \times S$ determined by $\sim$, then MS^{-1} is an RS^{-1}-module if addition and the RS^{-1}-action are defined on MS^{-1} by

$$(x/s) + (y/t) = (xt + ys)/st \quad \text{and} \quad (x/s)(a/t) = xa/st$$

for all $x/s, y/t \in MS^{-1}$ and $a/t \in RS^{-1}$.

Problem Set 8.2

In the following exercises, S is a right denominator set in R.

1. (a) If M is a right RS^{-1}-module, prove that the canonical map $\varphi : M \to MS^{-1}$ is an isomorphism.

   (b) If U is also a right denominator set in R and $U \subseteq S$, show that $MS^{-1} \cong (MU^{-1})S^{-1} \cong (MS^{-1})U^{-1}$. Conclude that if M is any R-module, then $MS^{-1} \cong (MS^{-1})S^{-1} \cong ((MS^{-1})S^{-1})S^{-1} \cong \cdots$.

   (c) If M is an (R, R)-bimodule and if S is also a left denominator set in R, then is it the case that $S^{-1}M \cong MS^{-1}$?

2. If M is an R-module, then an element $x \in M$ is said to be an S-*torsion element* of M if there is an element $s \in S$ such that $xs = 0$.

   (a) If $t(M)$ denotes the set of all S-torsion elements of M, show that $t(M)$ is a submodule of M.

   (b) An R-module M is said to be an S-*torsion module* if $t(M) = M$ and an S-*torsion free module* if $t(M) = 0$. Show that the R-module $M/t(M)$ is S-torsion free and that the induced mapping $\bar{\varphi} : M/t(M) \to MS^{-1}$ is an R-linear embedding. Conclude that the canonical map $\varphi : M \to MS^{-1}$ is a monomorphism if and only if $t(M) = 0$.

3. Make MS^{-1} into an R-module by pullback along the canonical map $\varphi_R : R \to RS^{-1}$. An R-module M is said to be S-*injective* if for each right ideal A of R that contains an element of S and for each R-linear mapping $f : A \to M$ there is an $x \in M$ such that $f(a) = xa$ for each $a \in A$. An R-module M is called S-*divisible* if for each $s \in S$ and each $y \in M$, there is an $x \in M$ such that $xs = y$, that is, if $Ms = M$ for each $s \in S$. Decide which, if any, of the

following implies the other(s). [Hint: Look at Baer's criteria, Exercise 2 and Proposition 5.1.9.]

(a) M is S-torsion free and S-injective.

(b) M is S-torsion free and S-divisible.

(c) $M \cong MS^{-1}$.

4. Let R be a ring, let X be a noncommuting indeterminate and let $\sigma : R \to R$ be a ring homomorphism. Suppose also that $R[[X, \sigma]]$ is the *ring of skew power series* over R. [Hint: Example 2 in Section 1.3.]

(a) Show that $R[[X, \sigma]]$ is noncommutative even if R is commutative.

(b) If $\sigma : R \to R$ is a ring isomorphism, show that $S = \{X^n \mid n$ is a nonnegative integer$\}$ is a left and a right denominator set in $R[[X, \sigma]]$.

(c) If $\sigma : R \to R$ is a ring isomorphism, describe the rings $S^{-1}R[[X, \sigma]]$ and $R[[X, \sigma]]S^{-1}$ and show that $S^{-1}R[[X, \sigma]] \cong R[[X, \sigma]]S^{-1}$. The ring $R[[X, \sigma]]S^{-1}$, often denoted by $R[[X, X^{-1}, \sigma]]$, is called the *ring of skew Laurent power series* over R.

5. Let $\sigma : R \to R$ be an injective ring homomorphism, where R is an integral domain. Prove that if R is a right Ore ring, then the skew polynomial ring $R[X, \sigma]$ is a right Ore ring. [Hint: Example 2 in Section 1.3.]

6. Let δ be a derivation on an integral domain R. Show that the differential polynomial ring $R[X, \delta]$ is a right Ore ring if and only if R is a right Ore ring. [Hint: Example 9, Section 1.1.]

7. **Common Denominator Property.** If $x_1/s_1, x_2/s_2, \ldots, x_n/s_n$ are in MS^{-1}, show that there are $x_1', x_2', \ldots, x_n' \in M$ and an $s \in S$ such that $x_1/s_1 + x_2/s_2 + \cdots + x_n/s_n = (x_1' + x_2' + \cdots + x_n')/s$.

In the following exercises, R is a commutative ring and S is a multiplicative system in R.

8. (a) Prove that the equivalence relation $\sim$ defined on $M \times S$ in Lemma 8.2.2 is equivalent to the relation $\sim$ on $M \times S$ defined by $(x, s) \sim (y, t)$ if and only if there is a $u \in S$ such that $(xt - ys)u = 0$.

(b) Under the conditions of (a) show that the operations defined in Proposition 8.2.3 become

$$x/s + y/t = (xt + ys)/st \quad \text{and} \quad (x/s)(a/t) = xa/st.$$

9. (a) If $f : M \to N$ is an R-linear mapping, show that the mapping $fS^{-1} : MS^{-1} \to NS^{-1}$ defined by $fS^{-1}(x/s) = f(x)/s$ is an RS^{-1}-module homomorphism.

(b) If $f : M_1 \to M$ and $g : M \to M_2$ are R-linear mappings, deduce that $(gf)S^{-1} = gS^{-1}fS^{-1}$.

(c) If $M_1 \to M \to M_2$ is exact, show that $M_1 S^{-1} \to M S^{-1} \to M_2 S^{-1}$ is exact. Conclude that if $0 \to M_1 \to M \to M_2 \to 0$ is a short exact sequence in $\mathbf{Mod}_R$, then $0 \to M_1 S^{-1} \to M S^{-1} \to M_2 S^{-1} \to 0$ is exact in $\mathbf{Mod}_{RS^{-1}}$ and that S^{-1} preserves monomorphisms and epimorphisms.

(d) Prove that $S^{-1} : \mathbf{Mod}_R \to \mathbf{Mod}_{RS^{-1}}$ is a functor (which (e) shows to be exact). Show also that if $f, g \in \mathrm{Hom}_R(M, N)$, then $(f + g)S^{-1} = f S^{-1} + g S^{-1}$ so that S^{-1} is an exact additive functor.

10. (a) If N is a submodule of M, show that $N S^{-1}$ is a submodule of $M S^{-1}$ and that $(M/N)S^{-1} \cong (M S^{-1})/(N S^{-1})$. [Hint: Consider (c) of Exercise 9.]

(b) If N_1 and N_2 are submodules of M, show that $(N_1 + N_2)S^{-1} = N_1 S^{-1} + N_2 S^{-1}$ and that $(N_1 \cap N_2)S^{-1} = N_1 S^{-1} \cap N_2 S^{-1}$.

(c) Show that if M is a flat R-module, then $M S^{-1}$ is a flat RS^{-1}-module. [Hint: Use (a) and Lemma 8.2.4.] Conclude that RS^{-1} is a flat RS^{-1}-module as well as a flat R-module.

11. (a) If I is an ideal of R, prove that $I S^{-1} = \{a/s \mid a \in I, s \in S\}$ is an ideal of RS^{-1}.

(b) Show that $\bar{I}$ is an ideal of RS^{-1} if and only if there is an ideal I of R such that $I S^{-1} = \bar{I}$. [Hint: Let $I = R \cap \bar{I}$.]

(c) Prove that the function $\mathfrak{p} \mapsto \mathfrak{p}S^{-1}$ is a bijective function from the prime ideals of R that are disjoint from S and the prime ideals of RS^{-1}. [Hint: If $\bar{\mathfrak{p}}$ is a prime ideal of RS^{-1}, let $\mathfrak{p} = R \cap \bar{\mathfrak{p}}$.]

(d) Let $\mathfrak{p}$ be a prime ideal of R. Show that $R_\mathfrak{p}$ is a local ring with unique maximal ideal $\mathfrak{p}S^{-1}$.

12. If $\mathfrak{p}$ is a prime ideal of R, then the module of quotients $M S^{-1}$ of an R-module M at $S = R - \mathfrak{p}$ is often denoted by $M_\mathfrak{p}$. With this notation in mind, prove that the following are equivalent.

(a) $M = 0$.

(b) $M_\mathfrak{p} = 0$ for every prime ideal $\mathfrak{p}$ of R.

(c) $M_\mathfrak{m} = 0$ for every maximal ideal $\mathfrak{m}$ of R.

[Hint: For (c) $\Rightarrow$ (a), let $x \in M$, $x \neq 0$, and suppose that $\mathfrak{m}$ is a maximal ideal that contains $\mathrm{ann}(x)$. Consider $x/1$ in $M_\mathfrak{m}$ and show that $M_\mathfrak{m} \neq 0$.]

13. Show that if M and N are R-modules, then $(M \otimes_R N)S^{-1} \cong M S^{-1} \otimes_{RS^{-1}} N S^{-1}$.

14. Answer each of the following for an R-module M.

(a) If M is projective, then is $M S^{-1}$ a projective RS^{-1}-module?

(b) If M is free, then is $M S^{-1}$ a free RS^{-1}-module?

(c) If M is finitely generated, then is $M S^{-1}$ a finitely generated RS^{-1}-module?

(d) If M is finitely generated, then can it be said that $M S^{-1} = 0$ if and only if there is an $s \in S$ such that $M s = 0$?

(e) If M is finitely presented, then is $M S^{-1}$ a finitely presented RS^{-1}-module?

8.3   Goldie's Theorem

We now consider rings that have a semisimple classical ring of quotients.

Definition 8.3.1. An R-module M is said to be (Goldie) *finite dimensional* if M does not contain an infinite collection of nonzero submodules whose sum is direct. If R is finite dimensional as an R-module, then R is a (right) *finite dimensional ring*. The right annihilator of a nonempty subset of M is called an *annihilator right ideal* of R or simply a *right annihilator*. *Annihilator left ideals* have a similar definition. If $A_1 \subseteq A_2 \subseteq A_3 \subseteq \cdots$ is an ascending chain of right annihilators in R, then we say that R satisfies the *ascending chain condition on right annihilators* if every such ascending chain terminates. A finite dimensional ring that satisfies the ascending chain condition on right annihilators is said to be a *right Goldie ring*.

The following lemmas play a central role in the development of conditions that are necessary and sufficient for R to have a semisimple classical ring of quotients.

Lemma 8.3.2. *If R is finite dimensional and $a \in R$ is such that* $\mathrm{ann}_r(a) = 0$, *then aR is an essential right ideal of R.*

Proof. If aR is not an essential right ideal of R, then there is a nonzero right ideal A of R such that $aR \cap A = 0$. Since $\mathrm{ann}_r(a) = 0$, we claim that this gives a direct sum $A \oplus aA \oplus a^2A \oplus a^3A \oplus \cdots$ of right ideals of R. Suppose that

$$a^{n_1}b_1 + a^{n_2}b_2 + \cdots + a^{n_j}b_j = 0$$

is an element of $A \oplus aA \oplus a^2A \oplus a^3A \oplus \cdots$, where $n_1 < n_2 < \cdots < n_j$. Then the fact that $\mathrm{ann}_r(a) = 0$ gives

$$b_1 + a^{n_2-n_1}b_2 + \cdots + a^{n_j-n_1}b_j = 0.$$

Therefore, $a^{n_2-n_1}b_2 + \cdots + a^{n_j-n_1}b_j = -b_1 \in aR \cap A = 0$ and we have $b_1 = 0$. Furthermore,

$$a^{n_2-n_1}b_2 + \cdots + a^{n_j-n_1}b_j = 0$$

implies that $a^{n_3-n_2}b_3 + \cdots + a^{n_j-n_2}b_{n_j} = -b_2 \in aR \cap A = 0$. Thus, $b_2 = 0$ and $a^{n_3-n_2}b_3 + \cdots + a^{n_j-n_2}b_{n_j} = 0$. Continuing in this way we finally arrive at $b_1 = b_2 = \cdots = b_j = 0$, so the sum is direct. But R is finite dimensional, so we have a contradiction. Hence, aR must be an essential right ideal of R. $\square$

Lemma 8.3.3. *If R is semiprime and if R satisfies the ascending condition on right annihilators, then R has no nonzero nil left ideals and no nonzero nil right ideals.*

Proof. Let A be a nil left ideal of R. If A has a nonzero element a, then $Ra \neq 0$ is a nil left ideal of R. Since R satisfies the ascending chain condition on right annihilators, then by the Remark immediately preceding Proposition 4.2.5, the set $\mathcal{A} = \{\text{ann}_r(b) \mid b \in Ra, b \neq 0\}$ has a maximal element when ordered by inclusion. So suppose that $b \in Ra, b \neq 0$, is such that $\text{ann}_r(b)$ is maximal in $\mathcal{A}$. For each $c \in R$, let k be the index of nilpotency of $cb \in Ra$. Then $(cb)^k = 0$ and $\text{ann}_r(b) \subseteq \text{ann}_r((cb)^{k-1})$. Thus, $\text{ann}_r(b) = \text{ann}_r((cb)^{k-1})$ by the maximality of $\text{ann}_r(b)$. Hence, $cb \in \text{ann}_r(b)$, so $bcb = 0$. Since c was chosen arbitrarily in R, $bRb = 0$ and since R is semiprime, Proposition 6.2.20 indicates that $b = 0$, a contraction. Hence, A cannot have nonzero elements, so $A = 0$. Thus, R has no nonzero nil left ideals. A similar argument shows that R also has no nonzero nil right ideals.   □

Lemma 8.3.4. *If M is an R-module, then*

$$Z(M) = \{x \in M \mid \text{ann}_r(x) \text{ is an essential right ideal of } R\}$$

is a submodule of M and $Z(R_R)$ is an ideal of R. Furthermore, if N is an essential submodule of M and $Z(N) = 0$, then $Z(M) = 0$ as well.

Proof. First, note that $Z(M) \neq \varnothing$ since $\text{ann}_r(0) = R$ is an essential right ideal of R. If $x, y \in Z(M)$, then $\text{ann}_r(x) \cap \text{ann}_r(y) \subseteq \text{ann}_r(x + y)$ and $\text{ann}_r(x) \cap \text{ann}_r(y)$ is an essential right ideal of R. Hence, $\text{ann}_r(x + y)$ is an essential right ideal of R, so $x + y \in Z(M)$. If $x \in Z(M)$ and $a \in R$, then we need to show that $\text{ann}_r(xa)$ is an essential right ideal of R. Suppose that there is a nonzero right ideal A of R such that $\text{ann}_r(xa) \cap A = 0$. Under this assumption, if $b \in A, b \neq 0$, then $xab \neq 0$, so $ab \neq 0$. Therefore, aA is a nonzero right ideal of R. But $\text{ann}_r(x)$ is an essential right ideal of R, so $\text{ann}_r(x) \cap aA \neq 0$. Let c be an element of A such that $ac \neq 0$ and $xac = 0$. Then $c \in \text{ann}_r(xa) \cap A = 0$, a contradiction. Thus, $\text{ann}_r(xa)$ is an essential right ideal of R, so $xa \in Z(M)$ and we have that $Z(M)$ is a submodule of M. Moreover, $Z(R_R)$ is an ideal of R since if $b \in R$ and $a \in Z(R_R)$, then $\text{ann}_r(a) \subseteq \text{ann}_r(ba)$.

Finally, suppose that N is an essential submodule of M such that $Z(N) = 0$. If $Z(M) \neq 0$, let $0 \neq x \in N \cap Z(M)$. Then $\text{ann}_r(x)$ is an essential right ideal of R, so $x \in Z(N) = 0$, a contradiction. Hence, $Z(M) = 0$.   □

Definition 8.3.5. The submodule $Z(M)$ is called the *singular submodule* of M and $Z(R_R)$ is the *right singular ideal* of R. If $Z(M) = 0$, then M is said to be a *nonsingular module* and if $Z(M) = M$, then M is said to be *singular*. If $Z(R_R) = 0$, then R is a *right nonsingular ring*. Left singularity and left nonsingularity have analogous definitions.

Remark. Earlier it was pointed out that Osofsky proved that a ring R is semisimple if and only if every cyclic R-module is injective. In [15], Goodearl described the rings over which every singular module is injective. Today these rings are called *SI-rings*.

Lemma 8.3.6. *If R satisfies the ascending chain condition on right annihilators, then the right singular ideal of R is nilpotent.*

Proof. Let $Z = Z(R_R)$. If $Z = 0$, there is nothing to prove, so suppose $Z \neq 0$ and that Z is not nilpotent. If $n \geq 2$ is an integer, then $Z^n \neq 0$, so there are nonzero $z \in Z$ such that $Z^{n-1}z \neq 0$. Let $\mathcal{A} = \{\operatorname{ann}_r(z) \mid z \in Z \text{ and } Z^{n-1}z \neq 0\}$. We claim that $\mathcal{A}$ contains a maximal element. If $\operatorname{ann}_r(z_1) \in \mathcal{A}$ is not maximal, then there is a $z_2 \in Z$ such that $Z^{n-1}z_2 \neq 0$ and $\operatorname{ann}_r(z_1) \subsetneqq \operatorname{ann}_r(z_2)$. If $\operatorname{ann}_r(z_2)$ is not maximal in $\mathcal{A}$, then there is a $z_3 \in Z$ such that $Z^{n-1}z_3 \neq 0$ and $\operatorname{ann}_r(z_2) \subsetneqq \operatorname{ann}_r(z_3)$ and so on. Hence, if $\mathcal{A}$ does not contain a maximal element, then we can construct a strictly increasing chain $\operatorname{ann}_r(z_1) \subsetneqq \operatorname{ann}_r(z_2) \subsetneqq \operatorname{ann}_r(z_3) \subsetneqq \cdots$ of right annihilators in R, a clear contradiction. Thus, $\mathcal{A}$ contains a maximal element.

Now suppose that $\operatorname{ann}_r(a)$ is a maximal element of $\mathcal{A}$. For each $z \in Z$, $\operatorname{ann}_r(z)$ is essential in R, so $\operatorname{ann}_r(z) \cap aR \neq 0$. Hence, there is a $b \in R$ such that $ab \neq 0$ and yet $zab = 0$. Therefore, we see that $\operatorname{ann}_r(a) \subsetneqq \operatorname{ann}_r(za)$, so it must be the case that $Z^{n-1}za = 0$. Since z was chosen arbitrarily in Z, we have $Z^n a = 0$. Hence, $\operatorname{ann}_r(Z^{n-1}) \subsetneqq \operatorname{ann}_r(Z^n)$ for each integer $n \geq 2$, so we have a strictly increasing chain $\operatorname{ann}_r(Z^1) \subsetneqq \operatorname{ann}_r(Z^2) \subsetneqq \operatorname{ann}_r(Z^3) \subsetneqq \cdots$ of right annihilators in R. This contradiction shows that Z must be nilpotent. $\square$

Lemma 8.3.7. *The left annihilator of any nilpotent ideal of R is an essential right ideal of R.*

Proof. Suppose that $\mathfrak{n}$ is a nilpotent ideal of R and let A be a right ideal of R such that $(\operatorname{ann}_\ell(\mathfrak{n})) \cap A = 0$. Then $a\mathfrak{n} \neq 0$ for each $a \in A$, $a \neq 0$. Let n be the index of nilpotency of $\mathfrak{n}$ and suppose that a is a nonzero element of A. Since $a\mathfrak{n} \neq 0$, choose $ab_1 \in a\mathfrak{n}$ to be such that $ab_1 \neq 0$. Then since $ab_1 \in A$, $ab_1\mathfrak{n} \neq 0$, so there must be a $b_2 \in \mathfrak{n}$ such that $ab_1b_2 \neq 0$. Likewise, there is a $b_3 \in \mathfrak{n}$ such that $ab_1b_2b_3 \neq 0$. Continuing in this way we finally arrive at $ab_1b_2 \cdots b_n \neq 0$, where $b_1, b_2, \ldots, b_n \in \mathfrak{n}$. This is a clear contradiction since $b_1b_2 \cdots b_n = 0$. Thus, such an a cannot exist, so we have $A = 0$. $\square$

Sufficient mathematical machinery has now been developed for a proof of the main result of this section known as Goldie's theorem [57].

Proposition 8.3.8 (Goldie). *The following are equivalent for a ring R.*

(1) *R is a semiprime right Goldie ring.*

(2) *A right ideal A of R is essential if and only A contains a regular element of R.*

(3) *R has a semisimple classical ring of quotients.*

Proof. (1) $\Rightarrow$ (2). Suppose that R is a semiprime right Goldie ring and let A be an essential right ideal of R. Then $A \neq 0$, so let

$$\mathcal{A} = \{\mathrm{ann}_r(a) \mid a \in A \text{ and } \mathrm{ann}_r(a^2) \subsetneq R\}.$$

If $\mathcal{A} = \varnothing$, then $\mathrm{ann}_r(a^2) = R$ for each $a \in A$ which indicates that $a^2 = 0$ for each $a \in A$. Thus, A is a nil right ideal of R which is impossible by Lemma 8.3.3. Hence, $\mathcal{A} \neq \varnothing$, so since R satisfies the ascending chain condition on right annihilators, there is a nonzero element $a_1 \in A$ such that $\mathrm{ann}_r(a_1) = \mathrm{ann}_r(a_1^2)$. Indeed, the set $\mathcal{A}$ has a maximal element, say $\mathrm{ann}_r(a_1)$. But $\mathrm{ann}_r(a_1) \subseteq \mathrm{ann}_r(a_1^2)$, so $\mathrm{ann}_r(a_1) = \mathrm{ann}_r(a_1^2)$. Likewise, if

$$\mathcal{A}_1 = \{\mathrm{ann}_r(a) \mid a \in \mathrm{ann}_r(a_1) \cap A \text{ and } \mathrm{ann}_r(a^2) \subsetneq R\}$$

and $\mathrm{ann}_r(a_1) \cap A \neq 0$, then we can find a nonzero element $a_2 \in \mathcal{A}_1$ such that $\mathrm{ann}_r(a_2) = \mathrm{ann}_r(a_2^2)$. Similarly, for each $n \geq 3$, if

$$\mathcal{A}_{n-1} = \{\mathrm{ann}_r(a) \mid a \in \mathrm{ann}_r(a_1) \cap \cdots \cap \mathrm{ann}_r(a_{n-1}) \cap A \text{ and } \mathrm{ann}_r(a^2) \subsetneq R\}$$

and $\mathrm{ann}_r(a_1) \cap \cdots \cap \mathrm{ann}_r(a_{n-1}) \cap A \neq 0$, then there is an $a_n \in \mathcal{A}_{n-1}$ such that $\mathrm{ann}_r(a_n) = \mathrm{ann}_r(a_n^2)$. We claim that the sum $a_1 R + a_2 R + \cdots + a_n R$ is direct for each $n \geq 1$. The case for $n = 1$ is obvious, so suppose that the sum $a_1 R + a_2 R + \cdots + a_{n-1} R$, $n \geq 2$, is direct and let

$$a_n b_n = a_1 b_1 + a_2 b_2 + \cdots + a_{n-1} b_{n-1} \in (a_1 R + a_2 R + \cdots + a_{n-1} R) \cap a_n R.$$

By construction, $a_i \in \mathrm{ann}_r(a_1) \cap \mathrm{ann}_r(a_2) \cap \cdots \cap \mathrm{ann}_r(a_{i-1})$, for each i such that $2 \leq i \leq n$, so it follows that $0 = a_i a_k$ whenever $i < k$, for $k = 1, 2, \ldots, n$. If the equation

$$a_n b_n = a_1 b_1 + a_2 b_2 + \cdots + a_{n-1} b_{n-1}$$

is multiplied through on the left, first by a_1, then by a_2 and finally by a_{n-1}, we get $b_1 \in \mathrm{ann}_r(a_1), b_2 \in \mathrm{ann}_r(a_2), \ldots, b_{n-1} \in \mathrm{ann}_r(a_{n-1})$. Hence,

$$a_1 b_1 + a_2 b_2 + \cdots + a_{n-1} b_{n-1} = 0,$$

so the sum $a_1 R + a_2 R + \cdots + a_n R$ is direct.

Since R is finite dimensional, the process just described must terminate. If it terminates at the nth stage, then we have

$$\mathrm{ann}_r(a_1) \cap \mathrm{ann}_r(a_2) \cap \cdots \cap \mathrm{ann}_r(a_n) \cap A = 0$$

which gives

$$\mathrm{ann}_r(a_1) \cap \mathrm{ann}_r(a_2) \cap \cdots \cap \mathrm{ann}_r(a_n) = 0$$

since A is essential in R. Since the sum $a_1 R + a_2 R + \cdots + a_n R$ is direct, we have

$$\text{ann}_r(a_1) \cap \text{ann}_r(a_2) \cap \cdots \cap \text{ann}_r(a_n) = \text{ann}_r(a_1 + a_2 + \cdots + a_n).$$

Therefore, if $a = a_1 + a_2 + \cdots + a_n$, then we have found an element $a \in A$ such that $\text{ann}_r(a) = 0$. If we can now show that $\text{ann}_\ell(a) = 0$, then a will be the desired regular element in A. First, note that since $\text{ann}_r(a) = 0$, Lemma 8.3.2 shows that aR is an essential right ideal of R. If $b \in \text{ann}_\ell(a)$, then $baR = 0$, so $aR \subseteq \text{ann}_r(b)$. Thus, $\text{ann}_r(b)$ is essential in R since aR is, so $b \in Z(R_R)$. Thus $\text{ann}_\ell(a) \subseteq Z(R_R)$. Lemma 8.3.4 and Lemma 8.3.6 show that $Z(R_R)$ is a nilpotent ideal of R, so $Z(R_R) = 0$ since a semiprime ring contains no nonzero nilpotent ideals. Hence, $\text{ann}_\ell(a) = 0$ and we can conclude that every essential right ideal of R contains a regular element of R.

Conversely, suppose that A is a right ideal of R that contains a regular element, say a. Then $\text{ann}_r(a) = 0$, so using Lemma 8.3.2 again we see that aR is an essential right ideal of R. Hence, $aR \subseteq A$ implies that A is an essential right ideal of R.

$(2) \Rightarrow (3)$. The first step is to show that R is a right Ore ring. For this, let $a, s \in R$ with s regular and set $(sR : a) = \{r \in R \mid ar \in sR\}$. Since sR is an essential right ideal of R, it follows that $(sR : a)$ is essential in R. This means there is a regular element $t \in (sR : a)$ such that $at = sb$ for some $b \in R$. Thus, the right Ore condition is satisfied and so R has a classical ring of quotients $Q_{cl}^r(R)$.

It remains to show that $Q_{cl}^r(R)$ is a semisimple ring. If A is an essential right ideal of $Q_{cl}^r(R)$, then in view of Proposition 8.1.13, it follows that A^c, the contraction of A to R, is an essential right ideal of R. Hence, A^c contains a regular element a of R. But then for any regular element s of R, a/s is a unit of $Q_{cl}^r(R)$ and $a/s \in A^{ce}$, the extension of A^c to $Q_{cl}^r(R)$. But from (3) of Proposition 8.1.12 we see that $A = A^{ce}$, and so A contains a unit of $Q_{cl}^r(R)$. Thus, $A = Q_{cl}^r(R)$ and so $Q_{cl}^r(R)$ has no proper essential right ideals. Hence, by Corollary 6.4.6, $Q_{cl}^r(R)$ is a semisimple ring.

$(3) \Rightarrow (1)$. We now identify R with its canonical image in $Q_{cl}^r(R)$ and treat R as a subring of $Q_{cl}^r(R)$. So suppose that R has a semisimple classical ring of quotients $Q_{cl}^r(R)$. If

$$A_1 \oplus A_2 \oplus A_3 \oplus \cdots$$

is an infinite direct sum of nonzero right ideals of R, then by (5) of Proposition 8.1.12

$$A_1^e \oplus A_2^e \oplus A_3^e \oplus \cdots$$

is an infinite direct sum of nonzero right ideals of $Q_{cl}^r(R)$. But $Q_{cl}^r(R)$ is semisimple, so, by Corollary 6.4.12, we see that $Q_{cl}^r(R)$ is right noetherian. Thus, $Q_{cl}^r(R)$ is finite dimensional, so there is a positive integer n such that $A_m^e = 0$ for each $m \geq n$. But this gives $A_m = 0$ for all $m \geq n$ which means that R is finite dimensional. Next, let

$$\text{ann}_r^R(S_1) \subseteq \text{ann}_r^R(S_2) \subseteq \text{ann}_r^R(S_3) \subseteq \cdots \tag{8.1}$$

be an ascending chain of right annihilators in R, where each S_i is a nonempty subset of R. Then

$$\operatorname{ann}_\ell^{Q_{cl}^r(R)}(\operatorname{ann}_r^R(S_1)) \supseteq \operatorname{ann}_\ell^{Q_{cl}^r(R)}(\operatorname{ann}_r^R(S_2)) \supseteq \operatorname{ann}_\ell^{Q_{cl}^r(R)}(\operatorname{ann}_r^R(S_3)) \supseteq \cdots$$

is a decreasing chain of left ideals in $Q_{cl}^r(R)$ which must stabilize since $Q_{cl}^r(R)$ is left artinian. Therefore, the chain

$$R \cap \operatorname{ann}_\ell^{Q_{cl}^r(R)}(\operatorname{ann}_r^R(S_1)) \supseteq R \cap \operatorname{ann}_\ell^{Q_{cl}^r(R)}(\operatorname{ann}_r^R(S_2))$$

$$\supseteq R \cap \operatorname{ann}_\ell^{Q_{cl}^r(R)}(\operatorname{ann}_r^R(S_3)) \supseteq \cdots$$

stabilizes, so it follows that the chain

$$\operatorname{ann}_r^R(R \cap \operatorname{ann}_\ell^{Q_{cl}^r(R)}(\operatorname{ann}_r^R(S_1))) \subseteq \operatorname{ann}_r^R(R \cap \operatorname{ann}_\ell^{Q_{cl}^r(R)}(\operatorname{ann}_r^R(S_2)))$$

$$\subseteq \operatorname{ann}_r^R(R \cap \operatorname{ann}_\ell^{Q_{cl}^r(R)}(\operatorname{ann}_r^R(S_3))) \subseteq \cdots$$

must stabilize too. But

$$\operatorname{ann}_r^R(S_n) = \operatorname{ann}_r^R(R \cap \operatorname{ann}_\ell^{Q_{cl}^r(R)}(\operatorname{ann}_r^R(S_n))),$$

for each integer $n \geq 1$, so (8.1) stabilizes. Hence, R satisfies the ascending chain condition on right annihilators and so R is a right Goldie ring.

Finally, to complete the proof, we need to prove that R is semiprime. Let $\mathfrak{n}$ be a nonzero nilpotent ideal of R. Then Lemma 8.3.7 shows that $\operatorname{ann}_\ell(\mathfrak{n})$ is an essential right ideal of R. But if $\operatorname{ann}_\ell(\mathfrak{n})$ is an essential right ideal of R, then $(\operatorname{ann}_\ell(\mathfrak{n}))^e$ is an essential right ideal of $Q_{cl}^r(R)$. Since $Q_{cl}^r(R)$ is semisimple, $(\operatorname{ann}_\ell(\mathfrak{n}))^e = Q_{cl}^r(R)$ and so we have

$$1_{Q_{cl}^r(R)} = r_1(a_1/s_1) + r_2(a_2/s_2) + \cdots + r_n(a_n/s_n),$$

where $r_i \in \operatorname{ann}_\ell(\mathfrak{n})$, $a_i \in R$ and s_i is a regular element of R for $i = 1, 2, \ldots, n$. Corollary 8.1.11 gives a regular element s of R such that $(a_1/s_1)s, (a_2/s_2)s, \ldots, (a_n/s_n)s \in R$, so

$$s\mathfrak{n} \subseteq r_1(a_1/s_1)s\mathfrak{n} + r_2(a_2/s_2)s\mathfrak{n} + \cdots + r_n(a_n/s_n)s\mathfrak{n}$$

$$\subseteq r_1\mathfrak{n} + r_2\mathfrak{n} + \cdots + r_n\mathfrak{n} = 0.$$

It follows that $\mathfrak{n} = 0$ which contradicts the assumption that $\mathfrak{n} \neq 0$. Hence, R can have no nonzero nilpotent ideals, so Proposition 6.2.27 shows that R is a semiprime ring. $\square$

We also have the following result due to Goldie [58] and Lesieur–Groisot [67].

Corollary 8.3.9 (Goldie, Lesieur–Groisot). *A ring R has a simple artinian classical ring of quotients if and only if R is a prime right Goldie ring.*

Proof. For this proof, we identify R with its canonical image $\varphi(R)$ in $Q_{cl}^r(R)$ and assume that R is a subring of $Q_{cl}^r(R)$. If R has a simple artinian ring of quotients $Q_{cl}^r(R)$, then by Goldie's theorem we need only show that R is prime. Let A and B be ideals of R such that $AB = 0$. Then $(BQ_{cl}^r(R)A) \cap R$ is an ideal of R such that $(BQ_{cl}^r(R)A \cap R)^2 = 0$. But from Goldie's theorem we know that R is semiprime, so $BQ_{cl}^r(R)A \cap R = 0$. This gives $BQ_{cl}^r(R)A = 0$, since R_R is essential in $Q_{cl}^r(R)_R$. Since $Q_{cl}^r(R)$ is a simple ring, $Q_{cl}^r(R)$ is prime, so either $BQ_{cl}^r(R) = 0$ or $Q_{cl}^r(R)A = 0$. Hence, either $A = 0$ or $B = 0$ and R is therefore prime.

Conversely, suppose that R is a prime right Goldie ring. Then R is a semiprime right Goldie ring, so $Q_{cl}^r(R)$ is semisimple and therefore artinian. Since a semisimple prime ring is simple, we need only show that $Q_{cl}^r(R)$ is prime. If I_1 and I_2 are ideals of $Q_{cl}^r(R)$ such that $I_1 I_2 = 0$, then $(I_1 \cap R)(I_2 \cap R) = 0$, so either $I_1 \cap R = 0$ or $I_2 \cap R = 0$ since R is prime. Hence, either $I_1 = 0$ or $I_2 = 0$, since R is essential in $Q_{cl}^r(R)$. Thus, $Q_{cl}^r(R)$ is a semisimple prime ring, so $Q_{cl}^r(R)$ is a simple artinian ring. $\qquad\square$

Problem Set 8.3

1. (a) Prove that $f(Z(M)) \subseteq Z(M)$ for any R-module M and each $f \in \mathrm{End}_R(M)$.
   (b) If $\{M_\alpha\}_\Delta$ is a family of R-modules, deduce that $Z(\bigoplus_\Delta M_\alpha) = \bigoplus_\Delta Z(M_\alpha)$. Suppose that N is a submodule of an R-module M and prove each of the following.
   (c) $Z(N) = N \cap Z(M)$.
   (d) If N is essential in M, then is $Z(N)$ an essential submodule of $Z(M)$?

2. (a) If K is a field, show that the matrix ring $\begin{pmatrix} K & K \\ 0 & K \end{pmatrix}$ is both left and right nonsingular.
   (b) Prove that the ring $R = \begin{pmatrix} \mathbb{Z} & \mathbb{Z}_2 \\ 0 & \mathbb{Z}_2 \end{pmatrix}$ is right nonsingular but not left nonsingular. [Hint: Let $x = \begin{pmatrix} 0 & [1] \\ 0 & 0 \end{pmatrix}$ and compute $\mathrm{ann}_\ell(x)$ and $\mathrm{ann}_r(x)$.]
   (c) Show that there are modules M such that $Z(M/Z(M)) \neq 0$. [Hint: For the ring R of (b), show that $Z(_RR) = \{0, x\}$. Next, show that $Z(_RR/Z(_RR)) = \{0, x'\}$, where $x' = \begin{pmatrix} 0 & 0 \\ 0 & [1] \end{pmatrix} + Z(_RR)$. Note that since (b) was at hand, this was used to give the result for left R-modules.]
   (d) If $Z(R_R) = 0$, then $Z(M/Z(M)) = 0$ for every R-module M. [Hint: If $x + Z(M) \in Z(M/Z(M))$, then $\mathrm{ann}_r(x + Z(M))$ is an essential right ideal of R. Thus, $(Z(M) : x)$ is an essential right ideal of R such that $x(Z(M) : x) \subseteq Z(M)$. If A is a right ideal of R. let $a \in A \cap (Z(M) : x)$, $a \neq 0$, consider

$xa \in x(Z(M) : x) \subseteq Z(M)$ and show that $\text{ann}_r(x)$ is an essential right ideal of R.]

3. Determine if the following are equivalent for a ring R.

   (a) Every R-module is nonsingular.

   (b) Every left R-module is nonsingular.

   (c) R is a semisimple ring.

   [$(a) \Leftrightarrow (c)$, Hint: If A is a right ideal of R, show that $R/(A \oplus A_c)$ is a singular right R-module, where A_c is a complement of A in R. Conversely, if R is semisimple and $x \in Z(M)$, then $\text{ann}_r(x)$ is an essential right ideal of R.]

4. Prove that every noetherian and every artinian R-module is finite-dimensional.

5. In each of the following, show that $Q = Q^r_{cl}(R)$.

   (a) $R = \begin{pmatrix} Z & Z \\ 0 & Z \end{pmatrix}$ and $Q = \begin{pmatrix} Q & Q \\ 0 & Q \end{pmatrix}$.

   (b) $R = \begin{pmatrix} K & K[X] \\ 0 & K[X] \end{pmatrix}$ and $Q = \begin{pmatrix} K & K(X) \\ 0 & K(X) \end{pmatrix}$, where K is a field and $K(X)$ is the field of fractions of the integral domain $K[X]$.

6. Prove that aR is a nil right ideal of R if and only if Ra is a nil left ideal of R. [Hint: Consider $(ab)^{n+1} = a(ba)^n b$.] Conclude that R has no nonzero nil right ideals if and only if R has no nonzero nil left ideals. Use this to complete the proof of Lemma 8.3.3.

7. If R is a semiprime (prime) right Goldie ring, then is eRe is a semiprime (prime) right Goldie ring for a nonzero idempotent e of R?

8.4 The Maximal Ring of Quotients

We will now develop a *maximal (right) ring of quotients* $Q^r_{max}(R)$ of R. This ring has the property that a ring of quotients RS^{-1} of R at S embeds in $Q^r_{max}(R)$ when R is S-torsion free, that is, when $a \in R$ and $s \in S$ is such that $as = 0$, then $a = 0$. However, if R is not S-torsion free, there is no assurance that there is even a nontrivial homomorphism from RS^{-1} to $Q^r_{max}(R)$. So in general, a right ring of quotients RS^{-1} is quite distinct from $Q^r_{max}(R)$. Another distinction is that every ring R has a maximal ring of quotients $Q^r_{max}(R)$ in contrast to RS^{-1} which exists if and only if S is a right denominator set in R.

Definition 8.4.1. An R-module N is said to be a *rational extension* of a submodule M if for all $x, y \in N$, $y \neq 0$, there is an $a \in R$ such that $xa \in M$ and $ya \neq 0$. If R is a subring of a ring S, then S is a rational extension of R, if S_R is a rational extension of R_R. Rational extensions of left R-modules are defined analogously.

One immediate consequence is that if N is a rational extension of M, then M is an essential submodule of N.

Lemma 8.4.2. *The following are equivalent for an R-module M.*

(1) *N is a rational extension of M.*

(2) *If N' is a submodule of N containing M and f : N' → N is an R-linear mapping such that f(M) = 0, then f = 0.*

Proof. (1) ⇒ (2). Let N be a rational extension of M, suppose that N' is a submodule of N containing M and let $f : N' \to N$ be an R-linear mapping such that $f(M) = 0$. If $f \neq 0$, then there is an $x \in N'$ such that $f(x) \neq 0$. Thus, $x, f(x) \in N$ with $f(x) \neq 0$, so there is an $a \in R$ such that $xa \in M$ and $f(x)a \neq 0$. But $f(x)a = f(xa) = 0$ gives a contradiction, so $f = 0$.

(2) ⇒ (1). Let $x, y \in N$, $y \neq 0$ and suppose that $y(M : x) = 0$, where $(M : x) = \{a \in R \mid xa \in M\}$. If $N' = M + xR$, then $f : N' \to N$ given by $f(x' + xa) = ya$ is a well-defined R-linear map such that $f(M) = 0$. Hence, $f = 0$, so $0 = f(x) = y$. This contradiction shows that $y(M : x) \neq 0$, so N is a rational extension of M.   □

Examples

1. The motivation for the concept of a rational extension is probably due to the fact that the field of rational numbers as a $\mathbb{Z}$-module is a rational extension of the ring of integers viewed as a $\mathbb{Z}$-module. Indeed, if $p/q, s/t \in \mathbb{Q}_{\mathbb{Z}}$, $s/t \neq 0$, then $(p/q)q \in \mathbb{Z}$ and $(s/t)q \neq 0$.

2. If S is a right denominator set in R and R is S-torsion free, then RS^{-1} is a rational extension of R and the canonical map $a \mapsto a/1$ from R to RS^{-1} is an injection. Hence, we can consider R to be a subring of RS^{-1} by identifying a with $a/1$ for each $a \in R$. Now let $a/s, b/t \in RS^{-1}$ with $b/t \neq 0$. If $(c, u) \in R \times S$ is such that $sc = tu \in S$, then $(a/s)(sc) = ac \in R$ and $(b/t)(tu) = bu \neq 0$ since $u \in S$ and R is S-torsion free. In particular, if R is a right Ore ring, then $Q_{cl}^r(R)$ is a rational extension of R.

Definition 8.4.3. If M is an R-module, then a rational extension $\xi(M)$ of M is said to be a *maximal rational extension* of M if whenever N is a rational extension of M, there is a unique monomorphism $f : N \to \xi(M)$ such that the diagram

$$
\begin{array}{ccc}
 & & N \\
 & i_N \nearrow & \big\downarrow f \\
M & & \\
 & i_{\xi(M)} \searrow & \\
 & & \xi(M)
\end{array}
$$

is commutative, where i_N and $i_{\xi(M)}$ are canonical injections.

The proof of the following lemma is left as an exercise.

Lemma 8.4.4. *The following are equivalent for an R-module M.*

(1) $\xi(M)$ *is a maximal rational extension of M.*

(2) *If the rational extensions of M are ordered by inclusion and N is a rational extension of M that contains $\xi(M)$, then $\xi(M) = N$.*

Proposition 8.4.5. *If M is an R-module, E(M) is an injective envelope of M and* $\Delta = \{h \in \mathrm{End}_R(E(M)) \mid h(M) = 0\}$, *then*

$$\xi(M) = \bigcap_{\Delta} \mathrm{Ker}\, h$$

is a maximal rational extension of M.

Proof. Let M be an R-module and suppose that $E(M)$, Δ and $\xi(M)$ are as stated in the proposition. Then $\xi(M)$ is a submodule of $E(M)$ containing M, so suppose that N is a submodule of $\xi(M)$ containing M. If $h : N \to \xi(M) \subseteq E(M)$ is an R-linear mapping such that $h(M) = 0$, then h can be extended to an $h' \in \mathrm{End}_R(E(M))$ and $h' \in \Delta$. Hence, $\xi(M) \subseteq \mathrm{Ker}\, h'$, so $h = 0$. Thus, Lemma 8.4.2 shows that $\xi(M)$ is a rational extension of M.

Next, we claim that $\xi(M)$ contains every rational extension N of M that is contained in $E(M)$. If N is a rational extension of M contained in $E(M)$ and $N \not\subseteq \xi(M)$, then there is an $h \in \Delta$ such that $h(N) \neq 0$. Let $N' = \{x \in N \mid h(x) \in M\}$. Since M is essential in $E(M)$, $M \cap h(N) \neq 0$, so $h(N') \neq 0$. Now N' is a submodule of N containing M and $h' : N' \to N$ defined by $h'(x) = h(x)$ is such that $h'(M) = 0$. Hence, we must have $h' = 0$, since N is a rational extension of M. This gives $h(N') = 0$, a contradiction which shows that if $h \in \Delta$, then $h(N) = 0$. It now follows easily that $N \subseteq \xi(M)$.

Now suppose that N is a rational extension of M not necessarily contained in $E(M)$. Using the injectivity of $E(M)$, we see that there is an R-linear mapping $h : N \to E(M)$ such that the diagram

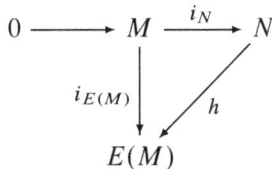

$$
\begin{array}{ccc}
0 \longrightarrow & M & \xrightarrow{\ i_N\ } N \\
 & \Big\downarrow{\scriptstyle i_{E(M)}} & \diagdown{\scriptstyle h} \\
 & E(M) &
\end{array}
$$

is commutative, where i_N and $i_{E(M)}$ are canonical injections. First, h is an injection for if $x \in M \cap \mathrm{Ker}\, h$, then $x = i_{E(M)}(x) = h i_N(x) = h(x) = 0$. Hence, $\mathrm{Ker}\, h = 0$, since M is essential in N. The next step is to show that $h(N)$ is a rational extension of M. Clearly $M \subseteq h(N)$, so let $h(x), h(y) \in h(N), h(y) \neq 0$, where $x, y \in N$ and

$y \neq 0$. Since N is a rational extension of M, there is an $a \in R$ such that $xa \in M$ and $ya \neq 0$. But h is the identity map on M, so we have $h(x)a = h(xa) = xa \in M$ and $h(y)a = h(ya) \neq 0$. Thus, $h(N)$ is a rational extension of M. But we have just seen that $\xi(M)$ contains every rational extension of M that is contained in $E(M)$. Therefore, $h(N) \subseteq \xi(M)$, so the diagram

$$0 \xrightarrow{} M \xrightarrow{\ i_N\ } N$$

$$i_{\xi(M)} \Big\downarrow \qquad \swarrow h$$

$$\xi(M)$$

is commutative.

Next, we show that the injective mapping $h : N \to \xi(M)$ is unique. Suppose that $h' : N \to \xi(M)$ also makes the diagram commute. Then $(h - h')(M) = 0$, so if $g : h'(N) \to \xi(M)$ is such that $g(h'(x)) = h'(x) - h(x)$, then $g(M) = 0$ and so (2) of Lemma 8.4.2 shows that $g = 0$. Therefore, $h' = h$ and we have that h is unique. Thus, $\xi(M)$ is a maximal rational extension of M. $\square$

Corollary 8.4.6. *If $\xi(M)$ is a maximal rational extension of M, then $\xi(M)$ is unique up to isomorphism via an R-linear mapping that extends the identity on M.*

Proof. This follows immediately since the existence of the unique R-linear map $h : N \to \xi(M)$ for each rational extension N of M shows that $i_{\xi(M)} : M \to \xi(M)$, where $i_{\xi(M)}$ is the canonical injection, has the universal mapping property. So an appropriate category can be constructed that has $i_{\xi(M)} : M \to \xi(M)$ as a final object. $\square$

We can conclude from Proposition 8.4.5 and its corollary that every module has a maximal rational extension that is unique up to isomorphism.

Remark. Let M be an R-module and suppose that $H = \operatorname{End}_R(E(M))$. Then $E(M)$ is an (H, R)-bimodule and the maximal rational extension of M is given by $\xi(M) = \operatorname{ann}_r^{E(M)}(\operatorname{ann}_\ell^H(M))$. This follows since

$$\operatorname{ann}_\ell^H(M) = \{h \in H \mid h(M) = 0\}, \text{ so}$$

$$\operatorname{ann}_r^{E(M)}(\operatorname{ann}_\ell^H(M)) = \{x \in E(M) \mid h(x) = 0 \text{ for all } h \in \operatorname{ann}_\ell^H(M)\}$$

$$= \bigcap_\Delta \operatorname{Ker} h, \quad \text{where } \Delta = \{h \in H \mid h(M) = 0\},$$

$$= \xi(M).$$

Definition 8.4.7. If D is a right ideal of R and if R is a rational extension of D as an R-module, then we say that D is a *dense right ideal* of R. If R is a subring of a ring Q, then Q is said to be a *general right ring of quotients* of R if Q is a rational extension of R as an R-module. The modifier "right" will be omitted with the understanding that general ring of quotients means a general right ring of quotients.

Examples

3. If $\{R_\alpha\}_\Delta$ is an arbitrary family of rings, then $\bigoplus_\Delta R_\alpha$ is a dense left ideal and a dense right ideal of the ring direct product $\prod_\Delta R_\alpha$.

4. An ideal I of R is a dense right ideal of R if and only if $\mathrm{ann}_\ell(I) = 0$. With this in mind it is easy to show that the ideal $I = \left(\begin{smallmatrix} [0] & [4]\mathbb{Z}_8 \\ [0] & \mathbb{Z}_8 \end{smallmatrix} \right)$ of the matrix ring $R = \left(\begin{smallmatrix} \mathbb{Z}_8 & [4]\mathbb{Z}_8 \\ [0] & \mathbb{Z}_8 \end{smallmatrix} \right)$ is not a dense right ideal of R.

Lemma 8.4.8. *The following hold in any ring R.*

(1) *If D is a dense right ideal of R, then D is essential in R.*

(2) *If $\{D\}_{i=2}^n$ is a family of dense right ideals of R, then $\bigcap_{i=2}^n D_i$ is a dense right ideal of R for each integer $n \geq 2$.*

(3) *If Q is a general ring of quotients of R, then $(R : q) = \{a \in R \mid qa \in R\}$ is a dense right ideal of R for each $q \in Q$.*

(4) *If $\mathbb{Z}(R_R) = 0$ and A is essential in R, then A is dense in R.*

Proof. (1) Let D be dense in R and suppose that A is a nonzero right ideal of R. If $a \in A, a \neq 0$, then, since R is a rational extension of D, there is a $b \in R$ such that $ab \in D, ab \neq 0$. Hence, $0 \neq ab \in D \cap A$.

(2) If $n = 2$, then we need to show that $D_1 \cap D_2$ is dense in R. If $a, a' \in R, a' \neq 0$, then since D_1 is dense in R, there is a $b \in R$ such that $ab \in D_1$ and $a'b \neq 0$. But D_2 is also dense in R, so there is a $c \in R$ such that $abc \in D_2$ and $a'bc \neq 0$. Hence, we have found an element $bc \in R$ such that $a(bc) \in D_1 \cap D_2$ and $a'(bc) \neq 0$, so $D_1 \cap D_2$ is dense in R. Now suppose that $\bigcap_{i=2}^n D_i$ is dense in R and consider $\bigcap_{i=2}^{n+1} D_i$. Since $\bigcap_{i=2}^{n+1} D_i = (\bigcap_{i=2}^n D_i) \cap D_{n+1}$, it follows from the case for $n = 2$ that $\bigcap_{i=2}^{n+1} D_i$ is dense in R. Hence, the result follows by induction.

(3) Suppose that Q is a general ring of quotients of R. If $q \in Q$, then we need to show that $(R : q)$ is dense in R. If $a, b \in R, b \neq 0$, then $qa, b \in Q$, so there is a $c \in R$ such that $qac \in R$ and $bc \neq 0$. Hence, $ac \in (R : q)$ and $bc \neq 0$.

(4) Suppose that $\mathbb{Z}(R_R) = 0$ and let A be an essential right ideal of R. If $a, b \in R$, $b \neq 0$, then $(A : a)$ is essential in R. However, $\mathrm{ann}_r(b)$ cannot be essential in R since $\mathbb{Z}(R_R) = 0$. It follows that $(A : a) \not\subseteq \mathrm{ann}_r(b)$, so there is an $c \in (A : a)$ such that $c \notin \mathrm{ann}_r(b)$. Hence, $ac \in A$ and $bc \neq 0$, so R is a rational extension of A. □

Parts (1) and (4) of the proposition above show that for a right nonsingular ring, the notions of "essential" and "dense" are equivalent.

Example

5. If S is a right denominator set in R and R is S-torsion free, then Example 2 shows that RS^{-1} is a general ring of quotients of R as is $Q_{cl}^r(R)$ when R is a right Ore ring.

Remark. If Q is a general ring of quotients of R, note that the elements of Q may not be fractions in the same sense that the elements $a/s \in RS^{-1}$ are fractions.

We saw in the opening remarks of Section 6.5 that if M is an R-module and $H = \mathrm{End}_R(M)$, then M is an (H, R)-bimodule and there is a canonical ring homomorphism $\varphi : R \to \mathrm{End}_H(M)$ given by $\varphi(a) = f_a$, where $f_a : {}_H M \to {}_H M$ is such that $(x) f_a = xa$ for all $x \in M$. Moreover, φ is an injective ring homomorphism when M is faithful. So since $E(R_R)$ is a faithful R-module, we have an embedding $\varphi : R \to \mathrm{End}_H(E(R_R))$.

We promised earlier to develop a maximal general ring of quotients. This begins with the next proposition.

Proposition 8.4.9. *There is a ring Q and a canonical ring embedding $\varphi : R \to Q$ such that if R is identified with its image in Q, then Q_R is a maximal rational extension of R_R.*

Proof. Let $E(R)$ be an injective envelope of R_R, suppose that $H = \mathrm{End}_R(E(R))$ and let $Q = \mathrm{End}_H(E(R))$. Then $E(R)$ is an (H, R)-bimodule and an (H, Q)-bimodule. The remarks immediately proceeding the statement of the proposition show that the canonical ring homomorphism $\varphi : R \to Q$ is an embedding, so we can consider R to be a subring of Q. The proof will be complete if we can show that $Q \cong \xi(R)$ as R-modules. First, note that if $q \in Q$, then $(1_R)q \in E(R)$, so let $\phi : Q \to E(R)$ be given by $\phi(q) = (1_R)q$. We claim that ϕ is an isomorphism from Q to $\xi(R)$. This map is clearly well defined and R-linear, so it remains only to show that it is a bijection.

ϕ is an injection: If $x \in E(R)$, then $g_x : R \to E(R)$ such that $g_x(a) = xa$ is an R-linear mapping that can be extended to an R-linear mapping $h_x \in H$. If $q \in \mathrm{Ker}\,\phi$, then $(1_R)q = 0$, so $0 = h_x((1_R)q) = (h_x(1_R))q = (x)q$. Since this is true for each $x \in E(R)$, we have $q = 0$ and so $\mathrm{Ker}\,\phi = 0$.

$\mathrm{Im}\,\phi \subseteq \xi(R)$: From the Remark immediately following Corollary 8.4.6, the maximal rational extension $\xi(R)$ of R is given by $\xi(R) = \mathrm{ann}_r^{E(R)}(\mathrm{ann}_\ell^H(R))$. Now let $(1_R)q \in \mathrm{Im}\,\phi$ and note that $h \in \mathrm{ann}_\ell^H(R)$ if and only if $h(R) = 0$ which in turn is true if and only if $h(1_R) = 0$. For such an h, we see that $0 = (h(1_R))q = h((1_R)q)$, so $(1_R)q \in \mathrm{ann}_r^{E(R)}(\mathrm{ann}_\ell^H(R)) = \xi(R)$.

ϕ is surjective: If $\bar{x} \in \xi(R)$, then we need to find a $q \in Q$ such that $\phi(q) = (1_R)q = \bar{x}$. Define $q : {}_H E(R) \to {}_H E(R)$ by $(h(1_R))q = h(\bar{x})$. The domain of q is $E(R)$ since we have previously seen that for each $x \in E(R)$, there is an $h_x \in H$

such that $h_x(1_R) = x$. Furthermore, the map q is clearly H-linear. Note also that $h(1_R) = 0$ implies that $h(R) = 0$ which in turn gives $h(\xi(R)) = 0$ since $\xi(R)$ is a rational extension of R. Hence, $h(\bar{x}) = 0$, so q is well defined. If $h = \mathrm{id}_{E(R)}$, then it follows that $(1_R)q = \bar{x}$, so ϕ is surjective.   □

Definition 8.4.10. The ring $Q = \mathrm{End}_H(E(R))$ is a *maximal general ring of quotients* of R that will be denoted by $Q^r_{\max}(R)$. We will refer to $Q^r_{\max}(R)$ as a *complete ring of quotients* of R.

The following proposition shows that $\varphi : R \to Q^r_{\max}(R)$ has the universal mapping property.

Proposition 8.4.11. *If Q is a general ring of quotients of R, then there is a unique injective ring homomorphism $h : Q \to Q^r_{\max}(R)$ such that the diagram*

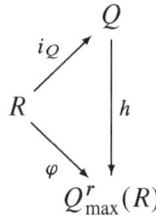

$$
\begin{array}{ccc}
 & & Q \\
 & \overset{i_Q}{\nearrow} & \big| \\
R & & \big| h \\
 & \underset{\varphi}{\searrow} & \big\downarrow \\
 & & Q^r_{\max}(R)
\end{array}
$$

is commutative, where i_Q and φ are the canonical ring embeddings.

Proof. Note first that i_Q and φ are R-linear mappings. Since $Q^r_{\max}(R)_R$ is a maximal rational extension of R_R, there is a unique R-linear injection $h : Q \to Q^r_{\max}(R)$ that will render the diagram

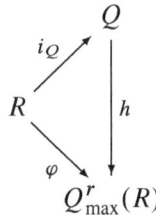

$$
\begin{array}{ccc}
 & & Q \\
 & \overset{i_Q}{\nearrow} & \big| \\
R & & \big| h \\
 & \underset{\varphi}{\searrow} & \big\downarrow \\
 & & Q^r_{\max}(R)
\end{array}
$$

commutative. We claim that h is a ring homomorphism. Let $q, q' \in Q$ and consider the R-linear mapping $g : Q^r_{\max}(R) \to Q^r_{\max}(R)$ given by $g(x) = (h(qq') - h(q)h(q'))x$. If $a \in (R : q') = \{a \in R \mid q'a \in R\}$, then

$$
\begin{aligned}
g(a) &= h(qq')a - h(q)h(q')a \\
&= h(qq'a) - h(q)h(q'a) \\
&= h(q)q'a - h(q)h(1_R)q'a \\
&= h(q)q'a - h(q)q'a \\
&= 0.
\end{aligned}
$$

But Q is a rational extension of R, so by (3) of Lemma 8.4.8, R is a rational extension of $(R : q')$. This means that $g(R) = 0$ which in turn gives $g = 0$. In particular, $g(1_R) = 0$, so $h(qq') = h(q)h(q')$.   □

Corollary 8.4.12. *The complete ring of quotients $Q^r_{max}(R)$ of R is unique up to ring isomorphism via a ring homomorphism that extends the identity map on R.*

Proof. It is easy to show that $\varphi : R \to Q^r_{max}(R)$ is a final object in an appropriately defined category.   □

Corollary 8.4.13. *If $Q^r_{cl}(R)$ exists, then there is a unique injective ring homomorphism $f : Q^r_{cl}(R) \to Q^r_{max}(R)$ such that $f|_R = \mathrm{id}_R$.*

Example

6. If S is a right denominator set in R and R is S-torsion free, then RS^{-1} is a general ring of quotients of R, so RS^{-1} embeds in $Q^r_{max}(R)$. If we identify RS^{-1} with its image in $Q^r_{max}(R)$, then we can consider RS^{-1} to be a subring of $Q^r_{max}(R)$. In particular, if R is a right Ore ring, then $Q^r_{cl}(R)$ can be viewed as a subring of $Q^r_{max}(R)$.

A *complete left ring of quotients* of R, denoted by $Q^\ell_{max}(R)$, can be developed in a similar fashion by beginning with the injective envelope of the left R-module $_R R$.

There is a connection between right nonsingular rings and rings that have a regular complete ring of quotients. To establish this connection, we need the following proposition. Recall that $J(R)$ denotes the Jacobson radical of a ring R.

Proposition 8.4.14. *If M is an injective R-module and $H = \mathrm{End}_R(M)$, then*

(1) $J(H) = \{h \in H \mid \mathrm{Ker}\, h \text{ is an essential submodule of } M\}$,

(2) $H/J(H)$ *is a regular ring, and*

(3) *idempotents of $H/J(H)$ can be lifted to H.*

Proof. First, we prove (1) and (2). Let $I = \{h \in H \mid \mathrm{Ker}\, h \text{ is an essential submodule of } M\}$. If $h \in H$, then we claim there is an element $f \in H$ such that $h - hfh \in I$. If K_c is a complement of $\mathrm{Ker}\, h$, then $\mathrm{Ker}\, h \oplus K_c$ is an essential submodule of M. Consider the mapping $g : h(K_c) \to K_c$ defined by $g(h(x)) = x$. If $h(x) = h(y)$, $x, y \in K_c$, then $x - y \in \mathrm{Ker}\, h \cap K_c = 0$, so $x = y$. Thus, g is a well-defined mapping that is easily shown to be R-linear. Hence, g can be extended to an R-linear mapping $f \in H$. If $x + y \in \operatorname{Ker} h \oplus K_c$, then

$$(h - hfh)(x + y) = h(x + y) - hfh(x + y)$$
$$= h(y) - hfh(y)$$
$$= h(y) - h(y)$$
$$= 0.$$

Therefore, $h - hfh \in I$ since $\operatorname{Ker} h \oplus K_c$ is an essential submodule of M.

The proof of (1) and (2) will be complete if we can show that $J(H) = I$. If $h \in I$ and $f \in H$, then $\operatorname{Ker} h \subseteq \operatorname{Ker} fh$, so $\operatorname{Ker} fh$ is an essential submodule of M. But $\operatorname{Ker} fh \cap \operatorname{Ker}(\operatorname{id}_M - fh) = 0$ and so we have $\operatorname{Ker}(\operatorname{id}_M - fh) = 0$. Hence, $\operatorname{id}_M - fh$ is an injective mapping and the injectivity of M can be used to produce a left inverse of $\operatorname{id}_M - fh$ in H. Therefore, by (3) of Proposition 6.1.8, $h \in J(H)$ and so $I \subseteq J(H)$.

Conversely, if $h \in J(H)$, then we have seen in the first paragraph of this proof that there is an $f \in H$ such that $h - hfh \in I$. But $J(H)$ is an ideal of H, so $hf \in J(H)$. Part (3) of Proposition 6.1.7 indicates that $\operatorname{id}_M - hf$ has an inverse in H, so $(\operatorname{id}_M - hf)^{-1}(h - hfh) = (\operatorname{id}_M - hf)^{-1}(\operatorname{id}_M - hf)h = h$. If I is a left ideal of H, then $h \in I$ and we will have $J(H) \subseteq I$. But I is a left ideal of H, since if $h, h' \in I$ and $f \in H$, then

$$\operatorname{Ker}(h + h') \supseteq \operatorname{Ker} h \cap \operatorname{Ker} h' \quad \text{and} \quad \operatorname{Ker}(fh) \supseteq \operatorname{Ker} h.$$

Therefore, $I = J(H)$ and so

$$h + J(H) = hfh + J(H) = (h + J(H))(f + J(H))(h + J(H)).$$

Thus, $H/J(H)$ is a regular ring.

(3) Suppose that $e + J(H)$ is an idempotent of $H/J(H)$. Then $e - e^2 \in J(H)$, so $K = \operatorname{Ker}(e - e^2)$ is an essential submodule of M. Since $E(eK)$ embeds in M, we can consider $E(eK)$ to be a submodule of M, so that $E(eK)$ is a direct summand of M. Thus, there is an idempotent endomorphism $h \in H$ such that $hM = E(eK)$. Since h is the identity map on eK, we see that $(he - e)K = 0$. But this gives $K \subseteq \operatorname{Ker}(he - e)$ and so it follows that $he - e \in J(H)$. Hence, $he + J(H) = e + J(H)$. If $f = h + he(\operatorname{id}_M - h)$, then $hf = f$, $fh = h$ and $f^2 = f$. In particular, f is an idempotent in H.

Finally, let $K' = (\operatorname{id}_M - h)M + eK$. Then K' is an essential submodule of M and $(f - he)K' = 0$, so $\operatorname{Ker}(f - he)$ is essential in M. Therefore, $f - he \in J(H)$ and we have $f + J(H) = he + J(H) = e + J(H)$. Hence, f lifts $e + J(H)$ to H. $\qquad\square$

Proposition 8.4.15. *The following are equivalent for any ring R.*

(1) *R is right nonsingular.*

(2) *$H = \operatorname{End}_R(E(R))$ is Jacobson semisimple.*

(3) *$Q^r_{\max}(R)$ is a regular ring.*

Proof. (1) $\Leftrightarrow$ (2). Suppose that $Z(R_R) = 0$ and let $h \in J(H)$. Then, by Proposition 8.4.14, $\operatorname{Ker} h$ is an essential submodule of $E(R)$. But $Z(R_R) = 0$ and Lemma 8.3.4 shows that $Z(E(R)) = 0$. Hence, $E(R)$ is a rational extension of $\operatorname{Ker} h$. Since $h(\operatorname{Ker} h) = 0$, (2) of Lemma 8.4.2 gives $h = 0$. Therefore, $J(H) = 0$, so H is Jacobson semisimple. Conversely, suppose that $J(H) = 0$ and let $a \in Z(R_R), a \neq 0$. Then $\operatorname{ann}_r(a)$ is an essential right ideal of R. The R-linear mapping $f : R \to R$ given by $f(b) = ab$ can be extended to an R-linear mapping $h : E(R) \to E(R)$. Since $\operatorname{ann}_r(a) = \operatorname{Ker} f \subseteq \operatorname{Ker} h$, $\operatorname{Ker} h$ is an essential submodule of $E(R)$. Thus, by Proposition 8.4.14, $h \in J(H) = 0$, so we see that $\operatorname{ann}_r(a) = 0$, a contradiction. Hence, $Z(R_R) = 0$.

(2) $\Rightarrow$ (3). If $J(H) = 0$, then because of the equivalence of (1) and (2), $Z(R_R) = 0$. But this gives $Z(E(R)) = 0$, so it follows that $E(R)$ is a rational extension of R. Thus, there is an injective R-linear mapping $f : E(R) \to Q^r_{\max}(R)$ that extends the identity mapping on R. Hence, $R \subseteq f(E(R)) \subseteq Q^r_{\max}(R)$. But $f(E(R))$ is an injective R-module, so there is an R-submodule N of $Q^r_{\max}(R)$ such that $N \oplus f(E(R)) = Q^r_{\max}(R)$. Since R is an essential R-submodule of $Q^r_{\max}(R)$ and since $R \cap N = 0$, we have $N = 0$. Therefore, $f(E(R)) = Q^r_{\max}(R)$ and we can identify the R-modules $E(R)$ and $f(E(R))$, so that $R_R \subseteq E(R)_R = Q^r_{\max}(R)_R$. It follows that the ring operations on $Q^r_{\max}(R)$ induce ring operations on $E(R)$ that extend the ring operations on R. Hence, $E(R)$ is a complete ring of quotients of R when $Z(R_R) = 0$.

Finally, we claim that H and $E(R)$ are isomorphic rings. As in the proof of Proposition 8.4.9, for each $x \in E(R)$, there is an $h_x \in H$ such that $h_x(1_R) = x$. If $h \in H$ is also such that $h(1_R) = x$, then $(h - h_x)(1_R) = 0$ implies that $(h - h_x)(R) = 0$. But $E(R)$ is a rational extension of R, so $h = h_x$. Thus, h_x is uniquely determined by x. Now define $\psi : E(R) \to H$ by $\psi(x) = h_x$. It is easy to show that ψ is an R-linear isomorphism, so H is also a complete ring of quotients R. Part (2) of Proposition 8.4.14 indicates that $Q^r_{\max}(R)$ is a regular ring.

(3) $\Rightarrow$ (1). Suppose that $Q^r_{\max}(R)$ is a regular ring. For each $a \in R \subseteq Q^r_{\max}(R)$, $a \neq 0$, there is an $q \in Q^r_{\max}(R)$ such that $a = aqa$, so $e = qa$ is a nonzero idempotent element of $Q^r_{\max}(R)$. Now $\operatorname{ann}_r^{Q^r_{\max}(R)}(e) = (1_{Q^r_{\max}(R)} - e)Q^r_{\max}(R)$ and $R \cap \operatorname{ann}_r^{Q^r_{\max}(R)}(e) = \operatorname{ann}_r^R(e)$. Since R_R is an essential submodule of $Q^r_{\max}(R)_R$, we have $R \cap (eQ^r_{\max}(R)) \neq 0$. This gives

$$R \cap (eQ^r_{\max}(R)) \cap \operatorname{ann}_r^R(e) \subseteq eQ^r_{\max}(R) \cap (1_{Q^r_{\max}(R)} - e)Q^r_{\max}(R) = 0,$$

so $\text{ann}_r^R(e)$ is not an essential right ideal of R. Since $\text{ann}_r^R(a) \subseteq \text{ann}_r^R(e)$, we have $a \notin Z(R_R)$, so $Z(R_R) = 0$.   □

Goldie's theorem gives the necessary and sufficient conditions on R that will ensure that $Q_{cl}^r(R)$ exists and that $Q_{cl}^r(R)$ is a semisimple ring. We will now develop a condition on R that is sufficient to give $Q_{cl}^r(R) = Q_{max}^r(R)$, whenever $Q_{cl}^r(R)$ exists.

Proposition 8.4.16. *If R is a right Ore ring and every dense right ideal of R contains a regular element of R, then $Q_{cl}^r(R) = Q_{max}^r(R)$.*

Proof. It suffices to show that if $q \in Q_{max}^r(R)$, then $q = a/s$ for some a in R and some regular element s of R. Since R_R is dense in $Q_{max}^r(R)_R$, we have, by (3) of Lemma 8.4.8, that $(R : q)$ is a dense right ideal of R. Therefore, $(R : q)$ contains a regular element of R, say s. If $a = qs$, then $q = a/s \in Q_{cl}^r(R)$, so we have the result.   □

Corollary 8.4.17. *If R is a semiprime right Goldie ring, then $Q_{max}^r(R)$ is a semisimple ring and, in fact, $Q_{max}^r(R) = Q_{cl}^r(R)$.*

Proof. If R is a semiprime right Goldie ring, then in view of the proposition and (2) of Goldie's theorem, Proposition 8.3.8, we have $Q_{cl}^r(R) = Q_{max}^r(R)$. The fact that $Q_{max}^r(R)$ is a semisimple ring is (3) of Goldie's theorem.   □

Goldie's theorem gives necessary and sufficient conditions for R to have a semisimple classical quotient ring. Our goal now is to develop necessary and sufficient conditions on R in order for $Q_{max}^r(R)$ to be semisimple even if $Q_{cl}^r(R)$ fails to exist. For this we need the concept of a uniform module.

Definition 8.4.18. A nonzero R-module M with the property that every nonzero submodule is essential in M is said to be a *uniform module*.

Recall that an R-module M is indecomposable if M cannot be written as a direct sum of nonzero submodules. It follows easily that the following implications hold for an R-module M.

$$\text{simple} \Rightarrow \text{uniform} \Rightarrow \text{indecomposable}$$

and

$$\text{injective} + \text{indecomposable} \Rightarrow \text{uniform}.$$

We now need the following lemma.

Lemma 8.4.19. *The following are equivalent for any R-module M.*

(1) *M is uniform.*

(2) *Any two nonzero submodules of M have nonzero intersection.*

(3) *The injective envelope of M is indecomposable.*

(4) *Every nonzero submodule of M is indecomposable.*

Proof. The equivalence of (1) and (2) is a direct consequence of the definition of a uniform module.

(2) $\Rightarrow$ (3). Let $E(M)$ be an injective envelope of M. If E_1 and E_2 are nonzero submodules of $E(M)$ such that $E(M) = E_1 \oplus E_2$, then $N_1 = M \cap E_1 \neq 0$ and $N_2 = M \cap E_2 \neq 0$, since M is an essential submodule of $E(M)$. But this gives $N_1 \cap N_2 = 0$ which contradicts (2).

(3) $\Rightarrow$ (4). Let N be a nonzero submodule of M that has nonzero submodules N_1 and N_2 such that $N = N_1 \oplus N_1$. If N_c is a complement of N in M, then $N \oplus N_c$ is an essential submodule of M, so, by Proposition 7.1.5, $E(M) = E(N \oplus N_c) = E(N) \oplus E(N_c)$. Therefore, $E(N_c) = 0$, since $E(M)$ is indecomposable. But this gives $N_c = 0$ and so we have $E(M) = E(N) = E(N_1) \oplus E(N_2)$. Hence, either $E(N_1)$ or $E(N_2)$ is zero, so either N_1 or N_2 is zero, and we have that N is indecomposable.

(4) $\Rightarrow$ (2). If (2) does not hold, then there exist nonzero submodules N_1 and N_2 of M such that $N_1 \cap N_2 = 0$. But then the nonzero submodule $N_1 \oplus N_2$ is not indecomposable, a contradiction. $\square$

Lemma 8.4.20. *If M is a finite dimensional nonzero R-module, then M contains a uniform submodule.*

Proof. Suppose that M does not contain a uniform submodule. Then if N is a nonzero submodule of M, then N is not uniform, so there must exist nonzero submodules N_1 and N_2' of N such that $N_1 \cap N_2' = 0$. Thus, we have a direct sum $N_1 \oplus N_2'$ and N_2' cannot be uniform. If N_2 and N_3' are nonzero submodules of N_2' such that $N_2 \cap N_3' = 0$, then we have a direct sum $N_1 \oplus N_2 \oplus N_3'$ and N_3' cannot be uniform. Continuing in this way, we can construct an infinite direct sum $N_1 \oplus N_2 \oplus N_3 \oplus \cdots$ which implies that M is not finite dimensional. $\square$

Proposition 8.4.21. *The following are equivalent for an R-module M.*

(1) *M is finite dimensional.*

(2) *M contains an essential submodule of the form $N_1 \oplus N_2 \oplus \cdots \oplus N_n$, where each N_i is uniform.*

(3) *The injective envelope of M is a direct sum of a finite number of indecomposable R-modules.*

Proof. (1) $\Rightarrow$ (2). By Lemma 8.4.20, M contains a uniform submodule N_1. If N_1 is essential in M, then we are done. If N_1 is not essential in M, let N_c be a complement of N_1 in M. Then $N_1 \oplus N_c$ is essential in M, so if N_c is uniform, then we are finished. If N_c is not uniform, then N_c contains a uniform submodule, say N_2. If $N_1 \oplus N_2$ is essential in M, then the proof is complete. If $N_1 \oplus N_2$ is not essential in M, let N_c' be a complement in M of $N_1 \oplus N_2$. Then $N_1 \oplus N_2 \oplus N_c'$ is essential in M. If N_c' is uniform, then $N_1 \oplus N_2 \oplus N_c'$ is the desired submodule of M. If N_c' is not uniform, then N_c' contains a uniform submodule N_3 and the sum $N_1 \oplus N_2 \oplus N_3$ is direct. Since M is finite dimensional, this process cannot be continued forever, so we must eventually arrive at a submodule $N_1 \oplus N_2 \oplus \cdots \oplus N_n$ that is essential in M, where each N_k is uniform.

(2) $\Rightarrow$ (3). Let $N_1 \oplus N_2 \oplus \cdots \oplus N_n$ be an essential submodule of M such that N_k is uniform for $k = 1, 2, \ldots, n$. Then in view of of Proposition 7.1.5, it follows that

$$E(M) = E(N_1 \oplus N_2 \oplus \cdots \oplus N_n) = E(N_1) \oplus E(N_2) \oplus \cdots \oplus E(N_n).$$

Furthermore, since each N_k is uniform, Lemma 8.4.19 shows that each $E(N_k)$ is indecomposable.

(3) $\Rightarrow$ (1). Suppose that $E(M) = \bigoplus_{k=1}^{n} E_k$, where each E_k is an indecomposable (and necessarily injective) R-module. If M is not finite dimensional, then there is an infinite family $\{N_\alpha\}_\Delta$ of nonzero submodules of M whose sum is direct. Clearly we can assume that $\Delta = \mathbb{N}$. Moreover, we can assume that $\bigoplus_\mathbb{N} N_k$ is essential in M. If not and N_c is a complement in M of $\bigoplus_\mathbb{N} N_k$, then $N_c \oplus (\bigoplus_\mathbb{N} N_k)$ is essential in M, so by reindexing we can produce a collection of nonzero submodules of M such that $\bigoplus_\mathbb{N} N_k'$ is essential in M. Proposition 7.1.5 now gives

$$\bigoplus_\mathbb{N} E(N_k) \subseteq E(\bigoplus_\mathbb{N} N_k) = \bigoplus_{k=1}^{n} E_k.$$

Therefore, the injective R-module $E(N_k)$ is a direct summand of $\bigoplus_{k=1}^{n} E_k$, for $k = 1, 2, 3, \ldots$. But each E_k is indecomposable, so it follows that each $E(N_k)$ is a direct sum of some of the E_k. Hence, only a finite number of the $E(N_k)$ can be distinct. Since the sum $\bigoplus_\mathbb{N} E(N_k)$ is direct, this means that $E(N_k) = 0$ for almost all k and this in turn gives $N_k = 0$ for almost all k. This contradiction means that M must be finite dimensional. $\square$

Recall that a nonzero idempotent e of R is said to be a *primitive idempotent* if e cannot be written as $e = f + g$, where f and g are nonzero orthogonal idempotents of R and that an idempotent e of R is a *local idempotent* if eRe is a local ring. We have previously seen that (1) if e is a local idempotent of R, then e is primitive and (2) if e is a primitive idempotent of R, then $e + I$ is primitive in R/I for any ideal I of R such that $I \subseteq J(R)$.

Remark. If R is a regular ring and e is a primitive idempotent of R, then eR is a minimal right ideal of R. To see this, suppose that $A \subseteq eR$, where A is a right ideal of R. If $a \in A$, $a \neq 0$, then since R is a regular ring, there is an idempotent $f \in R$ such that $fR = aR \subseteq eR$. Now $R = fR \oplus (1 - f)R$, so the modular law (Example 10 in Section 1.4) gives $eR = fR \oplus [eR \cap (1 - f)R]$. But e is primitive, so eR is indecomposable. Hence, $eR \cap (1 - f)R = 0$ and so $eR = fR$. Thus, $A = eR$, so eR is a minimal right ideal of R.

To prove the next proposition, we need the following lemma.

Lemma 8.4.22. *If M is an indecomposable injective R-module, then $\operatorname{End}_R(M)$ is a local ring.*

Proof. Let $f \in \operatorname{End}_R(M)$ be a nonunit in $\operatorname{End}_R(M)$. If $\operatorname{Ker} f = 0$, then $f(M) \subsetneq M$ and $f(M) \cong M$ indicates that $f(M)$ is an injective submodule of M. Hence, $M = f(M) \oplus N$ for some submodule N of M. But M is indecomposable and so $N = 0$. Thus, $M = f(M)$ which means that f is a bijection and so f unit in $\operatorname{End}_R(M)$, a contradiction. Thus, $\operatorname{Ker} f \neq 0$. Likewise, if $g \in \operatorname{End}_R(M)$ is also a nonunit in $\operatorname{End}_R(M)$, then $\operatorname{Ker} g \neq 0$. Since M is injective and indecomposable, M is uniform. Hence, we have $0 \neq \operatorname{Ker} f \cap \operatorname{Ker} g \subseteq \operatorname{Ker}(f + g)$ and so $f + g$ is not a unit in $\operatorname{Hom}_R(M)$. With this at hand, Proposition 7.2.11 shows that $\operatorname{End}_R(M)$ is a division ring. Consequently, $\operatorname{End}_R(M)$ is a local ring. $\square$

Proposition 8.4.23. *If M is a finite dimensional injective R-module, then $H = \operatorname{End}_R(M)$ is a semiperfect ring.*

Proof. If M is finite dimensional, then, by Proposition 8.4.21, we see that $M = \bigoplus_{k=1}^{n} E_k$, where each E_k is an indecomposable injective R-module. It follows that $H \cong \bigoplus_{k=1}^{n} \operatorname{End}_R(E_k)$ and since each E_k is injective and indecomposable, Lemma 8.4.22 shows that each $\operatorname{End}_R(E_k)$ is a local ring. Hence, there is a set $\{e_k\}_{k=1}^{n}$ of local idempotents of H such that $\operatorname{id}_H = e_1 + e_2 + \cdots + e_n$. But local idempotents are primitive, so each e_k is primitive and each $e_k + J(H)$ is primitive in $H/J(H)$. Thus, $\operatorname{id}_H + J(H) = e_1 + J(H) + e_2 + J(H) + \cdots + e_n + J(H)$ and it follows that $H/J(H)$ is a direct sum of a finite number of indecomposable right ideals in $H/J(H)$. But Proposition 8.4.14 indicates that $H/J(H)$ is a regular ring, so, by the preceding Remark, these right ideals must be minimal right ideals of $H/J(H)$. Hence, $H/J(H)$ is a semisimple ring. Proposition 8.4.14 also shows that idempotents of $H/J(H)$ can be lifted to H, so we have, by Proposition 7.2.23, that H is a semiperfect ring. $\square$

Remark. One can actually show that if M is a finite dimensional R-module and if $N_1 \oplus N_2 \oplus \cdots \oplus N_n$ and $N_1' \oplus N_2' \oplus \cdots \oplus N_m'$ are essential submodules of M with uniform summands, then $n = m$. For such a module, the *Goldie dimension* of M, denoted by $G.\dim M$, is defined to be the unique integer n. If M is not finite dimensional, then $G.\dim M$ is defined to be ∞. Additional information on Goldie dimension can be found in [2], [13] and [26].

We conclude this chapter by fulfilling an earlier promise to develop conditions on R that are necessary and sufficient for $Q^r_{max}(R)$ to be a semisimple ring.

Proposition 8.4.24. *The following are equivalent for any ring* R.

(1) R *is a finite dimensional, right nonsingular ring.*

(2) $Q^r_{max}(R)$ *is a semisimple ring.*

Proof. (1) $\Rightarrow$ (2). If R is finite dimensional, then we know from Proposition 8.4.23 that $H = \operatorname{End}_R(E(R))$ is a semiperfect ring. Hence, by Bass' theorem, Proposition 7.2.23, $H/J(H)$ is a semisimple ring. But R is right nonsingular and so Proposition 8.4.15 gives $J(H) = 0$. Thus, H is semisimple. Moreover, as we have seen in the proof of Proposition 8.4.15, $Q^r_{max}(R) \cong H$, so $Q^r_{max}(R)$ is a semisimple ring.

(2) $\Rightarrow$ (1). If $Q^r_{max}(R)$ is a semisimple ring, then by Proposition 6.6.2 we have that $Q^r_{max}(R)$ is a regular ring, so Proposition 8.4.15 indicates that R is a right nonsingular ring. Since R is right nonsingular, then, as before, we see that $Q^r_{max}(R) \cong H$. Thus, there is a complete set $\{e_i\}_{i=1}^n$ of orthogonal primitive idempotents of H such that $\operatorname{id}_H = e_1 + e_2 + \cdots + e_n$. It follows that $E(R) = \bigoplus_{i=1}^n e_i E(R)$, where each $e_i E(R)$ is an indecomposable R-module. Proposition 8.4.21 shows that R is finite dimensional. $\square$

Problem Set 8.4

1. (a) If D is a dense right ideal of R, show that $(D : a)$ is dense in R for each $a \in R$.

   (b) If $f : R \to R$ is an R-linear mapping and D is a dense right ideal of R, prove that $f^{-1}(D)$ is dense in R.

   (c) If D is a dense right ideal of R, show that $(D : q)$ is dense in R for each $q \in Q^r_{max}(R)$.

   (d) If D is a dense right ideal of R and $q \in Q^r_{max}(R)$ is such that $qD = 0$, show that $q = 0$.

2. Verify Examples 3 and 4.

3. (a) If N is a rational extension of M, show that M is an essential submodule of N.

   (b) If M is an essential submodule of N and $Z(M) = 0$, show that N is a rational extension of M.

   (c) If N is a rational extension of M, show that for any submodule N' of N, N is a rational extension of $N' \cap M$.

(d) If N' is a rational extension of N and N is a rational extension of M, prove that N' is a rational extension of M.

(e) Construct a category $\mathcal{C}$ such that the maximal rational extension $(\xi(M), i_{\xi(M)})$ is a final object in $\mathcal{C}$.

4. Prove Lemma 8.4.4.

5. A right R-module M is said to be *rationally complete* if $M = \xi(M)$.

(a) Prove that $\xi(M)$ is rationally complete.

(b) If R is an integral domain, prove that R_R is rationally complete if and only if R is a field.

(c) If M is a nonsingular R-module, show that $\xi(M)$ is an injective envelope of M. [Hint: Use (b) of Exercise 3.] Conclude that if M is a nonsingular R-module that is rationally complete, then M is an injective R-module. Note that an injective R-module M is rationally complete regardless of whether or not M is nonsingular.

(d) If $\{M_\alpha\}_\Delta$ is a family of R-modules each of which is rationally complete, then is $\prod_\Delta M_\alpha$ ($\bigoplus_\Delta M_\alpha$) rationally complete?

6. An element x of an R-module M is said to be *left fixed by an R-linear mapping* $h : M \to M$ if $h(x) = x$. Show that $\xi(M)$ is the set of all elements of $E(M)$ that are left fixed by all $h \in \mathrm{End}_R(E(M))$ that leave each element of M fixed.

7. Let S be a right denominator set in R and set $t(R) = \{a \in R \mid as = 0 \text{ for some } s \in S\}$.

(a) Show that $\bar{S} = \{s + t(R) \mid s \in S\}$ is a right denominator set in $\bar{R} = R/t(R)$.

(b) Prove that $\bar{R}\bar{S}^{-1}$ embeds in $Q^r_{\max}(R/t(R))$ as a ring.

Conclude that if R is S-torsion free, then RS^{-1} embeds in $Q^r_{\max}(R)$ as a ring and that if R is a right Ore ring, then $Q^r_{\max}(R)$ contains a copy of $Q^r_{\mathrm{cl}}(R)$.

8. Fill in the details of the proof of Corollary 8.4.6 and of Corollary 8.4.12.

9. Let $\mathcal{D}$ be the collection of all dense right ideals of R and consider the set $\bigcup_\mathcal{D} \mathrm{Hom}_R(D, R)$.

(a) If $f \in \bigcup_\mathcal{D} \mathrm{Hom}_R(D, R)$, then the dense right ideal of R that is the domain of f will be denoted by D_f. Define the relation $\sim$ on $\bigcup_\mathcal{D} \mathrm{Hom}_R(D, R)$ by $f \sim g$ if and only if f and g agree on $D_f \cap D_g$. Prove that $\sim$ is an equivalence relation on $\bigcup_\mathcal{D} \mathrm{Hom}_R(D, R)$.

(b) Let Q denote the set of equivalence classes $[f]$ in $\bigcup_\mathcal{D} \mathrm{Hom}_R(D, R)$ determined by $\sim$. Prove that Q is a ring if addition and multiplication are defined on Q by

$$[f] + [g] = [f + g], \quad \text{where } D_{f+g} = D_f \cap D_g \quad \text{and}$$
$$[f][g] = [fg], \quad \text{where } D_{fg} = g^{-1}(D_f).$$

(c) Prove that $Q \cong Q^r_{max}(R)$. [Hint: If $[f] \in Q$ and $f : D_f \to R$, let $h_f : E(R) \to E(R)$ be an R-linear mapping such that $h_f|_{D_f} = f$. Consider $[f] \mapsto h_f$.]

10. For any $a \in R$, prove that the inner derivation δ_a defined on R by $\delta_a(b) = ab - ba$ for each $b \in R$ has a unique extension to $Q^r_{max}(R)$. Use this fact to show that if R is commutative, then so is $Q^r_{max}(R)$. [Hint: See Example 9 in Section 1.1 and Exercise 3 in Problem Set 1.1.]

Chapter 9

Graded Rings and Modules

In this chapter we begin an investigation of graded rings and modules. One use of rings and modules with gradings is in describing certain topics in algebraic geometry. We will touch upon several concepts presented in the previous chapters, but they will now be reformulated for graded rings and modules and studied in this "new" setting. Proofs of the lemmas and theorems involving these reformulated concepts are often similar to the proofs of the ungraded results, so they may be left as exercises. Graded topics corresponding to ungraded concepts may also appear in the exercises.

Graded rings and modules can be defined with a group G as the set of degrees. The group G can be written either additively or multiplicatively and G may or may not be commutative. We have chosen to limit our discussion to an introduction to graded rings and module with the additive abelian group $\mathbb{Z}$ as the set of degrees. A complete development of graded rings and modules with a group G as the set of degrees would warrant a text in itself. The interested reader can consult [35] and [36] for a more extensive account of this subject.

9.1   Graded Rings and Modules

Graded Rings

As a way of introducing graded rings, consider the polynomial ring $R[X]$. For each $n \in \mathbb{Z}$, let $X^n R = \{X^n a \mid a \in R\}$, where $X^n R = 0$ for each $n < 0$ and $X^0 = 1$. Then each $X^n R$ is a subgroup of the additive group of $R[X]$ and it follows easily $R[X] = \bigoplus_{\mathbb{Z}} X^n R$. Thus, each $p(X) = a_0 + X a_1 + X^2 a_2 + \cdots + X^n a_n$ in $R[X]$ can be written uniquely as a finite sum of elements in the set of subgroups $\{X^n R\}_{\mathbb{Z}}$ and each summand $X^k a_k \in X^k R$ in $p(X)$ is a homogeneous polynomial of degree k. Finally, we see that $(X^m R)(X^n R) \subseteq X^{m+n} R$ for each $m, n \in \mathbb{Z}$. These observations regarding the polynomial ring $R[X]$ provide the motivation for the following definition.

Definition 9.1.1. A *graded ring* is a ring R together with a set $\{R_n\}_{\mathbb{Z}}$ of subgroups of the additive group of R such that $R = \bigoplus_{\mathbb{Z}} R_n$ and such that $R_m R_n \subseteq R_{m+n}$ for all $m, n \in \mathbb{Z}$. The family of subgroups $\{R_n\}_{\mathbb{Z}}$ is said to be a *grading* of R. Such a ring R is said to have $\mathbb{Z}$ as its *set of degrees*. A nonzero element $a \in R_n$ is referred to as a *homogeneous element* of R of degree n, denoted $\deg(a) = n$. If $\deg(a) = n$, then we will often write a_n for a to indicate that $a \in R_n$. Also, if $\{R_n\}_{\mathbb{Z}}$ is a grading of R, then not all of the R_n need be nonzero. If $R_m R_n = R_{m+n}$ for all $m, n \in \mathbb{Z}$, then R is said to be *strongly graded* by $\{R_n\}_{\mathbb{Z}}$. A graded ring R is *positively graded* if $R_n = 0$ for all $n < 0$. In this case, the zero subgroups R_n, $n < 0$, are suppressed and we write $\{R_n\}_{n \geq 0}$ for the grading of R. Similarly, R is *negatively graded* if $R_n = 0$ for all $n > 0$ and, in this case, we express the grading of R by $\{R_n\}_{n \leq 0}$.

If S is a subring of a graded ring R and if $S_n = S \cap R_n$ for each $n \in \mathbb{Z}$, then S is said to be a *graded subring* of R, if $S = \bigoplus_{\mathbb{Z}} S_n$. A right ideal (A left ideal, An ideal) A of a graded ring R is a *graded right ideal* (*graded left ideal, graded ideal*) of R, if $A = \bigoplus_{\mathbb{Z}} A_n$, where $A_n = A \cap R_n$ for each $n \in \mathbb{Z}$.

If R is a graded ring and S is a graded subring of R, then $(S \cap R_m)(S \cap R_n) \subseteq S \cap R_{m+n}$ for all $m, n \in \mathbb{Z}$. Hence, a graded subring S of R is a graded ring. Moreover, if R is a graded ring, then each nonzero $a \in R$ has a unique expression as a finite sum of nonzero homogeneous elements of R. If $a = \sum_{\mathbb{Z}} a_n$, then each a_n is said to be a *homogeneous component* of a. Note also that if A is a graded right ideal, then each $a \in A$ can be written as a finite sum of nonzero homogeneous components of R, each belonging to A.

Remark. If R is a graded ring, then $0 \in R_n$ for each $n \in \mathbb{Z}$, so no degree is assigned to 0. Furthermore, the zero ideal is to be viewed as a graded ideal of R that is a graded subideal of every graded right or graded left ideal of R.

Examples

1. **Trivial Grading.** If $R_0 = R$ and $R_n = 0$ for all $n \neq 0$, then $\{R_n\}_{\mathbb{Z}}$ is a grading of R called the *trivial grading* of R. Thus, every ring can be viewed as a graded ring.

   From this point forward, if we say that R is a graded ring, then we mean, unless stated otherwise, that the grading $\{R_n\}_{\mathbb{Z}}$ of R is nontrivial.

2. **Polynomial Rings.** As we saw in the opening remarks of this section, every polynomial ring $R[X]$ is positively graded with grading $\{X^n R\}_{n \geq 0}$. More generally, if $p(X_1, X_2, \ldots, X_k)$ is a polynomial in $R[X_1, X_2, \ldots, X_k]$, then a term $X_1^{j_1} X_2^{j_2} \cdots X_k^{j_k} a$ of $p(X_1, X_2, \ldots, X_k)$ is said to have degree n if $j_1 + j_2 + \cdots j_k = n$. A polynomial $p(X_1, X_2, \ldots, X_k)$ in $R[X_1, X_2, \ldots, X_k]$ is said to be a homogeneous polynomial of degree n if every term of $p(X_1, X_2, \ldots, X_k)$ has degree n. If $n \geq 0$ and P_n is the set of all homogeneous polynomials of $R[X_1, X_2, \ldots, X_k]$ of degree n together with the zero polynomial, then P_n is a subgroup of the additive group of $R[X_1, X_2, \ldots, X_n]$. Moreover, $R[X_1, X_2, \ldots, X_n] = \bigoplus_{n \geq 0} P_n$, so $R[X_1, X_2, \ldots X_k]$ is a positively graded ring.

3. **Not every right ideal of a graded ring must be graded.** Consider the positively graded ring $R[X]$ with grading $\{X^n R\}_{n \geq 0}$. Since $1 + X \in (1 + X)R[X]$ and $1 \notin (1 + X)R[X]$, $1 + X$ cannot be written as a sum of homogeneous elements of $(1 + X)R[X]$. Thus, the right ideal $(1 + X)R[X]$ of $R[X]$ is not graded.

4. **Laurent Polynomials.** Let R be a commutative ring and consider the polynomial ring $R[X]$. If $S = \{X^n \mid n = 0, 1, 2, \ldots\}$, then S is a multiplicative system and elements of $R[X]S^{-1}$ look like

$$\frac{a_0 + Xa_1 + X^2a_2 + \cdots + X^na_n}{X^m}, \quad m, n = 0, 1, 2, \ldots.$$

Thus, with a change of notation, we see that elements of $R[X]S^{-1}$ are "polynomials" of the form

$$p(X) = X^{-m}a_{-m} + \cdots + X^{-1}a_{-1} + a_0 + Xa_1 + \cdots + X^na_n,$$

$$m, n = 0, 1, 2, \ldots.$$

Each $p(X)$ in $R[X]S^{-1}$ is called a *Laurent polynomial* and $R[X]S^{-1}$ is referred to as the *ring of Laurent polynomials,* usually denoted by $R[X, X^{-1}]$. $R[X, X^{-1}]$ is a graded ring graded by the subgroups $\{X^n R\}_{\mathbb{Z}}$, where $X^n R = \{X^n a \mid a \in R\}$ for each $n \in \mathbb{Z}$. Since

$$\frac{a_0 + Xa_1 + X^2a_2 + \cdots + X^na_n}{1} = a_0 + Xa_1 + X^2a_2 + \cdots + X^na_n,$$

$R[X]$ can be viewed as a graded subring of $R[X, X^{-1}]$.

5. **$\mathbb{Z}_2$ as a set of degrees.** In the opening remarks of this chapter, it was indicated that a ring can be graded using an arbitrary group as a set of degrees. For instance, if $\mathbb{S} = \mathbb{M}_3(R)$ is the ring of 3×3 matrices over a ring R, then

$$\mathbb{S}_{[0]} = \begin{pmatrix} R & R & 0 \\ R & R & 0 \\ 0 & 0 & R \end{pmatrix}, \quad \mathbb{S}_{[1]} = \begin{pmatrix} 0 & 0 & R \\ 0 & 0 & R \\ R & R & 0 \end{pmatrix},$$

is a grading of $\mathbb{S}$ with the group $\mathbb{Z}_2 = \{[0], [1]\}$ as the set of degrees.

Recall that $U(R)$ denotes the multiplicative group of units of a ring R.

Proposition 9.1.2. *If R is a graded ring, then*

(1) *R_0 is a subring of R,*

(2) *if $a_n \in R_n \cap U(R)$, then $a_n^{-1} \in R_{-n} \cap U(R)$, and*

(3) *R is strongly graded if and only if $1 \in R_n R_{-n}$ for each $n \in \mathbb{Z}$.*

Proof. (1) Since $R_0 R_0 \subseteq R_0$, we need only show that $1 \in R_0$. Let $1 = \sum_{\mathbb{Z}} e_n$ be the unique decomposition of 1 into its homogeneous components. If $a_m \in R_m$, then $a_m = \sum_{\mathbb{Z}} e_n a_m$. Now each term on the left side of $a_m = \sum_{\mathbb{Z}} e_n a_m$ must correspond to a term on the right side of $a_m = \sum_{\mathbb{Z}} e_n a_m$ with the same degree. Since $\deg(e_n a_m) = n + m$ for each $n \in \mathbb{Z}$, it follows that the only possibility is that $a_m = e_0 a_m$. Similarly, $a_m = a_m e_0$, so e_0 acts as an identity element for R_m for each $m \in \mathbb{Z}$. But if this is the case, then e_0 acts as an identity for all elements of R and so $1 = e_0 \in R_0$.

(2) Suppose that $a_n \in R_n \cap U(R)$. If $a_n^{-1} = \sum_{\mathbb{Z}} b_m$, then $1 = a_n a_n^{-1} = \sum_{\mathbb{Z}} a_n b_m$. Now $\deg(1) = 0$, so by comparing degrees on the left and the right side of $1 = \sum_{\mathbb{Z}} a_n b_m$, it follows that it must be the case that $b_m = 0$ for all $m \neq -n$. Hence, $a_n^{-1} = b_{-n} \in R_{-n}$.

(3) If R is strongly graded, then $R_m R_n = R_{m+n}$ for all $m, n \in \mathbb{Z}$. In particular, $1 \in R_0 = R_n R_{-n}$. Conversely, suppose that $1 \in R_n R_{-n}$ for each $n \in \mathbb{Z}$. Since $R_m R_n \subseteq R_{m+n}$, it suffices to show that $R_{m+n} \subseteq R_m R_n$ for all $m, n \in \mathbb{Z}$. Now $1 \in R_m R_{-m}$ gives $R_0 = R_m R_{-m}$, so

$$R_{m+n} = R_0 R_{m+n} = (R_m R_{-m}) R_{m+n} = R_m (R_{-m} R_{m+n}) \subseteq R_m R_n. \qquad \square$$

Corollary 9.1.3. *If R is a graded ring, then R_n is a left and a right R_0-module for each $n \in \mathbb{Z}$.*

Definition 9.1.4. If R and S are graded rings, then a ring homomorphism $f : R \to S$ is a *graded ring homomorphism* if $f(R_n) \subseteq S_n$ for each $n \in \mathbb{Z}$. If R and S are graded rings, then $R \cong^{\mathrm{gr}} S$ will indicate that there is a graded ring isomorphism $f : R \to S$.

Clearly, if R, S and T are graded rings and if $f : R \to S$ and $g : S \to T$ are graded ring homomorphisms, then $gf : R \to T$ is a graded ring homomorphism. It follows that we can form the category **Gr-Ring** of graded rings whose objects are graded rings and whose morphisms are graded ring homomorphisms.

Proposition 9.1.5. *If R is a positively graded ring, then*

(1) $R_+ = \bigoplus_{n \geq 1} R_n$ *is an ideal of R and*

(2) R/R_+ *and R_0 are isomorphic rings.*

Proof. (1) R_+ is clearly closed under addition, so let $\sum_{n \geq 1} a_n \in R_+$ and suppose that $\sum_{n \geq 0} b_n \in R$. Then $(\sum_{n \geq 1} a_n) b_k = \sum_{n \geq 1} a_n b_k$ and $a_n b_k \in R_{n+k}$ for $n \geq 1$ and $k \geq 0$. Thus, $(\sum_{n \geq 1} a_n) b_k \in \bigoplus_{n \geq 1} R_{n+k} \subseteq R_+$ for each $k \geq 0$, so it follows that $(\sum_{n \geq 1} a_n)(\sum_{n \geq 0} b_n) \in R_+$. Similarly, for $(\sum_{n \geq 0} b_n)(\sum_{n \geq 1} a_n)$.

(2) By Proposition 9.1.2, R_0 is a ring and the obvious surjective ring homomorphism $R \to R_0$ has kernel R_+. $\qquad \square$

We can now prove an analog for graded rings that corresponds to parts of Propositions 1.3.3 and 1.3.5. For the proof, we will need the following lemma.

Lemma 9.1.6. *If R is a graded ring and I, $I \neq R$, is a graded ideal of R, then R/I is a graded ring with grading $\{(I + R_n)/I\}_{\mathbb{Z}}$ and the natural map $\eta : R \to R/I$ is a surjective graded ring homomorphism.*

Proof. If $\{R_n\}_{\mathbb{Z}}$ is the grading of R and if $\eta : R \to R/I$ is the canonical ring homomorphism, then $R/I = \eta(R) = \eta(\sum_{\mathbb{Z}} R_n) = \sum_{\mathbb{Z}} \eta(R_n) = \sum_{\mathbb{Z}} (I + R_n)/I$. If $\sum_{\mathbb{Z}}(a_n + I) \in \sum_{\mathbb{Z}}(I + R_n)/I$, where $a_n \in R_n$ for each $n \in \mathbb{Z}$, and $\sum_{\mathbb{Z}}(a_n + I) = 0$, then $(\sum_{\mathbb{Z}} a_n) + I = 0$. Hence, $\sum_{\mathbb{Z}} a_n \in I = \bigoplus_{\mathbb{Z}}(I \cap R_n) = \bigoplus_{\mathbb{Z}} I_n$, so it follows that $a_n \in I_n$ for each $n \in \mathbb{Z}$. Thus, $a_n + I = 0$ for each $n \in \mathbb{Z}$ and so the sum $\sum_{\mathbb{Z}}(I + R_n)/I$ is direct. Thus, R/I is graded by $\{(I + R_n)/I\}_{\mathbb{Z}}$. Since we clearly have $\eta(R_n) \subseteq (I + R_n)/I$, $\eta : R \to R/I$ is a surjective graded ring homomorphism.   □

The ring R/I with grading $\{(I + R_n)/I\}_{\mathbb{Z}}$ is referred to as a *graded factor ring* of R. Now for the proposition.

Proposition 9.1.7. *The following hold for a nonzero graded ring homomorphism $f : R \to S$.*

(1) *Im f is a graded subring of S.*

(2) *$K = \mathrm{Ker}\, f$ is a graded ideal of R.*

(3) *The induced map $\bar{f} : R/K \to \mathrm{Im}\, f$ such that $\bar{f}(a + K) = f(a)$ is a graded ring isomorphism.*

Proof. Let $\{R_n\}_{\mathbb{Z}}$ and $\{S_n\}_{\mathbb{Z}}$ be gradings of R and S, respectively.

(1) Set $T_n = \mathrm{Im}\, f \cap S_n$ for each $n \in \mathbb{Z}$. Let $f(a) \in \mathrm{Im}\, f$ and suppose that $a = \sum_{\mathbb{Z}} a_n$, where $a_n \in R_n$ for each $n \in \mathbb{Z}$. Then $f(a) = \sum_{\mathbb{Z}} f(a_n)$ and $f(a_n) \in \mathrm{Im}\, f \cap f(R_n) \subseteq \mathrm{Im}\, f \cap S_n = T_n$ for each $n \in \mathbb{Z}$. Thus, $\mathrm{Im}\, f \subseteq \sum_{\mathbb{Z}} T_n$. On the other hand, if $b \in \sum_{\mathbb{Z}} T_n$ and $b = \sum_{\mathbb{Z}} b_n$, where $b_n \in T_n = \mathrm{Im}\, f \cap S_n$, then for each n there is an $a' \in R$ such that $f(a') = b_n$. If $a' = \sum_{\mathbb{Z}} a'_n$, where $a'_n \in R_n$ for each $n \in \mathbb{Z}$, then $f(a') = \sum_{\mathbb{Z}} f(a'_n)$. But $\deg(f(a')) = n$ and $\deg(f(a'_k)) = k$ for each $k \in \mathbb{Z}$. Hence, it follows that $f(a') = f(a'_n)$, so there is an $\bar{a}_n \in R_n$ such that $f(\bar{a}_n) = b_n$ for each n. If $\bar{a} = \sum_{\mathbb{Z}} \bar{a}_n$, then $f(\bar{a}) = \sum_{\mathbb{Z}} f(\bar{a}_n) = \sum_{\mathbb{Z}} b_n = b$, so $\sum_{\mathbb{Z}} T_n \subseteq \mathrm{Im}\, f$. Thus, $\mathrm{Im}\, f = \sum_{\mathbb{Z}} T_n$.

(2) Suppose that $K_n = R_n \cap K$ for each $n \in \mathbb{Z}$. Clearly $\sum_{\mathbb{Z}} K_n \subseteq K$, and if $a \in K$, then $a \in R$, so $a = \sum_{\mathbb{Z}} a_n$, where $a_n \in R_n$ for each $n \in \mathbb{Z}$. Hence, $\sum_{\mathbb{Z}} f(a_n) = f(a) = 0$, so it follows that $f(a_n) = 0$ for each n. Thus, $a_n \in K_n$ for each n and we have $a \in \sum_{\mathbb{Z}} K_n$. Therefore, $K = \sum_{\mathbb{Z}} K_n$.

(3) We know from Lemma 9.1.6 that R/K is a graded ring with grading $\{(K + R_n)/K)\}_{\mathbb{Z}}$. The induced map $\bar{f} : M/K \to \mathrm{Im}\, f$ is an isomorphism, so we need

only show that $\bar{f}$ is graded. Since $\bar{f}((K + R_n)/K) = f(R_n) \subseteq S_n$, it follows from
the proof of (1) that $\bar{f}((K + R_n)/K) \subseteq T_n$.   □

Graded Modules

Definition 9.1.8. If R is a graded ring, then an R-module M is said to be a *graded
R-module* if there is a family $\{M_n\}_\mathbb{Z}$ of subgroups of M such that $M = \bigoplus_\mathbb{Z} M_n$
and $M_m R_n \subseteq M_{m+n}$ for all $m, n \in \mathbb{Z}$. If M is a graded R-module and if N is
a submodule of M, then N is a *graded submodule* of M, if $N = \bigoplus_\mathbb{Z} N_n$, where
$N_n = N \cap M_n$ for each $n \in \mathbb{Z}$. The additive abelian group $\mathbb{Z}$ is said to be the *set
of degrees* of M. If M is a graded R-module, then M is said to be *positively graded*
if $M_n = 0$ for all $n < 0$. Similarly, M is *negatively graded* if $M_n = 0$ for all
$n > 0$. A nonzero element $x \in M_n$ is referred to as a *homogeneous element* of M of
degree n, denoted by $\deg(x) = n$. We often write x_n for x to indicate that $x \in M_n$.

If M is a graded R-module and N is a graded submodule of M, then $(N \cap
M_m) R_n \subseteq N \cap M_{m+n}$ for all $m, n \in \mathbb{Z}$. Hence, N is a graded R-module with
grading $\{N \cap M_n\}_\mathbb{Z}$. Also, each nonzero element x of a graded R-module M can be
expressed uniquely as a finite sum $x = \sum_\mathbb{Z} x_n$ of nonzero homogeneous elements
of M and each summand is referred to as a *homogeneous component* of x. Clearly,
if M is a graded R-module, then each M_n is an R_0-module and if R has the trivial
grading, then each M_n is an R-submodule of M. It is also easy to see that if R is a
graded ring, then R is a graded R-module. Furthermore, if A is a right ideal of R,
then A is a graded right ideal of R if and only if A is a graded submodule of R_R.

Remark.

(1) If $M = \bigoplus_\mathbb{Z} M_n$ is a graded R-module, then $0 \in M_n$ for every $n \in \mathbb{Z}$, so no
   degree is assigned to 0. Moreover, the zero submodule is considered to be a
   graded submodule of every graded module. As with rings, if $x \in M$ and $x \neq 0$,
   then when x is written as a finite sum of its homogeneous components, 0 is not
   a member of this decomposition of x.

(2) If M is a graded R-module, then M has a set of *homogeneous generators*. If
   $\{x_\alpha\}_\Delta$ is a set of generators of M in $\mathbf{Mod}_R$, then x_α can be written as a finite
   sum $x_\alpha = \sum_\mathbb{Z} x_{\alpha,n}$, where $x_{\alpha,n} \in M_n$ for each $\alpha \in \Delta$ and each $n \in \mathbb{Z}$.
   If $\{x_{\alpha,n}\}_{(\alpha,n) \in \Delta \times \mathbb{Z}}$ is the set of homogeneous components of the set $\{x_\alpha\}_\Delta$ of
   generators of M, then $\{x_{\alpha,n}\}_{(\alpha,n) \in \Delta \times \mathbb{Z}}$ is a set of homogeneous generators of
   M. Furthermore, if M is finitely generated, then $\{x_{\alpha,n}\}_{(\alpha,n) \in \Delta \times \mathbb{Z}}$ contains only
   a finite number of nonzero homogeneous elements, so in this case, M is finitely
   generated by homogeneous elements.

(3) If R is a trivially graded ring, then an R-module M is said to be trivially graded
   by $\{M_n\}_\mathbb{Z}$ if $M_0 = M$ and $M_n = 0$ for each $n \in \mathbb{Z}$, $n \neq 0$. However, if R
   is not trivially graded, then M should not be viewed as trivially graded. For if

$\{R_n\}_{\mathbb{Z}}$ is a nontrivial grading of R and M is trivially graded, then for $R_k \neq 0$ we would have $MR_k = M_0 R_k \subseteq M_k = 0$. However, in general, $MR_k \neq 0$ for an R-module M.

> If M is a graded R-module, we now assume, unless stated otherwise, that the grading $\{R_n\}_{\mathbb{Z}}$ of R and the grading $\{M_n\}_{\mathbb{Z}}$ of M are nontrivial.

Proposition 9.1.9. *If M is a graded R-module and if N is a submodule of M, then the following are equivalent.*

(1) N *is a graded submodule of* M.

(2) *The homogeneous components in M of elements of N belong to N.*

(3) N *has a set of homogeneous generators.*

Proof. Suppose that N is a graded submodule of M. If $x \in N$, then we can write $x = \sum_{\mathbb{Z}} x_n \in \bigoplus_{\mathbb{Z}} M_n$, where $x_n \in M_n$ for each $n \in \mathbb{Z}$. Now $N = \sum_{\mathbb{Z}}(N \cap M_n)$, so $x = \sum_{\mathbb{Z}} y_n \in \sum_{\mathbb{Z}}(N \cap M_n)$, where $y_n \in N \cap M_n$ for each n. Thus, $\sum_{\mathbb{Z}} x_n = \sum_{\mathbb{Z}} y_n$, so it follows that $x_n = y_n \in N$ for each n, so we have that $(1) \Rightarrow (2)$. The fact that $(2) \Rightarrow (3)$ is obvious, so the proof will be complete if we can show that $(3) \Rightarrow (1)$. If $\{x_\alpha\}_\Delta$ is a set of nonzero homogeneous generators of N, then $\{x_\alpha\}_\Delta \subseteq N$ and every $x \in N$ can be expressed as $x = \sum_\Delta x_\alpha a_\alpha$, where $a_\alpha \in R$ and $a_\alpha = 0$ for almost all $\alpha \in \Delta$. For each $\alpha \in \Delta$, let $a_\alpha = \sum_{\mathbb{Z}} a_{\alpha,m} \in \bigoplus_{\mathbb{Z}} R_m$, where $a_{\alpha,m}$ is the homogeneous component of a_α of degree m. Thus, we have $x = \sum_\Delta x_\alpha a_\alpha = \sum_\Delta (\sum_{\mathbb{Z}} x_\alpha a_{\alpha,m})$. Now $\deg(x_\alpha a_{\alpha,m}) = \deg(x_\alpha) + m$, so if we collect the terms of $\sum_\Delta x_\alpha (\sum_{\mathbb{Z}} a_{\alpha,m})$ that have degree n for each $n \in \mathbb{Z}$, then

$$ x = \sum_\Delta x_\alpha \left(\sum_{\mathbb{Z}} a_{\alpha,m} \right) = \sum_{\mathbb{Z}} \left(\sum_{\Delta, \deg(x_\alpha)+m=n} x_\alpha a_{\alpha,m} \right) \in \bigoplus_{\mathbb{Z}} (N \cap M_n). $$

Hence, $N \subseteq \bigoplus_{\mathbb{Z}}(N \cap M_n)$ and so $N = \bigoplus_{\mathbb{Z}}(N \cap M_n)$. Consequently, N is a graded submodule of M. □

Definition 9.1.10. If M and N are graded R-modules and $f : M \to N$ is an R-linear mapping, then f is said to be a *graded R-module homomorphism of degree* k, if $f(M_n) \subseteq N_{n+k}$ for each $n \in \mathbb{Z}$. An injective (A surjective, A bijective) graded module homomorphism of degree zero is a *graded module monomorphism* (*graded module epimorphism, graded module isomorphism*). If M and N are graded R-modules and if there exists a graded module isomorphism $f : M \to N$ of degree zero, then we write $M \cong^{\mathrm{gr}} N$. Graded homomorphisms without an indication of degree are understood to have degree zero.

As with graded rings and graded ring homomorphisms, we can form the category $\mathbf{Gr}_R$ of graded R-modules whose objects are graded R-modules and whose morphisms are graded module homomorphisms (of degree zero). If M and N are objects in $\mathbf{Gr}_R$, then $\operatorname{Hom}_{\mathbf{Gr}_R}(M, N)$ will denote the additive abelian group of morphisms in $\mathbf{Gr}_R$ from M to N. Note that if M is an R-module with gradings $\{M_n\}_\mathbb{Z}$ and $\{M_n'\}_\mathbb{Z}$, then the graded module M with grading $\{M_n\}_\mathbb{Z}$ and the graded module M with grading $\{M_n'\}_\mathbb{Z}$ are distinct objects in $\mathbf{Gr}_R$. Note also that the definitions and the results obtained previously in this chapter hold for left R-modules. Thus, we can also form the category $_R\mathbf{Gr}$ of graded left R-modules. This observation also holds for subsequent definitions and results concerning graded R-modules when applied to graded left R-modules.

Examples

6. If M is a graded R-module and $a \in \operatorname{cent}(R)$, the center of R, is homogeneous of degree k, then $f : M \to M$ such that $f(x) = xa$ is a graded module homomorphism of degree k.

7. If $f : M \to M'$ and $g : M' \to M''$ are graded module homomorphisms with degrees k_1 and k_2, respectively, then $gf : M \to M''$ is a graded module homomorphism of degree $k_1 + k_2$.

8. Let M be a graded R-module with grading $\{M_n\}_\mathbb{Z}$ and k a fixed integer. If $M_n(k) = M_{n+k}$ for each $n \in \mathbb{Z}$, then $M = \bigoplus_\mathbb{Z} M_n(k)$ and $\{M_n(k)\}_\mathbb{Z}$ is a grading of M. The grading $\{M_n(k)\}_\mathbb{Z}$ is said to be obtained by *shifting the grading* $\{M_n\}_\mathbb{Z}$ by a factor of k.

    A graded R-module M with the grading of M shifted by a factor of k will be denoted by $M(k)$.

    If $N = \bigoplus_\mathbb{Z} N_n$ is a graded R-module, then, as above, $N(k)$ is graded by $\{N_n(k)\}_\mathbb{Z}$. Consequently, if $f : M \to N$ is a graded module homomorphism of degree k, then $f(M_n) \subseteq N_{n+k} = N_n(k)$, so $f : M \to N(k)$ is a graded R-module homomorphism of degree zero. Conversely, if $f : M \to N(k)$ is a graded module homomorphism of degree zero, then f can be considered to be a graded module homomorphism $f : M \to N$ of degree k.

9. By definition, $M \cong^{\mathrm{gr}} N$ if and only if there is a bijective graded module homomorphism $f : M \to N$. If $f : M \to N$ is a bijective graded module homomorphism of degree k, then we do not write $M \cong^{\mathrm{gr}} N$ since f shifts the grading of N by a factor of k and N and $N(k)$ are distinct objects in $\mathbf{Gr}_R$. However, it does follow that $M \cong^{\mathrm{gr}} N(k)$ since f gives a bijective graded module homomorphism $f : M \to N(k)$ of degree zero.

The proofs of the following lemma and proposition are left as exercises. The proofs are similar to the corresponding results for graded rings given in Lemma 9.1.6 and Proposition 9.1.7.

Lemma 9.1.11. *If M is a graded R-module and N is a graded submodule of M, then M/N is a graded R-module with grading $\{(N + M_n)/N\}_{\mathbb{Z}}$ and the natural map $\eta : M \to M/N$ is a graded module epimorphism.*

Proposition 9.1.12. *If $f : M \to N$ is a graded module homomorphism, then*

(1) *Im f is a graded submodule of N,*

(2) *$K = \mathrm{Ker}\, f$ is a graded submodule of M, and*

(3) *The induced map $\bar{f} : M/K \to \mathrm{Im}\, f$ such that $\bar{f}(x + K) = f(x)$ is a graded module isomorphism.*

The R-module M/N with grading $\{(N + M_n)/N\}_{\mathbb{Z}}$ is said to be a *graded factor module* of M.

Schur's lemma indicates that if S is a simple R-module, then $\mathrm{End}_R(S)$ is a division ring. There is an analog to Schur's lemma for graded simple modules. For this we need the following definition.

Definition 9.1.13. A graded nonzero R-module S is said to be a *graded simple R-module* (or a *simple object in* $\mathbf{Gr}_R$), *if the only graded submodules of S are 0 and S*. A graded ring D is said to be a *graded division ring* if every nonzero homogeneous element of D is a unit in D. If a graded division ring is commutative, then it is a *graded field*.

Example

10. If $K[X, X^{-1}]$ is the ring of Laurent polynomials of Example 4, where K is a field, then $K[X, X^{-1}]$ is a graded field, since it follows easily that each non-zero homogeneous element of $K[X, X^{-1}]$ is a unit. However, $K[X, X^{-1}]$ is not a field, so a graded field need not be a field.

If M and N are graded R-modules and if we let $\mathrm{HOM}_R(M, N)_k$ be the set of graded module homomorphisms from M to N of degree k, then $\mathrm{HOM}_R(M, N)_k$, is a subgroup of the additive abelian $\mathrm{Hom}_R(M, N)$. Furthermore,

$$\mathrm{HOM}_R(M, N)_0 = \mathrm{Hom}_{\mathbf{Gr}_R}(M, N) \quad \text{and}$$

$$\mathrm{HOM}_R(M, N)_k = \mathrm{HOM}_R(M, N(k))_0 = \mathrm{Hom}_{\mathbf{Gr}_R}(M, N(k))$$

$$= \mathrm{HOM}_R(M(-k), N)_0 = \mathrm{Hom}_{\mathbf{Gr}_R}(M(-k), N).$$

Next, let $\mathrm{HOM}_R(M, N) = \bigoplus_{\mathbb{Z}} \mathrm{HOM}_R(M, N)_k$. Then $\mathrm{HOM}_R(M, N)$ is a graded $\mathbb{Z}$-module with grading $\{\mathrm{HOM}_R(M, N)_k\}_{\mathbb{Z}}$ and $\mathrm{HOM}_R(M, N)$ is a subgroup of

$\text{Hom}_R(M, N)$. Moreover, $\text{END}_R(M) = \bigoplus_{\mathbb{Z}} \text{END}_R(M)_k$ is a graded ring and $\text{END}_R(M)$ is a subring of $\text{End}_R(M)$.

The following lemma is a restatement of Schur's lemma (Lemma 6.4.13) in the setting of graded rings and modules. Note that if S is a graded simple R-module, then so is $S(k)$ for each $k \in \mathbb{Z}$. (See Exercise 9.)

Lemma 9.1.14 (Schur's lemma for graded modules). *Let M be a nonzero graded R-module and suppose that S is a graded simple R-module.*

(1) *If $f : S \to M$ is a nonzero graded module homomorphism, then f is a monomorphism.*

(2) *If $f : M \to S$ is a nonzero graded module homomorphism, then f is an epimorphism.*

(3) $\text{END}_R(S) = \bigoplus_{\mathbb{Z}} \text{END}_R(S)_k$ *is a graded division ring.*

Proof. If $f : S \to M$ is a nonzero graded module homomorphism, then $\text{Ker } f \ne S$ and, by (2) of the previous proposition, $\text{Ker } f$ is a graded submodule of S. Hence, $\text{Ker } f = 0$, so f is a monomorphism and therefore (1) holds. If $f : M \to S$ is a nonzero graded module homomorphism, then $f(M)$ is, by (1) of the proposition above, a nonzero graded submodule of S. Hence, $f(M) = S$ and we have (2). To prove (3), it suffices to show that each nonzero element of $\text{END}_R(S)_k$ has an inverse in $\text{END}_R(S)_{-k}$. If $f \in \text{END}_R(S)_k$, $f \ne 0$, then by shifting the grading of S by a factor of k gives a graded R-module homomorphism $f : S \to S(k)$ of degree zero. So since $S(k)$ is also a graded simple R-module, it follows from (1) and (2) that $f : S \to S(k)$ is a graded isomorphism. This gives a graded module isomorphism, $f^{-1} : S(k) \to S$ such that $f^{-1}f = \text{id}_S$ and $ff^{-1} = \text{id}_{S(k)}$. But f^{-1} can be viewed as a map $S \to S$ of degree $-k$, so f has an inverse in $\text{END}_R(S)_{-k}$. $\quad\square$

Definition 9.1.15. If M is a graded R-module, then a graded submodule N of M is said to be a *maximal graded submodule* of M, if M/N is a graded simple module. A right ideal of a graded ring R is said to be *maximal graded right ideal* if it is a maximal graded submodule of R_R.

The proof of the following lemma is an exercise.

Lemma 9.1.16. *The following hold for a graded R-module M and a graded submodule N of M.*

(1) *N is a maximal graded submodule of M if and only if $N \ne M$ and $N + xR = M$ for each homogeneous element $x \in M - N$.*

(2) *If N is a maximal graded submodule of M, then N is a maximal element among the graded submodules of M.*

Previously we have seen that S is a simple R-module if and only if there is a maximal right ideal $\mathfrak{m}$ in R such that $R/\mathfrak{m} \cong S$. The following proposition links maximal graded right ideals in a graded ring R to graded simple R-modules.

Proposition 9.1.17. *The following hold for a graded simple R-module S.*

(1) *For each $n \in \mathbb{Z}$ such that $S_n \neq 0$, S_n is a simple R_0-module.*

(2) *There is at least one integer k such that $R/\mathfrak{m}_k \cong^{\mathrm{gr}} S(k)$, where $\mathfrak{m}_k$ is a maximal graded right ideal of R.*

Proof. (1) Suppose that $S_n \neq 0$ and let $x_n \in S_n$, $x_n \neq 0$. Then $x_n R$ is a nonzero submodule of S with a homogeneous generator. Thus, Proposition 9.1.9 shows that $x_n R$ is a graded submodule of S, so $x_n R = S$. If $y_n \in S_n$, $y_n \neq 0$, then $y_n \in S = x_n R$. Consequently, if $y_n = x_n a = x_n \sum_{\mathbb{Z}} a_k = \sum_{\mathbb{Z}} x_n a_k$, where each a_k is in R_k and $a_k = 0$ for almost all $k \in \mathbb{Z}$, then, by comparing degrees on the left and the right side of $y_n = \sum x_n a_k$, we see that $x_n a_k = 0$ if $k \neq 0$. Hence, $y_n = x_n a_0$. Therefore, $S_n \subseteq x_n R_0$, so $x_n R_0 = S_n$. Finally, if N is a nonzero R_0-submodule of S_n and $x_n \in N$, $x_n \neq 0$, then $x_n \in S_n$ and as above $x_n R_0 = S_n$. Therefore, if $y_n \in S_n$, then $y_n = x_n a_0$ for some $a_0 \in R_0$, so $y_n \in N$. Hence, $S_n \subseteq N$ and so $N = S_n$. Thus, S_n is a simple R_0-module.
(2) Let $x \in S$, $x \neq 0$, and suppose that $\deg(x) = k$. Then $f : R \to S$ defined by $f(a) = xa$ is such that $\mathrm{Im}(f) = xR = S$. Moreover, $f(R_n) \subseteq S_{n+k}$, so f is a nonzero graded module homomorphism of degree k. Shifting the grading of S by a factor of k gives a nonzero graded module homomorphism $f : R \to S(k)$ of degree zero. Hence, Proposition 9.1.12 indicates that $\mathfrak{m}_k = \mathrm{Ker}\, f$ is a graded right ideal of R. Using Proposition 9.1.12 again, we see that $\bar{f} : R/\mathfrak{m}_k \to S(k)$ such that $\bar{f}(a + \mathfrak{m}_k) = f(a)$ is a graded isomorphism. Hence, $R/\mathfrak{m}_k \cong^{\mathrm{gr}} S(k)$ and since $S(k)$ is a graded simple R-module, $\mathfrak{m}_k$ is a maximal graded right ideal of R. $\square$

Corollary 9.1.18. *If S is a graded simple R-module, then S is a semisimple R_0-module.*

Example

11. A submodule of a graded R-module of M may be a maximal graded submodule of M and yet not be a maximal submodule of M. For example, if $R = K[X, X^{-1}]$ is the graded ring of Example 10, then the zero submodule of the module R_R is a maximal graded submodule of R_R and yet zero is not a maximal submodule of R_R.

Problem Set 9.1

1. If R is a graded ring and $U^{gr}(R) = \bigcup_{\mathbb{Z}}(U(R) \cap R_n)$, then $U^{gr}(R)$ is the set of units of R that are homogeneous. Prove each of the following.

   (a) $U^{gr}(R)$ is a subgroup of $U(R)$.

   (b) The map $d : U^{gr}(R) \to \mathbb{Z}$ defined by $d(x) = \deg(x)$ is a group homomorphism from the multiplicative group $U^{gr}(R)$ to the additive group $\mathbb{Z}$. Moreover, $\mathrm{Ker}(d) = U(R_0)$.

   (c) d is an epimorphism if and only if $U(R) \cap R_n \neq 0$ for each $n \in \mathbb{Z}$.

2. Let R be a graded ring graded by $\{R_n\}_{\mathbb{Z}}$. Prove that $\{R_n[X]\}_{\mathbb{Z}}$ is a grading of the polynomial ring $R[X]$. Show also that if R is strongly graded, then $R[X]$ is strongly graded by $\{R_n[X]\}_{\mathbb{Z}}$.

3. Show that the only possible grading of the ring $\mathbb{Z}$ is the trivial grading. [Hint: Show that $\deg(1) = 0$.]

4. Let M be an (R, S)-bimodule and N an (S, R)-bimodule and consider $\mathbb{M} = \left(\begin{smallmatrix} R & M \\ N & S \end{smallmatrix}\right)$. Then $\mathbb{M}$ is a ring, if addition of matrices is defined in the usual fashion and multiplication is defined by

$$\begin{pmatrix} r_1 & m_1 \\ n_1 & s_1 \end{pmatrix}\begin{pmatrix} r_2 & m_2 \\ n_2 & s_2 \end{pmatrix} = \begin{pmatrix} r_1 r_2 & r_1 m_2 + m_1 s_2 \\ n_1 r_2 + s_1 n_2 & s_1 s_2 \end{pmatrix}.$$

   (a) Show that $\mathbb{M}$ is a graded ring with grading

$$\mathbb{M}_{-1} = \begin{pmatrix} 0 & 0 \\ N & 0 \end{pmatrix}, \quad \mathbb{M}_0 = \begin{pmatrix} R & 0 \\ 0 & S \end{pmatrix}, \quad \mathbb{M}_1 = \begin{pmatrix} 0 & M \\ 0 & 0 \end{pmatrix} \quad \text{and}$$

$$\mathbb{M}_n = \begin{pmatrix} 0 & 0 \\ 0 & 0 \end{pmatrix} \quad \text{if } n \neq -1, 0, 1.$$

   (b) Prove that $\left(\begin{smallmatrix} A & M' \\ N' & B \end{smallmatrix}\right)$ is a graded right ideal of $\mathbb{M}$, where M' is an R-submodule of M, N' is an S-submodule of N and A and B are right ideals of R and S, respectively, such that $AM \subseteq M'$ and $BN \subseteq N'$.

5. (a) Verify Examples 2, 4 and 5.

   (b) Verify Examples 8 and 9.

6. Suppose that N is a graded R-module, that L is a graded submodule of M and that M is a graded submodule of N. Is L a graded submodule of N? That is, decide if "graded submodule" is a transitive concept.

7. Prove Lemma 9.1.11. [Hint: Lemma 9.1.6.]

8. Prove Proposition 9.1.12. [Hint: Proposition 9.1.7.]

9. (a) If S is a graded simple R-module, prove that $S(k)$ is graded simple for each $k \in \mathbb{Z}$.

   (b) If D is a division ring and if $D[X, X^{-1}]$ is graded by $\{X^n D\}_{\mathbb{Z}}$, then is $D[X, X^{-1}]$ a graded division ring?

   (c) If R is a graded division ring, prove that R has no divisors of zero and that R_0 is a field.

10. Prove that the following are equivalent for a graded ring R.

   (a) R has no proper nonzero graded right ideals.

   (b) R has no proper nonzero graded left ideals.

   (c) R is a graded division ring.

11. (a) If K is a field with grading $K = \bigoplus_{\mathbb{Z}} K_n$, prove that $\{K_n\}_{\mathbb{Z}}$ is the trivial grading of K. [Hint: Suppose that $\{K_n\}_{\mathbb{Z}}$ is a nontrivial grading of K and let $x \in K_n$, $x \neq 0$ and $n > 0$. Consider $y = (1+x)^{-1} = \sum_{\mathbb{Z}} y_k$, where $\sum_{\mathbb{Z}} y_k$ is the homogeneous decomposition of y, and show that this gives a contradiction.]

   (b) Show that $K[X, X^{-1}]$ is a graded field, but that $K[X, X^{-1}]$ is not a field.

   Conclude from (a) and (b) that, in general, a graded field and a field that is graded are distinct graded objects.

12. Prove Lemma 9.1.16.

13. Let M be a graded R-module and suppose that N is a not necessarily graded submodule of M. Prove that $\sum_{\mathbb{Z}} (N \cap M_n)$ is a graded submodule of M.

14. Let $f : R \to S$ be a ring homomorphism and view S as an R-module by pullback along f. Prove that if R and S are graded rings, then $f : R \to S$ is a graded ring homomorphism if and only if the grading of S is an R-module grading.

15. (a) Let $f : M \to N$ be a morphism in $\mathbf{Gr}_R$ and suppose that X is a graded R-module. If

$$\mathrm{Hom}_{\mathbf{Gr}_R}(f, X) = f^* : \mathrm{Hom}_{\mathbf{Gr}_R}(N, X) \to \mathrm{Hom}_{\mathbf{Gr}_R}(M, X)$$

   is such that $f^*(g) = gf$, show that $\mathrm{Hom}_{\mathbf{Gr}_R}(-, X)$ is a left exact contravariant functor from $\mathbf{Gr}_R$ to $\mathbf{Ab}$. Prove also that if

$$\mathrm{Hom}_{\mathbf{Gr}_R}(X, f) = f_* : \mathrm{Hom}_{\mathbf{Gr}_R}(X, M) \to \mathrm{Hom}_{\mathbf{Gr}_R}(X, N)$$

   and $f_*(g) = fg$, then $\mathrm{Hom}_{\mathbf{Gr}_R}(X, -)$ is a left exact covariant functor from $\mathbf{Gr}_R$ to $\mathbf{Ab}$.

   (b) Define $\mathrm{HOM}_R(-, X)$ and $\mathrm{HOM}_R(X, -)$ as in (a) and (b) for a graded R-module X and a morphism $f : M \to N$ in $\mathbf{Gr}_R$. Show that $\mathrm{HOM}_R(-, X)$ and $\mathrm{HOM}_R(X, -)$ are left exact contravariant and covariant functors, respectively, from $\mathbf{Gr}_R$ to $\mathbf{Gr}_{\mathbb{Z}}$.

16. (a) If $\{N_\alpha\}_\Delta$ is a family of graded submodules of M, prove that $\bigcap_\Delta N_\alpha$ and $\sum_\Delta N_\alpha$ are graded submodules of M. [Hint: If $\{N_{\alpha,n}\}_\mathbb{Z}$ is the grading of N_α for each $\alpha \in \Delta$, consider $(\bigcap_\Delta N_\alpha)_n = \bigcap_\Delta N_{\alpha,n}$ and $(\sum_\Delta N_\alpha)_n = \sum_\Delta N_{\alpha,n}$, respectively, for each $n \in \mathbb{Z}$.]

(b) If x is a homogeneous element of a graded R-module M, show that $\mathrm{ann}_r(x)$ is a graded right ideal of R.

(c) If N is a graded submodule of a graded R-module M, prove that $(N : M) = \{a \in R \mid Ma \subseteq N\}$ is a graded ideal of R.

(d) If N is a graded submodule of a graded R-module M, prove that $\mathrm{ann}_r(N)$ is a graded ideal of R.

17. Let M be a graded R-module and suppose that N is a submodule of M, N not necessarily graded.

(a) Prove that there is a largest graded submodule N^{gr} of M such that $N^{\mathrm{gr}} \subseteq N$. [Hint: Use (a) of Exercise 16.]

(b) Prove that there is a smallest graded submodule N_{gr} of M such that $N \subseteq N_{\mathrm{gr}}$. [Hint: Use (a) of Exercise 16.]

Conclude that N is graded if and only if $N^{\mathrm{gr}} = N = N_{\mathrm{gr}}$.

18. (a) Let R be a graded ring and suppose that a is a homogeneous element of R. Prove that $C_a(R) = \{b \in R \mid ab = ba\}$ is a graded subring of R.

(b) Let R be a graded ring and suppose that S is a graded subring of R. Show that $\mathrm{cent}(S)$, the center of S, is a graded subring of R.

19. If M is a graded R-module and I is a graded ideal of R, prove that MI is a graded submodule of M and that M/MI is a graded R/I-module.

20. If $\{M_\alpha\}_\Delta$ is a family of graded R-modules, decide which, if any, of the following hold.

(a)
$$\mathrm{Hom}_{\mathbf{Gr}R}\left(\bigoplus_\Delta M_\alpha, N\right) \cong \prod_\Delta \mathrm{Hom}_{\mathbf{Gr}R}(M_\alpha, N)$$

(b)
$$\mathrm{Hom}_{\mathbf{Gr}R}\left(N, \prod_\Delta M_\alpha\right) \cong \prod_\Delta \mathrm{Hom}_{\mathbf{Gr}R}(N, M_\alpha)$$

(c)
$$\mathrm{HOM}_R\left(\bigoplus_\Delta M_\alpha, N\right) \cong \prod_\Delta \mathrm{HOM}_R(M_\alpha, N)$$

(d)
$$\mathrm{HOM}_R\left(N, \prod_\Delta M_\alpha\right) \cong \prod_\Delta \mathrm{HOM}_R(N, M_\alpha)$$

for a graded R-module N. [Hint: Proposition 2.1.12.]

21. **Nakayama's Graded Lemma.** Let R be a positively graded ring. Prove that each of the following hold for a finitely generated positively graded R-module M. [Hint: Lemma 6.1.10 and Proposition 9.1.5.]

(a) If $MR_+ = M$, then $M = 0$.

(b) If $N + MR_+ = M$ for a graded submodule N of M, then $N = M$.

22. Recall that if S is a multiplicative system in R, then S is a right denominator set if the following two conditions are satisfied.

D$_1$ If $(a, s) \in R \times S$, then there is a $(b, t) \in R \times S$ such that $at = sb$.

D$_2$ If $(a, s) \in R \times S$ is such that $sa = 0$, then there is a $t \in S$ such that $at = 0$.

If R is a graded ring, let $h(R)$ denote the set of homogeneous elements of R. Suppose also that $S \subseteq h(R)$ is a multiplicatively system that satisfies the following two conditions.

D$_1^{gr}$ If $(a, s) \in h(R) \times S$, then there is a $(b, t) \in h(R) \times S$ such that $at = sb$.

D$_2^{gr}$ If $(a, s) \in h(R) \times S$ is such that $sa = 0$, then there is a $t \in S$ such that $at = 0$.

Prove that if R is a graded ring and $S \subseteq h(R)$ is a multiplicative system in R, then D$_1$ and D$_2$ hold if and only if D$_1^{gr}$ and D$_2^{gr}$ hold. Conclude that S is a right denominator set in R if and only if D$_1^{gr}$ and D$_2^{gr}$ are satisfied.

23. Let R be a graded ring and suppose that $S \subseteq h(R)$ is a right denominator set in R.

(a) Prove that $(RS^{-1})_n = \{a/s \mid a \in h(R) \text{ and } \deg(a) - \deg(s) = n\}$ is a subgroup of the additive group of RS^{-1} for each $n \in \mathbb{Z}$ and that RS^{-1} is a graded ring with grading $\{(RS^{-1})_n\}_{\mathbb{Z}}$.[Hint: If a/s and b/t are in $(RS^{-1})_n$, then by Exercise 22 there is a pair $(c, u) \in h(R) \times S$ such that $a/s + b/t = (au + bc)/su$, where $su = tc$. Thus, $\deg(a/s + b/t) = \deg(au + bc)/su$. Now show that $\deg(au) = n + \deg(su)$ and $\deg(bc) = n + \deg(tc)$.]

(b) Let $(MS^{-1})_n = \{x/s \mid x \in h(M) \text{ and } \deg(x) - \deg(s) = n\}$ for each $n \in \mathbb{Z}$, where M is a graded R-module and $h(M)$ is the set of homogeneous elements of M. Show that each $(MS^{-1})_n$ is a subgroup of the additive group of MS^{-1} and that MS^{-1} is a graded RS^{-1}-module with grading $\{(MS^{-1})_n\}_{\mathbb{Z}}$.[Hint: Similar to the proof of (a).] If MS^{-1} is viewed as an R-module by pullback along the canonical map $a \mapsto a/1$ from R to RS^{-1}, then is MS^{-1} graded as an R-module?

9.2   Fundamental Concepts

In Chapter 2 we gave several fundamental constructions for rings and modules. In this section these same constructions are reformulated in the setting of graded rings and modules.

Graded Direct Products and Sums

Suppose that $\{M_\alpha\}_\Delta$ is a family of graded R-modules and let $\{M_{\alpha,n}\}_\mathbb{Z}$ be the grading of M_α for each $\alpha \in \Delta$. If $P_n = \prod_\Delta M_{\alpha,n}$, then P_n is a subgroup of $\prod_\Delta M_\alpha$ for each $n \in \mathbb{Z}$ and

$$\prod_\Delta M_\alpha = \prod_\Delta \bigoplus_\mathbb{Z} M_{\alpha,n} = \bigoplus_\mathbb{Z} \prod_\Delta M_{\alpha,n} = \bigoplus_\mathbb{Z} P_n.$$

Hence, $\prod_\Delta M_\alpha$ is a graded R-module graded by $\{P_n\}_\mathbb{Z}$. Now let $S_n = \bigoplus_\Delta M_{\alpha,n}$. Then

$$\bigoplus_\Delta M_\alpha = \bigoplus_\Delta \bigoplus_\mathbb{Z} M_{\alpha,n} = \bigoplus_\mathbb{Z} \bigoplus_\Delta M_{\alpha,n} = \bigoplus_\mathbb{Z} S_n,$$

so $\bigoplus_\Delta M_\alpha$ is a graded R-module graded by the subgroups $\{S_n\}_\mathbb{Z}$. These observations lead to the following definition.

Definition 9.2.1. If $\{M_\alpha\}_\Delta$ is a family of graded R-modules, then the R-module $\prod_\Delta M_\alpha$, graded by the family of subgroups $\{P_n\}_\mathbb{Z}$, where $P_n = \prod_\Delta M_{\alpha,n}$ for each $n \in \mathbb{Z}$, is said to be the *graded direct product* (or a *direct product in $\mathbf{Gr}_R$*) of the family $\{M_\alpha\}_\Delta$. The notation $\prod_\Delta^{\mathrm{gr}} M_\alpha$ will indicate that the R-module $\prod_\Delta M_\alpha$ is being considered as a module in $\mathbf{Gr}_R$ with the grading $\{P_n\}_\mathbb{Z}$. Likewise, $\bigoplus_\Delta^{\mathrm{gr}} M_\alpha$ indicates that $\bigoplus_\Delta M_\alpha$ is being viewed as a module in $\mathbf{Gr}_R$ with grading $\{S_n\}_\mathbb{Z}$, where $S_n = \bigoplus_\Delta M_{\alpha,n}$ for each $n \in \mathbb{Z}$. The graded module $\bigoplus_\Delta^{\mathrm{gr}} M_\alpha$ will be referred to as the *graded direct sum* (or a *direct sum in $\mathbf{Gr}_R$*) of $\{M_\alpha\}_\Delta$. Finally, if M is a graded R-module, then a graded submodule N of M is said to be a *graded direct summand* of M (or a *direct summand of M in $\mathbf{Gr}_R$*) if there is a graded submodule N' of M such that $M = N \oplus^{\mathrm{gr}} N'$. Note that if $\{N_n\}_\mathbb{Z}$ is the grading of N and $\{N'_n\}_\mathbb{Z}$ is the grading of N', then M is graded by $\{N_n \oplus N'_n\}_\mathbb{Z}$.

Remark.

(1) Let $\mathcal{U} : \mathbf{Gr}_R \to \mathbf{Mod}_R$ be such that $\mathcal{U}(M) = \underline{M}$, where $\underline{M}$ is the R-module M with the grading of M forgotten. If $f : M \to N$ is a morphism in $\mathbf{Gr}_R$, then $\mathcal{U}(f) = \underline{f}$, where $\underline{f} : \underline{M} \to \underline{N}$ is such that $\underline{f}(x) = f(x)$ for each $x \in M$. It's easy to show that $\mathcal{U}$ is a (forgetful) functor, called the *ungrading functor*. In what follows, if M is a graded R-module, then $\underline{M}$ will indicate that the grading of M has been forgotten. Similarly, if $f : M \to N$ is a a graded module homomorphism, then $\underline{f} : \underline{M} \to \underline{N}$ will be as above.

(2) If R is a graded ring, then we will often write R rather than $\underline{R}$, leaving it to the reader to determine whether R is being considered as a graded or as an ungraded ring. For example, if M is a graded R-module, then we will say that $\underline{M}$ is an

R-module rather than $\underline{M}$ is an $\underline{R}$-module. Moreover, if R is a graded ring then $\mathbf{Mod}_R$ will continue to denote the category of R-modules as opposed to $\mathbf{Mod}_{\underline{R}}$.

(3) The notation $M = N \oplus^{\mathrm{gr}} N'$ will indicate that the direct sum is taking place in $\mathbf{Gr}_R$ while $\underline{M} = \underline{N} \oplus \underline{N}'$ will mean that the direct sum is formed in $\mathbf{Mod}_{\underline{R}}$.

The following proposition points out the connection between direct summands of $\mathbf{Gr}_R$ and those in $\mathbf{Mod}_{\underline{R}}$.

Proposition 9.2.2. *If M is a graded R-module and N is a graded submodule of M, then N is a direct summand of M in $\mathbf{Gr}_R$ if and only $\underline{N}$ is a direct summand of $\underline{M}$ in $\mathbf{Mod}_{\underline{R}}$.*

Proof. If N is a direct summand of M in $\mathbf{Gr}_R$, then there is a graded submodule N' of M such that $M = N \oplus^{\mathrm{gr}} N'$. It follows from the definition of a graded direct sum that $\underline{M} = \underline{N} \oplus \underline{N}'$. Conversely, suppose that N is a graded submodule of a graded R-module M such that there is a submodule X of $\underline{M}$ such that $\underline{M} = \underline{N} \oplus X$. We claim that there is a graded submodule N' of M such that $M = N \oplus^{\mathrm{gr}} N'$. If $\{N_n = N \cap M_n\}_{\mathbb{Z}}$ is the grading of N, let Y_n be the subgroup $\bigoplus_{k \neq n} N_k$ and set $N'_n = (Y_n + X) \cap M_n$ for each $n \in \mathbb{Z}$. Then as groups $M = N_n \oplus Y_n \oplus X$ and $N_n \subseteq M_n$. The modular law of Example 10 in Section 1.4 gives

$$M_n = M \cap M_n = (N_n \oplus Y_n \oplus X) \cap M_n$$
$$= N_n \oplus ((Y_n \oplus X) \cap M_n)$$
$$= N_n \oplus N'_n.$$

Thus, if we set $N' = \bigoplus_{\mathbb{Z}} N'_n$, then $\underline{M} = \underline{N} \oplus \underline{N}'$ in $\mathbf{Mod}_{\mathbb{Z}}$. Moreover, $Y_m R_n \subseteq Y_{m+n}$, so $N'_m R_n \subseteq N'_{m+n}$. It follows that N' is a submodule of M that is in fact a graded submodule of M such that $M = N \oplus^{\mathrm{gr}} N'$.   □

Corollary 9.2.3. *A short exact sequence $0 \rightarrow M_1 \xrightarrow{f} M \xrightarrow{g} M_2 \rightarrow 0$ in $\mathbf{Gr}_R$ splits if and only if one of the following two equivalent conditions holds.*

(1) *There is a graded R-module homomorphism $f' : M \rightarrow M_1$ such that $f'f = \mathrm{id}_{M_1}$.*

(2) *There is a graded R-module homomorphism $g' : M_2 \rightarrow M$ such that $gg' = \mathrm{id}_{M_2}$.*

Corollary 9.2.4. *A short exact sequence $0 \rightarrow M_1 \xrightarrow{f} M \xrightarrow{g} M_2 \rightarrow 0$ in $\mathbf{Gr}_R$ splits in $\mathbf{Gr}_R$ if and only if one of the following three equivalent conditions holds.*

(1) *Im f is a graded direct summand of M.*

(2) *Ker g is a graded direct summand of M.*

(3) *$M \cong^{\mathrm{gr}} M_1 \oplus^{\mathrm{gr}} M_2$.*

Graded Tensor Products

Let M be a graded R-module with grading $\{M_i\}_{\mathbb{Z}}$ and N a graded left R-module with grading $\{N_j\}_{\mathbb{Z}}$. Consider the abelian group $M \otimes_R N$. Since each M_i and each N_j is a $\mathbb{Z}$-module and since Corollary 2.3.8 shows that tensor products and direct sums commute, it follows that as a $\mathbb{Z}$-module we have

$$M \otimes_{\mathbb{R}} N = \left(\bigoplus_{i\in\mathbb{Z}} M_i\right) \otimes_{\mathbb{Z}} \left(\bigoplus_{j\in\mathbb{Z}} N_j\right) \cong \bigoplus_{i\in\mathbb{Z}} \left(M_i \otimes_{\mathbb{Z}} \left(\bigoplus_{j\in\mathbb{Z}} N_j\right)\right)$$

$$\cong \bigoplus_{i\in\mathbb{Z}} \bigoplus_{j\in\mathbb{Z}} (M_i \otimes_{\mathbb{Z}} N_j) \cong \bigoplus_{n\in\mathbb{Z}} \left[\bigoplus_{i+j=n} M_i \otimes_{\mathbb{Z}} N_j\right]$$

$$= \bigoplus_{\mathbb{Z}} (M \otimes_{\mathbb{Z}} N)_n,$$

where $(M \otimes_{\mathbb{Z}} N)_n = \bigoplus_{i+j=n} M_i \otimes_{\mathbb{Z}} N_j$ for each $n \in \mathbb{Z}$. Hence, if the ring $\mathbb{Z}$ is given the trivial grading, then $M \otimes_{\mathbb{R}} N$ is a graded $\mathbb{Z}$-module with grading $\{(M \otimes_{\mathbb{Z}} N)_n\}_{\mathbb{Z}}$. (Note that we have seen in Exercise 3 in Problem Set 9.1 that the only possible grading of the ring $\mathbb{Z}$ is the trivial grading.)

Definition 9.2.5. If M is a graded R-module and N is a graded left R-module, then the object $M \otimes_R N$ in $\mathbf{Gr}_{\mathbb{Z}}$ with grading $\{(M \otimes_{\mathbb{Z}} N)_n\}_{\mathbb{Z}}$ will be referred to as the *graded tensor product* of M and N (or the *tensor product of M and N in $\mathbf{Gr}_R$*) and denoted by $M \otimes_R^{\mathrm{gr}} N$.

If $f : M \to N$ is a morphism in $\mathbf{Gr}_R$ and $g : M' \to N'$ is a morphism in $_R\mathbf{Gr}$, then Proposition 2.3.5 shows that there is a unique group homomorphism $f \otimes g : M \otimes_R N \to M' \otimes_R N'$ defined on generators by $(f \otimes g)(x \otimes y) = f(x) \otimes g(y)$. We claim that $f \otimes g$ is a morphism in $\mathbf{Gr}_{\mathbb{Z}}$. This follows easily since $f(M_i) \subseteq M'_i$ and $g(N_j) \subseteq N'_j$ gives $(f \otimes g)(M_i \otimes_{\mathbb{Z}} N_j) \subseteq M'_i \otimes_{\mathbb{Z}} N'_j$. So if $i + j = n$, then it follows that $(f \otimes g)((M \otimes_{\mathbb{Z}} N)_n) \subseteq (M' \otimes_{\mathbb{Z}} N')_n$ and we have that $f \otimes g$ is a morphism in $\mathbf{Gr}_{\mathbb{Z}}$.

Proposition 9.2.6. *If M is a graded left R-module, then $R \otimes_R^{\mathrm{gr}} M \cong^{\mathrm{gr}} M$ and if M is a graded R-module, then $M \otimes_R^{\mathrm{gr}} R \cong^{\mathrm{gr}} M$.*

Proof. If $f : R \otimes_R M \to M$ is such that $f(a \otimes x) = ax$, then we have seen in the proof of Proposition 2.3.4 that f is a well-defined R-linear isomorphism. Hence, $R \otimes_R M \cong M$ in $_R\mathbf{Mod}$. If R_i and M_j are subgroups in the grading of R and M, respectively, then $f(R_i \otimes_{\mathbb{Z}} M_j) \subseteq R_i M_j \subseteq M_{i+j}$. If $i + j = n$, then it follows that $f((R \otimes_{\mathbb{Z}} M)_n) \subseteq M_n$, so f is a graded isomorphism. Therefore, $R \otimes_R^{\mathrm{gr}} M \cong^{\mathrm{gr}} M$. Similarly, $M \otimes_R^{\mathrm{gr}} R \cong^{\mathrm{gr}} M$. $\square$

The following proposition follows from Proposition 3.3.4. The proof is left as an exercise.

Proposition 9.2.7. *If* $M_1 \xrightarrow{f} M \xrightarrow{g} M_2 \to 0$ *is an exact sequence in* $\mathbf{Gr}_R$*, then for any graded left R-module X, the sequence*

(1) $$M_1 \otimes_R^{gr} X \xrightarrow{f \otimes \mathrm{id}_X} M \otimes_R^{gr} X \xrightarrow{g \otimes \mathrm{id}_X} M_2 \otimes_R^{gr} X \to 0$$

is exact in $\mathbf{Gr}_{\mathbb{Z}}$*. Similarly, if* $M_1 \xrightarrow{f} M \xrightarrow{g} M_2 \to 0$ *is an exact sequence in* $_R\mathbf{Gr}$*, then for any graded R-module X, the sequence*

(2) $$X \otimes_R^{gr} M_1 \xrightarrow{\mathrm{id}_X \otimes f} X \otimes_R^{gr} M \xrightarrow{\mathrm{id}_X \otimes g} X \otimes_R^{gr} M_2 \to 0$$

is exact in $\mathbf{Gr}_{\mathbb{Z}}$*.*

Graded Free Modules

Definition 9.2.8. A graded R-module F is called a *graded free module* (or a *free module in* $\mathbf{Gr}_R$) if F has a basis consisting of homogeneous elements.

Example

1. Let $\{k_\alpha\}_\Delta$ be a set of distinct integers, suppose that $R = \bigoplus_{\mathbb{Z}} R_n$ is a graded ring and let $R'_n = R_n(-k_\alpha) = R_{n-k_\alpha}$ for each $\alpha \in \Delta$. Then $R(-k_\alpha) = \bigoplus_{\mathbb{Z}} R'_n = R$ for each $\alpha \in \Delta$ and $a \in R'_n$ is such that $\deg(a) = n$ with respect to the grading $\{R'_n\}_{\mathbb{Z}}$ of R. Thus, $1 \in R_0 = R'_{k_\alpha}$, so $\deg(1) = k_\alpha$ in this context and we write 1_{k_α} to indicate that $1 \in R'_{k_\alpha}$. Now $R_n = 1_{k_\alpha} R'_n$ and so we have $R(-k_\alpha) = R = \bigoplus_{\mathbb{Z}} R_n = \bigoplus_{\mathbb{Z}} 1_{k_\alpha} R'_n = 1_{k_\alpha} \bigoplus_{\mathbb{Z}} R'_n = 1_{k_\alpha} R$ for each $\alpha \in \Delta$. Therefore, $\bigoplus_\Delta^{gr} R(-k_\alpha) = \bigoplus_\Delta 1_{k_\alpha} R$, so $\{1_{k_\alpha}\}_\Delta$ is a homogeneous basis for $\bigoplus_\Delta^{gr} R(-k_\alpha)$. We will refer to $\{1_{k_\alpha}\}_\Delta$ as the *canonical basis* for the free R-module $\bigoplus_\Delta^{gr} R(-k_\alpha)$.

Remark. A graded free R-module is clearly a free R-module. However, the converse need not hold. For example, form the ring direct product $R = \mathbb{Z} \times \mathbb{Z}$ and let R have the trivial grading. Consider the R-module $M = \mathbb{Z} \times \mathbb{Z}$ with grading $M_0 = \mathbb{Z} \times 0$, $M_1 = 0 \times \mathbb{Z}$ and $M_n = 0$ if $n \neq 0, 1$. Then $\underline{M}$ is a free R-module with basis $\{(1, 1)\}$ but M is not graded free. For this, suppose that $\{(x, 0), (0, y)\}$ is a homogeneous basis for M. Then for any $(0, a), (b, 0) \in R$, with $(a, b) \neq (0, 0)$, we have $(x, 0)(0, a) + (0, y)(b, 0) = (0, 0)$, a contradiction. Thus, M cannot have a homogeneous basis and so M is not a graded free R-module.

We know that every R-module is the homomorphic image of a free R-module and that F is a free R-module if and only if there is a set Δ such that $F \cong \bigoplus_\Delta R_\alpha$, where R_α is a copy of R for each $\alpha \in \Delta$. Similar results hold for graded modules.

Proposition 9.2.9. *The following hold in* $\mathbf{Gr}_R$.

(1) $F = \bigoplus_{\mathbb{Z}} F_n$ *is a graded free R-module if and only if there is a set* $\{k_\alpha\}_\Delta$ *of integers such that* $\bigoplus_\Delta^{\mathrm{gr}} R(-k_\alpha) \cong^{\mathrm{gr}} F$.

(2) *Every graded R-module M is a graded homomorphic image of a graded free module F. Furthermore, if M is finitely generated, then F can be selected to be finitely generated.*

Proof. (1) Let $F = \bigoplus_{\mathbb{Z}} F_n$ be a graded free module with homogeneous basis $\{x_{k_\alpha}\}_{\alpha \in \Delta}$, where $\deg(x_{k_\alpha}) = k_\alpha$ for each $\alpha \in \Delta$ and $F_n = \sum_\Delta x_{k_\alpha} R_n(-k_\alpha)$ for each n. We have seen in Example 1 that $\{1_{k_\alpha}\}$ is a homogeneous basis for $\bigoplus_\Delta^{\mathrm{gr}} R(-k_\alpha)$, so if $f : \{1_{k_\alpha}\} \to \{x_{k_\alpha}\}_\Delta$ is such that $f(1_{k_\alpha}) = x_{k_\alpha}$ for each $\alpha \in \Delta$ and f is extended linearly to $f : \bigoplus_\Delta^{\mathrm{gr}} R(-k_\alpha) \to F$, then f is a bijective R-module homomorphism. If we can show that f is graded, then $\bigoplus_\Delta^{\mathrm{gr}} R(-k_\alpha) \cong^{\mathrm{gr}} F$. Now

$$\bigoplus_\Delta^{\mathrm{gr}} R(-k_\alpha) = \bigoplus_\Delta \bigoplus_{\mathbb{Z}} R_n(-k_\alpha) = \bigoplus_{\mathbb{Z}} \bigoplus_\Delta R_n(-k_\alpha) = \bigoplus_{\mathbb{Z}} S_n,$$

where $S_n = \bigoplus_\Delta R_n(-k_\alpha)$ for each n, so if $x = \sum_\Delta 1_{k_\alpha} a_{n-k_\alpha} \in S_n$, then $f(x) = \sum_\Delta f(1_{k_\alpha}) a_{n-k_\alpha} = \sum_\Delta x_{k_\alpha} a_{n-k_\alpha} \in F_n$. Hence, $f(S_n) \subseteq F_n$, so f is graded.

Conversely, if $f : \bigoplus_\Delta^{\mathrm{gr}} R(-k_\alpha) \to F$ is a graded isomorphism and $f(1_{k_k}) = x_{k_\alpha}$ for each $\alpha \in \Delta$, then $\{x_{k_\alpha}\}_\Delta$ is a homogeneous basis for F. Thus, F is a graded free module.

(2) Let M be a graded R-module and suppose that $\{x_{k_\alpha}\}_\Delta$ is a set of homogeneous generators of M. Then $f : \bigoplus_\Delta^{\mathrm{gr}} R(-k_\alpha) \to M$ such that $f(1_{k_\alpha}) = x_{k_\alpha}$ gives a graded R-module epimorphism. Finally, if M is finitely generated, then M has a finite set of homogeneous generators, so Δ can be selected to be a finite set. In this case, $\bigoplus_\Delta^{\mathrm{gr}} R(-k_\alpha)$ will have a finite basis $\{1_{k_\alpha}\}_\Delta$. $\quad\square$

Problem Set 9.2

1. Suppose that M is a graded R-module and that there is a family $\{M_\alpha\}_\Delta$ of graded R-modules such that $M \cong^{\mathrm{gr}} \bigoplus_\Delta^{\mathrm{gr}} M_\alpha$. Is there is a family $\{N_\alpha\}_\Delta$ of graded submodules of M such that $M = \bigoplus_\Delta^{\mathrm{gr}} N_\alpha$?

2. Prove Corollary 9.2.3. [Hint: Proposition 3.2.6.]

3. Prove Corollary 9.2.4. [Hint: Proposition 3.2.7.]

4. Prove Proposition 9.2.7. [Hint: Proposition 3.3.4.]

5. Let M be a graded R-module and suppose that N is a graded left R-module. Prove that $M(m) \otimes_R^{\mathrm{gr}} N(n) \cong^{\mathrm{gr}} (M \otimes_R^{\mathrm{gr}} N)(m + n)$ for any $m, n \in \mathbb{Z}$.

6. A short exact sequence $0 \to M_1 \to M \to M_2 \to 0$ in $\mathbf{Gr}_R$ is said to be *graded pure* if for each graded left R-module N,

$$0 \to M_1 \otimes_R^{\mathrm{gr}} N \to M \otimes_R^{\mathrm{gr}} N \to M_2 \otimes_R^{\mathrm{gr}} N \to 0$$

is exact in $\mathbf{Gr}_R$. Show that a short exact sequence in $\mathbf{Gr}_R$ is graded pure if and only the corresponding ungraded sequence is pure in $\mathbf{Mod}_R$. [Hint: Exercise 6 in Problem Set 5.3.]

7. Proposition 2.3.6 indicates that if M is an R-module and $\{N_\alpha\}_\Delta$ is a family of left R-modules, then $\varphi : M \otimes_R (\bigoplus_\Delta N_\alpha) \to \bigoplus_\Delta (M \otimes_R N_\alpha)$ such that $\varphi[x \otimes (y_\alpha)] = (x \otimes y_\alpha)$ is an isomorphism in $\mathbf{Mod}_{\mathbb{Z}}$. If M is a graded R-module and if $\{N_\alpha\}_\Delta$ is a family of graded left R-modules, then is there an isomorphism $\varphi : M \otimes_R^{\mathrm{gr}} (\bigoplus_\Delta^{\mathrm{gr}} N_\alpha) \to \bigoplus_\Delta^{\mathrm{gr}} (M \otimes_R^{\mathrm{gr}} N_\alpha)$ in $\mathbf{Gr}_{\mathbb{Z}}$?

9.3   Graded Projective, Graded Injective and Graded Flat Modules

Graded Projective and Graded Injective Modules

Definition 9.3.1. A graded R-module M is said to be a *graded projective module* (or a *projective module in* $\mathbf{Gr}_R$) if every row exact diagram

of graded R-modules and graded R-module homomorphisms can be completed commutatively by a graded R-module homomorphism $h : M \to L$.

Modules that are projective in $\mathbf{Gr}_R$ are closely connected to projective modules in $\mathbf{Mod}_R$. To establish this connection we need the following lemma.

Lemma 9.3.2. *Suppose that L, M and N are graded R-modules.*

(1) *Let $f : M \to N$ and $g : L \to N$ be graded R-module homomorphisms and consider the diagram*

If $h : \underline{M} \to \underline{L}$ is an R-linear mapping such that $\underline{f} = \underline{g}h$, then there is a graded R-module homomorphism $h' : M \to L$ such that the diagram

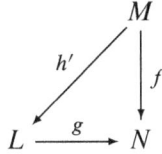

$$
\begin{array}{ccc}
 & & M \\
 & {\scriptstyle h'}\nearrow & {\big\downarrow}{\scriptstyle f} \\
L & \xrightarrow{\ g\ } & N
\end{array}
$$

is commutative.

(2) Let $f : N \to M$ and $g : N \to L$ be graded R-module homomorphisms and consider the diagram

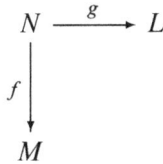

$$
\begin{array}{ccc}
N & \xrightarrow{\ g\ } & L \\
{\scriptstyle f}{\big\downarrow} & & \\
M & &
\end{array}
$$

If $h : \underline{L} \to \underline{M}$ is an R-linear mapping such that $\underline{f} = h\underline{g}$, then there is a graded R-module homomorphism $h' : L \to M$ such that the diagram

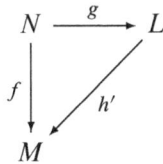

$$
\begin{array}{ccc}
N & \xrightarrow{\ g\ } & L \\
{\scriptstyle f}{\big\downarrow} & \swarrow{\scriptstyle h'} & \\
M & &
\end{array}
$$

is commutative.

Proof. We prove (1) and leave the proof of (2) as an exercise. If $x_n \in M_n$, then $f(x_n) \in N_n$, so $gh(x_n) = \underline{f}(x_n) \in N_n$. Now $h(x_n) \in L$, so if $h(x_n) = \sum_{\mathbb{Z}} y_n$, where $y_n \in L_n$ for each $n \in \mathbb{Z}$, then each y_n is unique and $gh(x_n) = \sum_{\mathbb{Z}} g(y_n) \in N_n$. But $\underline{g}(y_n) = g(y_n)$ for each $n \in \mathbb{Z}$, so it must be the case that $g(y_k) = 0$ for $k \neq n$. Hence, if $x_n \in M_n$, then $gh(x_n) = g(y_n)$. Next, define $h' : M \to L$ by $h'(\sum_{\mathbb{Z}} x_n) = \sum_{\mathbb{Z}} y_n$, where, for each n, y_n is that unique $y_n \in L_n$, such that $gh(x_n) = g(y_n)$. It follows that h' is a well-defined graded R-module homomorphism such that $\underline{f} = \underline{g}h'$. □

The following proposition is easily recognizable as a reformulation in $\mathbf{Gr}_R$ of results previously presented in an ungraded setting.

Proposition 9.3.3. *The following hold in the category* $\mathbf{Gr}_R$.

(1) A graded direct sum of a family $\{M_\alpha\}_\Delta$ of graded R-modules is a graded projective R-module if and only if each M_α is a graded projective R-module.

(2) If R is a graded ring, then $R(k)$ is a graded projective R-module for each $k \in \mathbb{Z}$.

(3) *Graded free R-modules are graded projective.*

(4) *A graded R-module M is graded projective if and only if M is graded isomorphic to a graded direct summand of a graded free R-module.*

Proof. We prove (4) and leave the proofs of (1), (2) and (3) as exercises. First, let M be a graded projective R-module. Then Proposition 9.2.9 gives a graded R-module epimorphism $f : F \rightarrow M$, where F is a graded free R-module. Since M is graded projective, we have a splitting map $g : M \rightarrow F$ in $\mathbf{Gr}_R$ such that $fg = \mathrm{id}_M$. But g is a graded monomorphism, so we have $M \cong^{\mathrm{gr}} g(M)$. Moreover, $\mathrm{Ker}\, f$ is a graded submodule of F, so it follows that $F = g(M) \oplus^{\mathrm{gr}} \mathrm{Ker}\, f$. Therefore, a graded projective R-module is isomorphic to a graded direct summand of a graded free R-module. The converse follows from (1) and (3) and from the fact that if $M \cong^{\mathrm{gr}} N$, then M is graded projective if and only if N is graded projective.   □

Proposition 9.3.4. *A graded R-module M is a projective module in $\mathbf{Gr}_R$ if and only if $\underline{M}$ is a projective module in $\mathbf{Mod}_R$.*

Proof. Let M be a graded R-module and suppose that $\underline{M}$ is a projective module in $\mathbf{Mod}_R$. A row exact diagram

$$
\begin{array}{ccc}
 & M & \\
 & \downarrow f & \\
L \xrightarrow{\ g\ } & N & \longrightarrow 0
\end{array}
$$

in $\mathbf{Gr}_R$ gives a row exact commutative diagram

$$
\begin{array}{ccc}
 & \underline{M} & \\
{}^{h}\nearrow & \downarrow \underline{f} & \\
\underline{L} \xrightarrow{\ \underline{g}\ } & \underline{N} & \longrightarrow 0
\end{array}
$$

in $\mathbf{Mod}_R$, where the R-linear mapping $h : \underline{M} \rightarrow \underline{L}$ is given by the projectivity of $\underline{M}$. The fact that M is a graded projective module now follows from (1) of Lemma 9.3.2.

Conversely, if M is a graded projective R-module, then, by (4) of the previous proposition, M is a direct summand in $\mathbf{Gr}_R$ of a graded free R-module F. But $\underline{F}$ is a free R-module, so $\underline{M}$ is a direct summand in $\mathbf{Mod}_R$ of a free R-module. But free R-modules are projective and direct summands of projective modules are projective, so $\underline{M}$ is a projective R-module.   □

Definition 9.3.5. A graded R-module M is said to be *graded injective* (or an *injective module in* $\mathbf{Gr}_R$) if every row exact diagram

$$0 \longrightarrow N \overset{g}{\longrightarrow} L$$

with $f : N \to M$ and $h : L \to M$

of graded R-modules and graded R-module homomorphisms can be completed commutatively by a graded R-module homomorphism $h : L \to M$.

The following proposition is an analogue of Baer's criteria for ungraded injective modules given in Section 5.1.

Proposition 9.3.6 (Baer's Criteria for Graded Modules). *The following are equivalent.*

(1) *M is a graded injective R-module.*

(2) *If A is a graded right ideal of R and $f : A \to M$ is a graded R-module homomorphism, there is a graded R-module homomorphism $g : R \to M$ such that $g|_A = f$.*

(3) *If A is a graded right ideal of R and $f : A \to M$ is a graded R-module homomorphism, there is an $x \in M$ such that $f(a) = xa$ for all $a \in A$.*

Proof. The implications $(1) \Rightarrow (2) \Rightarrow (3)$ are obvious. The proof of $(3) \Rightarrow (1)$ is a modification of the proof of $(3) \Rightarrow (1)$ in Proposition 5.1.3.   □

Proposition 9.3.7. *If M is a graded R-module such that $\underline{M}$ is an injective module in $\mathbf{Mod}_R$, then M is an injective module in $\mathbf{Gr}_R$.*

Proof. Use (2) of Lemma 9.3.2.   □

The converse of the proposition above fails as shown by the following example.

Example

1. If K is a field and $R = K[X, X^{-1}]$ is the ring of Laurent polynomials of Example 10 in Section 9.1, then R is a graded field. Hence, the only graded ideals of R are 0 and R. It follows from Baer's criteria for graded modules that R_R is graded injective. However, R_R is not an injective module in $\mathbf{Mod}_R$.

Graded Flat Modules

It has just been established in Chapter 5 that projective modules are flat. A similar result holds for graded projective modules.

Definition 9.3.8. A graded R-module M is said to be *graded flat* (or a *flat module in* $\mathbf{Gr}_R$) if

$$0 \to M \otimes_R^{\mathrm{gr}} N_1 \xrightarrow{\mathrm{id}_M \otimes f} M \otimes_R^{\mathrm{gr}} N_2$$

is exact in $\mathbf{Gr}_{\mathbb{Z}}$ whenever $0 \to N_1 \xrightarrow{f} N_2$ is an exact sequence in $_R\mathbf{Gr}$.

Proposition 9.3.9. *A graded R-module M is graded flat if and only if $\underline{M}$ is a flat R-module.*

Proof. Let M be a graded R-module, suppose that $\underline{M}$ is flat in $\mathbf{Mod}_R$ and let $f :
N_1 \to N_2$ be a graded R-module monomorphism, where N_1 and N_2 are graded left R-modules graded by $\{N_{1,n}\}_{\mathbb{Z}}$ and $\{N_{2,n}\}_{\mathbb{Z}}$, respectively. Then

$$0 \to \underline{M} \otimes_R \underline{N}_1 \xrightarrow{\mathrm{id}_{\underline{M}} \otimes f} \underline{M} \otimes_R \underline{N}_2$$

is exact in $\mathbf{Mod}_{\mathbb{Z}}$, so let M_i and $N_{1,j}$ be such that $i + j = n$. Now $\mathrm{id}_M(M_i) = M_i$
and $f(N_{1,j}) \subseteq N_{2,j}$, so $(\mathrm{id}_M \otimes f)(M_i \otimes_{\mathbb{Z}} N_{1,j}) = \mathrm{id}_M(M_i) \otimes_{\mathbb{Z}} f(N_{1,j}) \subseteq
M_i \otimes_{\mathbb{Z}} N_{2,j}$. Hence, $(\mathrm{id}_M \otimes f)((M \otimes_R N_1)_n) \subseteq (M \otimes_R N_2)_n$. Therefore,

$$\mathrm{id}_M \otimes f : M \otimes_R N_1 \to M \otimes_R N_2$$

is a graded R-module monomorphism, so M is graded flat. One proof of the converse uses the concept of an inductive limit, a concept not covered in this text. We outline such a proof and leave it to the interested reader to consult [10], [12] or [43] for the details on inductive limits. If M is graded flat, then M is an inductive limit of graded free R-modules. But if F is a graded free R-module, then $\underline{F}$ free, so $\underline{F}$ is a flat R-module. Thus, $\underline{M}$ is an inductive limit of flat R-modules and so $\underline{M}$ is flat.   □

Corollary 9.3.10. *Graded projective modules and graded free modules are graded flat modules.*

Proof. According to Proposition 9.3.4, a graded projective R-module is projective and a projective R-module is flat. Hence the proposition shows that graded projective R-modules are graded flat. From (3) of Proposition 9.3.3, we see that graded free R-modules are graded projective, so graded free R-modules are also graded flat.   □

Problem Set 9.3

1. Prove (2) of Lemma 9.3.2. [Hint: Dualize the proof of (1).]

2. Prove (1), (2) and (3) of Proposition 9.3.3. [(1), Hint: Proposition 5.2.3.] [(3), Hint: Propositions 5.2.6 and 9.3.4.]

3. If $\{M_\alpha\}_\Delta$ is a family of graded R-modules, prove that $\bigoplus_\Delta^{\mathrm{gr}} M_\alpha$ is a graded submodule of $\prod_\Delta^{\mathrm{gr}} M_\alpha$.

4. If M is a graded R-module, prove that the functors $\mathrm{Hom}_{\mathbf{Gr}_R}(M, -)$ and $\mathrm{HOM}_R(M, -)$ are exact if and only if M is a graded projective module. [Hint: Exercise 15 in Problem Set 9.1 and Propositions 3.3.2 and 5.2.11.]

5. Prove (3) $\Rightarrow$ (1) in Proposition 9.3.6. [Hint: Proposition 5.1.3.]

6. Prove Proposition 9.3.7. [Hint: (2) of Lemma 9.3.2.]

7. Verify Example 1.

8. If M is a graded R-module, show that M is graded injective if and only if $M(n)$ is graded injective for each $n \in \mathbb{Z}$.

9. Let M and N be graded R-modules and suppose that N is a graded submodule of M. If N is graded injective, show that N is a graded direct summand of M. [Hint: Corollary 9.2.4.]

10. Prove that each of the functors $\mathrm{Hom}_{\mathbf{Gr}_R}(-, M)$ and $\mathrm{HOM}_R(-, M)$ is exact if and only if M is a graded injective R-module. [Hint: Exercise 15 in Problem Set 9.1 and Propositions 3.3.2 and 5.1.11.]

11. If $0 \to M_1 \xrightarrow{f} M \xrightarrow{g} M_2 \to 0$ is a split short exact sequence in $\mathbf{Gr}_R$ and X is a graded R-module, then are

$$0 \to \mathrm{Hom}_{\mathbf{Gr}_R}(M_2, X) \xrightarrow{g^*} \mathrm{Hom}_{\mathbf{Gr}_R}(M, X) \xrightarrow{f^*} \mathrm{Hom}_{\mathbf{Gr}_R}(M_1, X) \to 0 \quad \text{and}$$

$$0 \to \mathrm{Hom}_{\mathbf{Gr}_R}(X, M_1) \xrightarrow{f_*} \mathrm{Hom}_{\mathbf{Gr}_R}(X, M) \xrightarrow{g_*} \mathrm{Hom}_{\mathbf{Gr}_R}(X, M_2) \to 0$$

split short exact sequences in $\mathbf{Ab}$?

12. If R and S are graded rings and $_R X_S$ is an (R, S)-bimodule, then X is a *graded (R, S)-bimodule* if $_R X$ is an object in $_R\mathbf{Gr}$ and X_S is an object in $\mathbf{Gr}_S$ and $R_i X_j S_k \subseteq X_{i+j+k}$ for all $i, j, k \in \mathbb{Z}$. Note also that if M is a graded R-module and if X is a graded (R, S)-bimodule, then $M \otimes_R^{\mathrm{gr}} X$ is a graded S-module under the operation $(x \otimes y)s = x \otimes ys$.

   (a) Let R and S be graded rings and suppose that N_S is an object in $\mathbf{Gr}_S$. If X is a graded (R, S)-bimodule, show that $\mathrm{HOM}_S(X, N)$ is a graded R-module.

(b) **Adjoint Associativity.** Let R and S be graded rings and suppose that M_R and N_S are objects in $\mathbf{Gr}_R$ and $\mathbf{Gr}_S$, respectively. If X is a graded (R, S)-bimodule, prove that there a graded isomorphism

$$\eta_{MN} : \mathrm{HOM}_S(M \otimes_R^{\mathrm{gr}} X, N) \to \mathrm{HOM}_R(M, \mathrm{HOM}_S(X, N))$$

of graded abelian groups that is natural in M_R and N_S. In particular, show that

$$\mathrm{Hom}_{\mathbf{Gr}_S}(M \otimes_R^{\mathrm{gr}} X, N) \cong^{\mathrm{gr}} \mathrm{Hom}_{\mathbf{Gr}_R}(M, \mathrm{HOM}_S(X, N)).$$

[Hint: Proposition 3.4.5.]

9.4   Graded Modules with Chain Conditions

A graded module can also satisfy chain conditions with respect to its submodules. Many classical results of modules with chain conditions can be reformulated and studied in the setting of graded modules.

Graded Noetherian and Graded Artinian Modules

Definition 9.4.1. A graded R-module M is said to be *graded noetherian* (*graded artinian*) or a *noetherian* (an *artinian*) *module* in $\mathbf{Gr}_R$ if every ascending (descending) chain of graded submodules of M terminates. If R_R is a *graded noetherian* (*graded artinian*) R-module, then we will say that R is a *right graded noetherian ring* (*right graded artinian ring*).

   The proofs of the following two propositions are similar to the proofs of Proposition 4.2.3, Proposition 4.2.4 and Exercise 8 in Problem Set 4.2.

Proposition 9.4.2. *The following are equivalent for a graded R-module M.*

(1) *M is graded noetherian.*

(2) *Every nonempty collection of graded submodules of M has a maximal element.*

(3) *Every graded submodule of M is finitely generated.*

Proposition 9.4.3. *The following are equivalent for a graded R-module M.*

(1) *M is graded artinian.*

(2) *Every nonempty collection of graded submodules of M has a minimal element.*

(3) *If $\{M_\alpha\}_\Delta$ is a family of graded submodules of M, there is a finite subset F of Δ such that $\bigcap_F M_\alpha = \bigcap_\Delta M_\alpha$.*

Remark. Let M be a graded R-module. The notation x_n indicating that $x_n \in M_n$ and $\deg(x_n) = n$ is convenient is some places and cumbersome in others. Consequently, we will sometimes write $x = x_1 + x_2 + \cdots + x_p$ for the homogeneous decomposition of x, where the subscript no longer denotes the degree of the x_i. The context of the discussion will indicate which notation is being used.

Let M be a graded R-module and suppose that N is a submodule of $\underline{M}$. Then each $x \in N$ can be expressed uniquely as $x = x_1 + x_2 + \cdots + x_p$, where each x_i is a homogeneous component in M of x and

$$\deg(x_1) < \deg(x_2) < \cdots < \deg(x_p).$$

Form the set $\{x_p\}_{x \in N}$ and let $N^{\#}$ be the submodule of M generated by $\{x_p\}_{x \in N}$. That is, $N^{\#}$ is the submodule of M generated by the components of highest degree in the homogeneous decomposition of the $x \in N$. Likewise, $N_{\#}$ will denote the submodule of M generated by the components of lowest degree in the homogeneous decomposition of the $x \in N$. Note that $N^{\#}$ and $N_{\#}$ are graded submodules of M since they are generated by homogeneous elements.

We now need the following notation. If R is a graded ring and M is a graded R-module, then for each $n_0 \in \mathbb{Z}$ let $R_{n \geq n_0} = \bigoplus_{n \geq n_0} R_n$, $R_{n \leq n_0} = \bigoplus_{n \leq n_0} R_n$, $M_{n \geq n_0} = \bigoplus_{n \geq n_0} M_n$ and $M_{n \leq n_0} = \bigoplus_{n \leq n_0} M_n$. The notation $R_{n > n_0}$, $R_{n < n_0}$, $M_{n > n_0}$, $M_{n < n_0}$ will have the obvious meaning.

Lemma 9.4.4. *Let M be a graded R-module and suppose that $n_0 \in \mathbb{Z}$. If X and Y are submodules of $\underline{M}$ such that $X \subseteq Y$, then the following are equivalent.*

(1) $X = Y$.

(2) $X^{\#} = Y^{\#}$ and $X \cap M_{n < n_0} = Y \cap M_{n < n_0}$.

(3) $X_{\#} = Y_{\#}$ and $X \cap M_{n > n_0} = Y \cap M_{n > n_0}$.

Proof. (1) $\Rightarrow$ (2) is obvious, so we show (2) $\Rightarrow$ (1). Suppose that $y \in Y$ and let $y = y_1 + y_2 + \cdots + y_{p_1}$ be the homogeneous decomposition of y in M, where $\deg(y_1) < \deg(y_2) < \cdots < \deg(y_{p_1})$. If $\deg(y_{p_1}) < n_0$, then $y \in Y \cap M_{n < n_0} = X \cap M_{n < n_0}$, so $y \in X$. If $\deg(y_{p_1}) = k \geq n_0$, then $y_{p_1} \in Y^{\#} = X^{\#}$ so there is an $x^1 \in X$ such that $x^1 = x_1^1 + x_2^1 + \cdots + x_{m_1}^1 + y_{p_1}$, where $\deg(x_1^1) < \deg(x_2^1) < \cdots < \deg(x_{m_1}^1) < k$. Thus,

$$y - x^1 = y_1 + y_2 + \cdots + y_{p_1 - 1} - (x_1^1 + x_2^1 + \cdots + x_{m_1}^1),$$

and the degree of each term in

$$y_1 + y_2 + \cdots + y_{p_1 - 1} - (x_1^1 + x_2^1 + \cdots + x_{m_1}^1)$$

is at most $k - 1$. Hence, $y - x^1$ has a homogeneous decomposition such that the degree of each component is at most $k - 1$. Now $x^1 \in X \subseteq Y$ and $y - x^1 \in Y$, so suppose that $y - x^1 = y_1' + y_2' + \cdots + y_{p_2}'$ is such that $\deg(y_{p_2}') \leq k - 1$. Then $y_{p_2}' \in Y^\# = X^\#$, so there is an $x^2 \in X$ such that $x^2 = x_1^2 + x_2^2 + \cdots + x_{m_2}^2 + y_{p_2}'$, where $\deg(x_1^2) < \deg(x_2^2) < \cdots < \deg(x_{m_2}^2) < k - 1$ and

$$y - (x^1 + x^2) = y_1' + y_2' + \cdots + y_{p_2-1}' - (x_1^2 + x_2^2 + \cdots + x_{m_2}^2).$$

Furthermore, the degree of each term in

$$y_1' + y_2' + \cdots + y_{p_2-1}' - (x_1^2 + x_2^2 + \cdots + x_{m_2}^2)$$

can be at most $k - 2$. Now let s be the positive integer for which $k - s = n_0 - 1$. If this process is repeated s times, then we arrive at

$$y - (x^1 + x^2 + \cdots + x^s) = \overline{y}_1 + \overline{y}_2 + \cdots + \overline{y}_{p_s-1} - (x_1^s + x_2^s + \cdots + x_{m_s}^s),$$

where the degree of each term of

$$\overline{y}_1 + \overline{y}_2 + \cdots + \overline{y}_{p_s-1} - (x_1^s + x_2^s + \cdots + x_{m_s}^s)$$

is at most $n_0 - 1$. Thus, $y - (x^1 + x^2 + \cdots + x^s)$ has a homogeneous decomposition in $M_{n<n_0}$. Now $x^1 + x^2 + \cdots + x^s \in X \subseteq Y$, so $y - (x^1 + x^2 + \cdots + x^s) \in Y$. Hence, $y - (x^1 + x^2 + \cdots + x^s) \in Y \cap M_{n<n_0} = X \cap M_{n<n_0}$ gives $y \in X$. Consequently, $X = Y$.

The proof of the equivalence of (1) and (3) is similar.   □

Corollary 9.4.5. *Let R be a positively graded ring (negatively graded ring) and suppose that M is a positively graded R-module (negatively graded R-module). If X and Y are submodules of $\underline{M}$, such that $X \subseteq Y$, then $X = Y$ if and only if $X^\# = Y^\# (X_\# = Y_\#)$.*

Lemma 9.4.6. *Suppose that M is a graded R-module and let $\{x_\alpha\}_\Delta$ be a set of homogeneous generators of $M_n R$. If x is a homogeneous element of M_n, then x can be written as $x = \sum_\Delta x_\alpha a_\alpha$, where each $a_\alpha \in R$ is homogeneous and $a_\alpha = 0$ for almost all $\alpha \in \Delta$.*

Proof. Let x be a homogeneous element of M_n. Then there are $a_\alpha \in R$, $\alpha \in \Delta$, such that $x = \sum_\Delta x_\alpha a_\alpha$, where $a_\alpha = 0$ for almost all $\alpha \in \Delta$. Since $\deg(x) = n$, by comparing degrees we see that $\deg(x_\alpha a_\alpha) = n$ for each $\alpha \in \Delta$. Suppose that $a_\alpha = \sum_{\mathbb{Z}} a_{\alpha,k}$ is the homogeneous decomposition in R of a_α for each $\alpha \in \Delta$ with $\deg(a_{\alpha,k}) = k$. Then $x = \sum_\Delta x_\alpha \sum_{\mathbb{Z}} a_{\alpha,k} = \sum_\Delta \sum_{\mathbb{Z}} x_\alpha a_{\alpha,k}$ and it follows that $\deg(x_\alpha a_{\alpha,k}) = n$ for each $\alpha \in \Delta$ and each $k \in \mathbb{Z}$. If α is fixed and if $\deg(x_\alpha) = m$, then $\deg(a_{\alpha,k}) = n - m$ for each k. Hence, $a_\alpha = \sum_{\mathbb{Z}} a_{\alpha,k} \in R_{n-m}$, so each a_α is homogeneous.   □

The proof of the following lemma is an exercise.

Lemma 9.4.7. *The following hold for a graded R-module M.*

(1) $R_{n\geq0}$ *and* $R_{n\leq0}$ *are graded subrings of R.*

(2) $M_{n\geq0}$ *is a graded* $R_{n\geq0}$*-module and* $M_{n\leq0}$ *is a graded* $R_{n\leq0}$*-module.*

Proposition 9.4.8. *The following hold for a graded noetherian R-module M.*

(1) *For each* $n \in \mathbb{Z}$, M_n *is a noetherian* R_0*-module.*

(2) $M_{n\geq0}$ *is a noetherian* $R_{n\geq0}$*-module.*

(3) $M_{n\leq0}$ *is a noetherian* $R_{n\leq0}$*-module.*

Conversely, if $M_{n\geq0}$ *and* $M_{n\leq0}$ *are noetherian* $R_{n\geq0}$ *and* $R_{n\leq0}$ *modules, respectively, then M is a graded noetherian R-module.*

Proof. (1) Lemma 9.4.7 indicates that $M_{n\geq0}$ is a graded $R_{n\geq0}$-module and that $M_{n\leq0}$ is a graded $R_{n\leq0}$-module. Moreover, each M_n is clearly an R_0-module, so we need only be concerned with proving that M_n is noetherian as an R_0-module.

Let N be an R_0-submodule of M_n. Since $NR = \bigoplus_{\mathbb{Z}}(NR \cap M_j)$, NR is a graded submodule of M. Hence, NR is finitely generated, so NR is finitely generated by homogeneous elements. Suppose that $\{x_1, x_2, \ldots, x_s\}$ is a set of homogeneous generators of NR. Now $x_k \in NR$ for $k = 1, 2, \ldots, s$, so each x_k can be written as $x_k = \sum_{i=1}^{p} x_{i_k} a_{i_k}$, where $x_{i_k} \in N$ and $a_{i_k} \in R_{i_k}$ for $i = 1, 2, \ldots, p$ with p depending on k. Hence, if $x \in NR$, then there are $a_1, a_2, \ldots, a_s \in R$ such that $x = \sum_{k=1}^{s} x_k a_k$, so $x = \sum_{k=1}^{s} \sum_{i=1}^{n} x_{i_k}(a_{i_k} a_k)$. Thus, NR is generated by the homogeneous elements x_{i_k} with each x_{i_k} belonging to N. Next, simplify notation and let $\{x'_1, x'_2, \ldots, x'_t\}$ be a set of homogeneous generators of NR with x'_i belonging to N for $i = 1, 2, \ldots, t$. If $y \in N$, then, by Lemma 9.4.6, there exist homogeneous elements $b_1, b_2, \ldots, b_t \in R$ such that $y = \sum_{i=1}^{t} x'_i b_i$. Since the $\deg(y) = n$ and $\deg(x'_i) = n$ for $i = 1, 2, \ldots, t$, the degree of each b_i must be zero. Thus, each b_i is an element of R_0, so N is generated as an R_0-module by $x'_1, x'_2, \ldots, x'_t$. Therefore, M_n is a noetherian R_0-module.

(2) Let $N = \bigoplus_{n\geq0} N_n$ be a graded $R_{n\geq0}$-submodule of $M_{n\geq0}$. Now $NR = \bigoplus_{\mathbb{Z}}(NR \cap M_j)$, so NR is a graded submodule of M. Hence, NR is finitely generated. It follows as in the proof of (1) that NR is generated by a set $\{x_1, x_2, \ldots, x_s\}$ of homogeneous elements of N. Thus, $NR = x_1 R + x_2 R + \cdots x_s R$, where $0 \leq \deg(x_1) < \deg(x_2) < \cdots < \deg(x_s)$. Let $n_0 = \deg(x_s)$. If $y \in N$ is homogeneous and $\deg(y) > n_0$, then there are homogeneous elements $a_1, a_2, \ldots, a_s \in R$ such that $y = x_1 a_1 + x_2 a_2 + \cdots + x_s a_s$ and $\deg(a_i) \geq 0$ for $i = 1, 2, \ldots, s$. Hence, each a_i is in $R_{n\geq0}$, so $N_{n\geq n_0}$ is a finitely generated $R_{n\geq0}$-module. Because of (1), each M_n is a noetherian R_0-module, so $M_0 \oplus M_1 \oplus \cdots \oplus M_{n_0-1}$ is a noetherian R_0-module. Thus, $N_0 \oplus N_1 \oplus \cdots \oplus N_{n_0-1}$ is a finitely generated R_0-module. If $\{y_1, y_2, \ldots, y_t\}$

is a set of generators of $N_0 \oplus N_1 \oplus \cdots \oplus N_{n_0-1}$, then $\{x_1, x_2, \ldots, x_s, y_1, y_2, \ldots, y_t\}$ is a set of generators of N as an $R_{n\geq0}$-module. Therefore, $M_{n\geq0}$ is a noetherian $R_{n\geq0}$-module.

(3) Similar to the proof of (2).

For the converse, suppose that $M_{n\geq0}$ and $M_{n\leq0}$ are noetherian $R_{n\geq0}$ and $R_{n\leq0}$ modules, respectively, and let $X_1 \subseteq X_2 \subseteq X_3 \subseteq \cdots$ be an ascending chain of graded submodules of M. Then

$$(X_1)_{n\geq0} \subseteq (X_2)_{n\geq0} \subseteq (X_3)_{n\geq0} \subseteq \cdots \quad \text{and}$$

$$(X_1)_{n\leq0} \subseteq (X_2)_{n\leq0} \subseteq (X_3)_{n\leq0} \subseteq \cdots$$

are ascending chains of graded submodules of $M_{n\geq0}$ and $M_{n\leq0}$, respectively. Since $M_{n\geq0}$ and $M_{n\leq0}$ are noetherian $R_{n\geq0}$ and $R_{n\leq0}$ modules, respectively, then $M_{n\geq0}$ and $M_{n\leq0}$ are graded noetherian $R_{n\geq0}$ and $R_{n\leq0}$ modules. Hence, there is an integer $n_0 \geq 0$ such that $(X_k)_{n\geq0} = (X_{n_0})_{n\geq0}$ and $(X_k)_{n\leq0} = (X_{n_0})_{n\leq0}$ for all $k \geq n_0$. But this gives $X_k = X_{n_0}$ for all $k \geq n_0$, so M is a graded noetherian R-module. $\square$

Proposition 9.4.9. *If M is a graded R-module, then M is graded noetherian if and only if $\underline{M}$ is a noetherian R-module.*

Proof. Let M be a graded R-module. If $\underline{M}$ is a noetherian R-module, then it is obvious that M is a graded noetherian R-module, so suppose that M is graded noetherian. If $X_1 \subseteq X_2 \subseteq X_3 \subseteq \cdots$ is an ascending chain of submodules of $\underline{M}$, then Exercise 8 shows that

$$(X_1)^\# \subseteq (X_2)^\# \subseteq (X_3)^\# \subseteq \cdots$$

is an ascending chain of graded submodules of M. Hence, there is an integer n_0 such that $(X_k)^\# = (X_{n_0})^\#$ for all $k \geq n_0$. We also have that

$$X_1 \cap M_{n\leq0} \subseteq X_2 \cap M_{n\leq0} \subseteq X_3 \cap M_{n\leq0} \subseteq \cdots$$

is an ascending chain of graded submodules of M, so there is an integer n_1 such that $X_k \cap M_{n\leq0} = X_{n_1} \cap M_{n\leq0}$ for all $k \geq n_1$. If $m = \max\{n_0, n_1\}$, then $(X_k)^\# = (X_m)^\#$ and $X_k \cap M_{n\leq0} = X_m \cap M_{n\leq0}$ for all $k \geq m$. But in view of Lemma 9.4.4 this means that $X_k = X_m$ for all $k \geq m$, so $\underline{M}$ is a noetherian R-module. $\square$

If M is a graded R-module and if $\underline{M}$ is an artinian R-module, then it follows easily that M is graded artinian. However, the converse does not hold. For example, the Laurent polynomial ring $R = K[X, X^{-1}]$, K a field, is a graded artinian R-module that is not artinian when the grading of R is forgotten.

Problem Set 9.4

1. Prove Proposition 9.4.2. [Hint: Proposition 4.2.3.]

2. Prove Proposition 9.4.3. [Hint: Proposition 4.2.4 and Exercise 8 in Problem Set 4.2.]

3. Prove the equivalence of (1) and (3) of Lemma 9.4.4.

4. Prove Corollary 9.4.5.

5. Prove Lemma 9.4.7.

6. Recall that direct sums of families of injective modules are injective if and only if R is a right noetherian ring. Does the graded version of this also hold?

7. In the proof of (2) of Proposition 9.4.8 it was pointed out that if M is a graded R-module and if $N = \bigoplus_{n \geq 0} N_n$ is a graded $R_{n \geq 0}$-submodule of $M_{n \geq 0}$, then $NR = \bigoplus_{\mathbb{Z}} (NR \cap M_j)$ so that NR is a graded submodule of M. Show that it is actually the case that $NR = \bigoplus_{\mathbb{Z}} (NR \cap M_j)$.

8. Prove that each of the following hold for a graded R-module M and submodules X and Y of $\underline{M}$.

   (a) $X^{\#}$ and $X_{\#}$ are graded submodules of M.

   (b) $X = X^{\#} = X_{\#}$ if and only if X is a graded submodule of M.

   (c) If $X \subseteq Y$, then $X^{\#} \subseteq Y^{\#}$ and $X_{\#} \subseteq Y_{\#}$.

   (d) If N is a graded submodule of M, then $(M/N)_{n \geq 0} \cong^{\mathrm{gr}} M_{n \geq 0}/N_{n \geq 0}$ and $(M/N)_{n \leq 0} \cong^{\mathrm{gr}} M_{n \leq 0}/N_{n \leq 0}$. [Hint: $M/N = \bigoplus_{\mathbb{Z}} (N + M_n/N)$, so $(M/N)_{n \geq 0} = \bigoplus_{n \geq 0} (N + M_n/N) \cong \bigoplus_{n \geq 0} M_n/(M_n \cap N)$.]

9. Show that the Laurent polynomial ring $R = K[X, X^{-1}]$, K a field, is not artinian as an ungraded R-module.

9.5 More on Graded Rings

The Graded Jacobson Radical

Definition 9.5.1. If M is a graded R-module, then the *graded Jacobson radical* of M, denoted by $\mathrm{Rad}^{\mathrm{gr}}(M)$, is the intersection of the maximal graded submodules of M. If M fails to have maximal graded submodules, then we set $\mathrm{Rad}^{\mathrm{gr}}(M) = M$. The graded Jacobson radical of the graded module $R_R({}_R R)$ will be denoted by $J^{\mathrm{gr}}(R_R)$ $(J^{\mathrm{gr}}({}_R R))$.

Lemma 9.5.2. *If M is a finitely generated nonzero graded R-module, then M has at least one maximal graded submodule.*

Proof. Let $\{x_1, x_2, \ldots, x_n\}$ be a minimal set of homogeneous generators of M. Then

$$N = x_2 R + x_3 R + \cdots + x_n R$$

is a proper graded submodule of M. Let $\mathscr{S}$ be the set of proper graded submodules N' of M that contain N. If $\mathscr{C}$ is a chain in $\mathscr{S}$, then

$$\bigcup_{\mathscr{C}} N' = \bigcup_{\mathscr{C}} \left(\bigoplus_{\mathbb{Z}} (N' \cap M_n) \right) = \bigoplus_{\mathbb{Z}} \left(\left(\bigcup_{\mathscr{C}} N' \right) \cap M_n \right),$$

so $\bigcup_{\mathscr{C}} N'$ is a graded submodule of M and $\bigcup_{\mathscr{C}} N' \neq M$, since $x_1 \notin \bigcup_{\mathscr{C}} N'$. Hence, $\mathscr{S}$ is inductive, so Zorn's lemma indicates that $\mathscr{S}$ has a maximal element, say N^*. Thus, $N \subseteq N^* \subsetneqq M$. It follows that M/N^* is a graded simple module, so N^* is a maximal graded submodule of M. $\qquad\square$

The proof of the following proposition closely follows that of the ungraded case.

Proposition 9.5.3. *The following hold for a nonzero graded R-module M.*

(1) *If M is finitely generated, then $\mathrm{Rad}^{\mathrm{gr}}(M) \subsetneqq M$.*

(2) $\mathrm{Rad}^{\mathrm{gr}}(M) = \bigcap \{\mathrm{Ker}\, f \mid f \in \mathrm{Hom}_{\mathbf{Gr}_R}(M, S), \ S \text{ a simple module in } \mathbf{Gr}_R\}$
  $= \bigcap \{\mathrm{Ker}\, f \mid f \in \mathrm{Hom}_{R}\mathbf{Gr}(M, S), \ S \text{ a simple module in } {}_R\mathbf{Gr}\}.$

(3) $J^{\mathrm{gr}}(R_R) = \bigcap \{\mathrm{ann}_r(S) \mid S \text{ a simple module in } \mathbf{Gr}_R\}$
  $= \bigcap \{\mathrm{ann}_\ell(S) \mid S \text{ a simple module in } {}_R\mathbf{Gr}\}.$

(4) $J^{\mathrm{gr}}(R_R)$ *is a graded ideal of R and* $J^{\mathrm{gr}}(R_R) = J^{\mathrm{gr}}({}_R R)$.

(5) *If M and N are graded R-modules and $f \in \mathrm{HOM}_R(M, N)$, then*

$$f(\mathrm{Rad}^{\mathrm{gr}}(M)) \subseteq \mathrm{Rad}^{\mathrm{gr}}(N).$$

(6) $J^{\mathrm{gr}}(R)$ *is the largest graded proper ideal of R such that if a is a homogeneous element of R and $a + J^{\mathrm{gr}}(R)$ is a unit in $R/J^{\mathrm{gr}}(R)$, then a is a unit in R.*

Proof. The proof of (1) follows immediately from Lemma 9.5.2 and the proofs of (2) through (5) are similar to the proofs of the classical results. We prove (6) and leave the proofs of (2) through (5) as exercises.

(6) Let a be a homogeneous element of R such that $a + J^{\mathrm{gr}}(R)$ is a unit in $R/J^{\mathrm{gr}}(R)$. Since a is homogeneous, aR is a graded right ideal of R, so if $aR \neq R$, then Zorn's lemma gives a maximal graded right ideal $\mathfrak{m}$ of R such that $aR \subseteq \mathfrak{m} \subsetneqq R$. If $b + J^{\mathrm{gr}}(R)$ is the inverse of $a + J^{\mathrm{gr}}(R)$ in $R/J^{\mathrm{gr}}(R)$, then $(a + J^{\mathrm{gr}}(R))(b + J^{\mathrm{gr}}(R)) = 1 + J^{\mathrm{gr}}(R)$. Hence, $1 - ab \in J^{\mathrm{gr}}(R) \subseteq \mathfrak{m}$ and $ab \in \mathfrak{m}$, so $1 \in \mathfrak{m}$. Thus, $\mathfrak{m} = R$, a contradiction. Therefore, $aR = R$, so there is a $c \in R$ such that $ac = 1$.

Similarly, the assumption that $Ra \neq R$ leads to a contradiction, so there is a $c' \in R$ such that $c'a = 1$. Hence, $c' = c$ and so a is a unit in R.

Finally, we claim that $J^{\text{gr}}(R)$ is the largest graded proper ideal of R with this property. Suppose that I is a proper graded ideal of R with the property that a homogenous element $a \in R$ is a unit in R whenever $a + I$ is a unit in R/I. If $I \nsubseteq J^{\text{gr}}(R)$, then there is a maximal graded right ideal $\mathfrak{m}$ of R such that $I \nsubseteq \mathfrak{m}$, so $I + \mathfrak{m} = R$. Let $a \in I$ and $b \in \mathfrak{m}$ be such that $a + b = 1$ with $b \notin I$. If $a = \sum_{\mathbb{Z}} a_n$ and $b = \sum_{\mathbb{Z}} b_n$, where $a_n, b_n \in R_n$ for each $n \in \mathbb{Z}$, then $\sum_{\mathbb{Z}} (a_n + b_n) = 1$. Now $\deg(1) = 0$, so $a_n + b_n = 0$ if $n \neq 0$ and $a_0 + b_0 = 1$ with $b_0 \notin I$. Thus, we see that we can select homogeneous elements $a \in I$ and $b \in \mathfrak{m}$ such that $a + b = 1$ with $b \notin I$. Consequently, $1 + I = b + I$, so $b + I$ is a unit in R/I. Therefore, b is a unit in R, so $\mathfrak{m} = R$, a contradiction. Hence, $I \subseteq J^{\text{gr}}(R)$ and we are done.   □

Because of (4) in the proposition above, we can unambiguously write $J^{\text{gr}}(R)$ for the graded Jacobson radical of R.

Graded Wedderburn–Artin Theory

Definition 9.5.4. If R is a graded ring, then R is said to be a *right (left) graded semisimple ring* if R is a direct sum of minimal graded right (left) ideals of R. Likewise, a graded (left) R-module M is said to be a *graded semisimple module* (or a *graded completely reducible module*) if M is a direct sum of graded simple submodules of M. If every module in $\mathbf{Gr}_R$ ($_R\mathbf{Gr}$) is graded semisimple, then we will refer to $\mathbf{Gr}_R$ ($_R\mathbf{Gr}$) as a *semisimple category*.

A development of graded semisimple rings can be carried out using methods that are similar to those used to investigate semisimple rings. Due to space limitations, we provide, without proof, the main results of such a development. For the reader who wishes to consider the details, we cite[35] and [36] as references.

Proposition 9.5.5. *The following are equivalent for a ring R.*

(1) *R is a right graded semisimple ring.*

(2) *R has a decomposition $R = A_1 \oplus A_2 \oplus \cdots \oplus A_n$, where each A_i is a minimal graded right ideal of R and there exist idempotents $e_i \in R$ such that $A_i = e_i R$ for $i = 1, 2, \ldots, n$.*

(3) $\mathbf{Gr}_R$ *is a semisimple category.*

Proposition 9.5.6. *A ring R is right graded semisimple if and only if R is left graded semisimple.*

Due to the proposition above, a right or left graded semisimple ring can be referred to simply as a *graded semisimple ring*.

Proposition 9.5.7. *If R is a graded semisimple ring, then R is a left and a right graded noetherian ring and a left and a right graded artinian ring.*

Definition 9.5.8. A graded semisimple ring $R = A_1 \oplus A_2 \oplus \cdots \oplus A_n$ is said to be *graded simple* if $\mathrm{HOM}_R(A_i, A_j) \neq 0$ for each pair of integers (i, j) with $1 \leq i, j \leq n$. That is, if there is an integer n_{ij} such that $A_j \cong^{\mathrm{gr}} A_i(n_{ij})$ whenever $1 \leq i, j \leq n$.

Proposition 9.5.9. *The following hold for a graded semisimple ring $R = A_1 \oplus A_2 \oplus \cdots \oplus A_n$.*

(1) *For each A_i, let $\overline{A_i} = \sum A_j$, where A_j is such that $A_j \cong^{\mathrm{gr}} A_i(n_{ij})$ for some integer n_{ij}. Then $\overline{A_i}$ is a graded ideal of R, for $i = 1, 2, \ldots, t$, with $t \leq n$.*

(2) *$R = \overline{A_1} \oplus \overline{A_2} \oplus \cdots \oplus \overline{A_t}$ and each $\overline{A_i}$ is a graded simple ring.*

Thus, we see that a graded semisimple ring is a finite direct product of graded simple rings. A graded simple ring can be expressed in the form of a graded matrix ring with entries from a graded division ring. To see, this let $\mathbb{M}_n(R)$ be the ring of $n \times n$ matrices over a graded ring R and suppose that $\overline{m} = (m_1, m_2, \ldots, m_n)$ is an n-tuple of integers. If

$$
\mathbb{M}_n(R_k)(\overline{m}) = \begin{pmatrix}
R_k & R_{k+m_2-m_1} & R_{k+m_3-m_1} & \cdots & R_{k+m_n-m_1} \\
R_{k+m_1-m_2} & R_k & R_{k+m_3-m_2} & \cdots & R_{k+m_n-m_2} \\
R_{k+m_1-m_3} & R_{k+m_2-m_3} & R_k & \cdots & R_{k+m_n-m_3} \\
\vdots & \vdots & \vdots & \ddots & \vdots \\
R_{k+m_1-m_n} & R_{k+m_2-m_n} & R_{k+m_3-m_n} & \cdots & R_k
\end{pmatrix},
$$

then we obtain a graded ring $\mathbb{M}_n(R)(\overline{m})$ with grading $\{\mathbb{M}_n(R_k)(\overline{m})\}_{k \in \mathbb{Z}}$. Consequently, there are an infinite number of distinct gradings of the $n \times n$ matrix ring $\mathbb{M}_n(R)$. With this notation in mind, we have the following proposition and corollary.

Proposition 9.5.10. *If R is a graded ring, then R is a graded simple ring if and only if there is a graded division ring D and an n-tuple of integers $\overline{m}$ such that $R \cong^{\mathrm{gr}} \mathbb{M}_n(D)(\overline{m})$.*

Corollary 9.5.11. *R is a graded semisimple ring if and only if there are graded division rings $D_1, D_2, \ldots, D_t$ and n_i-tuples of integers $\overline{m_i}$ such that*

$$
R \cong^{\mathrm{gr}} \mathbb{M}_{n_1}(D_1)(\overline{m_1}) \times \mathbb{M}_{n_2}(D_2)(\overline{m_2}) \times \cdots \times \mathbb{M}_{n_t}(D_t)(\overline{m_t}).
$$

The following graded version of a result of Hopkins, Corollary 6.6.6, also holds.

Proposition 9.5.12. *If R is a right graded artinian ring, then R is right graded noetherian.*

Problem Set 9.5

1. In the proof of Lemma 9.5.2 it was pointed out that if $\mathcal{C}$ is a chain of graded submodules N of a graded R-module M, then $\bigcup_{\mathcal{C}} N$ is a graded submodule of M. Prove that this is the case by showing that $\bigcup_{\mathcal{C}} N = \bigcup_{\mathcal{C}} (\bigoplus_{\mathbb{Z}} (N \cap M_n)) = \bigoplus_{\mathbb{Z}} ((\bigcup_{\mathcal{C}} N) \cap M_n)$.

2. Prove (2) through (5) of Proposition 9.5.3. [Hint: Consider the corresponding results for the ungraded cases given in Section 6.1.]

3. If R is a graded ring, prove that if A is a graded right ideal of R, then there is a maximal graded right ideal $\mathfrak{m}$ of R that contains A. [Hint: Exercise 1.]

4. For a ring R, we know that $J(R/J(R)) = 0$. Is it the case that $J^{\mathrm{gr}}(R/J^{\mathrm{gr}}(R)) = 0$ when R is a graded ring? [Hint: If f is a graded epimorphism and $\mathrm{Ker}\, f \subseteq \mathrm{Rad}(M)$, is there is a one-to-one correspondence among the maximal graded submodules of N and the maximal graded submodules of M that contain $\mathrm{Ker}\, f$?]

5. If R is a graded semisimple ring, show that $J^{\mathrm{gr}}(R) = 0$.

6. Let $\mathbb{M}_n(R)$ be the ring of $n \times n$ matrices over a graded ring R and suppose that $\overline{m} = (m_1, m_2, \ldots, m_n)$ is an n-tuple of integers. If

$$\mathbb{M}_n(R_k)(\overline{m}) = \begin{pmatrix} R_k & R_{k+m_2-m_1} & R_{k+m_3-m_1} & \cdots & R_{k+m_n-m_1} \\ R_{k+m_1-m_2} & R_k & R_{k+m_3-m_2} & \cdots & R_{k+m_n-m_2} \\ R_{k+m_1-m_3} & R_{k+m_2-m_3} & R_k & \cdots & R_{k+m_n-m_3} \\ \vdots & \vdots & \vdots & \ddots & \vdots \\ R_{k+m_1-m_n} & R_{k+m_2-m_n} & R_{k+m_3-m_n} & \cdots & R_k \end{pmatrix},$$

verify that we obtain a graded ring $\mathbb{M}_n(R)(\overline{m})$ with grading $\{\mathbb{M}_n(R_k)(\overline{m})\}_{k \in \mathbb{Z}}$. [Hint: Verify this for, say $n = 3$, and then generalize.]

7. A proper graded ideal $\mathfrak{p}$ of a graded ring R is said to be a *graded prime ideal* of R if whenever A and B are graded ideals of R such that $AB \subseteq \mathfrak{p}$, then either $A \subseteq \mathfrak{p}$ or $B \subseteq \mathfrak{p}$.

   (a) Prove that a proper graded ideal $\mathfrak{p}$ is a graded prime ideal of R if and only if whenever a and b are homogeneous elements of R such that $aRb \subseteq \mathfrak{p}$, then either $a \in \mathfrak{p}$ or $b \in \mathfrak{p}$. If 0 is a graded prime ideal of R, then R is said to be a *graded prime ring*.

   (b) A graded prime ideal of R is said to be a *minimal graded prime ideal* of R if it is a minimal element among the graded prime ideals of R. Prove that every graded prime ideal of R contains a minimal graded prime ideal of R.

(c) Let $\mathrm{rad}^{\mathrm{gr}}(R)$ denote the intersection of the graded prime ideals of R. $\mathrm{rad}^{\mathrm{gr}}(R)$ is called the *graded prime radical* of R. Prove that $\mathrm{rad}^{\mathrm{gr}}(R)$ is the intersection of the minimal prime graded ideals of R and that $\mathrm{rad}^{\mathrm{gr}}(R) \subseteq J^{\mathrm{gr}}(R)$.

(d) If $\mathrm{rad}^{\mathrm{gr}}(R) = 0$, prove that R has no nilpotent graded ideals. If $\mathrm{rad}^{\mathrm{gr}}(R) = 0$, then R is said to be a *graded semiprime ring*.

[Hint: Solutions for the parts of Exercise 7 can be modeled after the corresponding results given in Section 6.2.]

Chapter 10

More on Rings and Modules

Reflexive modules (defined below) arise in several areas of mathematics. In particular, in functional analysis, where the modules under consideration are usually vector spaces. As we will see, vector spaces are reflexive if and only if they are finite dimensional. If we pass from vector spaces to modules over an arbitrary ring, then finitely generated projective modules are reflexive. However, reflexivity often fails for finitely generated modules if the module is not projective. Our goal is to find conditions on a ring R that will ensure that reflexivity holds for finitely generated modules. To do this, we introduce a special class of rings known as quasi-Frobenius rings.

We begin our discussion of reflexive modules with definitions and basic concepts. Let M be an R-module and set

$$M^* = \operatorname{Hom}_R(M_R, R_R).$$

Thus, if $f \in M^*$ and we define $(af)(x) = af(x)$ for $x \in M$ and $a \in R$, then $af \in M^*$ and M^* is a left R-module. Likewise, we can form

$$M^{**} = \operatorname{Hom}_R({}_R M^*, {}_R R),$$

so if $(fa)(g) = f(g)a$ for $f \in M^{**}$, $g \in M^*$ and $a \in R$, then $fa \in M^{**}$ and this makes M^{**} into an R-module. If $x \in M$ and $f_x : M^* \to R$ is such that $f_x(g) = g(x)$ for each $g \in M^*$, then $f_x \in M^{**}$, so let

$$\varphi_M : M \to M^{**} \quad \text{be such that } \varphi_M(x) = f_x.$$

If $x, y \in M$ and $a \in R$, then it follows easily that $f_{x+y} = f_x + f_y$ and $f_{xa} = f_x a$, so φ_M is R-linear. If $f : M \to N$ is an R-linear mapping of R-modules, then

$$f^* : \operatorname{Hom}_R(N_R, R_R) \to \operatorname{Hom}_R(M_R, R_R)$$

such that $f^*(h) = hf$ is an R-linear mapping of left R-modules and

$$f^{**} : \operatorname{Hom}_R({}_R M^*, {}_R R) \to \operatorname{Hom}_R({}_R N^*, {}_R R)$$

given by $f^{**}(h) = hf^*$ is an R-linear mapping of R-modules. This produces diagrams

$$
\begin{array}{ccc}
M & \xrightarrow{\varphi_M} & M^{**} \\
\downarrow{\scriptstyle h} & & \downarrow{\scriptstyle h^{**}} \\
N & \xrightarrow{\varphi_N} & N^{**}
\end{array}
\qquad \text{and} \qquad
\begin{array}{ccc}
x & \longrightarrow & f_x \\
\downarrow & & \downarrow \\
h(x) & \longrightarrow & f_{h(x)}
\end{array}
$$

where $x \in M$. Note that $h^{**}(f_x) = f_x h^*$, so if $g \in M^*$, then

$$h^{**}(f_x)(g) = f_x h^*(g) = f_x(gh) = gh(x) = f_{h(x)}(g).$$

Hence, $h^{**}(f_x) = f_{h(x)}$ and so the diagrams are commutative.

If M is an R-module, then the left R-module M^* is said to be the *dual* of M and the R-module M^{**} is the *double dual* of M. The left exact contravariant functor $(-)^* = \mathrm{Hom}_R(-, R)$ is often referred to as the *duality functor*. The R-linear mapping φ_M is said to be the *canonical map* from M to M^{**}. If φ_M is a monomorphism, then we say that M is a *torsionless module* and if φ_M is an isomorphism, then M is *reflexive*. If M is a left R-module, then the preceding observations and definitions can easily be adapted to M. In this case, M^* is an R-module and M^{**} is a left R-module.

Remark. If F is a free R-module with basis $\mathcal{B} = \{x_\alpha\}_\Delta$, $\{a_\alpha\}_\Delta$ is any set of elements of R and

$$f : \mathcal{B} \to R \quad \text{such that } f(x_\alpha) = a_\alpha \text{ for each } \alpha \in \Delta$$

is extended linearly to F, then $f : F \to R$ is an R-linear mapping belonging to the left R-module F^*. In particular, if

$$x_\alpha^* : \mathcal{B} \to R \quad \text{is such that } x_\alpha^*(x_\beta) = \delta_{\alpha\beta}, \text{ where}$$

$$\delta : \Delta \times \Delta \to R \quad \text{is the Kronecker delta function,}$$

and each x_α^* is extended linearly to F, then $\mathcal{B}^* = \{x_\alpha^*\}_\Delta$ is a linearly independent set of elements of F^*. Indeed, if $\sum_\Delta a_\alpha x_\alpha^* = 0$ in F^*, where $a_\alpha = 0$ for almost all $\alpha \in \Delta$, then $0 = (\sum_\Delta a_\alpha x_\alpha^*)(x_\beta) = \sum_\Delta a_\alpha x_\alpha^*(x_\beta) = a_\beta$ for all $\beta \in \Delta$. As we will see, $\mathcal{B}^*$ may not be a basis for F^*, so, in general, $\bigoplus_\Delta Rx_\alpha^* \subsetneqq F^*$. However, if it turns out that $\{x_\alpha^*\}_\Delta$ is a basis for F^*, then this process can be repeated and we can, for each $\alpha \in \Delta$, set $x_\alpha^{**}(x_\beta^*) = \delta_{\alpha\beta}$ for all $\beta \in \Delta$ and then extend x_α^{**} linearly to F^{**}. In this case, $\mathcal{B}^{**} = \{x_\alpha^{**}\} \subseteq F^{**}$ and we have

$$\varphi_F(x_\alpha)(x_\beta^*) = f_{x_\alpha}(x_\beta^*) = x_\beta^*(x_\alpha) = x_\alpha^{**}(x_\beta^*), \quad \text{so } \varphi_F(x_\alpha) = f_{x_\alpha} = x_\alpha^{**}.$$

If $\mathcal{B} = \{x_\alpha\}_\Delta$ is a basis for F and $\mathcal{B}^* = \{x_\alpha^*\}_\Delta$ is a basis of F^*, then $\mathcal{B}^*$ is said to be the *basis dual to* $\mathcal{B}$ or the *dual basis* of $\mathcal{B}$.

10.1   Reflexivity and Vector Spaces

Proposition 10.1.1. *If V be a finite dimensional vector space over a division ring D, then V^* and V^{**} are finite dimensional vector spaces with bases $\mathcal{B}^*$ and $\mathcal{B}^{**}$, respectively. Furthermore, $\dim(V) = \dim(V^*) = \dim(V^{**})$ and V is reflexive.*

Proof. It follows as in the Remark above that $\mathcal{B}^* = \{x_i^*\}_{i=1}^n$ is a set of linearly independent vectors in V^*. We claim that $\mathcal{B}^*$ spans V^*. Let $f \in V^*$ and $x \in V$. If $x = x_1 b_1 + x_2 b_2 + \cdots + x_n b_n$ is a representation of x with elements of $\mathcal{B}$, then $f(x) = f(x_1)b_1 + f(x_2)b_2 + \cdots + f(x_n)b_n$. If $f(x_i) = a_i$ for $i = 1, 2, \ldots, n$, then $f(x) = a_1 b_1 + a_2 b_2 + \cdots + a_n b_n$. We claim that $f = a_1 x_1^* + a_2 x_2^* + \cdots + a_n x_n^*$. If $g = a_1 x_1^* + a_2 x_2^* + \cdots + a_n x_n^*$, then

$$g(x) = (a_1 x_1^* + a_2 x_2^* + \cdots + a_n x_n^*)(x)$$
$$= (a_1 x_1^* + a_2 x_2^* + \cdots + a_n x_n^*)(x_1 b_1 + x_2 b_2 + \cdots + x_n b_n)$$
$$= a_1 b_1 + a_2 b_2 + \cdots + a_n b_n$$
$$= f(x).$$

Hence, $f = a_1 x_1^* + a_2 x_2^* + \cdots + a_n x_n^*$ and so $\mathcal{B}^*$ is a basis for V^*. Since $\mathcal{B}^*$ is a basis for V^*, a similar proof shows that $\mathcal{B}^{**} = \{x_i^{**}\}_{i=1}^n$ is a basis for V^{**} and so $\dim(V) = \dim(V^*) = \dim(V^{**}) = n$. Finally, the canonical map $\varphi_V : V \to V^{**}$ is such that $\varphi_V(x_i) = x_i^{**}$ for $i = 1, 2, \ldots, n$, so φ_V is an isomorphism. Hence, V is reflexive. $\square$

Proposition 10.1.2. *A vector space over a division ring D is reflexive if and only it is finite dimensional.*

Proof. We have just seen in the previous proposition that if a vector space V over a division ring D is finite dimensional, then V is reflexive. So it remains to show that if V is reflexive, then V is finite dimensional. Suppose that V is an infinite dimensional with basis $\mathcal{B} = \{x_\alpha\}_\Delta$. Then $\operatorname{card}(\Delta) \geq \aleph_0$, so let $\Gamma \subseteq \Delta$ be such that $\operatorname{card}(\Gamma) \geq \aleph_0$ and for each $\alpha \in \Gamma$ let a_α be a nonzero element of D. Then $\{a_\alpha\}_\Gamma$ is an infinite set of nonzero elements of D. Define f on elements of $\mathcal{B}^*$ by $f(x_\alpha^*) = a_\alpha$ for $\alpha \in \Gamma$ and $f(x_\alpha^*) = 0$ when $\alpha \in \Delta - \Gamma$. If f is extended linearly to V^*, then $f \in V^{**}$. If x is an arbitrary element of V, let $x = \sum_\Delta x_\beta b_\beta$, where $b_\beta = 0$ for almost all $\beta \in \Delta$ and $x_\beta \in \mathcal{B}$ for each $\beta \in \Delta$. Then $\varphi_V(x) = f_x$ and $f_x(x_\alpha^*) = x_\alpha^*(\sum_\Delta x_\beta b_\beta) = b_\alpha$ for each $\alpha \in \Delta$, so $f_x(x_\alpha^*) = 0$ for almost all $\alpha \in \Delta$. But $f(x_\alpha^*) = a_\alpha$ and $a_\alpha \neq 0$ for an infinite number of $\alpha \in \Delta$, so there are $\alpha \in \Delta$ such that $f_x(x_\alpha^*) \neq f(x_\alpha^*)$. Hence, it must be the case that $\varphi_V(x) = f_x \neq f$ for all $x \in V$, so φ_V is not surjective and φ_V cannot be an isomorphism. Consequently, if V is not finite dimensional, then V is not reflexive. $\square$

In the Remark given in the introduction to this chapter, it was pointed out that if $\{x_\alpha\}_\Delta$ is a basis for a free R-module F, then $\{x_\alpha^*\}_\Delta$ may not be a basis for F^*. To see this, suppose it is always the case that if $\{x_\alpha\}_\Delta$ is a basis for a free R-module F, then $\{x_\alpha^*\}_\Delta$ is a basis for F^*. If V is an infinite dimensional vector space over a field K, then V is a free K-module, so, due to our assumption, if $\{x_\alpha\}_\Delta$ is a basis for V, then $\{x_\alpha^*\}_\Delta$ is a basis for V^*. Hence, V^* is a free K-module with basis $\{x_\alpha^*\}_\Delta$ and so

V^{**} is a free K-module with basis $\{x_\alpha^{**}\}_\Delta$. But this means that the canonical map $\varphi_V : V \rightarrow V^{**}$, defined on basis elements by $\varphi_V(x_\alpha) = x_\alpha^{**}$ for each $\alpha \in \Delta$, is an isomorphism which indicates that V is reflexive. But this contradicts the fact established in the previous proposition that V must be finite dimensional in order to be reflexive. Hence, if $\{x_\alpha\}_\Delta$ is a basis for a free R-module F, then $\{x_\alpha^*\}_\Delta$ need not be a basis for F^*.

Problem Set 10.1

1. Prove each of the following for a vector space V over a division ring D, where U, U_1 and U_2 are subspaces of V, X, X_1 and X_2 are subspaces of V^* and

$$\text{ann}_\ell^{V^*}(U) = \{f \in V^* \mid f(x) = 0 \text{ for all } x \in U\} \quad \text{and}$$
$$\text{ann}_r^V(X) = \{x \in V \mid f(x) = 0 \text{ for all } f \in X\}.$$

   (a) $\text{ann}_\ell^{V^*}(U)$ is a subspace of the left D-vector space V^* and $\text{ann}_r^V(X)$ is a subspace of V.
   (b) If $U_1 \subseteq U_2$, then $\text{ann}_\ell^{V^*}(U_2) \subseteq \text{ann}_\ell^{V^*}(U_1)$ and if $X_1 \subseteq X_2$, then $\text{ann}_r^V(X_2) \subseteq \text{ann}_r^V(X_1)$.
   (c) $U \subseteq \text{ann}_r^V(\text{ann}_\ell^{V^*}(U))$ and $X \subseteq \text{ann}_\ell^{V^*}(\text{ann}_r^V(X))$.
   (d) $\text{ann}_r^{V^*}(U) = \text{ann}_r^{V^*}(\text{ann}_\ell^V(\text{ann}_r^{V^*}(U)))$ and $\text{ann}_r^V(X) = \text{ann}_r^V(\text{ann}_\ell^{V^*}(\text{ann}_r^V(X)))$.
   (e) $\text{ann}_r^{V^*}(U_1 + U_2) = \text{ann}_r^{V^*}(U_1) \cap \text{ann}_r^{V^*}(U_2)$ and $\text{ann}_r^V(X_1 + X_2) = \text{ann}_r^V(X_1) \cap \text{ann}_r^V(X_2)$.

2. Let V be a finite dimensional vector space over a division ring D. Prove that each basis of V^* is dual to some basis of V. [Hint: If $\{x_i\}_{i=1}^n$ is a basis for V^*, then $\{x_i^*\}_{i=1}^n$ is a basis for V^{**} and $V \cong V^{**}$.]

3. Let V be an n dimensional vector space over a division ring D. If $f, g \in V^*$, assume that $f \neq g$, that neither f nor g is zero, and that $\text{Ker } f \neq \text{Ker } g$. Compute the dimensions of each of the following.
   (a) $\text{Ker } f$ and $\text{Ker } g$
   (b) $\text{Ker } f \cap \text{Ker } g$.
   (c) $\text{Ker } f + \text{Ker } g$.

4. Consider the $\mathbb{R}$-vector space $\mathbb{R}^3$. Show that $\mathcal{B} = \{x_1 = (1, 0, 1), x_2 = (-1, 1, 0), x_3 = (0, 1, 2)\}$ is a basis for $\mathbb{R}^3$ and compute $\mathcal{B}^*$. [Hint: Show that the representation of the vector (x, y, z) relative to this basis is $x_1(2x + 2y - z) + x_2(x + 2y - z) + x_3(-x - y + z)$.]

10.2 Reflexivity and R-modules

Since only finitely generated vector spaces are reflexive, if we are to study reflexive R-modules, it would appear that we will have to restrict our investigation to finitely generated free modules. However, if the module is finitely generated and projective, then we can show that it is reflexive.

Lemma 10.2.1. *A finitely generated free R-module is reflexive.*

Proof. If F is a finitely generated free R-module, then F has a finite basis, so there is a positive integer n such that $F \cong (R_R)^{(n)}$. This gives $F^* \cong (_R R)^{(n)}$ which in turn shows that $F^{**} \cong (R_R)^{(n)} \cong F$. $\square$

Proposition 10.2.2. *Let M be a projective R-module generated by n elements. Then*

(1) *M^* is a projective left R-module generated by n elements, and*

(2) *M is reflexive.*

Proof. (1) If M is a finitely generated projective R-module, then there is a split short exact sequence $0 \to N \xrightarrow{f} F \xrightarrow{g} M \to 0$, where F is a free R-module with a finite basis. Now $0 \to M^* \xrightarrow{g^*} F^* \xrightarrow{f^*} N^* \to 0$ is split exact, since the contravariant functor $\operatorname{Hom}_R(-, R)$ preserves split short exact sequences. Since there is an integer n such that $F \cong (R_R)^{(n)}$, $F^* \cong (_R R)^{(n)}$, so F^* is a free left R-module with a basis of n elements. Now $F^* \cong M^* \oplus N^*$, so if $\{x_1, x_2, \ldots, x_n\}$ is a basis for $M^* \oplus N^*$, then each x_i can be written as $x_i = (x_{M_i^*}, x_{N_i^*})$, where $x_{M_i^*} \in M^*$ and $x_{N_i^*} \in N^*$ for $i = 1, 2, \ldots, n$. It follows that $\{x_{M_1^*}, x_{M_2^*}, \ldots, x_{M_n^*}\}$ is a set of generates for M^*, so M^* is generated by n elements. Since M^* is isomorphic to a direct summand of the projective left R-module F^*, M^* is projective.

(2) As in (1) there is a split short exact sequence $0 \to N \xrightarrow{f} F \xrightarrow{g} M \to 0$, so $F \cong M \oplus N$. It follows that $F^{**} \cong M^{**} \oplus N^{**}$. Moreover, we have a commutative diagram

$$
\begin{array}{ccc}
M \oplus N & \xrightarrow{\ \varphi_F\ } & M^{**} \oplus N^{**} \\
\Big\downarrow{\scriptstyle \pi} & & \Big\downarrow{\scriptstyle \pi^{**}} \\
M & \xrightarrow{\ \varphi_M\ } & M^{**}
\end{array}
$$

where $\pi : M \oplus N \to M$ is the canonical projection. Now F is finitely generated and free, so by Lemma 10.2.1, F is reflexive. Since $\varphi_F|_M = \varphi_M$, it follows that φ_M is an isomorphism. $\square$

Observe that the proof of (2) of Proposition 10.2.2 shows that a direct summand of a reflexive module is reflexive. If the condition that the module is projective is removed from the preceding proposition, then the proposition, in general, fails. Recall that over a principal ideal domain a module is projective if and only if it is free. (See Proposition 5.2.16.) A simple cardinality argument shows that $\mathbb{Z}_n$, $n \geq 2$, as a $\mathbb{Z}$-module is not free and hence cannot be projective. Moreover, $\mathbb{Z}_n^* = \mathbb{Z}_n^{**} = 0$, so in this case, Proposition 10.2.2 does, in fact, fail. Thus, if the projective condition is to be removed, and if finitely generated modules are to remain reflexive, then the class of rings under consideration will have to be restricted. Toward this end, we investigate self-injective rings, rings that contain a copy of every simple R-module, and semiprimary rings.

Self-injective Rings

Definition 10.2.3. A ring R is said to be *right self-injective* if R_R is an injective R-module. *Left self-injective* rings are defined similarly.

Lemma 10.2.4. *The following properties of a ring R are equivalent.*

(1) *If A is a finitely generated right ideal of R and $f : A \to R$ is an R-linear mapping, then there is an $x \in R$ such that $f(a) = xa$ for all $a \in A$.*

(2) (a) *If $\{A_i\}_{i=1}^n$ is a family of finitely generated right ideals of R, then $\operatorname{ann}_\ell(\bigcap_{i=1}^n A_i) = \sum_{i=1}^n \operatorname{ann}_\ell(A_i)$ and*
(b) *for any $a \in R$, $\operatorname{ann}_\ell(\operatorname{ann}_r(Ra)) = Ra$.*

Proof. (1) $\Rightarrow$ (2a). If A and B are right ideals of R, then $\operatorname{ann}_\ell(A) + \operatorname{ann}_\ell(B) \subseteq \operatorname{ann}_\ell(A \cap B)$. So suppose that A and B are finitely generated, that $c \in \operatorname{ann}_\ell(A \cap B)$ and let $f : A + B \to R$ be such that

$$f(a) = a \text{ if } a \in A \quad \text{and} \quad f(b) = (1 + c)b \text{ if } b \in B.$$

(Since the two expressions for f agree on $A \cap B$, this map is well defined and R-linear.) Due to (1) there is an $x \in R$ such that $f(r) = xr$ for all $r \in A + B$. If $a \in A$, then $xa = a$, so $(x - 1)a = 0$. Hence, $x - 1 \in \operatorname{ann}_\ell(A)$. If $b \in B$, then

$$(1 + c - x)b = (1 + c)b - xb = f(b) - xb = xb - xb = 0,$$

so $1 + c - x \in \operatorname{ann}_\ell(B)$. Thus,

$$c = (x - 1) + (1 + c - x) \in \operatorname{ann}_\ell(A) + \operatorname{ann}_\ell(B)$$

and we have $\operatorname{ann}_\ell(A \cap B) \subseteq \operatorname{ann}_\ell(A) + \operatorname{ann}_\ell(B)$. Therefore,

$$\operatorname{ann}_\ell(A \cap B) = \operatorname{ann}_\ell(A) + \operatorname{ann}_\ell(B).$$

It now follows by induction that $\operatorname{ann}_\ell(\bigcap_{i=1}^n A_i) = \sum_{i=1}^n \operatorname{ann}_\ell(A)$ for any family $\{A_i\}_{i=1}^n$ of finitely generated right ideals of R.

(1) $\Rightarrow$ (2b). If $a \in R$, then we immediately have $Ra \subseteq \operatorname{ann}_\ell(\operatorname{ann}_r(Ra))$. On the other hand, if $b \in \operatorname{ann}_\ell(\operatorname{ann}_r(Ra))$, then $f : aR \to bR \subseteq R$ such that $f(ar) = br$ is a well-defined R-linear map. By (1), there is an $x \in R$ such that $f(ar) = xar$ for all $ar \in aR$. Hence, $br = xar$ for all $r \in R$. In particular, for $r = 1$, we have $b = xa$ and this gives $b \in Ra$. Hence, $\operatorname{ann}_\ell(\operatorname{ann}_r(a)) \subseteq Ra$, so

$$\operatorname{ann}_\ell(\operatorname{ann}_r(Ra)) = Ra.$$

(2) $\Rightarrow$ (1). Let A be a finitely generated right ideal of R and suppose that $f : A \to R$ is R-linear. We proceed by induction on n, the number of generators of A. If $n = 1$, then $A = aR$, so suppose that $f : aR \to R$ is a module homomorphism and let $c = f(a)$. Then $c \cdot \operatorname{ann}_r(Ra) = 0$, so (b) gives $c \in \operatorname{ann}_\ell(\operatorname{ann}_r(Ra)) = Ra$. Hence, $c = xa$ for some $x \in R$ and the map $f : aR \to R$ given by $f(ar) = xar$ for all $ar \in aR$ is a well-defined module homomorphism. Next, let $A = a_1 R + a_2 R + \cdots + a_n R$ and make the induction hypothesis that (1) holds for all finitely generated right ideals of R with k generators, where $1 \le k < n$. If $f : A \to R$ is an R-linear mapping, then there are $x', x'' \in R$ such that $f(a) = x'a$ if $a \in A' = a_1 R + a_2 R + \cdots + a_{n-1} R$ and $f(a) = x''a$ if $a \in a_n R$. If $a \in A' \cap a_n R$, then $(x' - x'')a = 0$, so it follows from (2a) that $x' - x'' \in \operatorname{ann}_\ell(A' \cap a_n R) = \operatorname{ann}_\ell(A') + \operatorname{ann}_\ell(a_n R)$. Hence, $x' - x'' = b' - b''$, where $b'A' = 0$ and $b''a_n R = 0$ and so $x' - b' = x'' - b''$. From this we see that if $a = c + a_n r \in A' + a_n R$, then $(x' - b')(c + a_n r) = (x'' - b'')(c + a_n r)$. Hence, if we let $x = x' - b'$, then $f(a) = xa$ for all $a \in A$ and we have (1). □

Proposition 10.2.5. *If R is a right self-injective ring, then*

(1) *if $\{A\}_{i=1}^n$ is a family of right ideals of R, then $\operatorname{ann}_\ell(\bigcap_{i=1}^n A_i) = \sum_{i=1}^n \operatorname{ann}_\ell(A_i)$ and*

(2) *if A is a finitely generated left ideal of R, then $\operatorname{ann}_\ell(\operatorname{ann}_r(A)) = A$.*

Conversely, if (1) and (2) hold and R is right noetherian, then R is right self-injective.

Proof. If R is right self-injective, then the proof of (1) $\Rightarrow$ (2a) of Lemma 10.2.4 with the finitely generated condition removed works and so (1). For the proof of (2), let $A = Ra_1 + Ra_2 + \cdots + Ra_n$ be a finitely generated left ideal of R. It follows easily that $\operatorname{ann}_r(\sum_{i=1}^n Ra_i) = \bigcap_{i=1}^n \operatorname{ann}_r(Ra_i)$ and so

$$\operatorname{ann}_\ell(\operatorname{ann}_r(A)) = \operatorname{ann}_\ell\left(\operatorname{ann}_r\left(\sum_{i=1}^n Ra_i\right)\right) = \operatorname{ann}_\ell\left(\bigcap_{i=1}^n \operatorname{ann}_r(Ra_i)\right).$$

Now $\{\operatorname{ann}_r(Ra_i)\}_{i=1}^n$ is a family of right ideals of R, so (1) gives

$$\operatorname{ann}_\ell\left(\bigcap_{i=1}^n \operatorname{ann}_r(Ra_i)\right) = \sum_{i=1}^n \operatorname{ann}_\ell(\operatorname{ann}_r(Ra_i))$$

and (2b) of Proposition 10.2.4 shows that $\text{ann}_\ell(\text{ann}_r(Ra_i)) = Ra_i$ for $i = 1, 2,$ $\ldots, n$. Hence,

$$\text{ann}_\ell(\text{ann}_r(A)) = Ra_1 + Ra_2 + \cdots + Ra_n = A.$$

Conversely, if R is right noetherian, then Lemma 10.2.4 and Baer's criteria show that R is right self-injective.   □

Proposition 10.2.6. *A right noetherian right self-injective ring is left noetherian.*

Proof. Let $J = J(R)$. Since R is right self-injective, Proposition 8.4.14 shows that R/J is a regular ring. We also see that R/J is right noetherian since R is, and so it follows from Proposition 6.6.2 that R/J is a semisimple ring. Also since R is right self-injective, Proposition 8.4.14 indicates that

$$J = \{a \in R \mid \text{ann}_r(a) \text{ is an essential right ideal of } R\}.$$

We claim that J is nilpotent. Consider the descending chain

$$J \supseteq J^2 \supseteq J^3 \supseteq \cdots .$$

Since R is right noetherian, the ascending chain

$$\text{ann}_r(J) \subseteq \text{ann}_r(J^2) \subseteq \text{ann}_r(J^3) \subseteq \cdots$$

of right ideals of R terminates. If

$$\text{ann}_r(J^n) = \text{ann}_r(J^{n+1}) = \cdots ,$$

then we claim that $J^n = 0$. Assume that $J^n \neq 0$, let

$$A = \{\text{ann}_r(y) \mid y \in R \text{ and } J^n y \neq 0\}$$

and choose a maximal element $\text{ann}_r(x)$ from A. Then $J^n x \neq 0$. If $a \in J$, then $\text{ann}_r(a) \cap xR \neq 0$, so $axy = 0$ for some $y \in R$ with $xy \neq 0$. Now $\text{ann}_r(x) \subsetneq \text{ann}_r(ax)$, so the maximality of $\text{ann}_r(x)$ means that $J^n ax = 0$. Since this holds for each $a \in J$, we see that $J^{n+1} x = 0$. Hence, $x \in \text{ann}_r(J^{n+1}) = \text{ann}_r(J^n)$. Therefore, $J^n x = 0$, so we have a contradiction. Thus, $J^n = 0$.

Hence, we have a decreasing chain

$$R \supseteq J \supseteq J^2 \supseteq \cdots \supseteq J^{n-1} \supseteq J^n = 0$$

and since R/J is semisimple each factor module J^{i-1}/J^i, $i = 1, 2, \ldots, n$, is a direct sum of simple R/J-modules. But this sum is finite by Lemma 6.6.4, so each factor module J^{i-1}/J^i has a composition series. Thus, it follows that J^{i-1}/J^i is right artinian for $i = 1, 2 \ldots, n$. It also follows, exactly as in the proof of Proposition 6.6.5, that R is right artinian.

Finally, let

$$A_1 \subseteq A_2 \subseteq A_3 \subseteq \cdots$$

be an ascending chain of finitely generated left ideals of R. Then

$$\mathrm{ann}_r(A_1) \supseteq \mathrm{ann}_r(A_2) \supseteq \mathrm{ann}_r(A_3) \supseteq \cdots$$

is a descending chain of right ideals which terminates, since R is right artinian. If

$$\mathrm{ann}_r(A_n) = \mathrm{ann}_r(A_{n+1}) = \cdots, \quad \text{then}$$
$$\mathrm{ann}_\ell(\mathrm{ann}_r(A_n)) = \mathrm{ann}_\ell(\mathrm{ann}_r(A_{n+1})) = \cdots.$$

But by Proposition 10.2.5, $A_n = A_{n+1} = \cdots$ and so R satisfies the ascending chain condition of finitely generated left ideals. It is straightforward to show that a ring R that satisfies this condition on finitely generated left ideals is R is left noetherian. □

Kasch Rings and Injective Cogenerators

Definition 10.2.7. A ring R is said to be a *right (left) Kasch ring* if R contains a copy of each simple (left) R-module. A left and right Kasch ring is referred to simply as a *Kasch ring*. An injective R-module M is an *injective cogenerator* for $\mathbf{Mod}_R$ if every R-module is cogenerated by M. Injective cogenerators for $_R\mathbf{Mod}$ are defined in the obvious way.

We now develop several conditions that will ensure that a right self-injective ring will be a cogenerator for $\mathbf{Mod}_R$. Left and right self-injective cogenerator rings are important in our investigation of rings over which every finitely generated (left) R-module is reflexive.

Lemma 10.2.8. *The following are equivalent for a maximal right (left) ideal $\mathfrak{m}$ of R.*

(1) $R/\mathfrak{m}$ *embeds in R.*

(2) $\mathfrak{m} = \mathrm{ann}_r(x)$ $(\mathfrak{m} = \mathrm{ann}_\ell(x))$ *for some $x \in R$, $x \neq 0$.*

(3) $\mathrm{ann}_\ell(\mathfrak{m}) \neq 0$ $(\mathrm{ann}_r(\mathfrak{m}) \neq 0)$.

(4) $\mathfrak{m} = \mathrm{ann}_r(\mathrm{ann}_\ell(\mathfrak{m}))$ $(\mathfrak{m} = \mathrm{ann}_\ell(\mathrm{ann}_r(\mathfrak{m})))$.

(5) *If $y \in R$ and $(\mathfrak{m} : y) = \{a \in R \mid ya \in \mathfrak{m}\}$ $((\mathfrak{m} : y) = \{a \in R \mid ay \in \mathfrak{m}\})$, then there is a $y \in R$ such that $\mathrm{ann}_\ell((\mathfrak{m} : y)) \neq 0$ $(\mathrm{ann}_\ell((\mathfrak{m} : y)) \neq 0)$.*

Proof. (1) $\Rightarrow$ (2). If $f : R/\mathfrak{m} \rightarrow R$ is an R-linear embedding and $f(1 + \mathfrak{m}) = x \neq 0$, then $\mathrm{ann}_r(x) = \mathfrak{m}$.

(2) $\Rightarrow$ (3). Since $x\mathfrak{m} = x\,\mathrm{ann}_r(x) = 0$, $0 \neq x \in \mathrm{ann}_\ell(\mathfrak{m})$.

(3) $\Rightarrow$ (4). Since $\text{ann}_\ell(\mathfrak{m}) \neq 0$, we see that $\mathfrak{m} \subseteq \text{ann}_r(\text{ann}_\ell(\mathfrak{m})) \subsetneqq R$. Hence, the maximality of $\mathfrak{m}$ gives $\mathfrak{m} = \text{ann}_r(\text{ann}_\ell(\mathfrak{m}))$.

(4) $\Rightarrow$ (5). If $\text{ann}_\ell((\mathfrak{m} : y)) = 0$ for all $y \in R$, then $\text{ann}_\ell((\mathfrak{m} : 1)) = \text{ann}_\ell(\mathfrak{m}) = 0$. But then $\text{ann}_r(\text{ann}_\ell(\mathfrak{m})) = R$ and this contradicts (4).

(5) $\Rightarrow$ (1). Let $y \in R$ be such that $\text{ann}_\ell((\mathfrak{m} : y)) \neq 0$. If $0 \neq x \in \text{ann}_\ell((\mathfrak{m} : y))$ and if $f : R/(\mathfrak{m} : y) \to R$ and $g : R/(\mathfrak{m} : y) \to R/\mathfrak{m}$ are such that $f(a + (\mathfrak{m} : y)) = xa$ and $g(a + (\mathfrak{m} : y)) = (y + \mathfrak{m})a$, then f and g are well-defined R-linear mappings. If $g(a + (\mathfrak{m} : y)) = 0$, then $0 = (y + \mathfrak{m})a = ya + \mathfrak{m}$, so $ya \in \mathfrak{m}$. Hence, $a \in (\mathfrak{m} : y)$, so $a + (\mathfrak{m} : y) = 0$. Thus, g is an injection. Since $(\mathfrak{m} : y) \neq R$, $R/(\mathfrak{m} : y) \neq 0$, so $g \neq 0$. Consequently, g must be an isomorphism since $R/\mathfrak{m}$ is a simple R-module and since $\text{Im}(g)$ is a nonzero submodule of $R/\mathfrak{m}$. Finally, we can write $R/\mathfrak{m} = (y + \mathfrak{m})R$, so $fg^{-1} : R/\mathfrak{m} \to R$ is such that $fg^{-1}((y + \mathfrak{m})a) = f(a + (\mathfrak{m} : y)) = xa$. Hence, if $fg^{-1} = 0$, then $xa = 0$ for each $a \in R$. In particular, we have $x = 0$, when $a = 1$. But this contradicts the fact that $x \neq 0$ and so $fg^{-1} \neq 0$. Hence, $\text{Ker}(fg^{-1})$ is a proper submodule of the simple R-module $R/\mathfrak{m}$, so $\text{Ker}(fg^{-1}) = 0$. Therefore, $fg^{-1} : R/\mathfrak{m} \to R$ is an embedding and we have (1).

The left-hand version of the proposition has a similar proof. $\square$

Proposition 10.2.9. *The following are equivalent for a right (left) self-injective ring R.*

(1) *R is a right (left) Kasch ring.*

(2) *R_R ($_R R$) is a cogenerator for $\mathbf{Mod}_R$ ($_R\mathbf{Mod}$).*

(3) *$\text{Hom}_R(M, R) \neq 0$ for every nonzero cyclic (left) R-module M.*

(4) *$\text{Hom}_R(M, R) \neq 0$ for every nonzero (left) R-module M.*

(5) *$\text{ann}_\ell(A) \neq 0$ ($\text{ann}_r(A) \neq 0$) for every proper right (left) ideal A of R.*

(6) *$\text{ann}_r(\text{ann}_\ell(A)) = A$ ($\text{ann}_\ell(\text{ann}_r(A)) = A$) for every right (left) ideal A of R.*

Proof. (1) $\Rightarrow$ (2). Suppose that R is a right Kasch ring. If M is a nonzero R-module, let $x \in M$, $x \neq 0$. Since xR is finitely generated, by considering Proposition 6.1.2, we see that xR has a maximal submodule, say N_x. The composition of the maps $xR \xrightarrow{\eta_x} xR/N_x \xrightarrow{f_x} R$, where η_x is the natural surjection and f_x is an embedding of the simple R-module xR/N_x into R, gives an R-linear map $g_x : xR \to R$ with kernel N_x such that $g_x(x) \neq 0$. For each $x \in M$, $x \neq 0$, choose exactly one such $g_x : xR \to R$. If H is this set of homomorphisms, then $\phi : M \to R^H$ such that $\phi(x) = (g_x(x))$ for each $x \in M$ is an R-linear mapping and we claim that $\text{Ker}\,\phi = \bigcap_H \text{Ker}\,g_x = 0$. If $z \in \bigcap_H \text{Ker}\,g_x$, $z \neq 0$, then $z \in \bigcap_{x \in M} N_x$. In particular, $z \in N_z$ which means that $g_z(z) = 0$, a contradiction. Hence, $\text{Ker}\,\phi = 0$, so R is a cogenerator for $\mathbf{Mod}_R$.

(2) $\Rightarrow$ (3). If $M = xR$ is a nonzero cyclic R-module, then as in (1) $\Rightarrow$ (2), the composition map $xR \xrightarrow{\eta_x} xR/N \xrightarrow{f_x} R$ shows that $\mathrm{Hom}_R(M, R) \neq 0$.

(3) $\Rightarrow$ (4). If M is a nonzero R-module and $x \in M$, $x \neq 0$, then $\mathrm{Hom}_R(xR, R)$ $\neq 0$. Since R is right self-injective, any nonzero map in $\mathrm{Hom}_R(xR, R)$ extends to a nonzero map in $\mathrm{Hom}_R(M, R)$.

(4) $\Rightarrow$ (5). If A is a proper right ideal of R, then R/A is a cyclic R-module with generator $1 + A$. If $0 \neq f \in \mathrm{Hom}_R(R/A, R)$ and $f(1 + A) = x$, then $x \neq 0$ and $f(r + A) = xr$ for all $r + A \in R/A$. Since $xA = 0$, $x \in \mathrm{ann}_\ell(A)$, so $\mathrm{ann}_\ell(A) \neq 0$.

(5) $\Rightarrow$ (1) If $\mathfrak{m}$ is a maximal right ideal of R, then $\mathrm{ann}_\ell(\mathfrak{m}) \neq 0$. Hence, we see from Lemma 10.2.8 that $R/\mathfrak{m}$ embeds in R. Thus, R contains a copy of each simple R-module, so R is a right Kasch ring.

(2) $\Rightarrow$ (6). Let $f : R/A \to R^\Delta$ be an embedding. If $f(1 + A) = (a_\alpha)$, then $A = \mathrm{ann}_r((a_\alpha))$, so

$$\mathrm{ann}_r(\mathrm{ann}_\ell(A)) = \mathrm{ann}_r(\mathrm{ann}_\ell(\mathrm{ann}_r((a_\alpha)))) = \mathrm{ann}_r((a_\alpha)) = A.$$

(6) $\Rightarrow$ (1) follows from Lemma 10.2.8.

The left-hand version of the proposition follows by symmetry. $\square$

Semiprimary Rings

We now prove two propositions that establish properties of rings that will be required in our investigation of rings over which every finitely generated module is reflexive. J continues to denote the Jacobson radical of R.

Definition 10.2.10. A ring R is said to be *semiprimary* if J is nilpotent and R/J is a semisimple ring.

Proposition 10.2.11. *If R is a semiprimary ring, then an R-module is artinian if and only if it is noetherian.*

Proof. A technique similar to that used in the proof of Proposition 6.6.5 can be used to prove the proposition, so this is left as an exercise. $\square$

Proposition 10.2.12. *A ring R is right artinian if and only if it is semiprimary and right noetherian.*

Proof. If R is semiprimary and right noetherian, then R is right artinian by the preceding Proposition.

Conversely, if R is right artinian, then Corollary 6.6.6 shows that R is right noetherian and Proposition 6.3.1 indicates that J is nilpotent. Since R/J is right artinian and Jacobson semisimple, Proposition 6.6.2 shows that R/J is semisimple. Thus, R is semiprimary. $\square$

Quasi-Frobenius Rings

Finally, we are in a position to investigate rings over which every finitely generated R-module is reflexive. Previously, we called a ring artinian (noetherian) if it was left and right artinian (left and right noetherian). The left and right modifiers will now be added for clarity and emphasis.

Definition 10.2.13. A ring R is said to be a *quasi-Frobenius* ring or simply a *QF-ring* if R is left and right artinian and

(a) $\operatorname{ann}_r(\operatorname{ann}_\ell(A)) = A$ for all right ideals A of R and

(b) $\operatorname{ann}_\ell(\operatorname{ann}_r(A)) = A$ for all left ideals A of R.

There are various conditions on a ring that will render it a QF-ring.

Proposition 10.2.14. *The following are equivalent for a ring R.*

(1) *R is a QF-ring.*

(2) *R is a right (left) noetherian ring that satisfies the conditions*
   (a) $\operatorname{ann}_r(\operatorname{ann}_\ell(A)) = A$ *for all right ideals A of R and*
   (b) $\operatorname{ann}_\ell(\operatorname{ann}_r(A)) = A$ *for all left ideals A of R.*

(3) *R is right (left) noetherian and right (left) self-injective.*

(4) *R is left (right) noetherian and right (left) self-injective.*

Proof. (1) $\Rightarrow$ (2). Since R is right artinian, Corollary 6.6.6 shows that R is right noetherian.

(2) $\Rightarrow$ (3). To show that R is right self-injective, let A and B be right ideals of R. Then using (2a) we have

$$\operatorname{ann}_r(\operatorname{ann}_\ell(A) + \operatorname{ann}_\ell(B)) = \operatorname{ann}_r(\operatorname{ann}_\ell(A)) \cap \operatorname{ann}_r(\operatorname{ann}_\ell(B)) = A \cap B,$$

so by taking left annihilators and using (2b), we get

$$\operatorname{ann}_\ell(A) + \operatorname{ann}_\ell(B) = \operatorname{ann}_\ell(A \cap B).$$

Induction and Proposition 10.2.5 shows that R is right self-injective.

(3) $\Rightarrow$ (4). This is Proposition 10.2.6.

(4) $\Rightarrow$ (1). This proof will be divided into several parts. For each part we assume that R is left noetherian and right self-injective.

Part I. Part (2) of Proposition 10.2.5 gives $\operatorname{ann}_\ell(\operatorname{ann}_r(A)) = A$ for each finitely generated left ideal A of R. Since R is left noetherian, we have

$$\operatorname{ann}_\ell(\operatorname{ann}_r(A)) = A \quad \text{for each left ideal } A \text{ of } R.$$

Part II. Note first that R/J is left noetherian and since R is right self-injective, R/J is a regular ring due to Proposition 8.4.14. Hence, the left-hand version of Proposition 6.6.2 shows that R/J is semisimple. We claim that J is nilpotent. Consider the descending chain $J \supseteq J^2 \supseteq J^3 \supseteq \cdots$ of ideals of R. Then $\operatorname{ann}_r(J) \subseteq \operatorname{ann}_r(J^2) \subseteq \operatorname{ann}_r(J^3) \subseteq \cdots$ is an ascending chain of ideals of R which terminates, since R is left noetherian. If $\operatorname{ann}_r(J^n) = \operatorname{ann}_r(J^{n+1}) = \cdots$, then $\operatorname{ann}_\ell(\operatorname{ann}_r(J^n)) = \operatorname{ann}_\ell(\operatorname{ann}_r(J^{n+1}))$, so by Part I, $J^n = J^{n+1}$. Hence, Nakayama's lemma (Lemma 6.1.10) gives $J^n = 0$. Thus, R is a semiprimary ring and the left-hand version of Proposition 10.2.11 shows that

$$R \text{ is left artinian.}$$

Part III. If we can show that R is a right Kasch ring, then Proposition 10.2.9 will give

$$\operatorname{ann}_r(\operatorname{ann}_\ell(A)) = A \quad \text{for each right ideal } A \text{ of } R.$$

For this, first note that as in Part II, J is nilpotent. So Lemma 8.3.7 indicates that $\operatorname{ann}_\ell(J)$ is an essential right ideal of R. If $H_1, H_2, \ldots, H_n$ are the homogeneous components of $\operatorname{Soc}(_R R)$, then the H_i are ideals of R and $A_i = \operatorname{ann}_\ell(J) \cap H_i \neq 0$ is an ideal of R for $i = 1, 2, \ldots, n$. From Part II we see that R/J is a semisimple ring, so since Proposition 6.1.7 gives $A_i J = 0$, each A_i is a semisimple R/J-module. Thus, A_i contains a simple R/J-module S_i which is also a simple R-module. If S is a simple R-module, then $SJ = 0$, so S is a simple R/J-module. Thus, $S \cong S_i$ for some i, $1 \leq i \leq n$, so R contains a copy of each simple R-module, and R is therefore a right Kasch ring.

Part IV. Finally, we need to show that R is right artinian. If

$$A_1 \supseteq A_2 \supseteq A_3 \supseteq \cdots$$

is a decreasing chain of right ideals of R, then

$$\operatorname{ann}_\ell(A_1) \subseteq \operatorname{ann}_\ell(A_2) \subseteq \operatorname{ann}_\ell(A_3) \subseteq \cdots$$

is an ascending chain of left ideals of R. This latter chain terminates since R is left noetherian. If $\operatorname{ann}_\ell(A_n) = \operatorname{ann}_\ell(A_{n+1}) = \cdots$, then $\operatorname{ann}_r(\operatorname{ann}_\ell(A_n)) = \operatorname{ann}_r(\operatorname{ann}_\ell(A_{n+1})) = \cdots$. But Part III gives $A_n = A_{n+1} = \cdots$, so

$$R \text{ is right artinian}$$

and the proof is complete. (A QF-ring is left-right symmetric, so the parenthetical versions of the proposition also hold.) $\square$

Remark. Corollary 6.6.6 shows that if a ring is right (left) artinian, then it is right (left) noetherian. It follows that a ring R is a QF-ring if and only if R is left and right noetherian and left and right self-injective.

Examples

1. If K is a field, then K is a QF-ring.
2. If R is a principal ideal domain, then $R/(a)$ is a QF-ring for each nonzero nonunit $a \in R$. Hence, $\mathbb{Z}_n$ is a QF-ring for each integer $n \geq 2$. Note $\mathbb{Z}$ is not self-injective, so $\mathbb{Z}$ is not a QF-ring.
3. If $R = \prod_{i=1}^{n} R_i$, then R is a QF-ring if and only if each R_i is a QF-ring.

We have one more proposition to prove before we can consider finitely generated modules over a QF-ring. One result of the following proposition is that over a QF-ring the class of projective modules and the class of injective modules coincide.

Proposition 10.2.15. *If R is a QF-ring, then the following are equivalent for an R-module M.*

(1) *M is injective.*

(2) *There is a family $\{e_\alpha\}_\Delta$ of idempotents of R such that $M \cong \bigoplus_\Delta e_\alpha R$.*

(3) *M is projective.*

Proof. (1) $\Rightarrow$ (2). Since R is right noetherian and M is injective, it follows from Proposition 4.2.10 and Corollary 5.1.14 that $M = \bigoplus_\Delta E_\alpha$, where each E_α is an indecomposable injective R-module. Let $x \in E_\alpha$, $x \neq 0$, and consider the R-linear mapping $f : R \to E_\alpha$ given by $f(a) = xa$ and the induced embedding $R/\operatorname{Ker} f \to E_\alpha$. If $\{A_\alpha\}_\Delta$ is the family of right ideals of R that properly contain $\operatorname{Ker} f$, then, since R is right artinian, $\{A_\alpha\}_\Delta$ has a minimal element, say A. But then $A/\operatorname{Ker} f$ is a simple R-module, so we see that each E_α contains a simple submodule S_α. But S_α is essential in its injective envelope $E(S_\alpha)$, so $E(S_\alpha)$ embeds in E_α. It follows that $E_\alpha \cong E(S_\alpha)$, since E_α is indecomposable. Now R is a right Kasch ring, so S_α is isomorphic to a minimal right ideal of R, and since R is right self-injective, E_α embeds in R. Thus, $E_\alpha \cong e_\alpha R$ for some idempotent e_α of R.

(2) $\Rightarrow$ (3). Each $e_\alpha R$ is a direct summand of R and so is projective. But direct sums of projectives are projective and this gives (3).

(3) $\Rightarrow$ (1). Since R is right noetherian, direct sums of injectives are injective. (See Proposition 7.1.6.) Hence, every free R-module is injective since R_R is injective. But every projective R-module is a direct summand of a free R-module, so it follows that M is injective. $\square$

We can now address reflexivity of finitely generated modules.

Proposition 10.2.16. *The following hold for a QF-ring R.*

(1) *Every (left) R-module is torsionless.*

(2) *Every finitely generated (left) R-module is reflexive.*

(3) *A right (left) R-module M is finitely generated if and only if the left (right) R-module M^* is finitely generated.*

Proof. (1) Every *R*-module is the homomorphic image of a free *R*-module, so there is a free *R*-module F and an epimorphism $f : F \rightarrow E(M)$, where $E(M)$ is the injective envelope of M. Since the preceding proposition indicates that $E(M)$ is projective, there is an *R*-linear mapping $g : E(M) \rightarrow F$ such that $fg = \mathrm{id}_{E(M)}$. It follows that g is a monomorphism, so we can assume that $E(M)$ is a submodule of F. Hence, $M \subseteq E(M) \subseteq F$. But Exercise 3 indicates that free modules are torsionless and submodules of torsionless modules are torsionless, so M is torsionless. A similar result holds for a left *R*-module, so over a QF-ring every left and right *R*-module is torsionless.

(2) Suppose that M is a finitely generated *R*-module. Then we have a short exact sequence $0 \rightarrow N \rightarrow F \xrightarrow{f} M \rightarrow 0$, where F is a finitely generated free *R*-module. Since R is left and right self-injective, Corollary 5.1.12 shows that the contravariant functors $\mathrm{Hom}_R(-, R_R)$ and $\mathrm{Hom}_R(-, {}_RR)$ are exact. Hence, we have a commutative diagram

$$
\begin{array}{ccccccccc}
0 & \longrightarrow & N & \longrightarrow & F & \xrightarrow{\ f\ } & M & \longrightarrow & 0 \\
 & & \Big\downarrow{\varphi_N} & & \Big\downarrow{\varphi_F} & & \Big\downarrow{\varphi_M} & & \\
0 & \longrightarrow & N^{**} & \longrightarrow & F^{**} & \xrightarrow{\ f^{**}\ } & M^{**} & \longrightarrow & 0
\end{array}
$$

and Proposition 10.2.2 indicates that φ_F is an isomorphism. But (1) indicates that M is torsionless, so φ_M is an injection. Furthermore, f^{**} is an epimorphism and a simple diagram chase shows that φ_M is also an epimorphism. Hence, φ_M is an isomorphism and so M is reflexive. Symmetry gives the same result for left *R*-modules.

(3) If M is a finitely generated *R*-module, then we have a surjective mapping $(R_R)^{(n)} \rightarrow M$ for some integer $n \geq 1$. So as in (1), we have an injective mapping $M^* \rightarrow (R_R)^{(n)*} \cong ({}_RR)^{(n)}$. But R is left noetherian so it follows that M^* is finitely generated. Conversely, suppose that M^* is finitely generated. What we have just proved also holds for left *R*-modules, so if the left *R*-module M^* is finitely generated, then M^{**} is a finitely generated *R*-module. Now (1) indicates that M is torsionless, so $\varphi_M : M \rightarrow M^{**}$ is an injection. Since R is right noetherian, we see that M is finitely generated.   □

Therefore, over a quasi-Frobenius ring every finitely generated left and right *R*-module is reflexive. It is also the case that if R is a left and a right noetherian ring and if every finitely generated left and right *R*-module is reflexive, then R is a QF-ring. A proof of this fact will be delayed until Chapter 12 where a proof will given using homological methods.

Remark. We also point out a nice result on quasi-Frobenius rings obtained by Faith and Walker. They proved that a ring R is quasi-Frobenius if and only if every projective R-module is injective if and only if every injective R-module is projective. Details can be found in [14]. Also, recent results as well as open problems on QF-rings and related rings can be found in [56].

Problem Set 10.2

1. Show that the following are equivalent for an injective R-module E.

   (a) E is a cogenerator for $\mathbf{Mod}_R$.

   (b) $\mathrm{Hom}_R(S, E) \neq 0$ for every simple R-module.

   (c) Every simple R-module embeds in E.

   (d) E cogenerates very simple R-module.

   [(b) $\Rightarrow$ (a), Hint: Let M be a right R-modules and suppose that x is a nonzero element of M. Then xR is finitely generated, so Proposition 6.1.2 indicates that xR contains a maximal submodule, say N. Thus, xR/N is a simple R-module and (b) gives $\mathrm{Hom}_R(xR/N, E) \neq 0$. Consider the commutative diagram

$$
\begin{array}{ccc}
0 \longrightarrow xR \longrightarrow M \\
\hspace{1.2cm}\downarrow \hspace{0.6cm}\nearrow \\
\hspace{1cm}xR/N \\
\hspace{1.2cm}\downarrow \nearrow \\
\hspace{1cm}E
\end{array}
$$

   and Exercise 3 in Problem Set 4.1.]

2. Verify Examples 1, 2 and 3. [2, Hint: Exercise 14 in Problem Set 5.1.] [3, Hint: Show that R is right noetherian if and only each R_i is right noetherian and that R is right self-injective if and only if each $(R_i)_R$ is injective if and only if each $(R_i)_{R_i}$ is injective.]

3. Answer each of the following questions or prove the given statement.

   (a) Are free modules are torsionless?

   (b) Submodules of torsionless modules are torsionless. [Hint: Let N be a submodule of the torsionless R-module M, let $i_N : N \to M$ be the canonical injection and consider the commutative diagram

$$
\begin{array}{ccc}
N & \xrightarrow{\varphi_N} & N^{**} \\
\downarrow{i_N} & & \downarrow{i_N^{**}} \\
M & \xrightarrow{\varphi_N} & M^{**}
\end{array} \ ]
$$

   (c) Factor modules of torsionless modules need not be torsionless.

   (d) If $\{M_i\}_{i=1}^n$ is a family of reflexive modules, then is $\bigoplus_{i=1}^n M_i$ reflexive?

4. (a) Let R be an integral domain and suppose that M is an R-module. If $t(M)$ denotes the torsion submodule of M and $\eta : M \rightarrow M/t(M)$ is the canonical surjection, prove that $\eta^* : (M/t(M))^* \rightarrow M^*$ is an isomorphism.

   (b) If R is a principal ideal domain and M is a finitely generated R-module, prove that M^{**} and $M/t(M)$ are isomorphic.

5. If R is a ring without zero divisors, prove that R is right self-injective if and only if R is a division ring.

6. Prove each of the following for an integer $n \geq 1$.

   (a) $(R_R)^{(n)*} \cong (_R R)^{(n)}$ and $(R_R)^{(n)**} \cong (R_R)^{(n)}$.

   (b) $(_R R)^{(n)*} \cong (R_R)^{(n)}$ and $(_R R)^{(n)**} \cong (_R R)^{(n)}$.

7. Prove Proposition 10.2.11. [Hint: Proposition 6.6.5.]

8. (a) Show that an R-module M is torsionless if and only if for each $x \in M$, $x \neq 0$, there is an $f \in M^*$ such that $f(x) \neq 0$. [Hint: $\varphi_M : M \rightarrow M^{**}$ must be an injection.]

   (b) If R is an integral domain, prove that if M is a finitely generated torsionless R-module, then M is torsion free. [Hint: Use (a).]

9. If D is a division ring, show that the ring $R = \begin{pmatrix} D & D \\ 0 & D \end{pmatrix}$ is not a QF-ring. [Hint: Consider the right ideal $\mathfrak{m} = \begin{pmatrix} 0 & D \\ 0 & D \end{pmatrix}$ of R and compute $\text{ann}_r(\text{ann}_\ell(\mathfrak{m}))$.]

10. Suppose that R is a commutative ring and let M and N be R-modules.

    (a) Show that $\phi : M^* \otimes_R N \rightarrow \text{Hom}_R(M, N)$ such that $\phi(f \otimes y)(z) = f(z)y$ is an R-linear mapping.

    (b) If M and N are free R-modules of rank m and n, respectively, show that $M^* \otimes_R N$ and $\text{Hom}_R(M, N)$ are free R-modules of rank mn. Conclude that $M^* \otimes_R N$ and $\text{Hom}_R(M, N)$ are isomorphic R-modules. In particular, show that ϕ, as defined in (a), is an isomorphism.

Chapter 11

Introduction to Homological Algebra

We know that every module is the homomorphic image of a projective module and that every module can be embedded in an injective module. These observations provide the tools necessary to build projective and injective resolutions that play a central role in homological algebra and in the theory of derived functors. We continue with the assumption from Chapter 3 that the term "functor" means covariant functor.

11.1   Chain and Cochain Complexes

A sequence $\mathbf{M} = \{M_n, \alpha_n\}_{\mathbb{Z}}$ of R-modules and R-module homomorphisms, also denoted by

$$\mathbf{M} : \cdots \to M_{n+1} \xrightarrow{\alpha_{n+1}} M_n \xrightarrow{\alpha_n} M_{n-1} \xrightarrow{\alpha_{n-1}} M_{n-2} \to \cdots,$$

is said to be a *chain complex* if $\alpha_n \alpha_{n+1} = 0$ for each $n \in \mathbb{Z}$. Each mapping α_n : $M_n \to M_{n-1}$ is said to be a *boundary mapping* (or a *differential operator*). Note that for a chain complex the subscripts decrease from left to right and the subscript on each boundary map agrees with the subscript on its domain. A chain complex $\mathbf{M}$ is said to be exact at M_n if $\mathrm{Im}(\alpha_{n+1}) = \mathrm{Ker}(\alpha_n)$ and $\mathbf{M}$ is *exact* if it is exact at M_n for each $n \in \mathbb{Z}$. A chain complex of the form

$$\mathbf{M} : \cdots \to M_n \xrightarrow{\alpha_n} M_{n-1} \to \cdots \to M_1 \xrightarrow{\alpha_1} M_0 \to 0,$$

where the additional zeroes to the right have been omitted, is said to be *positive*. Similarly, a sequence $\mathbf{M} = \{M^n, \alpha^n\}_{\mathbb{Z}}$ of R-modules and R-module homomorphisms, also denoted by

$$\mathbf{M} : \cdots \to M^{n-1} \xrightarrow{\alpha^{n-1}} M^n \xrightarrow{\alpha^n} M^{n+1} \xrightarrow{\alpha^{n+1}} M^{n+2} \to \cdots,$$

such that $\alpha^{n+1} \alpha^n = 0$ for all $n \in \mathbb{Z}$, is a *cochain complex*. For a cochain complex, the superscripts increase from left to right and the superscripts on the boundary maps agree with the superscripts on their domains. A cochain complex $\mathbf{M}$ is exact at M^n if $\mathrm{Im}(\alpha_{n-1}) = \mathrm{Ker}(\alpha_n)$ and it is *exact* if it is exact at M^n for each $n \in \mathbb{Z}$. A cochain complex $\mathbf{M}$ is said to be *positive* if $M^n = 0$ for all $n < 0$, that is, if $\mathbf{M}$ is of the form

$$\mathbf{M} : 0 \to M^0 \xrightarrow{\alpha^0} M^1 \to \cdots \to M^{n-1} \xrightarrow{\alpha^{n-1}} M^n \to \cdots,$$

where the additional zeroes to the left have been suppressed. Using descending subscripts for chain complexes and ascending superscripts for cochain complexes cannot always be strictly adhered to. For example, if $\mathcal{F} : \mathbf{Mod}_R \to \mathbf{Mod}_S$ is an additive contravariant functor and $\mathbf{M} = \{M_n, \alpha_n\}_{\mathbb{Z}}$ is a chain complex, then $\mathcal{F}(\mathbf{M}) = \{\mathcal{F}(M_n), \mathcal{F}(\alpha_n)\}_{\mathbb{Z}}$ is a cochain complex

$$\mathcal{F}(\mathbf{M}) : \cdots \to \mathcal{F}(M_{n-2}) \xrightarrow{\mathcal{F}(\alpha_{n-1})} \mathcal{F}(M_{n-1}) \xrightarrow{\mathcal{F}(\alpha_n)} \mathcal{F}(M_n) \to \cdots$$

with lower indices. Moreover, the subscripts on the mappings $\mathcal{F}(\alpha_n)$, $n \in \mathbb{Z}$, no longer agree with the subscripts on their domains. Both of these notational anomalies could be easily corrected by a change in notation, but such a change would only introduce additional notation. It is more efficient to forgo any notational change and to simply interpret the notation correctly from the context of the discussion.

Finally, if $\mathbf{M} = \{M_n, \alpha_n\}_{\mathbb{Z}}$ is a chain complex, then $\mathbf{M}$ can be converted to a cochain complex $\mathbf{N} = \{N^n, \beta^n\}_{\mathbb{Z}}$ by *raising indices*, that is, by setting $N^n = M_{-n}$ and $\beta^n = \alpha_{-n}$ for all $n \in \mathbb{Z}$. Similarly, a cochain complex can be converted to a chain complex by *lowering indices*. Thus, we see that the only difference between a chain complex and a cochain complex is in the notation used. For this reason, a result obtained for a chain complex can often be obtained for a cochain complex simply by raising indices and conversely.

Definition 11.1.1. If $\mathbf{M}$ is a chain complex, then $H_n(\mathbf{M}) = \operatorname{Ker}\alpha_n / \operatorname{Im}\alpha_{n+1}$ is called the *nth homology module* of $\mathbf{M}$ and if $\mathbf{M}$ is a chain complex of abelian groups, then we call $H_n(\mathbf{M})$ the *nth homology group* of $\mathbf{M}$. Similarly, if M is a cochain complex, then $H^n(\mathbf{M}) = \operatorname{Ker}\alpha^n / \operatorname{Im}\alpha^{n-1}$ is the *nth cohomology module* of $\mathbf{M}$.

Another important concept in homological algebra is that of a (co)chain map.

Definition 11.1.2. If $\mathbf{M}$ and $\mathbf{N}$ are chain complexes, then a *chain map* $\mathbf{f} : \mathbf{M} \to \mathbf{N}$ *of degree* k is a family $\mathbf{f} = \{f_n : M_n \to N_{n+k}\}_{\mathbb{Z}}$ of R-linear mappings such that the diagram

$$\begin{array}{ccccc}
\cdots \xrightarrow{\alpha_{n+1}} & M_n & \xrightarrow{\alpha_n} & M_{n-1} & \xrightarrow{\alpha_{n-1}} \cdots \\
& \downarrow{\scriptstyle f_n} & & \downarrow{\scriptstyle f_{n-1}} & \\
\cdots \xrightarrow{\beta_{n+k+1}} & N_{n+k} & \xrightarrow{\beta_{n+k}} & N_{n+k-1} & \xrightarrow{\beta_{n+k-1}} \cdots
\end{array}$$

is commutative for each $n \in \mathbb{Z}$. *Cochain maps* $\mathbf{f} = \{f^n : M^n \to N^{n+k}\}_{\mathbb{Z}}$ and the *degree of a cochain map* are defined in the obvious way. If $\mathbf{f} : \mathbf{M} \to \mathbf{N}$ is a (co)chain map and no degree is specified, then it will be assumed that $\mathbf{f}$ has degree zero. If a (co)chain map $\mathbf{f}$ is such that f_n (f^n) is a monomorphism (an epimorphism, an isomorphism) for each $n \in \mathbb{Z}$, then we will refer to $\mathbf{f}$ as a monomorphism (an epimorphism, an isomorphism).

If $\mathbf{M}$ is a chain complex, then it is obvious that $\mathbf{M}$ is exact at M_n if and only if $H_n(\mathbf{M}) = 0$. A similar observation holds if $\mathbf{M}$ is a cochain complex.

Remark. Whenever possible the notation for boundary maps will be in matching alphabetical order with the order of the (co)chain complexes. For example, if $\mathbf{L}, \mathbf{M}$ and $\mathbf{N}$ are chain complexes, then the boundary maps for these complexes will be denoted by α_n, β_n and γ_n, respectively.

If $\mathbf{M}$ and $\mathbf{N}$ are chain complexes and $\mathbf{f} : \mathbf{M} \to \mathbf{N}$ is a chain map, then for each $n \in \mathbb{Z}$ there is an induced *nth homology mapping* $H_n(\mathbf{f}) : H_n(\mathbf{M}) \to H_n(\mathbf{N})$ established by the following proposition.

Proposition 11.1.3. *If $\mathbf{f} : \mathbf{M} \to \mathbf{N}$ is a chain map, then for each $n \in \mathbb{Z}$ there is an R-linear mapping $H_n(\mathbf{f}) : H_n(\mathbf{M}) \to H_n(\mathbf{N})$ defined by*

$$H_n(\mathbf{f})(x + \operatorname{Im}\alpha_{n+1}) = f_n(x) + \operatorname{Im}\beta_{n+1}$$

for all $x + \operatorname{Im}\alpha_{n+1} \in H_n(\mathbf{M})$.

Proof. For each $n \in \mathbb{Z}$, we have a commutative diagram

$$\cdots \xrightarrow{\alpha_{n+1}} M_n \xrightarrow{\alpha_n} M_{n-1} \xrightarrow{\alpha_{n-1}} \cdots$$
$$\quad\quad \Big\downarrow f_n \quad\quad \Big\downarrow f_{n-1}$$
$$\cdots \xrightarrow{\beta_{n+1}} N_n \xrightarrow{\beta_n} N_{n-1} \xrightarrow{\beta_{n-1}} \cdots$$

so let $H_n(\mathbf{f})$ be as stated in the proposition. If $x \in \operatorname{Ker}\alpha_n$, then

$$\beta_n f_n(x) = f_{n-1}\alpha_n(x) = 0,$$

so $f_n(x) \in \operatorname{Ker}\beta_n$. Hence, $H_n(\mathbf{f})$ maps $\operatorname{Ker}\alpha_n / \operatorname{Im}\alpha_{n+1}$ to $\operatorname{Ker}\beta_n / \operatorname{Im}\beta_{n+1}$ as required. Next, let $x, x' \in \operatorname{Ker}\alpha_n$ and suppose that $x + \operatorname{Im}\alpha_{n+1} = x' + \operatorname{Im}\alpha_{n+1}$. Then $x - x' \in \operatorname{Im}\alpha_{n+1}$, so there is a $y \in M_{n+1}$ such that $\alpha_{n+1}(y) = x - x'$. Thus,

$$f_n(x) - f_n(x') = f_n(x - x') = f_n\alpha_{n+1}(y) = \beta_{n+1}f_{n+1}(y),$$

so $f_n(x) - f_n(x') \in \operatorname{Im}\beta_{n+1}$. Therefore, $H_n(\mathbf{f})$ is well defined. It is immediate that $H_n(\mathbf{f})$ is R-linear since f_n is, so the proof is complete. $\qquad\square$

There is also an *nth cohomology mapping* $H^n(\mathbf{f}) : H^n(\mathbf{M}) \to H^n(\mathbf{N})$ defined by

$$H^n(\mathbf{f})(x + \operatorname{Im}\alpha^{n-1}) = f^n(x) + \operatorname{Im}\beta^{n-1}$$

for all $x + \operatorname{Im}\alpha^{n-1} \in H^n(\mathbf{M})$, where $\mathbf{f} : \mathbf{M} \to \mathbf{N}$ is any cochain map.

An important question in homological algebra is, "When do two (co)chain maps $\mathbf{f}, \mathbf{g} : \mathbf{M} \to \mathbf{N}$ induce the same nth (co)homology map from the nth (co)homology module of $\mathbf{M}$ to the nth (co)homology module of $\mathbf{N}$?" To answer this question, we need the following definition.

Definition 11.1.4. If $\mathbf{f}, \mathbf{g} : \mathbf{M} \to \mathbf{N}$ are chain maps, then a *homotopy* φ from $\mathbf{f}$ to $\mathbf{g}$, denoted by $\varphi : \mathbf{f} \to \mathbf{g}$, is a chain map $\varphi = \{\varphi_n : M_n \to N_{n+1}\}_{\mathbb{Z}}$ of degree $+1$ such that $f_n - g_n = \beta_{n+1}\varphi_n + \varphi_{n-1}\alpha_n$ for each $n \in \mathbb{Z}$. The following diagram illustrates a homotopy for chain maps.

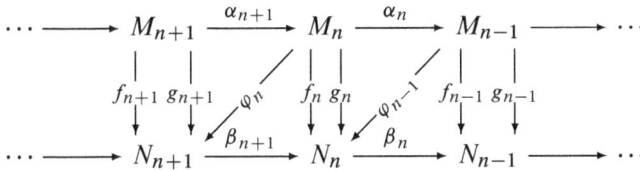

If there is a homotopy $\varphi : \mathbf{f} \to \mathbf{g}$, then $\mathbf{f}$ and $\mathbf{g}$ are said to be *homotopic chain maps*. If $\mathbf{f}$ and $\mathbf{g}$ are cochain maps, then a homotopy $\varphi : \mathbf{f} \to \mathbf{g}$ is a cochain map of degree -1 such that $f^n - g^n = \beta^{n-1}\varphi^n + \varphi^{n+1}\alpha^n$ for each $n \in \mathbb{Z}$. A cochain homotopy is illustrated in the following diagram.

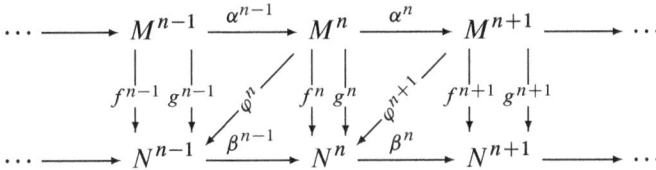

The notation $\mathbf{f} \approx \mathbf{g}$ will indicate that $\mathbf{f}$ and $\mathbf{g}$ are homotopic (co)chain maps. Two (co)chain complexes $\mathbf{M}$ and $\mathbf{N}$ are said to be of the *same homotopy type* if there exist (co)chain maps $\mathbf{f} : \mathbf{M} \to \mathbf{N}$ and $\mathbf{g} : \mathbf{N} \to \mathbf{M}$ such that $\mathbf{gf} \approx \mathbf{id_M}$ and $\mathbf{fg} \approx \mathbf{id_N}$, where $\mathbf{id_M}$ and $\mathbf{id_N}$ are the identity (co)chain maps on $\mathbf{M}$ and $\mathbf{N}$, respectively. Such a (co)chain map $\mathbf{f}$ (or $\mathbf{g}$) is called a *homotopy equivalence*.

Remark. The diagrams of Definition 11.1.4 are no longer commutative. For example, in the first diagram of the definition, there is no reason to expect that the triangle formed by the maps φ_n, β_{n+1} and f_n is commutative.

Proposition 11.1.5. *If* $\mathbf{f}, \mathbf{g} : \mathbf{M} \to \mathbf{N}$ *are homotopic chain maps, then* $H_n(\mathbf{f}) = H_n(\mathbf{g})$ *for each* $n \in \mathbb{Z}$.

Proof. If φ is a homotopy from $\mathbf{f}$ to $\mathbf{g}$, then $f_n - g_n = \beta_{n+1}\varphi_n + \varphi_{n-1}\alpha_n$ for each $n \in \mathbb{Z}$. If $x + \operatorname{Im}\alpha_{n+1} \in H_n(\mathbf{M})$, where $x \in \operatorname{Ker}\alpha_n$, then we see that $f_n(x) - g_n(x) = \beta_{n+1}\varphi_n(x) + \varphi_{n-1}\alpha_n(x) = \beta_{n+1}\varphi_n(x) \in \operatorname{Im}\beta_{n+1}$. Therefore, $f_n(x) + \operatorname{Im}\beta_{n+1} = g_n(x) + \operatorname{Im}\beta_{n+1}$. The fact that $f_n(x)$ and $g_n(x)$ are elements of $\operatorname{Ker}\beta_n$ was demonstrated in the proof of Proposition 11.1.3. Hence, $H_n(\mathbf{f}) = H_n(\mathbf{g})$. $\square$

If $\mathbf{f}, \mathbf{g} : \mathbf{M} \to \mathbf{N}$ are homotopic cochain maps, then it follows by an argument similar to that given in the proof of Proposition 11.1.5 that $H^n(\mathbf{f}) = H^n(\mathbf{g})$.

Chain complexes and chain maps form an additive category that we denote by $\mathbf{Chain}_R$. If $\mathbf{f} : \mathbf{L} \to \mathbf{M}$ and $\mathbf{g} : \mathbf{M} \to \mathbf{N}$ are chain maps, then $\mathbf{gf}$ is defined in

Chain$_R$ by $\mathbf{gf} = \{g_n f_n : L_n \to N_n\}_{\mathbb{Z}}$. Furthermore, if $\mathbf{f}, \mathbf{g} : \mathbf{L} \to \mathbf{M}$ are chain maps, then $\mathbf{f} + \mathbf{g} : \mathbf{L} \to \mathbf{M}$ is also a chain map if $\mathbf{f} + \mathbf{g}$ is defined in the obvious way.

The following proposition presents a useful relation among chain maps.

Proposition 11.1.6. *Let* $\mathbf{M}$ *and* $\mathbf{N}$ *be chain complexes. Then the relation* $\approx$ *on* $\mathrm{Mor}(\mathbf{M}, \mathbf{N})$ *given by* $\mathbf{f} \approx \mathbf{g}$ *if there is a homotopy* $\varphi : \mathbf{f} \to \mathbf{g}$ *is an equivalence relation on* $\mathrm{Mor}(\mathbf{M}, \mathbf{N})$ *in* **Chain**$_R$.

Proof. $\approx$ *is reflexive.* If $\mathbf{f}$ is a chain map in $\mathrm{Mor}(\mathbf{M}, \mathbf{N})$, then the zero homotopy $0 : \mathbf{f} \to \mathbf{f}$ shows that $\mathbf{f} \approx \mathbf{f}$.

$\approx$ *is symmetric.* Suppose that $\mathbf{f}$ and $\mathbf{g}$ are chain maps in $\mathrm{Mor}(\mathbf{M}, \mathbf{N})$ such that $\mathbf{f} \approx \mathbf{g}$. If $\varphi : \mathbf{f} \to \mathbf{g}$ is a homotopy, then $f_n - g_n = \beta_{n+1}\varphi_n + \varphi_{n-1}\alpha_n$ implies that $g_n - f_n = \beta_{n+1}(-\varphi_n) + (-\varphi_{n-1})\alpha_n$ for each $n \in \mathbb{Z}$. This gives a homotopy $\varphi : \mathbf{g} \to \mathbf{f}$, so $\mathbf{g} \approx \mathbf{f}$.

$\approx$ *is transitive.* Let $\mathbf{f}, \mathbf{g}$ and $\mathbf{h}$ be chain maps in $\mathrm{Mor}(\mathbf{M}, \mathbf{N})$ and suppose that $\mathbf{f} \approx \mathbf{g}$ and $\mathbf{g} \approx \mathbf{h}$. If $\varphi : \mathbf{f} \to \mathbf{g}$ and $\psi : \mathbf{g} \to \mathbf{h}$ are homotopies, then for each $n \in \mathbb{Z}$ we have

$$f_n - g_n = \beta_{n+1}\varphi_n + \varphi_{n-1}\alpha_n \quad \text{and}$$
$$g_n - h_n = \beta_{n+1}\psi_n + \psi_{n-1}\alpha_n.$$

So

$$f_n - h_n = \beta_{n+1}(\varphi_n + \psi_n) + (\varphi_{n-1} + \psi_{n-1})\alpha_n$$

gives a homotopy $\varphi + \psi : \mathbf{f} \to \mathbf{h}$ and we have $\mathbf{f} \approx \mathbf{h}$. $\square$

If $\mathbf{f} : \mathbf{M} \to \mathbf{N}$ is a chain map in **Chain**$_R$, then the equivalence class $[\mathbf{f}]$ determined by the equivalence relation of Proposition 11.1.6 is called the *homotopy class* of $\mathbf{f}$.

The proofs of the following two propositions are straightforward and are left as exercises.

Proposition 11.1.7. *For each* $n \in \mathbb{Z}$, $H_n :$ **Chain**$_R \to$ **Mod**$_R$ *is an additive functor.*

If we form the category **Cochain**$_R$ of cochain complexes and cochain maps, then $H^n :$ **Cochain**$_R \to$ **Mod**$_R$ is also an additive functor. The functors H_n and H^n are called the *nth homology functor* and the *nth cohomology functor*, respectively.

Proposition 11.1.8. *Suppose that* $\mathcal{F} :$ **Mod**$_R \to$ **Mod**$_S$ *is an additive functor.*

(1) *If* $\mathbf{M}$ *is a chain complex in* **Chain**$_R$, *then* $\mathcal{F}(\mathbf{M})$ *is a chain complex in* **Chain**$_S$.

(2) *If* $\mathbf{f} : \mathbf{M} \to \mathbf{N}$ *is a chain map of degree* k *in* **Chain**$_R$, *then*

$$\mathcal{F}(\mathbf{f}) : \mathcal{F}(\mathbf{M}) \to \mathcal{F}(\mathbf{N})$$

is a chain map of degree k *in* **Chain**$_S$.

Definition 11.1.4. If $f, g : M \to N$ are chain maps, then a *homotopy* φ from f to g, denoted by $\varphi : f \to g$, is a chain map $\varphi = \{\varphi_n : M_n \to N_{n+1}\}_{\mathbb{Z}}$ of degree $+1$ such that $f_n - g_n = \beta_{n+1}\varphi_n + \varphi_{n-1}\alpha_n$ for each $n \in \mathbb{Z}$. The following diagram illustrates a homotopy for chain maps.

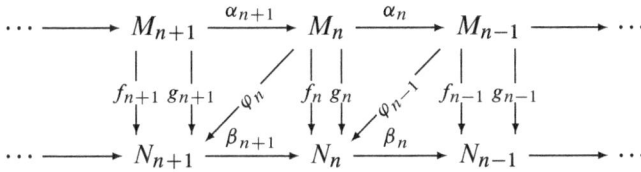

$$
\begin{array}{ccccccc}
\cdots \longrightarrow & M_{n+1} & \xrightarrow{\ \alpha_{n+1}\ } & M_n & \xrightarrow{\ \alpha_n\ } & M_{n-1} & \longrightarrow \cdots \\
 & \big\downarrow{\scriptstyle f_{n+1}\, g_{n+1}}\ \ {\scriptstyle \varphi_n} & & \big\downarrow{\scriptstyle f_n\, g_n}\ \ {\scriptstyle \varphi_{n-1}} & & \big\downarrow{\scriptstyle f_{n-1}\, g_{n-1}} & \\
\cdots \longrightarrow & N_{n+1} & \xrightarrow{\ \beta_{n+1}\ } & N_n & \xrightarrow{\ \beta_n\ } & N_{n-1} & \longrightarrow \cdots
\end{array}
$$

If there is a homotopy $\varphi : f \to g$, then f and g are said to be *homotopic chain maps*. If f and g are cochain maps, then a homotopy $\varphi : f \to g$ is a cochain map of degree -1 such that $f^n - g^n = \beta^{n-1}\varphi^n + \varphi^{n+1}\alpha^n$ for each $n \in \mathbb{Z}$. A cochain homotopy is illustrated in the following diagram.

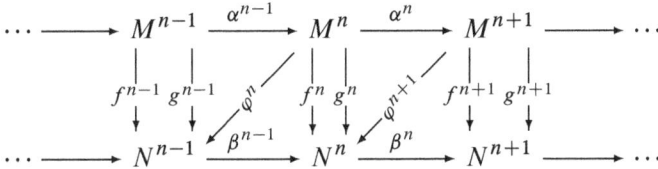

$$
\begin{array}{ccccccc}
\cdots \longrightarrow & M^{n-1} & \xrightarrow{\ \alpha^{n-1}\ } & M^n & \xrightarrow{\ \alpha^n\ } & M^{n+1} & \longrightarrow \cdots \\
 & \big\downarrow{\scriptstyle f^{n-1}\, g^{n-1}}\ \ {\scriptstyle \varphi^n} & & \big\downarrow{\scriptstyle f^n\, g^n}\ \ {\scriptstyle \varphi^{n+1}} & & \big\downarrow{\scriptstyle f^{n+1}\, g^{n+1}} & \\
\cdots \longrightarrow & N^{n-1} & \xrightarrow{\ \beta^{n-1}\ } & N^n & \xrightarrow{\ \beta^n\ } & N^{n+1} & \longrightarrow \cdots
\end{array}
$$

The notation $f \approx g$ will indicate that f and g are homotopic (co)chain maps. Two (co)chain complexes M and N are said to be of the *same homotopy type* if there exist (co)chain maps $f : M \to N$ and $g : N \to M$ such that $gf \approx id_M$ and $fg \approx id_N$, where id_M and id_N are the identity (co)chain maps on M and N, respectively. Such a (co)chain map f (or g) is called a *homotopy equivalence*.

Remark. The diagrams of Definition 11.1.4 are no longer commutative. For example, in the first diagram of the definition, there is no reason to expect that the triangle formed by the maps φ_n, β_{n+1} and f_n is commutative.

Proposition 11.1.5. *If* $f, g : M \to N$ *are homotopic chain maps, then* $H_n(f) = H_n(g)$ *for each* $n \in \mathbb{Z}$.

Proof. If φ is a homotopy from f to g, then $f_n - g_n = \beta_{n+1}\varphi_n + \varphi_{n-1}\alpha_n$ for each $n \in \mathbb{Z}$. If $x + \operatorname{Im}\alpha_{n+1} \in H_n(M)$, where $x \in \operatorname{Ker}\alpha_n$, then we see that $f_n(x) - g_n(x) = \beta_{n+1}\varphi_n(x) + \varphi_{n-1}\alpha_n(x) = \beta_{n+1}\varphi_n(x) \in \operatorname{Im}\beta_{n+1}$. Therefore, $f_n(x) + \operatorname{Im}\beta_{n+1} = g_n(x) + \operatorname{Im}\beta_{n+1}$. The fact that $f_n(x)$ and $g_n(x)$ are elements of $\operatorname{Ker}\beta_n$ was demonstrated in the proof of Proposition 11.1.3. Hence, $H_n(f) = H_n(g)$. $\square$

If $f, g : M \to N$ are homotopic cochain maps, then it follows by an argument similar to that given in the proof of Proposition 11.1.5 that $H^n(f) = H^n(g)$.

Chain complexes and chain maps form an additive category that we denote by **Chain**$_R$. If $f : L \to M$ and $g : M \to N$ are chain maps, then gf is defined in

Chain$_R$ by $\mathbf{gf} = \{g_n f_n : L_n \to N_n\}_{\mathbb{Z}}$. Furthermore, if $\mathbf{f}, \mathbf{g} : \mathbf{L} \to \mathbf{M}$ are chain maps, then $\mathbf{f} + \mathbf{g} : \mathbf{L} \to \mathbf{M}$ is also a chain map if $\mathbf{f} + \mathbf{g}$ is defined in the obvious way.

The following proposition presents a useful relation among chain maps.

Proposition 11.1.6. *Let* $\mathbf{M}$ *and* $\mathbf{N}$ *be chain complexes. Then the relation* $\approx$ *on* $\mathrm{Mor}(\mathbf{M}, \mathbf{N})$ *given by* $\mathbf{f} \approx \mathbf{g}$ *if there is a homotopy* $\varphi : \mathbf{f} \to \mathbf{g}$ *is an equivalence relation on* $\mathrm{Mor}(\mathbf{M}, \mathbf{N})$ *in* **Chain**$_R$.

Proof. $\approx$ *is reflexive.* If $\mathbf{f}$ is a chain map in $\mathrm{Mor}(\mathbf{M}, \mathbf{N})$, then the zero homotopy $\mathbf{0} : \mathbf{f} \to \mathbf{f}$ shows that $\mathbf{f} \approx \mathbf{f}$.

$\approx$ *is symmetric.* Suppose that $\mathbf{f}$ and $\mathbf{g}$ are chain maps in $\mathrm{Mor}(\mathbf{M}, \mathbf{N})$ such that $\mathbf{f} \approx \mathbf{g}$. If $\varphi : \mathbf{f} \to \mathbf{g}$ is a homotopy, then $f_n - g_n = \beta_{n+1}\varphi_n + \varphi_{n-1}\alpha_n$ implies that $g_n - f_n = \beta_{n+1}(-\varphi_n) + (-\varphi_{n-1})\alpha_n$ for each $n \in \mathbb{Z}$. This gives a homotopy $\varphi : \mathbf{g} \to \mathbf{f}$, so $\mathbf{g} \approx \mathbf{f}$.

$\approx$ *is transitive.* Let $\mathbf{f}, \mathbf{g}$ and $\mathbf{h}$ be chain maps in $\mathrm{Mor}(\mathbf{M}, \mathbf{N})$ and suppose that $\mathbf{f} \approx \mathbf{g}$ and $\mathbf{g} \approx \mathbf{h}$. If $\varphi : \mathbf{f} \to \mathbf{g}$ and $\psi : \mathbf{g} \to \mathbf{h}$ are homotopies, then for each $n \in \mathbb{Z}$ we have

$$f_n - g_n = \beta_{n+1}\varphi_n + \varphi_{n-1}\alpha_n \quad \text{and}$$
$$g_n - h_n = \beta_{n+1}\psi_n + \psi_{n-1}\alpha_n.$$

So

$$f_n - h_n = \beta_{n+1}(\varphi_n + \psi_n) + (\varphi_{n-1} + \psi_{n-1})\alpha_n$$

gives a homotopy $\varphi + \psi : \mathbf{f} \to \mathbf{h}$ and we have $\mathbf{f} \approx \mathbf{h}$. $\square$

If $\mathbf{f} : \mathbf{M} \to \mathbf{N}$ is a chain map in **Chain**$_R$, then the equivalence class $[\mathbf{f}]$ determined by the equivalence relation of Proposition 11.1.6 is called the *homotopy class* of $\mathbf{f}$.

The proofs of the following two propositions are straightforward and are left as exercises.

Proposition 11.1.7. *For each* $n \in \mathbb{Z}$, $H_n :$ **Chain**$_R \to$ **Mod**$_R$ *is an additive functor.*

If we form the category **Cochain**$_R$ of cochain complexes and cochain maps, then $H^n :$ **Cochain**$_R \to$ **Mod**$_R$ is also an additive functor. The functors H_n and H^n are called the *nth homology functor* and the *nth cohomology functor*, respectively.

Proposition 11.1.8. *Suppose that* $\mathscr{F} :$ **Mod**$_R \to$ **Mod**$_S$ *is an additive functor.*

(1) *If* $\mathbf{M}$ *is a chain complex in* **Chain**$_R$, *then* $\mathscr{F}(\mathbf{M})$ *is a chain complex in* **Chain**$_S$.

(2) *If* $\mathbf{f} : \mathbf{M} \to \mathbf{N}$ *is a chain map of degree k in* **Chain**$_R$, *then*

$$\mathscr{F}(\mathbf{f}) : \mathscr{F}(\mathbf{M}) \to \mathscr{F}(\mathbf{N})$$

is a chain map of degree k in **Chain**$_S$.

(3) *If* $\mathbf{f}, \mathbf{g} : \mathbf{M} \rightarrow \mathbf{N}$ *are homotopic chain maps in* **Chain**$_R$, *then*

$$\mathscr{F}(\mathbf{f}), \mathscr{F}(\mathbf{g}) : \mathscr{F}(\mathbf{M}) \rightarrow \mathscr{F}(\mathbf{N})$$

are homotopic in **Chain**$_S$.

(4) *If* $\mathbf{f}, \mathbf{g} : \mathbf{M} \rightarrow \mathbf{N}$ *are homotopic chain maps in* **Chain**$_R$, *then*

$$H_n(\mathscr{F}(\mathbf{f})) = H_n(\mathscr{F}(\mathbf{g})) : H_n(\mathscr{F}(\mathbf{M})) \rightarrow H_n(\mathscr{F}(\mathbf{N}))$$

in **Mod**$_S$ *for each* $n \in \mathbb{Z}$.

Clearly there is a dual version of Proposition 11.1.8 that holds for cochain complexes and cochain maps. Part (4) of the preceding proposition shows that each homotopy class [**f**] of a chain map $\mathbf{f} : \mathbf{M} \rightarrow \mathbf{N}$ in **Chain**$_R$ produces exactly one homology mapping from $H_n(\mathscr{F}(\mathbf{M}))$ to $H_n(\mathscr{F}(\mathbf{N}))$.

Homology and Cohomology Sequences

The category **Chain**$_R$ of chain complexes and chain maps enjoys many of the properties of the category **Mod**$_R$. For example, we can form a sequence $\mathbf{L} \xrightarrow{\mathbf{f}} \mathbf{M} \xrightarrow{\mathbf{g}} \mathbf{N}$ of chain complexes and chain maps, we can form subchains of chain complexes, and we can form factor chains of chain complexes. (These constructions will be addressed in the exercises.) A sequence $\mathbf{L} \xrightarrow{\mathbf{f}} \mathbf{M} \xrightarrow{\mathbf{g}} \mathbf{N}$ of chain complexes and chain maps is said to be *exact* if $L \xrightarrow{f_n} M \xrightarrow{g_n} N$ is exact for each $n \in \mathbb{Z}$. A *short exact sequence of chain complexes* $0 \rightarrow \mathbf{L} \xrightarrow{\mathbf{f}} \mathbf{M} \xrightarrow{\mathbf{g}} \mathbf{N} \rightarrow 0$ is actually a 2-dimensional commutative diagram of R-modules and R-module homomorphisms, where the columns are chain com-plexes and $0 \to L_n \xrightarrow{f_n} M_n \xrightarrow{g_n} N_n \to 0$ is a short exact sequence in $\mathbf{Mod}_R$ for each $n \in \mathbb{Z}$.

Corresponding to each short exact sequence of chain complexes $0 \to \mathbf{L} \xrightarrow{f} \mathbf{M} \xrightarrow{g} \mathbf{N} \to 0$, there is a long exact sequence

$$\cdots \xrightarrow{\Phi_{n+1}} H_n(\mathbf{L}) \xrightarrow{H_n(f)} H_n(\mathbf{M}) \xrightarrow{H_n(g)} H_n(\mathbf{N}) \xrightarrow{\Phi_n}$$

$$\xrightarrow{\Phi_n} H_{n-1}(\mathbf{L}) \xrightarrow{H_{n-1}(f)} H_{n-1}(\mathbf{M}) \xrightarrow{H_{n-1}(g)} H_{n-1}(\mathbf{N}) \xrightarrow{\Phi_{n-1}} \cdots$$

of homology modules. The mapping Φ_n is said to be a *connecting homomorphism* for each $n \in \mathbb{Z}$.

To establish the existence of the *long exact sequence in homology*, we begin with the following lemma. But first note that if the diagram

of R-modules and R-module homomorphisms is commutative, then there are induced mappings $\bar{f} : \operatorname{Ker}\alpha \to \operatorname{Ker}\beta$ and $\bar{g} : \operatorname{Coker}\alpha \to \operatorname{Coker}\beta$. Indeed, let $\bar{f}(x) = f(x)$ for each $x \in \operatorname{Ker}\alpha$ and $\bar{g}(x + \operatorname{Im}\alpha) = g(x) + \operatorname{Im}\beta$ for all $x + \operatorname{Im}\alpha \in \operatorname{Coker}\alpha$.

Lemma 11.1.9 (Snake Lemma). *Let*

be a row exact commutative diagram of R-modules and R-module homomorphisms. Then there is an R-linear mapping $\Phi : \operatorname{Ker}\gamma \to \operatorname{Coker}\alpha$ such that the sequence

$$\operatorname{Ker}\alpha \xrightarrow{\bar{f}_1} \operatorname{Ker}\beta \xrightarrow{\bar{g}_1} \operatorname{Ker}\gamma \xrightarrow{\Phi} \operatorname{Coker}\alpha \xrightarrow{\bar{f}_2} \operatorname{Coker}\beta \xrightarrow{\bar{g}_2} \operatorname{Coker}\gamma$$

is exact.

Proof. It is not difficult to show that each of the sequences

$$\operatorname{Ker}\alpha \xrightarrow{\bar{f}_1} \operatorname{Ker}\beta \xrightarrow{\bar{g}_1} \operatorname{Ker}\gamma \quad \text{and}$$

$$\operatorname{Coker}\alpha \xrightarrow{\bar{f}_2} \operatorname{Coker}\beta \xrightarrow{\bar{g}_2} \operatorname{Coker}\gamma$$

formed from the induced maps is exact. So it only remains to establish the existence of Φ and to show that the sequence is exact at $\operatorname{Ker} \gamma$ and at $\operatorname{Coker} \alpha$.

The map $\Phi : \operatorname{Ker} \gamma \to \operatorname{Coker} \alpha$ is defined as follows. If $z \in \operatorname{Ker} \gamma$, let $x \in M$ be such that $g_1(x) = z$. Then $\gamma g_1(x) = 0$, so $g_2 \beta(x) = 0$. Hence, $\beta(x) \in \operatorname{Ker} g_2 = \operatorname{Im} f_2$. Since f_2 is an injection, there is a unique $y \in N_1$ such that $f_2(y) = \beta(x)$. Now define Φ by $\Phi(z) = y + \operatorname{Im} \alpha$. To show that Φ is well defined, we need to show that the definition of Φ is independent of the choice of x. Let x' be another element of M such that $g_1(x') = z$ and suppose that $y' \in N_1$ is such that $f_2(y') = \beta(x')$. Then $x - x' \in \operatorname{Ker} g_1 = \operatorname{Im} f_1$, so there is a $w \in M_1$ such that $f_1(w) = x - x'$. We now have $f_2(y) - f_2(y') = \beta(x) - \beta(x') = \beta f_1(w) = f_2 \alpha(w)$ which gives $y - y' - \alpha(w) \in \operatorname{Ker} f_2 = 0$. Hence, $y - y' = \alpha(w) \in \operatorname{Im} \alpha$ and so $y + \operatorname{Im} \alpha = y' + \operatorname{Im} \alpha$. Each mapping involved in the construction of Φ is R-linear, so a direct computation will verify that Φ is also R-linear. Thus, we have an R-linear mapping $\Phi : \operatorname{Ker} \gamma \to \operatorname{Coker} \alpha$.

Next, let us show that the sequence is exact at $\operatorname{Ker} \gamma$. Let $z \in \operatorname{Ker} \gamma$ be such that $z \in \operatorname{Ker} \Phi$ and suppose that $x \in M$ and $y \in N_1$ are as above. Then $0 = \Phi(z) = y + \operatorname{Im} \alpha$, so $y \in \operatorname{Im} \alpha$. Hence, there is a $u \in M_1$ such that $\alpha(u) = y$. Thus, $\beta f_1(u) = f_2 \alpha(u) = f_2(y) = \beta(x)$, so we have $x - f_1(u) \in \operatorname{Ker} \beta$. If $x - f_1(u) = w \in \operatorname{Ker} \beta$, then $z = g_1(x) = g_1(x) - g_1 f_1(u) = g_1(w)$. Therefore, $z \in \operatorname{Im} \bar{g}_1$, so $\operatorname{Ker} \Phi \subseteq \operatorname{Im} \bar{g}_1$. Conversely, let $z \in \operatorname{Ker} \gamma$ be such that $z \in \operatorname{Im} \bar{g}_1 \subseteq \operatorname{Im} g_1$. Then there is an $x \in \operatorname{Ker} \beta$ such that $g_1(x) = z$ and a $y \in N_1$ such that $f_2(y) = \beta(x) = 0$. Thus, $y = 0$ since f_2 is injective and this gives $\Phi(z) = y + \operatorname{Im} \alpha = 0$. Hence, $\operatorname{Im} \bar{g}_1 \subseteq \operatorname{Ker} \Phi$ and we have $\operatorname{Im} \bar{g}_1 = \operatorname{Ker} \Phi$.

The proof that sequence is exact at $\operatorname{Coker} \alpha$ is an exercise. $\square$

We also need the following lemma.

Lemma 11.1.10. *If* $\mathbf{M}$ *is a chain complex, then the map* $\alpha_n : M_n \to M_{n-1}$ *induces an R-linear mapping* $\bar{\alpha}_n : \operatorname{Coker} \alpha_{n+1} \to \operatorname{Ker} \alpha_{n-1}$. *Moreover,* $H_n(\mathbf{M}) = \operatorname{Ker} \bar{\alpha}_n$ *and* $H_{n-1}(\mathbf{M}) = \operatorname{Coker} \bar{\alpha}_n$.

Proof. $\operatorname{Im} \alpha_{n+1} \subseteq \operatorname{Ker} \alpha_n$ gives an epimorphism $\bar{\alpha}_n : M_n / \operatorname{Im} \alpha_{n+1} \to M_n / \operatorname{Ker} \alpha_n$ such that $x + \operatorname{Im} \alpha_{n+1} \mapsto x + \operatorname{Ker} \alpha_n$. But $M_n / \operatorname{Ker} \alpha_n \cong \operatorname{Im} \alpha_n \subseteq \operatorname{Ker} \alpha_{n-1}$, so we have an R-linear mapping $\bar{\alpha}_n : \operatorname{Coker} \alpha_{n+1} \to \operatorname{Ker} \alpha_{n-1}$. The proofs that $H_n(\mathbf{M}) = \operatorname{Ker} \bar{\alpha}_n$ and $H_{n-1}(\mathbf{M}) = \operatorname{Coker} \bar{\alpha}_n$ are straightforward computations. $\square$

In the Snake Lemma, if f_1 is injective and g_2 is surjective, then $\bar{f}_1$ is injective and $\bar{g}_2$ is surjective. Thus, if $0 \to \mathbf{L} \overset{f}{\to} \mathbf{M} \overset{g}{\to} \mathbf{N} \to 0$ is a short exact sequence of chain complexes, then for each $n \in \mathbb{Z}$ we have a row and column exact commutative

diagram

$$
\begin{array}{ccccccccc}
 & & 0 & & 0 & & 0 & & \\
 & & \downarrow & & \downarrow & & \downarrow & & \\
0 & \longrightarrow & \operatorname{Ker}\alpha_n & \longrightarrow & \operatorname{Ker}\beta_n & \longrightarrow & \operatorname{Ker}\gamma_n & \longrightarrow & 0 \\
 & & \downarrow & & \downarrow & & \downarrow & & \\
0 & \longrightarrow & L_n & \longrightarrow & M_n & \longrightarrow & N_n & \longrightarrow & 0 \\
 & & \alpha_n \downarrow & & \beta_n \downarrow & & \gamma_n \downarrow & & \\
0 & \longrightarrow & L_{n-1} & \longrightarrow & M_{n-1} & \longrightarrow & N_{n-1} & \longrightarrow & 0 \\
 & & \downarrow & & \downarrow & & \downarrow & & \\
0 & \longrightarrow & \operatorname{Coker}\alpha_n & \longrightarrow & \operatorname{Coker}\beta_n & \longrightarrow & \operatorname{Coker}\gamma_n & \longrightarrow & 0 \\
 & & \downarrow & & \downarrow & & \downarrow & & \\
 & & 0 & & 0 & & 0 & &
\end{array}
$$

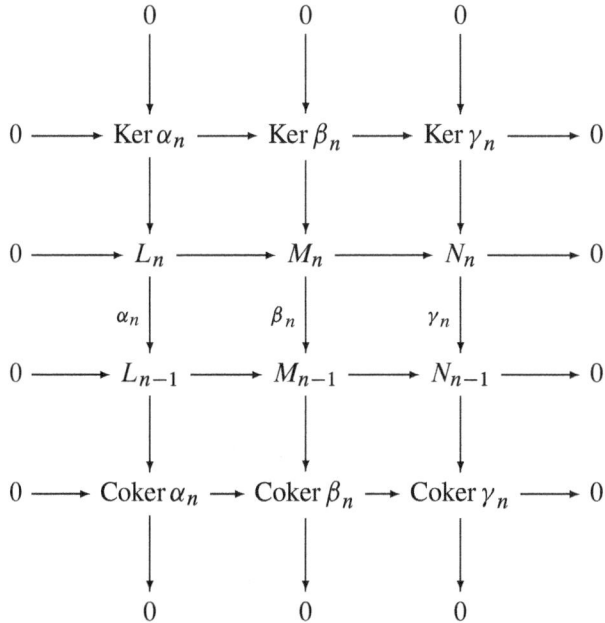

Using this diagram and Lemma 11.1.10, we get a commutative diagram

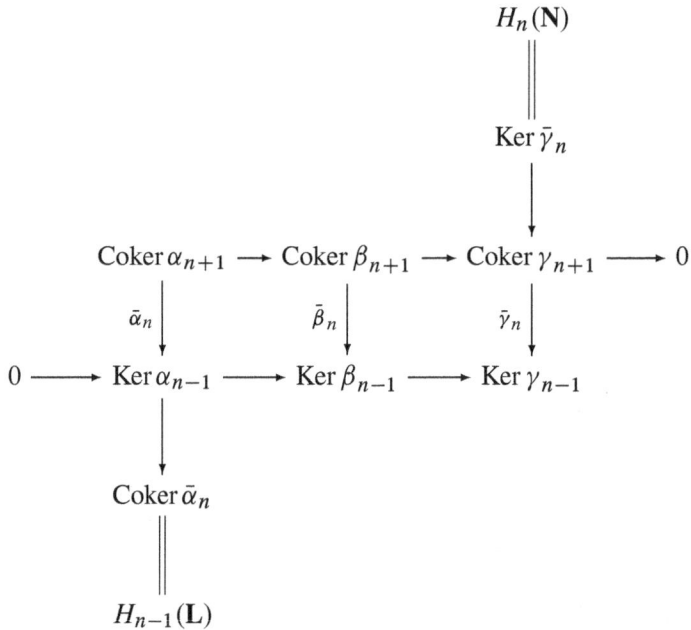

$$
\begin{array}{ccccccc}
 & & & & H_n(\mathbf{N}) & & \\
 & & & & \| & & \\
 & & & & \operatorname{Ker}\bar{\gamma}_n & & \\
 & & & & \downarrow & & \\
\operatorname{Coker}\alpha_{n+1} & \longrightarrow & \operatorname{Coker}\beta_{n+1} & \longrightarrow & \operatorname{Coker}\gamma_{n+1} & \longrightarrow & 0 \\
\bar{\alpha}_n \downarrow & & \bar{\beta}_n \downarrow & & \bar{\gamma}_n \downarrow & & \\
\operatorname{Ker}\alpha_{n-1} & \longrightarrow & \operatorname{Ker}\beta_{n-1} & \longrightarrow & \operatorname{Ker}\gamma_{n-1} & & \\
\downarrow & & & & & & \\
\operatorname{Coker}\bar{\alpha}_n & & & & & & \\
\| & & & & & & \\
H_{n-1}(\mathbf{L}) & & & & & &
\end{array}
$$

(with $0 \longrightarrow \operatorname{Ker}\alpha_{n-1}$ at the left of that row)

for each $n \in \mathbb{Z}$. Hence, for each $n \in \mathbb{Z}$, the Snake Lemma gives a connecting homomorphism $\Phi_n : H_n(\mathbf{N}) \to H_{n-1}(\mathbf{L})$ and so we have the following proposition.

Proposition 11.1.11. *Corresponding to each short exact sequence* $0 \to L \xrightarrow{f} M \xrightarrow{g} N \to 0$ *of chain complexes, there is a long exact sequence in homology*

$$\cdots \xrightarrow{\Phi_{n+1}} H_n(L) \xrightarrow{H_n(f)} H_n(M) \xrightarrow{H_n(g)} H_n(N) \xrightarrow{\Phi_n}$$

$$\xrightarrow{\Phi_n} H_{n-1}(L) \xrightarrow{H_{n-1}(f)} H_{n-1}(M) \xrightarrow{H_{n-1}(g)} H_{n-1}(N) \xrightarrow{\Phi_{n-1}} \cdots,$$

where Φ_n *is a connecting homomorphism for each* $n \in \mathbb{Z}$.

There is also a *long exact sequence in cohomology* that corresponds to each short exact sequence of cochain complexes.

Proposition 11.1.12. *Corresponding to each short exact sequence of cochain complexes* $0 \to L \xrightarrow{f} M \xrightarrow{g} N \to 0$, *there is a long exact sequence in cohomology*

$$\cdots \xrightarrow{\Phi^{n-1}} H^n(L) \xrightarrow{H^n(f)} H^n(M) \xrightarrow{H^n(g)} H^n(N) \xrightarrow{\Phi^n}$$

$$\xrightarrow{\Phi^n} H^{n+1}(L) \xrightarrow{H^{n+1}(f)} H^{n+1}(M) \xrightarrow{H^{n+1}(g)} H^{n+1}(N) \xrightarrow{\Phi^{n+1}} \cdots,$$

where Φ^n *is a connecting homomorphism for each* $n \in \mathbb{Z}$.

Problem Set 11.1

1. The *nth cohomology mapping* $H^n(f) : H^n(M) \to H^n(N)$, where $f : M \to N$ is a cochain map, is defined by $H^n(f)(x + \operatorname{Im} \alpha^{n-1}) = f^n(x) + \operatorname{Im} \beta^{n-1}$ for all $x + \operatorname{Im} \alpha^{n-1} \in H^n(M)$. Prove that $H^n(f)$ is a well-defined R-linear mapping for all $n \in \mathbb{Z}$.

2. If $f, g : M \to N$ are homotopic cochain maps, show that $H^n(f) = H^n(g)$ for each $n \in \mathbb{Z}$.

3. (a) If $M = \{M_n, \alpha_n\}_{\mathbb{Z}}$ is a chain complex, deduce that α_n can be changed to $-\alpha_n$ in M for any integer n and the result will remain a chain complex. Show that the same is true for a cochain complex.
   (b) Let $M = \{M_n, \alpha_n\}_{\mathbb{Z}}$ be a chain complex. If α_n is changed to $-\alpha_n$ in M and N is the resulting complex, compare $H_n(M)$ and $H_n(N)$.
   (c) If $M = \{M_n, \alpha_n\}_{\mathbb{Z}} (M = \{M^n, \alpha^n\}_{\mathbb{Z}})$ is a chain complex (cochain complex), verify that M can be converted to a cochain complex (chain complex) by lowering (raising) indices.

4. Let $M \xrightarrow{0} N$ denote the zero mapping. If $\{M_n\}_{\mathbb{Z}}$ is any family of R-modules, verify that

$$\cdots \to M_{n+1} \xrightarrow{0} M_n \xrightarrow{0} M_{n-1} \to \cdots$$

is a chain complex and compute $H_n(M_n)$.

5. Verify that **Chain**$_R$ and **Cochain**$_R$ are additive categories and prove Proposition 11.1.7.

6. (a) Prove Proposition 11.1.8.

   (b) State the analogue of Proposition 11.1.8 for chain complexes and an additive contravariant functor $\mathcal{F} : \mathbf{Mod}_R \to \mathbf{Mod}_S$.

   (c) State the analogue of Proposition 11.1.8 for cochain complexes and an additive functor $\mathcal{F}$.

7. Show that each of the following hold in the category **Chain**$_R$.

   (a) If $\mathbf{f}, \mathbf{f}' : \mathbf{L} \to \mathbf{M}$ are homotopic chain maps and $\mathbf{g} : \mathbf{M} \to \mathbf{N}$ is any chain map, then $\mathbf{gf} \approx \mathbf{gf}'$. [Hint: If $\varphi = \{\varphi_n : L_n \to M_{n+1}\}_{\mathbb{Z}}$ is a homotopy from $\mathbf{f}$ to $\mathbf{f}'$, show that $\chi = \{g_{n+1}\varphi_n : L_n \to N_{n+1}\}_{\mathbb{Z}}$ is a homotopy from $\mathbf{gf}$ to $\mathbf{gf}'$.] Likewise, if $\mathbf{h} : \mathbf{L} \to \mathbf{M}$ is a chain map and $\mathbf{f}, \mathbf{f}' : \mathbf{M} \to \mathbf{N}$ are homotopic chain maps, then $\mathbf{fh} \approx \mathbf{f'h}$.

   (b) If $\mathbf{f}, \mathbf{f}' : \mathbf{L} \to \mathbf{M}$ and $\mathbf{g}, \mathbf{g}' : \mathbf{M} \to \mathbf{N}$ are chain maps such that $\mathbf{f} \approx \mathbf{f}'$ and $\mathbf{g} \approx \mathbf{g}'$, then $\mathbf{gf} \approx \mathbf{g'f'}$.

8. If
$$\mathbf{M} : \cdots \to M_{n+1} \xrightarrow{\alpha_{n+1}} M_n \xrightarrow{\alpha_n} M_{n-1} \to \cdots$$
is a chain complex, then a *subchain* of $\mathbf{M}$ is a chain complex
$$\mathbf{N} : \cdots \to N_{n+1} \xrightarrow{\beta_{n+1}} N_n \xrightarrow{\beta_n} N_{n-1} \to \cdots$$
such that N_n is a submodule of M_n and $\beta_n = \alpha_n|_{N_n}$ for each $n \in \mathbb{Z}$. If $\mathbf{N}$ is a subchain of $\mathbf{M}$, then the *factor* (or *quotient*) *chain* of $\mathbf{M}$ by $\mathbf{N}$ is given by
$$\mathbf{M/N} : \cdots \to M_{n+1}/N_{n+1} \xrightarrow{\bar{\alpha}_{n+1}} M_n/N_n \xrightarrow{\bar{\alpha}_n} M_{n-1}/N_{n-1} \to \cdots,$$
where $\bar{\alpha}_n(x + N_n) = \alpha_n(x) + N_{n-1}$ for all $x + N_n$ and $n \in \mathbb{Z}$.

   (a) If $\mathbf{f} : \mathbf{M} \to \mathbf{N}$ is a chain map, define the subchains Ker $\mathbf{f}$ and Im $\mathbf{f}$.

   (b) If $\mathbf{f} : \mathbf{M} \to \mathbf{N}$ is a chain map, prove that there is a chain isomorphism from $\mathbf{M}/$ Ker $\mathbf{f}$ to Im $\mathbf{f}$. Can we write $\mathbf{M}/$ Ker $\mathbf{f} \cong$ Im $\mathbf{f}$ in the category **Chain**$_R$?

   Observe that all of the definitions given in this exercise can be dualized to cochain complexes and cochain maps. Does the dual of (b) hold for cochain maps?

9. Prove that each of the following hold for the Snake Lemma.

   (a) The connecting homomorphism Φ is an R-linear mapping.

   (b) The sequences
$$\text{Ker}\,\alpha \xrightarrow{\bar{f}_1} \text{Ker}\,\beta \xrightarrow{\bar{g}_1} \text{Ker}\,\gamma \quad \text{and}$$
$$\text{Coker}\,\alpha \xrightarrow{\bar{f}_2} \text{Coker}\,\beta \xrightarrow{\bar{g}_2} \text{Coker}\,\gamma$$
are exact.

(c) The sequence given in the lemma is exact at $\operatorname{Coker}\alpha$.

(d) If f_1 is injective, then the map $\bar{f}_1 : \operatorname{Ker}\alpha \to \operatorname{Ker}\beta$ is also injective.

(e) If g_2 is surjective, then so is $\bar{g}_2 : \operatorname{Coker}\beta \to \operatorname{Coker}\gamma$.

10. Let $\mathcal{F} : \mathbf{Mod}_R \to \mathbf{Mod}_S$ be an additive functor. Prove that $\mathcal{F}$ is an exact functor if and only if for each chain complex $\mathbf{M}$ of R-modules and R-module homomorphisms $\mathcal{F}(H_n(\mathbf{M})) = H_n(\mathcal{F}(\mathbf{M}))$ for each $n \in \mathbb{Z}$.

11. Let

$$
\begin{array}{ccccccccc}
0 & \longrightarrow & \mathbf{L} & \longrightarrow & \mathbf{M} & \longrightarrow & \mathbf{N} & \longrightarrow & 0 \\
 & & \downarrow f & & \downarrow g & & \downarrow h & & \\
0 & \longrightarrow & \mathbf{L'} & \longrightarrow & \mathbf{M'} & \longrightarrow & \mathbf{N'} & \longrightarrow & 0
\end{array}
$$

be a commutative diagram of short exact sequences of chain complexes, where $\mathbf{f}, \mathbf{g}$ and $\mathbf{h}$ are chain maps.

(a) Prove that there is a chain map from the long exact sequence of homology modules arising from

$$0 \to \mathbf{L} \to \mathbf{M} \to \mathbf{N} \to 0$$

to the long exact sequence of homology modules arising from

$$0 \to \mathbf{L'} \to \mathbf{M'} \to \mathbf{N'} \to 0$$

and that the resulting diagram is commutative.

(b) Show that if any two of the chain maps induce isomorphisms in homology, then so does the third.

11.2 Projective and Injective Resolutions

We now lay the groundwork to investigate the left and right derived functors of an additive (contravariant) functor $\mathcal{F}$. Central to the development of these functors are projective and injective resolutions of a module.

If $\mathbf{C}$ is the positive chain complex

$$\mathbf{C} : \cdots \to M_n \to M_{n-1} \to \cdots \to M_1 \to M_0 \to M \to 0,$$

then $\mathbf{C}_M$ will denote the chain complex

$$\mathbf{C}_M : \cdots \to M_n \to M_{n-1} \to \cdots \to M_1 \to M_0 \to 0,$$

where M has been removed from $\mathbf{C}$. Similarly, given a positive cochain complex

$$\mathbf{C} : 0 \to M \to M^0 \to M^1 \to \cdots \to M^{n-1} \to M^n \to \cdots ,$$

$\mathbf{C}^M$ will denote the cochain complex

$$\mathbf{C}^M : 0 \to M^0 \to M^1 \to \cdots \to M^{n-1} \to M^n \to \cdots$$

with M removed.

Definition 11.2.1. An exact positive chain complex

$$\mathbf{P} : \cdots \to P_n \xrightarrow{\alpha_n} P_{n-1} \to \cdots \to P_1 \xrightarrow{\alpha_1} P_0 \xrightarrow{\alpha_0} M \to 0$$

is said to be a *projective resolution* of M if P_n is a projective module for $n = 0, 1, 2, \ldots$. If $\mathbf{P}$ is a projective resolution of M, then $\mathbf{P}_M$ is said to be a *deleted projective resolution* of M. Dually, an exact positive cochain complex

$$\mathbf{E} : 0 \to M \xrightarrow{\alpha^{-1}} E^0 \xrightarrow{\alpha^0} E^1 \to \cdots \to E^{n-1} \xrightarrow{\alpha^{n-1}} E^n \to \cdots,$$

where each E^n is an injective R-module, is an *injective resolution* of M. If $\mathbf{E}$ is an injective resolution of M, then $\mathbf{E}^M$ is a *deleted injective resolution* of M.

Proposition 11.2.2. *Every R-module M has a projective and an injective resolution.*

Proof. If M is an R-module, then we know that there is a projective module P_0 and an epimorphism $\alpha_0 : P_0 \to M$. Hence, we have an exact sequence $0 \to K_0 \xrightarrow{i_0} P_0 \xrightarrow{\alpha_0} M \to 0$, where K_0 is the kernel of α_0 and i_0 is the canonical injection. Continuing in this fashion, there are short exact sequences of R-modules and R-module homomorphisms

$$0 \to K_0 \xrightarrow{i_0} P_0 \xrightarrow{\alpha_0} M \to 0$$

$$0 \to K_1 \xrightarrow{i_1} P_1 \xrightarrow{p_1} K_0 \to 0$$

$$0 \to K_2 \xrightarrow{i_2} P_2 \xrightarrow{p_2} K_1 \to 0$$

$$\vdots$$

$$0 \to K_n \xrightarrow{i_n} P_n \xrightarrow{p_n} K_{n-1} \to 0$$

$$\vdots$$

where P_n is projective, K_n is the kernel of α_n and i_n is the canonical injection, for $n = 0, 1, 2, \ldots$ with $p_0 = \alpha_0$. If these short exact sequences are spliced together as

shown in the following diagram

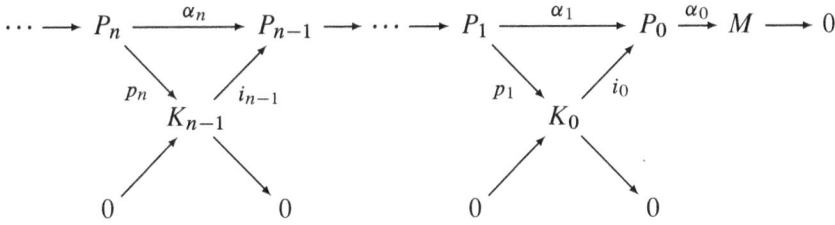

by letting $\alpha_n = i_{n-1} p_n$ for $n = 1, 2, 3, \ldots$, then $\operatorname{Im} \alpha_n = \operatorname{Ker} \alpha_{n-1} = K_{n-1}$ for $n = 1, 2, 3, \ldots$ and

$$\mathbf{P} : \cdots \to P_n \xrightarrow{\alpha_n} P_{n-1} \to \cdots \to P_1 \xrightarrow{\alpha_1} P_0 \xrightarrow{\alpha_0} M \to 0$$

is a projective resolution of M.

An injective resolution

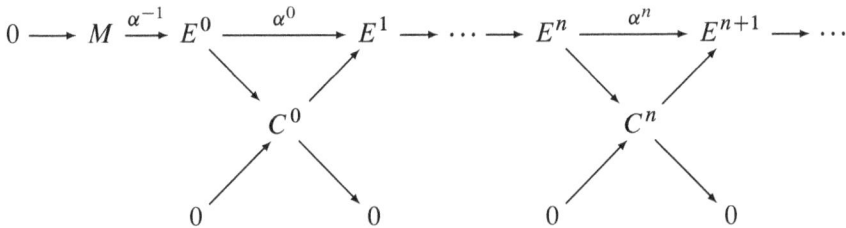

of M can be constructed by using the fact that every module can be embedded in an injective module and by using cokernels C^n in a manner dual to how kernels were used in the development of a projective resolution of M. If

$$\mathbf{E} : 0 \to M \xrightarrow{\alpha^{-1}} E^0 \xrightarrow{\alpha^0} E^1 \to \cdots \to E^n \xrightarrow{\alpha^n} E^{n+1} \to \cdots$$

is such an injective resolution of M, then $\operatorname{Im} \alpha^{n-1} = \operatorname{Ker} \alpha^n = C^n$ for $n = 0, 1, 2, \ldots$.   □

Remark. If

$$\mathbf{P} : \cdots \to P_n \xrightarrow{\alpha_n} P_{n-1} \to \cdots \to P_1 \xrightarrow{\alpha_1} P_0 \xrightarrow{\alpha_0} M \to 0$$

is a projective resolution of M, then

$$\mathbf{P}_M : \cdots \to P_n \xrightarrow{\alpha_n} P_{n-1} \to \cdots \to P_1 \xrightarrow{\alpha_1} P_0 \to 0$$

and $H_0(\mathbf{P}_M) = \operatorname{Ker}(P_0 \to 0)/\operatorname{Im} \alpha_1 = P_0/\operatorname{Im} \alpha_1 = P_0/\operatorname{Ker} \alpha_0 \cong M$. Similarly, if $\mathbf{E}$ is an injective resolution of M, then $H^0(\mathbf{E}^M) \cong M$. Thus, there is essentially no loss of information when M is deleted from a projective resolution or from an injective resolution since M can be recovered, up to isomorphism, by forming the 0th homology or the 0th cohomology module of $\mathbf{P}_M$ or $\mathbf{E}^M$, respectively.

Examples

1. $\cdots \to \mathbb{Z}_4 \xrightarrow{\alpha_n} \mathbb{Z}_4 \xrightarrow{\alpha_{n-1}} \mathbb{Z}_4 \to \cdots \to \mathbb{Z}_4 \xrightarrow{\alpha_0} \mathbb{Z}_2 \to 0$ is a projective resolution of the $\mathbb{Z}_4$-module $\mathbb{Z}_2$, where $\alpha_n([a]) = [2a]$ for $n \geq 1$ and $\alpha_0([a]) = [a]$.

2. $\cdots \to 0 \to 0 \to \mathbb{Z} \xrightarrow{\alpha_1} \mathbb{Z} \xrightarrow{\alpha_0} \mathbb{Z}_k \to 0$ is a projective resolution of the $\mathbb{Z}$-module $\mathbb{Z}_k$, where $k \geq 2$ and $\alpha_1(a) = ka$ and $\alpha_0(a) = [a]$ for all $a \in \mathbb{Z}$.

3. $0 \to \mathbb{Z} \xrightarrow{\alpha^{-1}} \mathbb{Q} \xrightarrow{\alpha^0} \mathbb{Q}/\mathbb{Z} \to 0 \to 0 \to \cdots$ is an injective resolution of the $\mathbb{Z}$-module $\mathbb{Z}$, where α^{-1} is the canonical injection and α^0 is the natural mapping.

Our goal now is to show that any two projective resolutions of a module are of the same homotopy type. For this, we need the following two lemmas.

Lemma 11.2.3. *Let*

$$\mathbf{P} : \cdots \to P_n \xrightarrow{\alpha_n} P_{n-1} \to \cdots \to P_1 \xrightarrow{\alpha_1} P_0 \xrightarrow{\alpha_0} M \to 0$$

be a chain complex such that P_n is projective for $n = 0, 1, 2, \ldots$ and suppose that

$$\mathbf{Q} : \cdots \to N_n \xrightarrow{\beta_n} N_{n-1} \to \cdots \to N_1 \xrightarrow{\beta_1} N_0 \xrightarrow{\beta_0} N \to 0$$

is exact. Then for any R-linear mapping $f : M \to N$ there is a chain map $\mathbf{f} : \mathbf{P}_M \to \mathbf{Q}_N$ such that the diagram

$$
\begin{array}{ccccccccccccc}
\cdots & \longrightarrow & P_n & \xrightarrow{\alpha_n} & P_{n-1} & \xrightarrow{\alpha_{n-1}} & \cdots & \xrightarrow{\alpha_1} & P_0 & \xrightarrow{\alpha_0} & M & \longrightarrow & 0 \\
& & \downarrow{f_n} & & \downarrow{f_{n-1}} & & & & \downarrow{f_0} & & \downarrow{f} & & \\
\cdots & \longrightarrow & N_n & \xrightarrow{\beta_n} & N_{n-1} & \xrightarrow{\beta_{n-1}} & \cdots & \xrightarrow{\beta_1} & N_0 & \xrightarrow{\beta_0} & N & \longrightarrow & 0
\end{array}
$$

is commutative.

Proof. Since P_0 is projective and since we have an R-linear mapping $P_0 \xrightarrow{f\alpha_0} N$, then the fact that β_0 is an epimorphism ensures that there is an R-linear mapping $f_0 : P_0 \to N_0$ such that the diagram

$$
\begin{array}{ccc}
P_0 & \xrightarrow{\alpha_0} & M \\
\downarrow{f_0} & & \downarrow{f} \\
N_0 & \xrightarrow{\beta_0} & N \longrightarrow 0
\end{array}
$$

is commutative. Now suppose that R-linear mappings $f_0, f_1, f_2, \ldots, f_{n-1}$ have been found such that the diagram

$$\cdots \longrightarrow P_n \xrightarrow{\alpha_n} P_{n-1} \xrightarrow{\alpha_{n-1}} P_{n-2} \xrightarrow{\alpha_{n-2}} \cdots \xrightarrow{\alpha_1} P_0 \xrightarrow{\alpha_0} M \longrightarrow 0$$

with vertical maps $f_n, f_{n-1}, f_{n-2}, f_0, f$ down to

$$\cdots \longrightarrow N_n \xrightarrow{\beta_n} N_{n-1} \xrightarrow{\beta_{n-1}} N_{n-2} \xrightarrow{\beta_{n-2}} \cdots \xrightarrow{\beta_1} N_0 \xrightarrow{\beta_0} N \longrightarrow 0$$

is commutative. If we can produce an R-linear mapping $f_n : P_n \rightarrow N_n$ such that $\beta_n f_n = f_{n-1}\alpha_n$, then the proposition will follow by induction. From the last diagram above, $\beta_{n-1}f_{n-1} = f_{n-2}\alpha_{n-1}$, so $\beta_{n-1}f_{n-1}\alpha_n = f_{n-2}\alpha_{n-1}\alpha_n = 0$ gives Im $f_{n-1}\alpha_n \subseteq$ Ker $\beta_{n-1} =$ Im β_n. Since P_n is projective, we have a diagram

$$\begin{array}{ccc} & & P_n \\ & {\scriptstyle f_n}\nearrow & \downarrow {\scriptstyle f_{n-1}\alpha_n} \\ N_n \xrightarrow{\beta_n} & \mathrm{Im}\,\beta_n & \longrightarrow 0 \end{array}$$

that can be completed commutatively by an R-linear map $f_n : P_n \rightarrow N_n$.   □

We will now refer to the chain map $\mathbf{f} : \mathbf{P}_M \rightarrow \mathbf{Q}_N$ produced in Lemma 11.2.3 as a *chain map generated by* $f : M \rightarrow N$. There is no assurance that such a chain map is unique. However, any two chain maps generated by f are homotopic.

Lemma 11.2.4. *Let* $\mathbf{P}$ *and* $\mathbf{Q}$ *be as in Lemma 11.2.3. If* $f : M \rightarrow N$ *is an R-linear mapping, then any pair of chain maps* $\mathbf{f}, \mathbf{g} : \mathbf{P}_M \rightarrow \mathbf{Q}_N$ *generated by* f *are homotopic.*

Proof. Suppose that $\mathbf{f}, \mathbf{g} : \mathbf{P}_M \rightarrow \mathbf{Q}_N$ are chain maps generated by f. We need to produce a family $\varphi = \{\varphi_n : P_n \rightarrow N_{n+1}\}_{\mathbb{Z}}$ of R-linear mappings such that $f_n - g_n = \beta_{n+1}\varphi_n + \varphi_{n-1}\alpha_n$ for each $n \in \mathbb{Z}$. Since $\mathbf{P}_M$ is positive, we can let $\varphi_n = 0$ for all $n < 0$. Thus, in the 0th position we need to find $\varphi_0 : P_0 \rightarrow N_1$ such that $f_0 - g_0 = \beta_1\varphi_0$. For this consider the diagram

$$\begin{array}{ccccccc} P_1 & \xrightarrow{\alpha_1} & P_0 & \xrightarrow{\alpha_0} & M & \longrightarrow & 0 \\ {\scriptstyle f_1\,g_1}\downarrow\downarrow & {\scriptstyle \varphi_0}\swarrow & \downarrow\downarrow {\scriptstyle f_0\,g_0} & & \downarrow {\scriptstyle f} & & \\ N_1 & \xrightarrow{\beta_1} & N_0 & \xrightarrow{\beta_0} & N & \longrightarrow & 0 \end{array}$$

Since $\beta_0(f_0 - g_0) = f(\alpha_0 - \alpha_0) = 0$, Im$(f_0 - g_0) \subseteq$ Ker $\beta_0 =$ Im β_1. It follows that the projectivity of P_0 gives an R-linear mapping $\varphi_0 : P_0 \rightarrow N_1$ such that $f_0 - g_0 =$

$\beta_1\varphi_0$. Next, suppose that R-linear maps $\varphi_k : P_k \to N_{k+1}$ have been found such that $f_k - g_k = \beta_{k+1}\varphi_k + \varphi_{k-1}\alpha_k$, for $k = 0, 1, 2, \dots, n-1$, and consider the diagram

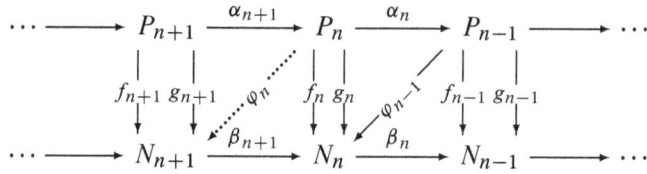

$$
\begin{array}{ccccccc}
\cdots \longrightarrow & P_{n+1} & \overset{\alpha_{n+1}}{\longrightarrow} & P_n & \overset{\alpha_n}{\longrightarrow} & P_{n-1} & \longrightarrow \cdots \\
& \downarrow\downarrow & \varphi_n & \downarrow\downarrow & \varphi_{n-1} & \downarrow\downarrow & \\
& {\scriptstyle f_{n+1}\, g_{n+1}} & & {\scriptstyle f_n\, g_n} & & {\scriptstyle f_{n-1}\, g_{n-1}} & \\
\cdots \longrightarrow & N_{n+1} & \overset{\beta_{n+1}}{\longrightarrow} & N_n & \overset{\beta_n}{\longrightarrow} & N_{n-1} & \longrightarrow \cdots
\end{array}
$$

Since $f_{n-1} - g_{n-1} - \beta_n\varphi_{n-1} = \varphi_{n-2}\alpha_{n-1}$, we have

$$
\begin{aligned}
\beta_n(f_n - g_n - \varphi_{n-1}\alpha_n) &= \beta_n f_n - \beta_n g_n - \beta_n\varphi_{n-1}\alpha_n \\
&= f_{n-1}\alpha_n - g_{n-1}\alpha_n - \beta_n\varphi_{n-1}\alpha_n \\
&= (f_{n-1} - g_{n-1} - \beta_n\varphi_{n-1})\alpha_n \\
&= \varphi_{n-2}\alpha_{n-1}\alpha_n = 0.
\end{aligned}
$$

Hence, $\operatorname{Im}(f_n - g_n - \varphi_{n-1}\alpha_n) \subseteq \operatorname{Ker}\beta_n = \operatorname{Im}\beta_{n+1}$ and so since P_n is projective, there is an R-linear mapping $\varphi_n : P_n \to N_{n+1}$ that completes the diagram

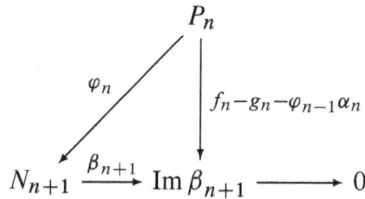

$$
\begin{array}{ccc}
& P_n & \\
{\scriptstyle \varphi_n}\swarrow & & \downarrow {\scriptstyle f_n - g_n - \varphi_{n-1}\alpha_n} \\
N_{n+1} \overset{\beta_{n+1}}{\longrightarrow} & \operatorname{Im}\beta_{n+1} & \longrightarrow 0
\end{array}
$$

commutatively. Therefore, $f_n - g_n = \beta_{n+1}\varphi_n + \varphi_{n-1}\alpha_n$, so the proposition follows by induction. $\square$

Proposition 11.2.5. *If P and Q are projective resolutions of M, then P_M and Q_M are of the same homotopy type.*

Proof. Suppose that the identity map $\operatorname{id}_M : M \to M$ generates chain maps $f : P_M \to Q_M$ and $g : Q_M \to P_M$. The proposition follows from the fact that Lemma 11.2.4 gives $gf \approx \operatorname{id}_{P_M}$ and $fg \approx \operatorname{id}_{Q_M}$, where $\operatorname{id}_{P_M} : P_M \to P_M$ and $\operatorname{id}_{Q_M} : Q_M \to Q_M$ are the identity chain maps on P_M and Q_M, respectively. Thus, P_M and Q_M are of the same homotopy type. $\square$

The dual versions of Lemma 11.2.3, Lemma 11.2.4 and Proposition 11.2.5 are Exercises 2 and 3 in the following problem set.

Problem Set 11.2

1. Verify Examples 1, 2 and 3. [3, Hint: Proposition 5.1.9.]

2. Let

$$\mathbf{D}: 0 \to N \to N^0 \to N^1 \to \cdots \to N^{n-1} \to N^n \to \cdots$$

be an exact cochain complex and suppose that

$$\mathbf{E}: 0 \to M \to E^0 \to E^1 \to \cdots \to E^{n-1} \to E^n \to \cdots$$

is a cochain complex such that E^n is injective for $n = 0, 1, 2, \ldots$.

(a) Show that for any R-linear mapping $f : N \to M$, there is a cochain map $\mathbf{f} : \mathbf{D}^N \to \mathbf{E}^M$ such that the diagram

is commutative.

(b) Deduce that any pair of cochain maps $\mathbf{f}, \mathbf{g} : \mathbf{D}^N \to \mathbf{E}^M$ generated by f are homotopic.

(c) If $\mathbf{D}$ and $\mathbf{E}$ are injective resolutions of M, prove that $\mathbf{D}^M$ and $\mathbf{E}^M$ are of the same homotopy type.

[Hint: The proofs of (a), (b) and (c) are duals of the proofs of results in this section.]

3. Let $\mathbf{M}$ be a chain complex and suppose that $\mathbf{id_M}$ and $\mathbf{0_M}$ are the identity and zero chain maps from $\mathbf{M}$ to $\mathbf{M}$, respectively. Prove that if $\mathbf{id_M}$ and $\mathbf{0_M}$ are homotopic, then $\mathbf{M}$ is exact.

4. If R is a right noetherian ring and M is a finitely generated R-module, show that M has a projective resolution in which the projective modules and the kernels of the boundary maps are finitely generated.

5. Exercise 11 in Problem Set 5.1 and Exercise 7 in Problem Set 5.2 gave Schanuel's lemma for injectives and projectives, respectively. Prove the following *long versions of Schanuel's lemmas.*

(a) If

$$0 \to K_n \to P_n \to P_{n-1} \to \cdots \to P_0 \xrightarrow{\alpha} M \to 0 \quad \text{and}$$

$$0 \to K'_n \to Q_n \to Q_{n-1} \to \cdots \to Q_0 \xrightarrow{\beta} M \to 0$$

are exact and P_k and Q_k are projective for $k = 0, 1, 2, \ldots, n$, then

$$K_n \oplus Q_n \oplus P_{n-1} \oplus Q_{n-2} \cdots \cong K_n' \oplus P_n \oplus Q_{n-1} \oplus P_{n-2} \cdots .$$

[Hint: If $K_\alpha = \operatorname{Ker}\alpha$ and $K_\beta = \operatorname{Ker}\beta$, then Schanuel's lemma for projectives gives $Q_0 \oplus K_\alpha \cong P_0 \oplus K_\beta$. Show that we have exact sequences

$$0 \to K_n \to P_n \to P_{n-1} \to \cdots \to P_2 \to Q_0 \oplus P_1 \to Q_0 \oplus K_\alpha \to 0 \quad \text{and}$$

$$0 \to K_n' \to Q_n \to Q_{n-1} \to \cdots \to Q_2 \to P_0 \oplus Q_1 \to P_0 \oplus K_\beta \to 0$$

and use induction.

(b) If

$$0 \to M \xrightarrow{\alpha} D^0 \to \cdots \to D^{n-1} \to D^n \to C^n \to 0 \quad \text{and}$$

$$0 \to M \xrightarrow{\beta} E^0 \to \cdots \to E^{n-1} \to E^n \to C^{n'} \to 0$$

are exact and D^k and E^k are injective for $k = 0, 1, 2, \ldots, n$, then

$$C^n \oplus E^n \oplus D^{n-1} \oplus E^{n-2} \oplus \cdots \cong C^{n'} \oplus D^n \oplus E^{n-1} \oplus D^{n-2} \oplus \cdots .$$

6. (a) If $\mathbf{P}:\ \cdots \to P_n \xrightarrow{\alpha_n} P_{n-1} \to \cdots \to P_1 \xrightarrow{\alpha_1} P_0 \xrightarrow{\alpha_0} M \to 0$ is a projective resolution of M such that $\operatorname{Ker}\alpha_n$ is a projective R-module, prove that if $\mathbf{Q}:\ \cdots \to Q_n \xrightarrow{\beta_n} Q_{n-1} \to \cdots \to Q_1 \xrightarrow{\beta_1} Q_0 \xrightarrow{\beta_0} M \to 0$ is a projective resolution of M, then $\operatorname{Ker}\beta_n$ is also projective.

(b) If $0 \to M \xrightarrow{\alpha^{-1}} D^0 \xrightarrow{\alpha^0} \cdots \to D^n \xrightarrow{\alpha^n} D^{n+1} \to \cdots$ is an injective resolution of M such that $\operatorname{Im}\alpha^n$ is injective, deduce that if $0 \to M \xrightarrow{\beta^{-1}} E^0 \xrightarrow{\beta^0} \cdots \to E^n \xrightarrow{\beta^n} E^{n+1} \to \cdots$ is an injective resolution of M, then $\operatorname{Im}\beta^n$ is injective as well.

7. If $\mathbf{P}_M$ is a deleted projective resolution of an R-module M, then we have seen that there is exactly one homotopy class of deleted projective resolutions of M, denoted by $[\mathbf{P}_M]$. Let $\mathbf{P}_M$ and $\mathbf{Q}_N$ be deleted projective resolutions of M and N, respectively, suppose that $f : M \to N$ is an R-linear mapping and let $[\mathbf{f}]$ denote the homotopy class of $\mathbf{f}$, where $\mathbf{f} : \mathbf{P}_M \to \mathbf{Q}_N$ is a chain map generated by f. Next, let $\mathcal{C}$ be the possible category whose objects are $[\mathbf{P}_M]$ and whose morphism sets are $[\mathbf{f}] \in \operatorname{Mor}([\mathbf{P}_M], [\mathbf{Q}_N])$ for each pair of R-modules M and N. If composition is defined in $\mathcal{C}$ in the obvious way, then is $\mathcal{C}$ a (an additive) category?

11.3 Derived Functors

A central theme of homological algebra is that of left and right derived functors. Since we are primarily interested in the left and right derived functors that can be developed from Hom and $\otimes$ in their first and second variables and since these functors take $\mathbf{Mod}_R$ to $\mathbf{Ab}$ or $_R\mathbf{Mod}$ to $\mathbf{Ab}$, we will provide a general treatment for an additive functor $\mathscr{F} : \mathbf{Mod}_R \to \mathbf{Ab}$. The parallel but dual case for an additive contravariant functor can be obtained from the covariant case simply by reversing the arrows and making dual arguments.

Recall that if

$$\mathbf{P} : \cdots \to P_n \xrightarrow{\alpha_n} P_{n-1} \to \cdots \xrightarrow{\alpha_1} P_0 \xrightarrow{\alpha_0} M \to 0$$

is a projective resolution of M and $\mathscr{F} : \mathbf{Mod}_R \to \mathbf{Ab}$ is a functor, then $\mathscr{F}(\mathbf{P})$ denotes the sequence

$$\mathscr{F}(\mathbf{P}) : \cdots \to \mathscr{F}(P_n) \xrightarrow{\mathscr{F}(\alpha_n)} \mathscr{F}(P_{n-1}) \to \cdots \xrightarrow{\mathscr{F}(\alpha_1)} \mathscr{F}(P_0) \xrightarrow{\mathscr{F}(\alpha_0)} \mathscr{F}(M) \to 0$$

of abelian groups and group homomorphisms. If $\mathscr{F}$ is also additive, then

$$\mathscr{F}(\alpha_n)\mathscr{F}(\alpha_{n+1}) = 0,$$

so $\mathscr{F}(\mathbf{P})$ is a chain complex in $\mathbf{Ab}$. However, $\mathscr{F}(\mathbf{P})$ may not be exact and there is no reason to expect that each $\mathscr{F}(P_n)$ is projective. Similar observations hold for $\mathscr{F}(\mathbf{E})$ if $\mathbf{E}$ is an injective resolution of M and for $\mathscr{F}(\mathbf{P})$ and $\mathscr{F}(\mathbf{E})$ when $\mathscr{F}$ is an additive contravariant functor.

If $\mathbf{P}$ and $\mathbf{Q}$ are projective resolutions of M and N, respectively, and $f : M \to N$ is an R-linear mapping, then there is a commutative diagram

$$
\begin{array}{ccccccccccc}
\cdots & \longrightarrow & P_{n+1} & \xrightarrow{\alpha_{n+1}} & P_n & \xrightarrow{\alpha_n} & P_{n-1} & \longrightarrow & \cdots \xrightarrow{\alpha_1} & P_0 & \xrightarrow{\alpha_0} & M & \longrightarrow & 0 \\
& & \downarrow{f_{n+1}} & & \downarrow{f_n} & & \downarrow{f_{n-1}} & & & \downarrow{f_0} & & \downarrow{f} & & \\
\cdots & \longrightarrow & Q_{n+1} & \xrightarrow{\beta_{n+1}} & Q_n & \xrightarrow{\beta_n} & Q_{n-1} & \longrightarrow & \cdots \xrightarrow{\beta_1} & Q_0 & \xrightarrow{\beta_0} & N & \longrightarrow & 0
\end{array}
$$

where $\mathbf{f} : \mathbf{P}_M \to \mathbf{Q}_N$ is a chain map generated by f. So if $\mathscr{F} : \mathbf{Mod}_R \to \mathbf{Ab}$ is an additive functor, then we get a commutative diagram

$$
\begin{array}{ccccccccc}
\cdots \longrightarrow & \mathscr{F}(P_{n+1}) & \xrightarrow{\mathscr{F}(\alpha_{n+1})} & \mathscr{F}(P_n) & \xrightarrow{\mathscr{F}(\alpha_n)} & \mathscr{F}(P_{n-1}) & \longrightarrow \cdots \xrightarrow{\mathscr{F}(\alpha_1)} & \mathscr{F}(P_0) & \longrightarrow 0 \\
& \downarrow{\mathscr{F}(f_{n+1})} & & \downarrow{\mathscr{F}(f_n)} & & \downarrow{\mathscr{F}(f_{n-1})} & & \downarrow{\mathscr{F}(f_0)} & \\
\cdots \longrightarrow & \mathscr{F}(Q_{n+1}) & \xrightarrow{\mathscr{F}(\beta_{n+1})} & \mathscr{F}(Q_n) & \xrightarrow{\mathscr{F}(\beta_n)} & \mathscr{F}(Q_{n-1}) & \longrightarrow \cdots \xrightarrow{\mathscr{F}(\beta_1)} & \mathscr{F}(Q_0) & \longrightarrow 0
\end{array}
$$

where the top and bottom row are chain complexes. Hence, for $n = 0, 1, 2, \ldots$, there is a homology mapping $H_n(\mathcal{F}(\mathbf{f}))$ that maps the nth homology group of $\mathcal{F}(\mathbf{P}_M)$ to the nth homology group of $\mathcal{F}(\mathbf{Q}_N)$. If $\mathcal{F}$ is additive and contravariant, then the arrows reverse in the preceding diagram and, for each n, there is an nth cohomology mapping $H^n(\mathcal{F}(\mathbf{f}))$ taking the nth cohomology group of $\mathcal{F}(\mathbf{Q}_N)$ to the nth cohomology group of $\mathcal{F}(\mathbf{P}_M)$.

Proposition 11.3.1. *Let $\mathcal{F} : \mathbf{Mod}_R \to \mathbf{Ab}$ be an additive functor.*

(1) *If $\mathbf{P}$ and $\mathbf{Q}$ are projective resolutions of M, then for $n \geq 0$ the nth homology group of $\mathcal{F}(\mathbf{P}_M)$ is isomorphic to the nth homology group of $\mathcal{F}(\mathbf{Q}_M)$. Moreover, if $f : M \to N$ is an R-linear mapping and $\mathbf{P}$ and $\mathbf{Q}$ are projective resolutions of M and N, respectively, then for $n \geq 0$ the group homomorphism $H_n(\mathcal{F}(\mathbf{f})) : H_n(\mathcal{F}(\mathbf{P}_M)) \to H_n(\mathcal{F}(\mathbf{Q}_N))$ does not depend on the choice of the chain map $\mathbf{f} : \mathbf{P}_M \to \mathbf{Q}_N$ generated by f.*

(2) *If $\mathbf{D}$ and $\mathbf{E}$ are injective resolutions of M, then for $n \geq 0$ the nth cohomology group of $\mathcal{F}(\mathbf{D}^M)$ is isomorphic to the nth cohomology group of $\mathcal{F}(\mathbf{E}^M)$. If $f : M \to N$ is an R-linear mapping and $\mathbf{D}$ and $\mathbf{E}$ are injective resolutions of M and N, respectively, then for $n \geq 0$ the group homomorphism $H^n(\mathcal{F}(\mathbf{f})) : H^n(\mathcal{F}(\mathbf{D}^M)) \to H^n(\mathcal{F}(\mathbf{E}^N))$ does not depend on the choice of the cochain map $\mathbf{f} : \mathbf{D}^M \to \mathbf{E}^N$ generated by f.*

Proof. We prove (1) and omit the proof of (2) since it is similar. If $\mathbf{P}$ and $\mathbf{Q}$ are projective resolutions of M and if the identity map $\mathrm{id}_M : M \to M$ generates chain maps $\mathbf{f} : \mathbf{P}_M \to \mathbf{Q}_M$ and $\mathbf{g} : \mathbf{Q}_M \to \mathbf{P}_M$, then Lemma 11.2.5 indicates that $\mathbf{gf}$ and $\mathbf{fg}$ are homotopic to the chain maps $\mathbf{id}_{\mathbf{P}_M}$ and $\mathbf{id}_{\mathbf{Q}_M}$, respectively. It follows from (3) of Proposition 11.1.8 that $\mathcal{F}(\mathbf{gf})$ and $\mathcal{F}(\mathbf{fg})$ are homotopic to $\mathcal{F}(\mathbf{id}_{\mathbf{P}_M}) = \mathbf{id}_{\mathcal{F}(\mathbf{P}_M)}$ and $\mathcal{F}(\mathbf{id}_{\mathbf{Q}_M}) = \mathbf{id}_{\mathcal{F}(\mathbf{Q}_M)}$, respectively. By applying (4) of Proposition 11.1.8 and using properties of the functors H_n and $\mathcal{F}$, we see that

$$H_n(\mathcal{F}(\mathbf{g}))H_n(\mathcal{F}(\mathbf{f})) = H_n(\mathcal{F}(\mathbf{g})\mathcal{F}(\mathbf{f}))$$

$$= H_n(\mathcal{F}(\mathbf{gf})) = H_n(\mathcal{F}(\mathbf{id}_{\mathbf{P}_M}))$$

$$= H_n(\mathbf{id}_{\mathcal{F}(\mathbf{P}_M)}) = \mathrm{id}_{H_n(\mathcal{F}(\mathbf{P}_M))}.$$

Similarly, $H_n(\mathcal{F}(\mathbf{f}))H_n(\mathcal{F}(\mathbf{g})) = \mathrm{id}_{H_n(\mathcal{F}(\mathbf{Q}_M))}$, so

$$H_n(\mathcal{F}(\mathbf{f})) : H_n(\mathcal{F}(\mathbf{P}_M)) \to H_n(\mathcal{F}(\mathbf{Q}_M))$$

is an isomorphism.

Finally, suppose that $\mathbf{f}, \mathbf{g} : \mathbf{P}_M \to \mathbf{Q}_N$ are chain maps generated by $f : M \to N$. Then $\mathbf{f}$ and $\mathbf{g}$ are homotopic, so by (3) of Proposition 11.1.8 we see that $\mathcal{F}(\mathbf{f})$, $\mathcal{F}(\mathbf{g}) : \mathcal{F}(\mathbf{P}_M) \to \mathcal{F}(\mathbf{Q}_N)$ are homotopic as well. Hence, (4) of Proposition 11.1.8 shows that

$$H_n(\mathcal{F}(\mathbf{f})) = H_n(\mathcal{F}(\mathbf{g})) : H_n(\mathcal{F}(\mathbf{P}_M)) \to H_n(\mathcal{F}(\mathbf{Q}_N)). \qquad \square$$

Proposition 11.3.1 provides the tools necessary to establish nth left and nth right derived functors. The development of these functors proceeds as follows for an additive functor $\mathcal{F} : \mathbf{Mod}_R \to \mathbf{Ab}$.

1. **The Functor $\mathcal{L}_n\mathcal{F} : \mathbf{Mod}_R \to \mathbf{Ab}$.** Choose and fix a projective resolution of each R-module. If

$$\mathbf{P} : \cdots \to P_n \to P_{n-1} \to \cdots \to P_0 \to M \to 0$$

is the chosen projective resolution of M, then

$$\mathcal{F}(\mathbf{P}) : \cdots \to \mathcal{F}(P_n) \to \mathcal{F}(P_{n-1}) \to \cdots \to \mathcal{F}(P_0) \to \mathcal{F}(M) \to 0$$

is a chain complex in $\mathbf{Ab}$. Form the homology groups $H_n(\mathcal{F}(\mathbf{P}_M))$, that is, *take homology* and set $\mathcal{L}_n\mathcal{F}(M) = H_n(\mathcal{F}(\mathbf{P}_M))$. For an R-linear mapping $f : M \to N$, let $H_n(\mathcal{F}(\mathbf{f})) : H_n(\mathcal{F}(\mathbf{P}_M)) \to H_n(\mathcal{F}(\mathbf{Q}_N))$ be the nth homology map, where $\mathbf{Q}$ is the chosen projective resolutions of N and $\mathbf{f} : \mathbf{P}_M \to \mathbf{Q}_N$ is a chain map generated by f. Part (1) of Proposition 11.3.1 shows that $H_n(\mathcal{F}(\mathbf{f}))$ depends only on f and not on the chain map $\mathbf{f}$ generated by f. If we let $\mathcal{L}_n\mathcal{F}(f) = H_n(\mathcal{F}(\mathbf{f}))$, then

$$\mathcal{L}_n\mathcal{F}(f) : \mathcal{L}_n\mathcal{F}(M) \to \mathcal{L}_n\mathcal{F}(N)$$

and we have an additive functor $\mathcal{L}_n\mathcal{F} : \mathbf{Mod}_R \to \mathbf{Ab}$ for each integer $n \geq 0$. However, $\mathcal{L}_n\mathcal{F}$ is not unique since its construction depends on the projective resolutions chosen for the modules. If the projective resolutions are chosen again and perhaps in a different way, then we obtain a second functor $\overline{\mathcal{L}}_n\mathcal{F} : \mathbf{Mod}_R \to \mathbf{Ab}$ constructed in exactly the same fashion as $\mathcal{L}_n\mathcal{F}$. The functors $\mathcal{L}_n\mathcal{F}$ and $\overline{\mathcal{L}}_n\mathcal{F}$ are, in general, distinct but the important point is that they are naturally equivalent. From this we can conclude that $\mathcal{L}_n\mathcal{F}$ and $\overline{\mathcal{L}}_n\mathcal{F}$ can be interchanged via isomorphisms.

2. **The Functor $\mathcal{R}^n\mathcal{F} : \mathbf{Mod}_R \to \mathbf{Ab}$.** Choose and fix an injective resolution of each R-module. If

$$\mathbf{D} : 0 \to M \to D^0 \to \cdots \to D^{n-1} \to D^n \to \cdots$$

is the chosen injective resolution of M, then

$$\mathcal{F}(\mathbf{D}) : 0 \to \mathcal{F}(M) \to \mathcal{F}(D^0) \to \cdots \to \mathcal{F}(D^{n-1}) \to \mathcal{F}(D^n) \to \cdots$$

is a cochain complex in $\mathbf{Ab}$. *Take cohomology*, that is, form the cohomology groups in $\mathcal{F}(\mathbf{D}^M)$ and let $\mathcal{R}^n\mathcal{F}(M) = H^n(\mathcal{F}(\mathbf{D}^M))$. For an R-linear mapping $f : M \to N$, let $H^n(\mathcal{F}(\mathbf{f})) : H^n(\mathcal{F}(\mathbf{D}^M)) \to H^n(\mathcal{F}(\mathbf{E}^N))$ be the nth cohomology map, where $\mathbf{E}$ is the chosen injective resolution of N and $\mathbf{f} : \mathbf{D}^M \to \mathbf{E}^N$

is a cochain map generated by f. Part (2) of Proposition 11.3.1 indicates that $H^n(\mathscr{F}(\mathbf{f}))$ does not depend on the cochain map $\mathbf{f}$ generated by f. Hence, if we let $\mathscr{R}^n \mathscr{F}(f) = H^n(\mathscr{F}(\mathbf{f}))$, then

$$\mathscr{R}^n \mathscr{F}(f) : \mathscr{R}^n \mathscr{F}(M) \to \mathscr{R}^n \mathscr{F}(N)$$

which establishes an additive functor $\mathscr{R}^n \mathscr{F} : \mathbf{Mod}_R \to \mathbf{Ab}$, for $n = 0, 1, 2, \ldots$. It follows that if injective resolutions of the modules are chosen in a different way, then a functor $\overline{\mathscr{R}}^n \mathscr{F} : \mathbf{Mod}_R \to \mathbf{Ab}$ can be formed exactly in the same manner as the functor $\mathscr{R}^n \mathscr{F}$ and the functors $\mathscr{R}^n \mathscr{F}$ and $\overline{\mathscr{R}}^n \mathscr{F}$ are naturally equivalent.

Proposition 11.3.2. *If $\mathscr{F} : \mathbf{Mod}_R \to \mathbf{Ab}$ is an additive functor, then the functors $\mathscr{L}_n \mathscr{F}$ and $\overline{\mathscr{L}}_n \mathscr{F}$ are naturally equivalent as are $\mathscr{R}^n \mathscr{F}$ and $\overline{\mathscr{R}}^n \mathscr{F}$.*

Proof. If $\mathbf{P}$ and $\mathbf{P}'$ are projective resolutions of M, then by (1) of Proposition 11.3.1 we see that $H_n(\mathbf{P}_M)$ and $H_n(\mathbf{P}'_M)$ are isomorphic and in fact the isomorphism is given by $H_n(\mathscr{F}(\mathbf{f}))$, where $\mathbf{f}$ is any chain map generated by $\mathrm{id}_M : M \to M$. Furthermore, (4) of Proposition 11.1.8 shows that the isomorphism $H_n(\mathscr{F}(\mathbf{f}))$ is unique, so let $\eta_M = H_n(\mathscr{F}(\mathbf{f}))$. Next, suppose that $\mathbf{Q}$ and $\mathbf{Q}'$ are projective resolutions of N. If $f : M \to N$ is an R-linear mapping, then we have a commutative diagram

$$
\begin{array}{ccc}
H_n(\mathscr{F}(\mathbf{P}_M)) & \xrightarrow{\;\eta_M\;} & H_n(\mathscr{F}(\mathbf{P}'_M)) \\[2pt]
\Big\downarrow{\scriptstyle H_n(\mathscr{F}(\mathbf{f}))} & & \Big\downarrow{\scriptstyle H_n(\mathscr{F}(\mathbf{g}))} \\[2pt]
H_n(\mathscr{F}(\mathbf{Q}_N)) & \xrightarrow{\;\eta_N\;} & H_n(\mathscr{F}(\mathbf{Q}'_N))
\end{array}
$$

where $\mathbf{f} : \mathbf{P}_M \to \mathbf{Q}_N$ and $\mathbf{g} : \mathbf{P}'_M \to \mathbf{Q}'_N$ are chain maps generated by f. Thus, we have a natural isomorphism $\eta : \mathscr{L}_n \mathscr{F} \to \overline{\mathscr{L}}_n \mathscr{F}$, so $\mathscr{L}_n \mathscr{F}$ and $\overline{\mathscr{L}}_n \mathscr{F}$ are naturally equivalent functors.

The proof that $\mathscr{R}^n \mathscr{F}$ and $\overline{\mathscr{R}}^n \mathscr{F}$ are naturally equivalent is just as straightforward. $\square$

Hence, the functors $\mathscr{L}_n \mathscr{F}$ and $\mathscr{R}^n \mathscr{F}$ do not depend on the projective and injective resolution chosen for their development. This fact underlies the construction of the right derived functors of Hom and the construction of the left derived functors of $\otimes$.

Definition 11.3.3. If $\mathscr{F} : \mathbf{Mod}_R \to \mathbf{Ab}$ is an additive functor, then $\mathscr{L}_n \mathscr{F}$ and $\mathscr{R}^n \mathscr{F}$ are called the *nth left derived functor* of $\mathscr{F}$ and the *nth right derived functor* of $\mathscr{F}$, respectively, for $n = 0, 1, 2, \ldots$.

To complete the development of left and right derived functors, we note that there are nth left and right derived functors that correspond to an additive contravariant functor both of which are unique up to natural isomorphism. If $\mathcal{F} : \mathbf{Mod}_R \to \mathbf{Ab}$ is an additive contravariant functor, then

(1) $\mathcal{L}_n \mathcal{F}$ is constructed using injective resolutions and

(2) $\mathcal{R}^n \mathcal{F}$ is constructed using projective resolutions.

One useful and general result is that if $\mathcal{F} : \mathbf{Mod}_R \to \mathbf{Ab}$ is a right exact additive functor, then $\mathcal{L}_0 \mathcal{F}$ and $\mathcal{F}$ are naturally equivalent with a similar result holding when $\mathcal{F}$ is left exact and additive.

Proposition 11.3.4. *Let* $\mathcal{F} : \mathbf{Mod}_R \to \mathbf{Ab}$ *be an additive functor.*

(1) *If* $\mathcal{F}$ *is right exact, then* $\mathcal{L}_0 \mathcal{F}(M) \cong \mathcal{F}(M)$, *for each R-module M, so that* $\mathcal{L}_0 \mathcal{F}$ *and* $\mathcal{F}$ *are naturally equivalent functors and if P is a projective R-module, then* $\mathcal{L}_n \mathcal{F}(P) = 0$, *for* $n = 1, 2, 3, \ldots$.

(2) *If* $\mathcal{F}$ *is left exact, then* $\mathcal{R}^0 \mathcal{F}(M) \cong \mathcal{F}(M)$, *for each R-module M, so that* $\mathcal{R}^0 \mathcal{F}$ *and* $\mathcal{F}$ *are naturally equivalent functors and if E is an injective R-module, then* $\mathcal{R}^n \mathcal{F}(E) = 0$, *for* $n = 1, 2, 3, \ldots$.

Proof. If $\mathbf{P}$ is a projective resolution of M, then

$$P_1 \xrightarrow{\alpha_1} P_0 \xrightarrow{\alpha_0} M \to 0$$

is exact. Since $\mathcal{F}$ is right exact, it follows that

$$\mathcal{F}(P_1) \xrightarrow{\mathcal{F}(\alpha_1)} \mathcal{F}(P_0) \xrightarrow{\mathcal{F}(\alpha_0)} \mathcal{F}(M) \to 0$$

is exact. By considering

$$\mathcal{F}(P_1) \xrightarrow{\mathcal{F}(\alpha_1)} \mathcal{F}(P_0) \to 0,$$

we see that

$$\mathcal{L}_0 \mathcal{F}(M) = H_0(\mathcal{F}(\mathbf{P}_M)) = \mathrm{Ker}(\mathcal{F}(P_0) \to 0) / \mathrm{Im}\, \mathcal{F}(\alpha_1)$$
$$= \mathcal{F}(P_0) / \mathrm{Ker}\, \mathcal{F}(\alpha_0) \cong \mathcal{F}(M).$$

Thus, there is an isomorphism $\eta_M : \mathcal{L}_0 \mathcal{F}(M) \to \mathcal{F}(M)$. We claim that the family $\{\eta_M : \mathcal{L}_0 \mathcal{F}(M) \to \mathcal{F}(M)\}$ of these isomorphisms produces a natural isomorphism $\eta : \mathcal{L}_0 \mathcal{F} \to \mathcal{F}$. Let $f : M \to N$ be an R-linear mapping and suppose that $\mathbf{Q}$ is a projective resolution of N. If $\mathbf{f} : \mathbf{P}_M \to \mathbf{Q}_N$ is a chain map generated by f, then we have a commutative diagram

$$
\begin{array}{ccc}
\mathcal{L}_0 \mathcal{F}(M) & \xrightarrow{\eta_M} & \mathcal{F}(M) \\
{\scriptstyle \mathcal{L}_0 \mathcal{F}(\mathbf{f})} \downarrow & & \downarrow {\scriptstyle \mathcal{F}(f)} \\
\mathcal{L}_0 \mathcal{F}(N) & \xrightarrow{\eta_N} & \mathcal{F}(N)
\end{array}
$$

where η_M and η_N are isomorphisms. Thus, $\mathcal{L}_0\mathcal{F}$ and $\mathcal{F}$ are naturally equivalent functors as asserted.

If P is a projective R-module, then

$$\mathbf{P}: \cdots \to 0 \to P \xrightarrow{\text{id}_P} P \to 0$$

is a projective resolution of P. This gives the chain complex

$$\mathcal{F}(\mathbf{P}_P): \cdots \to 0 \to \mathcal{F}(P) \to 0$$

and it follows immediately that $\mathcal{L}_n\mathcal{F}(P) = 0$, for $n = 1, 2, 3, \ldots$.

The proof of (2) is similar. $\square$

The proof of the following proposition is an exercise.

Proposition 11.3.5. *Let $\mathcal{F}: \mathbf{Mod}_R \to \mathbf{Ab}$ be an additive contravariant functor.*

(1) *If $\mathcal{F}$ is left exact, then $\mathcal{R}^0\mathcal{F}(M) \cong \mathcal{F}(M)$, for each R-module M, so that $\mathcal{R}^0\mathcal{F}$ and $\mathcal{F}$ are naturally equivalent functors and if P is a projective R-module, then $\mathcal{R}^n\mathcal{F}(P) = 0$, for $n = 1, 2, 3, \ldots$.*

(2) *If $\mathcal{F}$ is right exact, then $\mathcal{L}_0\mathcal{F}(M) \cong \mathcal{F}(M)$, for each R-module M, so that $\mathcal{L}_0\mathcal{F}$ and $\mathcal{F}$ are naturally equivalent functors and if E is an injective R-module, then $\mathcal{L}_n\mathcal{F}(E) = 0$, for $n = 1, 2, 3, \ldots$.*

Problem Set 11.3

1. Let $\mathcal{F}: \mathbf{Mod}_R \to \mathbf{Ab}$ be an additive contravariant functor.

   (a) If $\mathbf{P}$ and $\mathbf{Q}$ are projective resolutions of M, show that the nth cohomology group of $\mathcal{F}(\mathbf{P}_M)$ is isomorphic to the nth cohomology group of $\mathcal{F}(\mathbf{Q}_M)$. Also if $f: M \to N$ is an R-linear mapping and $\mathbf{P}$ and $\mathbf{Q}$ are projective resolutions of M and N, respectively, prove that the group homomorphism $H^n(\mathcal{F}(\mathbf{f}))$: $H^n(\mathcal{F}(\mathbf{Q}_N)) \to H^n(\mathcal{F}(\mathbf{P}_M))$ does not depend on the choice of the chain map $\mathbf{f}: \mathbf{P}_M \to \mathbf{Q}_N$ generated by f.

   (b) If $\mathbf{D}$ and $\mathbf{E}$ are injective resolutions of M, prove that the nth homology group of $\mathcal{F}(\mathbf{D}^M)$ is isomorphic to the nth homology group of $\mathcal{F}(\mathbf{E}^M)$. Furthermore, if $f: M \to N$ is an R-linear mapping and $\mathbf{D}$ and $\mathbf{E}$ are injective resolutions of M and N, respectively, show that the group homomorphism $H_n(\mathcal{F}(\mathbf{f}))$: $H_n(\mathcal{F}(\mathbf{E}^N)) \to H_n(\mathcal{F}(\mathbf{D}^M))$ does not depend on the choice of the chain map $\mathbf{f}: \mathbf{D}^M \to \mathbf{E}^N$ generated by f.

2. Outline the development of the left and right derived functors $\mathcal{L}_n\mathcal{F}$ and $\mathcal{R}^n\mathcal{F}$ of $\mathcal{F}$, if $\mathcal{F}: \mathbf{Mod}_R \to \mathbf{Ab}$ is an additive contravariant functor.

3. (a) Let $\mathcal{F} : \mathbf{Mod}_R \to \mathbf{Ab}$ be an additive functor. Prove that the functors $\mathcal{L}_n\mathcal{F}$ and $\mathcal{R}_n\mathcal{F}$ are additive. [Hint: Proposition 11.1.7.]

   (b) Let $\mathcal{F} : \mathbf{Mod}_R \to \mathbf{Ab}$ be an additive contravariant functor. Show that the functors $\mathcal{L}_n\mathcal{F}$ and $\mathcal{R}_n\mathcal{F}$ are additive.

4. Prove Proposition 11.3.5. [Hint: Dualize the proof of Proposition 11.3.4.]

11.4 Extension Functors

If X is a fixed R-module, then $\operatorname{Hom}_R(-, X) : \mathbf{Mod}_R \to \mathbf{Ab}$ is a left exact additive contravariant functor. Indeed, if $f : M \to N$ is an R-linear mapping, then $\operatorname{Hom}_R(f, X) : \operatorname{Hom}_R(N, X) \to \operatorname{Hom}_R(M, X)$ is such that if $g, h \in \operatorname{Hom}_R(N, X)$, then

$$\operatorname{Hom}_R(f, X)(g + h) = f^*(g + h) = (g + h)f$$
$$= gf + hf = f^*(g) + f^*(h)$$
$$= \operatorname{Hom}_R(f, X)(g) + \operatorname{Hom}_R(f, X)(h).$$

$\operatorname{Hom}_R(-, X)$ is clearly contravariant and we saw in Chapter 3 that $\operatorname{Hom}_R(-, X)$ is left exact. Likewise, $\operatorname{Hom}_R(X, -)$ is a left exact additive functor from $\mathbf{Mod}_R$ to $\mathbf{Ab}$.

Right Derived Functors of $\operatorname{Hom}_R(-, X)$

If $\mathbf{P} : \cdots \to P_n \xrightarrow{\alpha_n} P_{n-1} \to \cdots \to P_1 \xrightarrow{\alpha_1} P_0 \xrightarrow{\alpha_0} M \to 0$ is a projective resolution of an R-module M, then for a fixed R-module X, we have a cochain complex

$$\operatorname{Hom}_R(\mathbf{P}_M, X) : \ 0 \to \operatorname{Hom}_R(P_0, X) \xrightarrow{\alpha_1^*} \cdots \xrightarrow{\alpha_n^*} \operatorname{Hom}_R(P_n, X) \xrightarrow{\alpha_{n+1}^*} \cdots .$$

Take cohomology in $\mathbf{Ab}$ and let

$$\operatorname{Ext}_R^n(M, X) = H^n(\operatorname{Hom}_R(\mathbf{P}_M, X)).$$

Next, for an R-linear mapping $f : M \to N$ and a projective resolution $\mathbf{Q}$ of N, let

$$\operatorname{Ext}_R^n(f, X) = H^n(\operatorname{Hom}_R(\mathbf{f}, X)),$$

where $\mathbf{f} : \mathbf{P}_M \to \mathbf{Q}_N$ is a chain map generated by f. Thus, we have

$$\operatorname{Ext}_R^n(f, X) : \operatorname{Ext}_R^n(N, X) \to \operatorname{Ext}_R^n(M, X)$$

and so for each $n \geq 0$

$$\operatorname{Ext}_R^n(-, X) : \mathbf{Mod}_R \to \mathbf{Ab}$$

is an additive contravariant functor. By construction, $\mathrm{Ext}_R^n(-, X) = \mathcal{R}^n \mathrm{Hom}_R(-, X)$ is the *nth right derived functor* of $\mathrm{Hom}_R(-, X)$. One immediate observation is that for any R-module M the group $\mathrm{Ext}_R^n(M, X)$ depends only on M and X and not on the projective resolution chosen for M. It also follows that $\mathrm{Ext}_R^n(-, X)$ is additive since $\mathrm{Hom}_R(-, X)$ and H^n are additive.

Definition 11.4.1. The contravariant functor $\mathrm{Ext}_R^n(-, X) : \mathbf{Mod}_R \rightarrow \mathbf{Ab}$ is called the *nth extension functor* of $\mathrm{Hom}_R(-, X)$, for $n = 0, 1, 2, \ldots$.

Remark. The name "extension functor" comes from the fact that it is possible to turn equivalence classes of n-fold extensions of M by X into an additive abelian group isomorphic to $\mathrm{Ext}_R^n(M, X)$. (An exact sequence of the form

$$0 \rightarrow X \rightarrow M_n \rightarrow \cdots \rightarrow M_1 \rightarrow M \rightarrow 0$$

is said to be an n-fold extension of M by X.) Such a development of $\mathrm{Ext}_R^n(-, X)$ is not required for our purposes, so we only mention it in passing. Additional details can be found in [8], [18] and [30].

For each short exact sequence $0 \rightarrow L \rightarrow M \rightarrow N \rightarrow 0$ of R-modules and R-module homomorphisms, there is a long exact sequence in cohomology corresponding to the contravariant functor $\mathrm{Ext}_R^n(-, X)$. To establish the existence of this sequence, we need the following lemma.

Lemma 11.4.2 (Horse Shoe Lemma for Projectives). *Consider the diagram* *where the bottom row is exact and* **P** *and* **R** *are projective resolutions of* L *and* N, *respectively. Then there is a projective resolution* **Q** *of* M *and chain maps* $\mathbf{f} : \mathbf{P}_L \to \mathbf{Q}_M$ *and* $\mathbf{g} : \mathbf{Q}_M \to \mathbf{R}_N$ *such that*

$$
\begin{array}{ccccccccc}
 & & \vdots & & \vdots & & \vdots & & \\
 & & \downarrow{\scriptstyle \alpha_2} & & \downarrow{\scriptstyle \beta_2} & & \downarrow{\scriptstyle \gamma_2} & & \\
0 & \longrightarrow & P_1 & \xrightarrow{f_1} & Q_1 & \xrightarrow{g_1} & R_1 & \longrightarrow & 0 \\
 & & \downarrow{\scriptstyle \alpha_1} & & \downarrow{\scriptstyle \beta_1} & & \downarrow{\scriptstyle \gamma_1} & & \\
0 & \longrightarrow & P_0 & \xrightarrow{f_0} & Q_0 & \xrightarrow{g_0} & R_0 & \longrightarrow & 0 \\
 & & \downarrow{\scriptstyle \alpha_0} & & \downarrow{\scriptstyle \beta_0} & & \downarrow{\scriptstyle \gamma_0} & & \\
0 & \longrightarrow & L & \xrightarrow{f} & M & \xrightarrow{g} & N & \longrightarrow & 0 \\
 & & \downarrow & & \downarrow & & \downarrow & & \\
 & & 0 & & 0 & & 0 & &
\end{array}
$$

is a commutative row exact diagram. Furthermore, $Q_n = P_n \oplus R_n$ *for each* $n \geq 0$.

Proof. Consider the diagram

$$
\begin{array}{ccccccccc}
 & & P_0 & & & & R_0 & & \\
 & & \downarrow{\scriptstyle \alpha_0} & & & & \downarrow{\scriptstyle \gamma_0} & & \\
0 & \longrightarrow & L & \xrightarrow{f} & M & \xrightarrow{g} & N & \longrightarrow & 0 \\
 & & \downarrow & & & & \downarrow & & \\
 & & 0 & & & & 0 & &
\end{array}
$$

and let $Q_0 = P_0 \oplus R_0$. Since R_0 is projective, there is an R-linear map $h : R_0 \to M$ such that $gh = \gamma_0$. So if $f_0 = i_0 : P_0 \to Q_0$ is the canonical injection and $g_0 = \pi_2 : Q_0 \to R_0$ is the canonical projection, let $\beta_0 : Q_0 \to M$ be defined by $\beta_0(x, y) = f\alpha_0(x) + h(y)$. We claim that β_0 is an epimorphism. If $z \in M$, then $g(z) \in N$, so there is a $y \in R_0$ such that $\gamma_0(y) = g(z)$. Now $g(z - h(y)) = 0$ and so $z - h(y) \in \operatorname{Ker} g = \operatorname{Im} f$. Hence, there is an $x \in P_0$ such that $f\alpha_0(x) = z - h(y)$. Thus, $\beta_0(x, y) = z$ and so β_0 is an epimorphism as asserted. Furthermore, the sequence $0 \to P_0 \xrightarrow{f_0} Q_0 \xrightarrow{g_0} R_0 \to 0$ is exact. Hence, we have a row and column exact diagram

$$
\begin{array}{ccccccccc}
0 & \longrightarrow & P_0 & \xrightarrow{\ f_0\ } & Q_0 & \xrightarrow{\ g_0\ } & R_0 & \longrightarrow & 0 \\
& & \downarrow{\scriptstyle \alpha_0} & & \downarrow{\scriptstyle \beta_0} & & \downarrow{\scriptstyle \gamma_0} & & \\
0 & \longrightarrow & L & \xrightarrow{\ f\ } & M & \xrightarrow{\ g\ } & N & \longrightarrow & 0 \\
& & \downarrow & & \downarrow & & \downarrow & & \\
& & 0 & & 0 & & 0 & &
\end{array}
$$

and a simple diagram chase shows that this latter diagram is commutative. Now suppose that the first n components

$$
Q_{n-1} = P_{n-1} \oplus R_{n-1} \xrightarrow{\ \beta_{n-1}\ } \cdots \to Q = P_0 \oplus R_0 \xrightarrow{\ \beta_0\ } M \to 0
$$

of a projective resolution $\mathbf{Q}$ of M and R-linear mappings $f_0, f_1, \ldots, f_{n-1}$ and $g_0, g_1,$ $\ldots, g_{n-1}$ have been found such that the diagram formed by filling in the middle column and rows of the original diagram with these components results in a commutative row and column exact diagram. At the nth position this gives a row and column exact diagram

$$
\begin{array}{ccccccccc}
& & P_n & & & & R_n & & \\
& & \downarrow{\scriptstyle p_n^\alpha} & & & & \downarrow{\scriptstyle p_n^\gamma} & & \\
0 & \longrightarrow & \operatorname{Ker}\alpha_{n-1} & \xrightarrow{\ \bar{f}_{n-1}\ } & \operatorname{Ker}\beta_{n-1} & \xrightarrow{\ \bar{g}_{n-1}\ } & \operatorname{Ker}\gamma_{n-1} & \longrightarrow & 0 \\
& & \downarrow & & & & \downarrow & & \\
& & 0 & & & & 0 & &
\end{array}
$$

with $p_n^\alpha = \alpha_n : P_n \to \operatorname{Im}\alpha_n = \operatorname{Ker}\alpha_{n-1}$ and $p_n^\gamma = \gamma_n : R_n \to \operatorname{Im}\gamma_n = \operatorname{Ker}\gamma_{n-1}$ and where $\bar{f}_{n-1}$ and $\bar{g}_{n-1}$ are the restrictions of $f_{n-1} : P_{n-1} \to Q_{n-1}$ and $g_{n-1} : Q_{n-1} \to R_{n-1}$ to $\operatorname{Ker}\alpha_{n-1}$ and $\operatorname{Ker}\beta_{n-1}$, respectively. An epimorphism $p_n^\beta : Q_n = P_n \oplus R_n \to \operatorname{Ker}\beta_{n-1}$ and maps $0 \to P_n \xrightarrow{\ f_n\ } Q_n$ and $Q_n \xrightarrow{\ g_n\ } R_n \to 0$ can now be constructed exactly as β_0, f_0 and g_0 were constructed and this gives a row and column exact commutative diagram

$$
\begin{array}{ccccccccc}
0 & \longrightarrow & P_n & \xrightarrow{f_n} & Q_n & \xrightarrow{g_n} & R_n & \longrightarrow & 0 \\
& & \downarrow{p_n^\alpha} & & \downarrow{p_n^\beta} & & \downarrow{p_n^\gamma} & & \\
0 & \longrightarrow & \operatorname{Ker}\alpha_{n-1} & \xrightarrow{\bar{f}_{n-1}} & \operatorname{Ker}\beta_{n-1} & \xrightarrow{\bar{g}_{n-1}} & \operatorname{Ker}\gamma_{n-1} & \longrightarrow & 0 \\
& & \downarrow & & \downarrow & & \downarrow & & \\
& & 0 & & 0 & & 0 & &
\end{array}
$$

If $i_{n-1}^\alpha : \operatorname{Ker}\alpha_{n-1} \to P_{n-1}$, $i_{n-1}^\beta : \operatorname{Ker}\beta_{n-1} \to Q_{n-1}$ and $i_{n-1}^\gamma : \operatorname{Ker}\gamma_{n-1} \to R_{n-1}$ are canonical injections, then $\alpha_n = i_{n-1}^\alpha\, p_n^\alpha$, $\beta_n = i_{n-1}^\beta\, p_n^\beta$ and $\gamma_n = i_{n-1}^\gamma\, p_n^\gamma$. Hence, the row and column exact diagram

$$
\begin{array}{ccccccccc}
0 & \longrightarrow & P_n & \xrightarrow{f_n} & Q_n & \xrightarrow{g_n} & R_n & \longrightarrow & 0 \\
& & \downarrow{\alpha_n} & & \downarrow{\beta_n} & & \downarrow{\gamma_n} & & \\
0 & \longrightarrow & P_{n-1} & \xrightarrow{f_{n-1}} & Q_{n-1} & \xrightarrow{g_{n-1}} & R_{n-1} & \longrightarrow & 0 \\
& & \downarrow & & \downarrow & & \downarrow & & \\
& & \vdots & & \vdots & & \vdots & & \\
& & \downarrow & & \downarrow & & \downarrow & & \\
0 & \longrightarrow & L & \xrightarrow{f} & M & \xrightarrow{g} & N & \longrightarrow & 0 \\
& & \downarrow & & \downarrow & & \downarrow & & \\
& & 0 & & 0 & & 0 & &
\end{array}
$$

is commutative and so the lemma follows by induction. $\square$

Proposition 11.4.3. *If* $0 \to L \xrightarrow{f} M \xrightarrow{g} N \to 0$ *is an exact sequence of R-modules and R-module homomorphisms, then for any R-module X, there is a long exact cohomology sequence*

$$
0 \to \operatorname{Hom}_R(N, X) \xrightarrow{g^*} \operatorname{Hom}_R(M, X) \xrightarrow{f^*} \operatorname{Hom}_R(L, X) \xrightarrow{\Phi^0}
$$

$$
\xrightarrow{\Phi^0} \operatorname{Ext}_R^1(N, X) \xrightarrow{\operatorname{Ext}_R^1(g,X)} \operatorname{Ext}_R^1(M, X) \xrightarrow{\operatorname{Ext}_R^1(f,X)} \operatorname{Ext}_R^1(L, X) \xrightarrow{\Phi^1} \cdots
$$

$$
\cdots \xrightarrow{\Phi^{n-1}} \operatorname{Ext}_R^n(N, X) \xrightarrow{\operatorname{Ext}_R^n(g,X)} \operatorname{Ext}_R^n(M, X) \xrightarrow{\operatorname{Ext}_R^n(f,X)} \operatorname{Ext}_R^n(L, X) \xrightarrow{\Phi^n} \cdots
$$

where Φ^n is a connecting homomorphism for each $n \geq 0$.

Proof. If **P** is a projective resolution of L and **R** is a projective resolution of N, then The Horse Shoe Lemma shows that there is a projective resolution **Q** of M and chain maps $\mathbf{f} : \mathbf{P} \to \mathbf{Q}$ and $\mathbf{g} : \mathbf{Q} \to \mathbf{R}$ such that $0 \to \mathbf{P} \xrightarrow{\mathbf{f}} \mathbf{Q} \xrightarrow{\mathbf{g}} \mathbf{R} \to \mathbf{0}$ is a short exact sequence of chain complexes. From the way **Q** was constructed in the Horse Shoe Lemma, $0 \to P_n \xrightarrow{f_n} Q_n \xrightarrow{g_n} R_n \to 0$ is a split short exact sequence with $Q_n = P_n \oplus R_n$ for each $n \geq 0$. Since $\mathrm{Hom}_R(-, X)$ preserves split short exact sequences, it follows that we have a commutative diagram

$$
\begin{array}{ccccccc}
& 0 & & 0 & & 0 & \\
& \downarrow & & \downarrow & & \downarrow & \\
0 \longrightarrow & \mathrm{Hom}_R(R_0, X) & \xrightarrow{\gamma_1^*} & \mathrm{Hom}_R(R_1, X) & \xrightarrow{\gamma_2^*} & \mathrm{Hom}_R(R_2, X) & \longrightarrow \cdots \\
& \downarrow{g_0^*} & & \downarrow{g_1^*} & & \downarrow{g_2^*} & \\
0 \longrightarrow & \mathrm{Hom}_R(P_0 \oplus R_0, X) & \xrightarrow{\beta_1^*} & \mathrm{Hom}_R(P_1 \oplus R_1, X) & \xrightarrow{\beta_2^*} & \mathrm{Hom}_R(P_2 \oplus R_2, X) & \longrightarrow \cdots \\
& \downarrow{f_0^*} & & \downarrow{f_1^*} & & \downarrow{f_2^*} & \\
0 \longrightarrow & \mathrm{Hom}_R(P_0, X) & \xrightarrow{\alpha_1^*} & \mathrm{Hom}_R(P_1, X) & \xrightarrow{\alpha_1^*} & \mathrm{Hom}_R(P_2, X) & \longrightarrow \cdots \\
& \downarrow & & \downarrow & & \downarrow & \\
& 0 & & 0 & & 0 &
\end{array}
$$

where the columns are exact and the rows give $\mathrm{Ext}_R^n(N, X)$, $\mathrm{Ext}_R^n(M, X)$ and $\mathrm{Ext}_R^n(L, X)$ for each $n \geq 0$. Thus,

$$0 \to \mathrm{Hom}_R(\mathbf{R}_N, X) \xrightarrow{\mathbf{g}^*} \mathrm{Hom}_R(\mathbf{Q}_M, X) \xrightarrow{\mathbf{f}^*} \mathrm{Hom}_R(\mathbf{P}_L, X) \to 0$$

is an exact sequence of cochain complexes. Since the contravariant functor $\mathrm{Hom}_R(-, X)$ is left exact and additive, (2) of Proposition 11.3.5 shows that we have

$$\mathrm{Ext}_R^0(N, X) = H^0(\mathrm{Hom}_R(\mathbf{P}_N, X)) = \mathrm{Hom}_R(N, X),$$
$$\mathrm{Ext}_R^0(M, X) = H^0(\mathrm{Hom}_R(\mathbf{Q}_M, X)) = \mathrm{Hom}_R(M, X) \quad \text{and}$$
$$\mathrm{Ext}_R^0(L, X) = H^0(\mathrm{Hom}_R(\mathbf{R}_L, X)) = \mathrm{Hom}_R(L, X).$$

This, together with Proposition 11.1.12, gives the result.   □

The sequence given in Proposition 11.4.3 is called the *long exact* Ext-*sequence in the first variable*. There is also a *long exact* Ext-*sequence in the second variable*. To see this, suppose that $0 \to L \xrightarrow{f} M \xrightarrow{g} N \to 0$ is a short exact sequence of

R-modules and R-module homomorphisms. Let **P** be a projective resolution of X and recall that if P is a projective R-module, then $\mathrm{Hom}_R(P, -)$ preserves short exact sequences. So for each $n \geq 0$, the sequence

$$0 \to \mathrm{Hom}_R(P_n, L) \xrightarrow{f_*} \mathrm{Hom}_R(P_n, M) \xrightarrow{g_*} \mathrm{Hom}_R(P_n, N) \to 0$$

is exact which leads to the short exact sequence of cochain complexes

$$\mathbf{0} \to \mathrm{Hom}_R(\mathbf{P}_X, L) \xrightarrow{\mathbf{f}_*} \mathrm{Hom}_R(\mathbf{P}_X, M) \xrightarrow{\mathbf{g}_*} \mathrm{Hom}_R(\mathbf{P}_X, N) \to \mathbf{0}.$$

Taking cohomology in this sequence and applying Propositions 11.1.12 and 11.3.5 establishes the following proposition.

Proposition 11.4.4. *If* $0 \to L \xrightarrow{f} M \xrightarrow{g} N \to 0$ *is an exact sequence of R-modules and R-module homomorphisms, then for any R-module X, there is an exact cohomology sequence*

$$0 \to \mathrm{Hom}_R(X, L) \xrightarrow{f_*} \mathrm{Hom}_R(X, M) \xrightarrow{g_*} \mathrm{Hom}_R(X, N) \xrightarrow{\Phi^0}$$

$$\xrightarrow{\Phi^0} \mathrm{Ext}^1_R(X, L) \xrightarrow{\mathrm{Ext}^1_R(X,f)} \mathrm{Ext}^1_R(X, M) \xrightarrow{\mathrm{Ext}^1_R(X,g)} \mathrm{Ext}^1_R(X, N) \xrightarrow{\Phi^1} \cdots$$

$$\cdots \xrightarrow{\Phi^{n-1}} \mathrm{Ext}^n_R(X, L) \xrightarrow{\mathrm{Ext}^n_R(X,f)} \mathrm{Ext}^n_R(X, M) \xrightarrow{\mathrm{Ext}^n_R(X,g)} \mathrm{Ext}^n_R(X, N) \xrightarrow{\Phi^n} \cdots,$$

where Φ^n *is a connecting homomorphism for each $n \geq 0$.*

Thus, we have a contravariant functor and a covariant functor

$$\mathrm{Ext}^n_R(-, X) \quad \text{and} \quad \mathrm{Ext}^n_R(X, -)$$

from $\mathbf{Mod}_R$ to $\mathbf{Ab}$, respectively, for each $n \geq 0$.

Proposition 11.4.5. *The following are equivalent for an R-module P.*

(1) *P is projective.*

(2) *$\mathrm{Ext}^n_R(P, X) = 0$ for each R-module X and every integer $n \geq 1$.*

(3) *$\mathrm{Ext}^1_R(P, X) = 0$ for every R-module X.*

Proof. (1) $\Rightarrow$ (2). Suppose that P is projective. Since $\mathrm{Hom}_R(-, X)$ is a left exact additive contravariant functor, it follows immediately from (1) of Proposition 11.3.5 that $\mathrm{Ext}^n_R(P, X) = 0$ for every integer $n \geq 1$.
(2) $\Rightarrow$ (3) is obvious.

(3) $\Rightarrow$ (1). If $\text{Ext}_R^1(P, X) = 0$ for each R-module X, then the long exact Ext-sequence in the second variable shows that

$$0 \to \text{Hom}_R(P, L) \to \text{Hom}_R(P, M) \to \text{Hom}_R(P, N) \to 0$$

is exact for each short exact sequence

$$0 \to L \to M \to N \to 0$$

of R-modules and R-module homomorphisms. Thus, we have by Corollary 5.2.12 that P is projective. $\square$

Proposition 11.4.6. *The following are equivalent for an R-module E.*

(1) *E is injective.*

(2) *$\text{Ext}_R^n(X, E) = 0$ for each R-module X and every integer $n \geq 1$.*

(3) *$\text{Ext}_R^1(X, E) = 0$ for every R-module X.*

(4) *$\text{Ext}_R^1(X, E) = 0$ for every cyclic R-module X.*

(5) *$\text{Ext}_R^1(R/A, E) = 0$ for every right ideal A of R.*

Proof. (1) $\Rightarrow$ (2). Suppose that E is injective and let

$$\mathbf{P}: \cdots \to P_{n+1} \xrightarrow{\alpha_{n+1}} P_n \xrightarrow{\alpha_n} P_{n-1} \to \cdots \to P_0 \xrightarrow{\alpha_0} X \to 0$$

be a projective resolution of X. Since E is injective, Corollary 5.1.12 shows that the functor $\text{Hom}_R(-, E)$ preserves short exact sequences and so it follows that $\text{Hom}_R(\mathbf{P}_X, E)$ is exact. Hence, $\text{Ext}_R^n(X, E) = H^n(\text{Hom}_R(\mathbf{P}_X, E)) = 0$ for $n \geq 1$.
(2) $\Rightarrow$ (3), (3) $\Rightarrow$ (4) and (4) $\Rightarrow$ (5) are obvious.
(5) $\Rightarrow$ (1). If A is a right ideal of R, then the short exact sequence $0 \to A \to R \to R/A \to 0$ gives rise to the long exact Ext-sequence

$$0 \to \text{Hom}_R(R/A, E) \to \text{Hom}_R(R, E) \to \text{Hom}_R(A, E) \to \text{Ext}_R^1(R/A, E) \to \cdots .$$

But (5) gives $\text{Ext}_R^1(R/A, E) = 0$ and this indicates that $\text{Hom}_R(R, E) \to \text{Hom}_R(A, E)$ is an epimorphism. Thus, Baer's criteria shows that E is injective. $\square$

The derived functors of Hom can be used to gain information about rings and modules. To foretell things to come, we will see in the next chapter that the functors $\text{Ext}_R^n(-, X)$ can be used to define a projective dimension of an R-module. This dimension will in some sense measure "how far" a module is from being projective.

Right Derived Functors of $\mathrm{Hom}_R(X, -)$

A parallel development to that of $\mathrm{Ext}_R^n(-, X)$ can be carried out using injective resolutions rather than projective resolutions. The result is a functor $\overline{\mathrm{Ext}}_R^n(X, -) : \mathbf{Mod}_R \to \mathbf{Ab}$ for each $n \geq 0$. We begin with a brief outline of a development of $\overline{\mathrm{Ext}}_R^n$. Proofs will be omitted since they are analogous to the proofs of propositions already given in the development of the functor $\mathrm{Ext}_R^n(-, X)$.

Let X and M be R-modules and suppose $\mathbf{D}$ is an injective resolution of M. Consider the cochain complex

$$\mathrm{Hom}_R(X, \mathbf{D}^M) : \ 0 \to \mathrm{Hom}_R(X, D^0) \xrightarrow{\alpha_*^0} \cdots \xrightarrow{\alpha_*^{n-1}} \mathrm{Hom}_R(X, D^n) \xrightarrow{\alpha_*^n} \cdots .$$

Take cohomology in $\mathbf{Ab}$ and let

$$\overline{\mathrm{Ext}}_R^n(X, M) = H^n(\mathrm{Hom}_R(X, \mathbf{D}^M))$$

for $n \geq 0$. Next, for an R-linear mapping $f : M \to N$ and an injective resolution $\mathbf{E}$ of N, let $\overline{\mathrm{Ext}}_R^n(X, f) = H^n(\mathrm{Hom}_R(X, \mathbf{f}))$, where $\mathbf{f} : \mathbf{D}^M \to \mathbf{E}^N$ is a cochain map generated by f. Then

$$\overline{\mathrm{Ext}}_R^n(X, f) : \overline{\mathrm{Ext}}_R^n(X, M) \to \overline{\mathrm{Ext}}_R^n(X, N),$$

so $\overline{\mathrm{Ext}}_R^n(X, -) : \mathbf{Mod}_R \to \mathbf{Ab}$ is a right derived functor of $\mathrm{Hom}_R(X, -)$ which is additive and covariant.

Definition 11.4.7. The covariant functor $\overline{\mathrm{Ext}}_R^n(X, -) : \mathbf{Mod}_R \to \mathbf{Ab}$ is (also) called the *nth extension functor* of $\mathrm{Hom}_R(X, -)$, for $n = 0, 1, 2, \ldots$.

Proposition 11.4.8 (Horse Shoe Lemma for Injectives). *If*

$$0 \to L \xrightarrow{f} M \xrightarrow{g} N \to 0$$

is a short exact sequence of R-modules and R-module homomorphisms and $\mathbf{D}$ and $\mathbf{F}$ are injective resolutions of L and N, respectively, then there is an injective resolution $\mathbf{E}$ of M and cochain maps $\mathbf{f} : \mathbf{D}^L \to \mathbf{E}^M$ and $\mathbf{g} : \mathbf{E}^M \to \mathbf{F}^N$ such that

$$0 \to \mathbf{D} \xrightarrow{\mathbf{f}} \mathbf{E} \xrightarrow{\mathbf{g}} \mathbf{F} \to 0$$

is a short exact sequence of cochain complexes, where $E^n = D^n \oplus F^n$ for each $n \geq 0$.

Applying $\mathrm{Hom}_R(X, -)$ to the short exact sequence

$$0 \to \mathbf{D} \xrightarrow{\mathbf{f}} \mathbf{E} \xrightarrow{\mathbf{g}} \mathbf{F} \to 0$$

of cochain complexes of injective resolutions of L, M and N and taking cohomology establishes the following proposition.

Proposition 11.4.9. *Corresponding to each short exact sequence* $0 \to L \xrightarrow{f} M \xrightarrow{g} N \to 0$ *of R-modules and R-module homomorphisms, there is a long exact cohomology sequence*

$$0 \to \mathrm{Hom}_R(X, L) \xrightarrow{f_*} \mathrm{Hom}_R(X, M) \xrightarrow{g_*} \mathrm{Hom}_R(X, N) \xrightarrow{\bar{\Phi}^0}$$

$$\xrightarrow{\bar{\Phi}^0} \overline{\mathrm{Ext}}_R^1(X, L) \xrightarrow{\overline{\mathrm{Ext}}_R^1(X,f)} \overline{\mathrm{Ext}}_R^1(X, M) \xrightarrow{\overline{\mathrm{Ext}}_R^1(X,g)} \overline{\mathrm{Ext}}_R^1(X, N) \xrightarrow{\bar{\Phi}^1} \cdots$$

$$\cdots \xrightarrow{\bar{\Phi}^{n-1}} \overline{\mathrm{Ext}}_R^n(X, L) \xrightarrow{\overline{\mathrm{Ext}}_R^n(X,f)} \overline{\mathrm{Ext}}_R^n(X, M) \xrightarrow{\overline{\mathrm{Ext}}_R^n(X,g)} \overline{\mathrm{Ext}}_R^n(X, N) \xrightarrow{\bar{\Phi}^n} \cdots,$$

where $\bar{\Phi}^n$ *is a connecting homomorphism for each* $n \geq 0$.

There is also a long exact $\overline{\mathrm{Ext}}$-sequence in the first variable corresponding to each short exact sequence $0 \to L \xrightarrow{f} M \xrightarrow{g} N \to 0$ of R-modules and R-module homomorphisms. If X is an R-module and $\mathbf{E}$ is an injective resolution of X, then since E^n is injective, $\mathrm{Hom}_R(-, E^n)$ preserves short exact sequences. Therefore, the sequence

$$0 \to \mathrm{Hom}_R(N, E^n) \xrightarrow{g^*} \mathrm{Hom}_R(M, E^n) \xrightarrow{f^*} \mathrm{Hom}_R(L, E^n) \to 0$$

is exact for each $n \geq 0$, so we have a short exact sequence

$$0 \to \mathrm{Hom}_R(N, \mathbf{E}^X) \xrightarrow{\mathbf{g}^*} \mathrm{Hom}_R(M, \mathbf{E}^X) \xrightarrow{\mathbf{f}^*} \mathrm{Hom}_R(L, \mathbf{E}^X) \to 0$$

of cochain complexes. Taking cohomology leads to the long exact cohomology sequence

$$0 \to \mathrm{Hom}_R(N, X) \xrightarrow{g^*} \mathrm{Hom}_R(M, X) \xrightarrow{f^*} \mathrm{Hom}_R(L, X) \xrightarrow{\bar{\Phi}^0}$$

$$\xrightarrow{\bar{\Phi}^0} \overline{\mathrm{Ext}}_R^1(N, X) \xrightarrow{\overline{\mathrm{Ext}}_R^1(g,X)} \overline{\mathrm{Ext}}_R^1(M, X) \xrightarrow{\overline{\mathrm{Ext}}_R^1(f,X)} \overline{\mathrm{Ext}}_R^1(L, X) \xrightarrow{\bar{\Phi}^1} \cdots$$

$$\cdots \xrightarrow{\bar{\Phi}^{n-1}} \overline{\mathrm{Ext}}_R^n(N, X) \xrightarrow{\overline{\mathrm{Ext}}_R^n(g,X)} \overline{\mathrm{Ext}}_R^n(M, X) \xrightarrow{\overline{\mathrm{Ext}}_R^n(f,X)} \overline{\mathrm{Ext}}_R^n(L, X) \xrightarrow{\bar{\Phi}^n} \cdots.$$

It can also be verified that $\overline{\mathrm{Ext}}_R^n(-, X) : \mathbf{Mod}_R \to \mathbf{Ab}$ is a contravariant functor, so

$$\overline{\mathrm{Ext}}_R^n(-, X) \quad \text{and} \quad \overline{\mathrm{Ext}}_R^n(X, -)$$

are contravariant and covariant functors from $\mathbf{Mod}_R$ to $\mathbf{Ab}$, respectively, for each $n \geq 0$.

Proposition 11.4.5 shows that an R-module P is projective if and only if $\text{Ext}_R^n(P, X) = 0$ for every R-module X and every $n \geq 1$. There is a similar result for injective modules. The implication $(5) \Rightarrow (1)$ in the following proposition is a result of Baer's criteria and the long exact $\overline{\text{Ext}}$-sequence in the first variable.

Proposition 11.4.10. *The following are equivalent for an R-module E.*

(1) E is injective.

(2) $\overline{\text{Ext}}_R^n(X, E) = 0$ for each R-module X and every $n \geq 1$.

(3) $\overline{\text{Ext}}_R^1(X, E) = 0$ for every R-module X.

(4) $\overline{\text{Ext}}_R^1(X, E) = 0$ for every cyclic R-module X.

(5) $\overline{\text{Ext}}_R^1(R/A, E) = 0$ for every right ideal A of R.

Proposition 11.4.11. *The following are equivalent for an R-module P.*

(1) P is projective.

(2) $\overline{\text{Ext}}_R^n(P, X) = 0$ for each R-module X and every $n \geq 1$.

(3) $\overline{\text{Ext}}_R^1(P, X) = 0$ for every R-module X.

Properties of the bifunctors Ext_R^n and $\overline{\text{Ext}}_R^n$ are strikingly similar. Actually, these functors are naturally equivalent. After this has been established, the distinction between these two functors can be ignored since they can be interchanged through the use of isomorphisms. No loss of generality will result if both functors are denoted by Ext_R^n.

Proposition 11.4.12. *The bifunctors Ext_R^n and $\overline{\text{Ext}}_R^n$ are naturally equivalent for $n \geq 0$.*

Proof. The proof is by induction. If M is a fixed R-module and $n = 0$, then we have seen that $\text{Ext}_R^0(M, N) = \overline{\text{Ext}}_R^0(M, N) = \text{Hom}_R(M, N)$, for each R-module N. Hence, we need only let

$$\eta_{MN}^0 : \text{Ext}_R^0(M, N) \to \overline{\text{Ext}}_R^0(M, N)$$

be the identity map on $\text{Hom}_R(M, N)$ to establish that $\text{Ext}_R^0(M, -)$ and $\overline{\text{Ext}}_R^0(M, -)$ are naturally equivalent functors. Next, let the short exact sequence $0 \to N \xrightarrow{u} E \xrightarrow{p} C \to 0$ represent an embedding of N into an injective R-module E. Then Proposition

11.4.4 gives the long exact Ext-sequence

$$0 \to \mathrm{Hom}_R(M, N) \xrightarrow{u_*} \mathrm{Hom}_R(M, E) \xrightarrow{p_*} \mathrm{Hom}_R(M, C) \xrightarrow{\Phi^0} \qquad (11.1)$$

$$\xrightarrow{\Phi^0} \mathrm{Ext}_R^1(M, N) \xrightarrow{\mathrm{Ext}_R^1(M,u)} \mathrm{Ext}_R^1(M, E) \xrightarrow{\mathrm{Ext}_R^1(M,p)} \mathrm{Ext}_R^1(M, C) \xrightarrow{\Phi^1} \cdots$$

$$\cdots \xrightarrow{\Phi^{n-1}} \mathrm{Ext}_R^n(M, N) \xrightarrow{\mathrm{Ext}_R^n(M,u)} \mathrm{Ext}_R^n(M, E) \xrightarrow{\mathrm{Ext}_R^n(M,p)} \mathrm{Ext}_R^n(M, C) \xrightarrow{\Phi^n} \cdots$$

and from Proposition 11.4.9 we get the long exact $\overline{\mathrm{Ext}}$-sequence

$$0 \to \mathrm{Hom}_R(M, N) \xrightarrow{u_*} \mathrm{Hom}_R(M, E) \xrightarrow{p_*} \mathrm{Hom}_R(M, C) \xrightarrow{\bar{\Phi}^0} \qquad (11.2)$$

$$\xrightarrow{\bar{\Phi}^0} \overline{\mathrm{Ext}}_R^1(M, N) \xrightarrow{\overline{\mathrm{Ext}}_R^1(M,u)} \overline{\mathrm{Ext}}_R^1(M, E) \xrightarrow{\overline{\mathrm{Ext}}_R^1(M,p)} \overline{\mathrm{Ext}}_R^1(M, C) \xrightarrow{\bar{\Phi}^1} \cdots$$

$$\cdots \xrightarrow{\bar{\Phi}^{n-1}} \overline{\mathrm{Ext}}_R^n(M, N) \xrightarrow{\overline{\mathrm{Ext}}_R^n(M,u)} \overline{\mathrm{Ext}}_R^n(M, E) \xrightarrow{\overline{\mathrm{Ext}}_R^n(M,p)} \overline{\mathrm{Ext}}_R^n(M, C) \xrightarrow{\bar{\Phi}^n} \cdots .$$

In view of Propositions 11.4.6 and 11.4.10, $\mathrm{Ext}_R^n(M, E) = \overline{\mathrm{Ext}}_R^n(M, E) = 0$ for $n = 1, 2, 3, \ldots$, so we have a row exact commutative diagram

$$\mathrm{Hom}_R(M, E) \xrightarrow{p_*} \mathrm{Hom}_R(M, C) \xrightarrow{\Phi^0} \mathrm{Ext}_R^1(M, N) \longrightarrow \mathrm{Ext}_R^1(M, E) = 0$$

$$\mathrm{Hom}_R(M, E) \xrightarrow{p_*} \mathrm{Hom}_R(M, C) \xrightarrow{\bar{\Phi}^0} \overline{\mathrm{Ext}}_R^1(M, N) \longrightarrow \overline{\mathrm{Ext}}_R^1(M, E) = 0.$$

with vertical map η_{MN}^1.

From this we see that there is a group isomorphism

$$\eta_{MN}^1 : \mathrm{Ext}_R^1(M, N) \to \overline{\mathrm{Ext}}_R^1(M, N)$$

for each R-module N. If the short exact sequence $0 \to N' \xrightarrow{u'} E' \xrightarrow{p'} C' \to 0$ also represents an embedding of an R-module N' into an injective module E', then we also have long exact sequences such as (11.1) and (11.2) with N, E, C, u and p replaced by N', E', C', u' and p', respectively. This gives a copy of the last row exact commutative diagram immediately above, but with primes in the appropriate places. If $f : N \to N'$ is an R-linear mapping, then by using the injectivity of E', we see that there is an induced map $g : C \to C'$. We have now established the groundwork necessary to show that η_{MN}^1 is not only a group isomorphism but also a natural transformation for each R-module N. The discussion to this point yields the

diagram

$$\begin{array}{ccc}
\operatorname{Hom}_R(M,C) & \xrightarrow{\quad g_* \quad} & \operatorname{Hom}_R(M,C')
\end{array}$$

$$\eta^0_{MC} \qquad \eta^0_{MC'}$$

$$\Phi^0 \qquad \operatorname{Hom}_R(M,C) \xrightarrow{\quad g_* \quad} \operatorname{Hom}_R(M,C')$$

$$\Phi'^0$$

$$\operatorname{Ext}^1_R(M,N) \xrightarrow{\quad \operatorname{Ext}_R(M,f) \quad} \operatorname{Ext}^1_R(M,N')$$

$$\eta^1_{MN} \qquad \overline{\Phi}^0 \qquad \eta^1_{MN'} \qquad \overline{\Phi}'^0$$

$$\overline{\operatorname{Ext}}^1_R(M,N) \xrightarrow{\quad \overline{\operatorname{Ext}}^1_R(M,f) \quad} \overline{\operatorname{Ext}}^1_R(M,N')$$

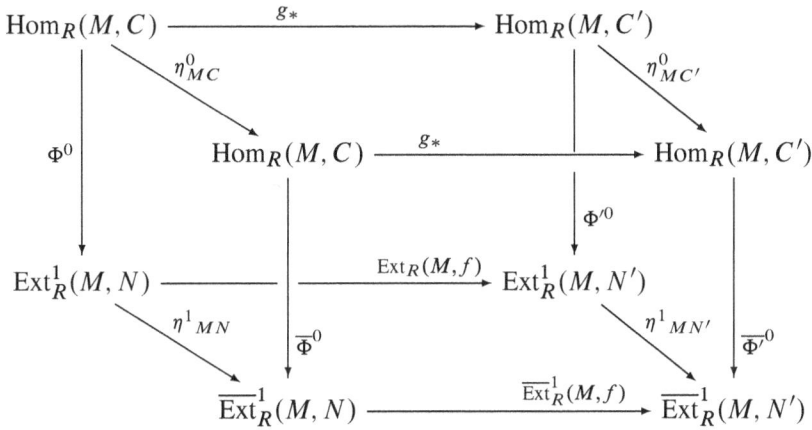

It follows that the top face of the diagram is commutative, as are the four faces on the sides. Since the map Φ^0 is an epimorphism, Exercise 6 in the Problem Set given in the chapter on preliminaries to the text shows that the bottom face is also commutative. Hence, we have that η^1_{MN} is a natural isomorphism for each R-module N, so the functors $\operatorname{Ext}^1_R(M,-)$ and $\overline{\operatorname{Ext}}^1_R(M,-)$ are naturally equivalent.

Finally, suppose that natural isomorphisms $\eta^0_{MN}, \eta^1_{MN}, \ldots, \eta^{n-1}_{MN}$ have been found that fit the requirements of the proposition. Considering the sequences (11.1) and (11.2) again, we see that there is a row exact commutative diagram

$$0 = \operatorname{Ext}^{n-1}_R(M,E) \longrightarrow \operatorname{Ext}^{n-1}_R(M,C) \xrightarrow{\Phi^n} \operatorname{Ext}^n_R(M,N) \longrightarrow \operatorname{Ext}^n_R(M,E) = 0$$

$$\eta^{n-1}_{MC} \qquad\qquad \eta^n_{MN}$$

$$0 = \overline{\operatorname{Ext}}^{n-1}_R(M,E) \longrightarrow \overline{\operatorname{Ext}}^{n-1}_R(M,C) \xrightarrow{\overline{\Phi}^n} \overline{\operatorname{Ext}}^n_R(M,N) \longrightarrow \overline{\operatorname{Ext}}^n_R(M,E) = 0$$

An argument parallel to that given for η^1_{MN} shows that η^n_{MN} is a natural isomorphism for each R-module N, so it follows by induction that $\operatorname{Ext}^n_R(M,-)$ and $\overline{\operatorname{Ext}}^n_R(M,-)$ are naturally equivalent functors for each $n \geq 0$. A similar argument can used to show that $\operatorname{Ext}^n_R(-,N)$ and $\overline{\operatorname{Ext}}^n_R(-,N)$ are naturally equivalent contravariant functors for each $n \geq 0$. Therefore, the bifunctors Ext^n_R and $\overline{\operatorname{Ext}}^n_R$ are naturally equivalent for each $n \geq 0$. $\qquad\square$

One result of Proposition 11.4.12 is that $\operatorname{Ext}^n_R(M,N)$ can be computed either by using a projective resolution of M or by using an injective resolution of N. For this reason, Ext^n_R is said to be a *balanced bifunctor*.

Example

4. Consider the projective resolution

$$\cdots \to 0 \to 0 \to \mathbb{Z} \xrightarrow{\alpha_1} \mathbb{Z} \xrightarrow{\alpha_0} \mathbb{Z}_k \to 0$$

of the $\mathbb{Z}$-module $\mathbb{Z}_k$, where $k \geq 2$ and $\alpha_1(a) = ka$ and $\alpha_0(a) = [a]$ for all $a \in \mathbb{Z}$. Let M be a $\mathbb{Z}$-module and form

$$0 \to \operatorname{Hom}_{\mathbb{Z}}(\mathbb{Z}, M) \xrightarrow{\alpha_1^*} \operatorname{Hom}_{\mathbb{Z}}(\mathbb{Z}, M) \to 0 \to 0 \to \cdots$$

using the deleted projective resolution of $\mathbb{Z}_k$. Since $\varphi : \operatorname{Hom}_{\mathbb{Z}}(\mathbb{Z}, M) \to M$ defined by $\varphi(f) = f(1)$ is an isomorphism, the diagram

is commutative, where $\xi(x) = kx$ for each $x \in M$. It follows that $\operatorname{Ext}_{\mathbb{Z}}^0(\mathbb{Z}_k, M) \cong \operatorname{Ker} \xi$ and that $\operatorname{Ext}_{\mathbb{Z}}^1(\mathbb{Z}_k, M) \cong M/kM$.

Since E is an injective R-module if and only if $\operatorname{Ext}_R^n(X, E) = 0$ for every R-module X and all $n \geq 1$, we will see in the next chapter that Ext_R^n in the second variable can be used to define an injective dimension for a module that will in some sense measure "how far" the module is from being injective.

Finally, one might inquire as to why the left derived functors of Hom in the first and second variable were not investigated. The answer is simply because the development of these functors would provide no useful information about Hom. We have seen that if $\mathcal{F} = \operatorname{Hom}_R(-, X)$ or if $\mathcal{F} = \operatorname{Hom}_R(X, -)$, then $\mathcal{R}^0\mathcal{F}$ and $\mathcal{F}$ are naturally equivalent functors. Because of this, $\operatorname{Hom}_R(-, X)$ and $\operatorname{Hom}_R(X, -)$ are linked to the long exact Ext-sequences. However, Exercise 6 shows that if $\mathcal{F}$ is an additive functor, then $\mathcal{L}_0\mathcal{F}$ is right exact. Hence, it cannot be the case that $\mathcal{L}_0\mathcal{F}$ and $\mathcal{F}$ are naturally equivalent when $\mathcal{F} = \operatorname{Hom}_R(X, -)$, since $\operatorname{Hom}_R(X, -)$ is not generally right exact. Similarly, $\mathcal{L}_0\mathcal{F}$ and $\mathcal{F}$ cannot be naturally equivalent when $\mathcal{F} = \operatorname{Hom}_R(-, X)$. Thus, the left derived functors of Hom are not connected to Hom as are its right derived functors.

Problem Set 11.4

1. Choose an injective resolution of each R-module, develop $\overline{\operatorname{Ext}}_R^n(X, -)$ and then choose another injective resolution of each R-module and construct $\widetilde{\operatorname{Ext}}_R^n(X, -)$. Prove that $\widetilde{\operatorname{Ext}}_R^n(X, -)$ and $\overline{\operatorname{Ext}}_R^n(X, -)$ are naturally equivalent.

2. Prove the *Horse Shoe Lemma for Injectives*. That is, given a column exact diagram

$$
\begin{array}{c}
0 \\
\downarrow \\
0 \longrightarrow L \xrightarrow{\alpha^{-1}} D^0 \xrightarrow{\alpha^0} D^1 \xrightarrow{\alpha^1} D^2 \xrightarrow{\alpha^2} \cdots \\
\downarrow{\scriptstyle f} \\
0 \longrightarrow M \\
\downarrow{\scriptstyle g} \\
0 \longrightarrow N \xrightarrow{\gamma^{-1}} F^0 \xrightarrow{\gamma^0} F^1 \xrightarrow{\gamma^1} F^2 \xrightarrow{\gamma^2} \cdots \\
\downarrow \\
0
\end{array}
$$

where the rows are injective resolutions of L and N respectively, show that there is an injective resolution $\mathbf{E}$ of M and cochain maps $\mathbf{f} : \mathbf{D}^L \to \mathbf{E}^M$ and $\mathbf{g} : \mathbf{E}^M \to \mathbf{F}^N$ such that the column exact diagram

$$
\begin{array}{ccccccccc}
0 & & 0 & & 0 & & 0 \\
\downarrow & & \downarrow & & \downarrow & & \downarrow \\
0 \longrightarrow L & \xrightarrow{\alpha^{-1}} & D^0 & \xrightarrow{\alpha^0} & D^1 & \xrightarrow{\alpha^1} & D^2 & \xrightarrow{\alpha^2} & \cdots \\
\downarrow{\scriptstyle f} & & \downarrow{\scriptstyle f_0} & & \downarrow{\scriptstyle f_1} & & \downarrow{\scriptstyle f_2} \\
0 \longrightarrow M & \xrightarrow{\beta^{-1}} & E^0 & \xrightarrow{\beta^0} & E^1 & \xrightarrow{\beta^1} & E^2 & \xrightarrow{\beta^2} & \cdots \\
\downarrow{\scriptstyle g} & & \downarrow{\scriptstyle g_0} & & \downarrow{\scriptstyle g_1} & & \downarrow{\scriptstyle g_2} \\
0 \longrightarrow N & \xrightarrow{\gamma^{-1}} & F^0 & \xrightarrow{\gamma^0} & F^1 & \xrightarrow{\gamma^1} & F^2 & \xrightarrow{\gamma^2} & \cdots \\
\downarrow & & \downarrow & & \downarrow & & \downarrow \\
0 & & 0 & & 0 & & 0
\end{array}
$$

is commutative, where $E^n = D^n \oplus F^n$ for each $n \geq 0$. [Hint: Dualize the proof of the Horse Shoe Lemma for Projectives.]

3. Prove Proposition 11.4.10. [(1) $\Rightarrow$ (2), Hint: Proposition 11.3.4.] [(5) $\Rightarrow$ (1), Hint: Baer's criteria.]

4. Prove Proposition 11.4.11. [Hint: Proposition 11.3.5.]

5. (a) Show that $\text{Ext}_R^n(-, -)$ is additive in each variable.

   (b) Suppose that $0 \to M_1 \to P \to M \to 0$ and $0 \to N \to E \to N_1 \to 0$ are short exact sequences where P and E are projective and injective R-modules, respectively. Prove that $\text{Ext}_R^1(M_1, N)$ and $\text{Ext}_R^1(M, N_1)$ are isomorphic. [Hint: Use the long exact Ext-sequences.]

6. If $\mathscr{F}$ is an additive functor, show that $\mathscr{L}_0\mathscr{F}$ is right exact. [Hint: Let $0 \to L \xrightarrow{f} M \xrightarrow{g} N \to 0$ be an exact sequence of R-modules and suppose that $\mathbf{P}$ and $\mathbf{R}$ are projective resolutions of L and N, respectively. Then, by the Horse Shoe Lemma for Projectives, there is a projective resolution $\mathbf{Q}$ of M and chain maps $\mathbf{f} : \mathbf{P}_L \to \mathbf{Q}_M$ and $\mathbf{g} : \mathbf{Q}_M \to \mathbf{R}_N$ such that $0 \to \mathbf{P}_L \xrightarrow{\mathbf{f}} \mathbf{Q}_M \xrightarrow{\mathbf{g}} \mathbf{R}_N \to 0$ is exact, where $Q_n = P_n \oplus R_n$ for each $n \geq 0$. This gives a long exact sequence

$$\cdots \to H_n(\mathscr{F}(\mathbf{P}_L)) \xrightarrow{H_n(\mathscr{F}(\mathbf{f}))} H_n(\mathscr{F}(\mathbf{Q}_M)) \xrightarrow{H_n(\mathscr{F}(\mathbf{g}))} H_n(\mathscr{F}(\mathbf{R}_N)) \to \cdots .$$

Show that this in turn gives

$$\cdots \to \mathscr{L}_n\mathscr{F}(L) \xrightarrow{\mathscr{L}_n\mathscr{F}(f)} \mathscr{L}_n\mathscr{F}(M) \xrightarrow{\mathscr{L}_n\mathscr{F}(g)} \mathscr{L}_n\mathscr{F}(N) \xrightarrow{\Phi_n} \mathscr{L}_{n-1}\mathscr{F}(L) \to \cdots$$

$$\cdots \xrightarrow{\Phi_1} \mathscr{L}_0\mathscr{F}(L) \xrightarrow{\mathscr{L}_0\mathscr{F}(f)} \mathscr{L}_0\mathscr{F}(M) \xrightarrow{\mathscr{L}_0\mathscr{F}(g)} \mathscr{L}_0\mathscr{F}(N) \to 0.]$$

7. Show that if $\text{Ext}_R^1(M, N) = 0$, then every short exact sequence of the form $0 \to N \xrightarrow{f} X \xrightarrow{g} M \to 0$ splits. [Hint: $0 \to \text{Hom}_R(M, N) \xrightarrow{g^*} \text{Hom}_R(X, N) \xrightarrow{f^*} \text{Hom}_R(N, N) \to \text{Ext}_R^1(M, N).]$

8. Prove that if $0 \to L \to M \to N \to 0$ is a split short exact sequence of R-modules and R-module homomorphisms, then

$$0 \to \text{Ext}_R^n(N, X) \to \text{Ext}_R^n(M, X) \to \text{Ext}_R^n(L, X) \to 0 \quad \text{and}$$
$$0 \to \text{Ext}_R^n(X, L) \to \text{Ext}_R^n(X, M) \to \text{Ext}_R^n(X, N) \to 0$$

are split short exact sequences for any R-module X and any $n \geq 0$.

9. If $0 \to K \xrightarrow{f} P \xrightarrow{g} M \to 0$ is a short exact sequence with P projective, then the sequence

$$0 \to \text{Hom}_R(M, N) \xrightarrow{g^*} \text{Hom}_R(P, N) \xrightarrow{f^*} \text{Hom}_R(K, N)$$

is exact. Show that $\text{Ext}_R^1(M, N) = \text{Coker } f^*$.

10. Prove each of the following for an R-module N and a family of R-modules $\{M_\alpha\}_\Delta$.

(a) For any $n \geq 0$, $\mathrm{Ext}_R^n(\bigoplus_\Delta M_\alpha, N) \cong \prod_\Delta \mathrm{Ext}_R^n(M_\alpha, N)$. [Hint: Use induction. The case for $n = 0$ is Proposition 2.1.12. For each $\alpha \in \Delta$, let $0 \to K_\alpha \to P_\alpha \to M_\alpha \to 0$ be a short exact sequence with P_α projective and consider the diagram

$$\mathrm{Hom}_R\left(\bigoplus P_\alpha, N\right) \longrightarrow \mathrm{Hom}_R\left(\bigoplus K_\alpha, N\right) \longrightarrow \mathrm{Ext}_R^1\left(\bigoplus M_\alpha, N\right) \longrightarrow \mathrm{Ext}_R^1\left(\bigoplus P_\alpha, N\right)$$

$$\prod_\Delta \mathrm{Hom}_R(P_\alpha, N) \longrightarrow \prod_\Delta \mathrm{Hom}_R(K_\alpha, N) \longrightarrow \prod_\Delta \mathrm{Ext}_R^1(M_\alpha, N) \longrightarrow \prod_\Delta \mathrm{Ext}_R^1(P_\alpha, N)$$

with vertical maps α, β, γ.

Show that the diagram is commutative, where α and β are isomorphisms, and then chase the diagram to establish that γ is an isomorphism. This establishes the case for $n = 1$.]

(b) For any $n \geq 0$, $\mathrm{Ext}_R^n(N, \prod_\Delta M_\alpha) \cong \prod_\Delta \mathrm{Ext}_R^n(N, M_\alpha)$.

11. Prove that the following are equivalent.

(a) $\mathrm{Ext}_R^{n+1}(M, N) = 0$ for every R-module N.

(b) There is a projective resolution of M of the form

$$\mathbf{P}: \cdots \to 0 \to 0 \to P_n \to \cdots \to P_0 \to M \to 0.$$

(c) For every projective resolution

$$\mathbf{Q}: \cdots \to Q_n \xrightarrow{\beta_n} Q_{n-1} \to \cdots \to Q_0 \to M \to 0$$

of M, $\mathrm{Ker}\,\beta_{n-1}$ is projective.

12. Prove that the following are equivalent.

(a) $\mathrm{Ext}_R^{n+1}(M, N) = 0$ for every R-module M.

(b) There is an injective resolution of N of the form

$$\mathbf{E}: 0 \to N \to E^0 \to \cdots \to E^n \to 0 \to 0 \to \cdots.$$

(c) For every injective resolution

$$\mathbf{D}: 0 \to N \to D^0 \to \cdots \to D^{n-1} \xrightarrow{\alpha^{n-1}} D^n \to \cdots$$

of N, $\mathrm{Im}\,\alpha^{n-1}$ is injective.

11.5   Torsion Functors

Left Derived Functors of $- \otimes_R X$ and $X \otimes_R -$

We outline a development of the left derived functors of $\otimes$ in both variables with the details left as exercises.

Let X be a left R-module and suppose that

$$\mathbf{P} : \cdots \to P_n \xrightarrow{\alpha_n} P_{n-1} \to \cdots \to P_1 \xrightarrow{\alpha_1} P_0 \xrightarrow{\alpha_0} M \to 0$$

is a projective resolution of an R-module M. Then we have a chain complex

$$\mathbf{P}_M \otimes_R X : \cdots \to P_n \otimes_R X \xrightarrow{\alpha_n \otimes \mathrm{id}_X} P_{n-1} \otimes_R X \to \cdots \xrightarrow{\alpha_1 \otimes \mathrm{id}_X} P_0 \otimes_R X \to 0.$$

Next, let N be an R-module and suppose that $\mathbf{Q}$ is a projective resolution of N. If $f : M \to N$ is an R-linear mapping, then we have a commutative diagram

$$
\begin{array}{ccccccccc}
\cdots & \longrightarrow & P_n \otimes_R X & \xrightarrow{\alpha_n \otimes \mathrm{id}_X} & P_{n-1} \otimes_R X & \longrightarrow & \cdots & \xrightarrow{\alpha_1 \otimes \mathrm{id}_X} & P_0 \otimes_R X & \longrightarrow & 0 \\
& & \downarrow{\scriptstyle f_n \otimes \mathrm{id}_X} & & \downarrow{\scriptstyle f_{n-1} \otimes \mathrm{id}_X} & & & & \downarrow{\scriptstyle f_0 \otimes \mathrm{id}_X} & & \\
\cdots & \longrightarrow & Q_n \otimes_R X & \xrightarrow{\beta_n \otimes \mathrm{id}_X} & Q_{n-1} \otimes_R X & \longrightarrow & \cdots & \xrightarrow{\beta_1 \otimes \mathrm{id}_X} & Q_0 \otimes_R X & \longrightarrow & 0
\end{array}
$$

of chain complexes, where $\mathbf{f} : \mathbf{P}_M \to \mathbf{Q}_N$ is a chain map generated by f. If we let $\mathrm{Tor}_n^R(M, X) = H_n(\mathbf{P}_M \otimes_R X)$ and $\mathrm{Tor}_n^R(f, X) = H_n(\mathbf{f} \otimes \mathrm{id}_X)$, then

$$\mathrm{Tor}_n^R(f, X) : \mathrm{Tor}_n^R(M, X) \to \mathrm{Tor}_n^R(N, X),$$

so we have a right exact additive functor $\mathrm{Tor}_n^R(-, X) : \mathbf{Mod}_R \to \mathbf{Ab}$ such that $\mathrm{Tor}_0^R(-, X) = - \otimes_R X$.

If $0 \to L \xrightarrow{f} M \xrightarrow{g} N \to 0$ is a short exact sequence of R-modules and R-module homomorphisms, then there is a short exact sequence

$$0 \to \mathbf{P}_L \xrightarrow{\mathbf{f}} \mathbf{Q}_M \xrightarrow{\mathbf{g}} \mathbf{R}_N \to 0$$

in $\mathbf{Chain}_R$, where $\mathbf{P}, \mathbf{Q}$ and $\mathbf{R}$ are projective resolutions of L, M and N, respectively. This gives a sequence

$$0 \to \mathbf{P}_M \otimes_R X \xrightarrow{\mathbf{f} \otimes \mathrm{id}_X} \mathbf{Q}_M \otimes_R X \xrightarrow{\mathbf{g} \otimes \mathrm{id}_X} \mathbf{R}_N \otimes_R X \to 0$$

in **Chain**$_\mathbb{Z}$, so taking homology produces a *long exact* Tor-*sequence in the first variable*

$$\cdots \to \operatorname{Tor}_n^R(L, X) \xrightarrow{\operatorname{Tor}_n^R(f,X)} \operatorname{Tor}_n^R(M, X) \xrightarrow{\operatorname{Tor}_n^R(g,X)} \operatorname{Tor}_n^R(N, X) \xrightarrow{\Phi_n} \cdots$$

$$\cdots \xrightarrow{\Phi_2} \operatorname{Tor}_1^R(L, X) \xrightarrow{\operatorname{Tor}_1^R(f,X)} \operatorname{Tor}_1^R(M, X) \xrightarrow{\operatorname{Tor}_1^R(g,X)} \operatorname{Tor}_1^R(N, X) \xrightarrow{\Phi_1}$$

$$\xrightarrow{\Phi_1} L \otimes_R X \xrightarrow{f \otimes \operatorname{id}_X} M \otimes_R X \xrightarrow{g \otimes \operatorname{id}_X} N \otimes_R X \to 0,$$

where Φ_n is a connecting homomorphism for each $n \geq 1$. There is also a *long exact* Tor-*sequence in the second variable*. For this, let $\mathbf{P}$ be a projective resolution of X and suppose that $0 \to L \xrightarrow{f} M \xrightarrow{g} N \to 0$ is a short exact sequence of left R-modules and left R-module homomorphisms. Since each P_n is projective and hence flat,

$$0 \to P_n \otimes_R L \xrightarrow{\operatorname{id}_{P_n} \otimes f} P_n \otimes_R M \xrightarrow{\operatorname{id}_{P_n} \otimes g} P_n \otimes_R N \to 0$$

is exact for each $n \geq 0$. Thus, we have a short exact sequence

$$\mathbf{0} \to \mathbf{P}_X \otimes_R L \xrightarrow{\operatorname{id}_{\mathbf{P}_X} \otimes f} \mathbf{P}_X \otimes_R M \xrightarrow{\operatorname{id}_{\mathbf{P}_X} \otimes g} \mathbf{P}_X \otimes_R N \to \mathbf{0}$$

of chain complexes, so by taking homology we have a long exact Tor-sequence

$$\cdots \to \operatorname{Tor}_n^R(X, L) \xrightarrow{\operatorname{Tor}_n^R(X,f)} \operatorname{Tor}_n^R(X, M) \xrightarrow{\operatorname{Tor}_n^R(X,g)} \operatorname{Tor}_n^R(X, N) \xrightarrow{\Phi_n} \cdots$$

$$\cdots \xrightarrow{\Phi_2} \operatorname{Tor}_1^R(X, L) \xrightarrow{\operatorname{Tor}_1^R(X,f)} \operatorname{Tor}_1^R(X, M) \xrightarrow{\operatorname{Tor}_1^R(X,g)} \operatorname{Tor}_1^R(X, N) \xrightarrow{\Phi_1}$$

$$\xrightarrow{\Phi_1} X \otimes_R L \xrightarrow{\operatorname{id}_X \otimes f} X \otimes_R M \xrightarrow{\operatorname{id}_X \otimes f} X \otimes_R N \to 0$$

in the second variable, where each Φ_n is a connecting homomorphism.

It is not difficult to show that $\operatorname{Tor}_n^R(X, -) : {}_R\mathbf{Mod} \to \mathbf{Ab}$ is an additive functor, where we let $\operatorname{Tor}_0^R(X, -) = X \otimes_R -$. Hence, we have right exact additive functors

$$\operatorname{Tor}_n^R(-, X) : \mathbf{Mod}_R \to \mathbf{Ab} \quad \text{and} \quad \operatorname{Tor}_n^R(X, -) : {}_R\mathbf{Mod} \to \mathbf{Ab}$$

for each $n \geq 0$. Furthermore, the bifunctor Tor_n^R is *balanced*, that is, $\operatorname{Tor}_n^R(M, N)$ can be computed by taking a projective resolution of M or by taking a projective resolution of N.

Example

1. Consider the $\mathbb{Z}_4$-module $\mathbb{Z}_2$. Then

$$\cdots \rightarrow \mathbb{Z}_4 \xrightarrow{\alpha_n} \mathbb{Z}_4 \xrightarrow{\alpha_{n-1}} \mathbb{Z}_4 \rightarrow \cdots \rightarrow \mathbb{Z}_4 \xrightarrow{\alpha_0} \mathbb{Z}_2 \rightarrow 0$$

is a projective resolution of $\mathbb{Z}_2$, where $\alpha_n([a]) = [2a]$ for each $n \geq 1$ and $\alpha_0([a]) = [a]$. Tensoring $\mathbb{Z}_2$ with the deleted projective resolution of $\mathbb{Z}_2$ gives

$$\cdots \rightarrow \mathbb{Z}_4 \otimes_{\mathbb{Z}_4} \mathbb{Z}_2 \xrightarrow{\alpha_n \otimes \mathrm{id}_{\mathbb{Z}_2}} \mathbb{Z}_4 \otimes_{\mathbb{Z}_4} \mathbb{Z}_2 \rightarrow \cdots \rightarrow \mathbb{Z}_4 \otimes_{\mathbb{Z}_4} \mathbb{Z}_2 \rightarrow 0.$$

It follows that this sequence can be replaced by

$$\cdots \rightarrow \mathbb{Z}_2 \xrightarrow{0} \mathbb{Z}_2 \xrightarrow{0} \mathbb{Z}_2 \rightarrow \cdots \rightarrow \mathbb{Z}_2 \xrightarrow{0} \mathbb{Z}_2 \rightarrow 0$$

and so $\mathrm{Tor}_n^{\mathbb{Z}_4}(\mathbb{Z}_2, \mathbb{Z}_2) \cong \mathbb{Z}_2$ for $n \geq 1$.

Remark. If G is an additive abelian group and if $t(G)$ is the torsion subgroup of G, then one can show that $\mathrm{Tor}_1^{\mathbb{Z}}(\mathbb{Q}/\mathbb{Z}, G) \cong t(G)$. Hence, the name *torsion functors* and the notation Tor for these functors.

Finally, we wish to point out that there is an important connection among the Tor functors and flat modules. To establish this connection, we need the following lemma.

Lemma 11.5.1. *If P is a projective left R-module, then $\mathrm{Tor}_n^R(M, P) = 0$ for every R-module M and all $n \geq 1$.*

Proof. Part (1) of Proposition 11.3.4 proves the lemma.   □

Now for the connection of flat modules to the Tor functors.

Proposition 11.5.2. *The following are equivalent for an R-module F.*

(1) *F is flat.*

(2) *$\mathrm{Tor}_1^R(F, N) = 0$ for every left R-module N.*

(3) *$\mathrm{Tor}_n^R(F, N) = 0$ for every left R-module N and all $n \geq 1$.*

Proof. It is obvious that $(3) \Rightarrow (2)$, so suppose that

$$0 \rightarrow L \rightarrow M \rightarrow N \rightarrow 0$$

is a short exact sequence of left R-modules and left R-module homomorphisms. Then the long exact Tor-sequence in the second variable gives the exact sequence

$$\cdots \rightarrow \mathrm{Tor}_1^R(F, M) \rightarrow \mathrm{Tor}_1^R(F, N) \rightarrow F \otimes_R L \rightarrow F \otimes_R M \rightarrow F \otimes_R N \rightarrow 0.$$

If $\text{Tor}_1^R(F, N) = 0$ for every left R-module N, then

$$0 \to F \otimes_R L \to F \otimes_R M \to F \otimes_R N \to 0$$

is exact, so F is flat. Hence, (2) $\Rightarrow$ (1). Finally, we need to show that (1) $\Rightarrow$ (3). Let F be a flat R-module and suppose that N is a left R-module. Then there is a short exact sequence $0 \to K \to P \to N \to 0$ of left R-modules and left R-module homomorphisms with P projective. For $n = 1$, we have the exact sequence

$$0 = \text{Tor}_1^R(F, P) \to \text{Tor}_1^R(F, N) \to F \otimes_R K \to F \otimes_R P,$$

where $\text{Tor}_1^R(F, P) = 0$ is given by Lemma 11.5.1. But F is flat, so $F \otimes_R K \to F \otimes_R P$ is an injective map. Hence, $\text{Tor}_1^R(F, N) = 0$. Now make the induction hypothesis that $\text{Tor}_k^R(F, N) = 0$ for $k = 1, 2, \ldots, n - 1$ and for every left R-module N. Then the long exact Tor-sequence in the second variable gives

$$\cdots \to \text{Tor}_n^R(F, P) \to \text{Tor}_n^R(F, N) \to \text{Tor}_{n-1}^R(F, K) \to \cdots.$$

Using Lemma 11.5.1 again shows that $\text{Tor}_n^R(F, P) = 0$ and the induction hypothesis gives $\text{Tor}_{n-1}^R(F, K) = 0$. Hence, $\text{Tor}_n^R(F, N) = 0$ and the fact that (1) $\Rightarrow$ (3) follows by induction. □

Proposition 11.5.2 is obviously symmetric, that is, F is a flat left R-module if and only if $\text{Tor}_n^R(M, F) = 0$ for every R-module M and all $n \geq 1$. It was previously mentioned that the functors Ext_R^n will be used in the following chapter to define a projective and an injective dimension of an R-module. A similar use will be made of the functors Tor_n^R with regard to flat modules.

We close with the following two propositions. The first is an immediate result due to the preceding proposition and Proposition 5.3.16, while the second follows directly from the preceding proposition and Proposition 5.3.18.

Proposition 11.5.3. *The following are equivalent for a ring R and all $n \geq 1$.*

(1) *R is left coherent.*

(2) *$\text{Tor}_n^R(\prod_\Delta F_\alpha, N) = 0$ for every family $\{F_\alpha\}_\Delta$ of flat R-modules and every left R-module N.*

(3) *$\text{Tor}_n^R(R^\Delta, N) = 0$ for every left R-module N and every set Δ.*

Proposition 11.5.4. *The following are equivalent for a ring R and all $n \geq 1$.*

(1) *R is a regular ring.*

(2) *$\text{Tor}_n^R(M, N) = 0$ for every R-module M and every left R-module N.*

(3) *$\text{Tor}_n^R(M, N) = 0$ for every cyclic R-module M and every left R-module N.*

Problem Set 11.5

1. Prove that F is a flat R-module if and only if $\text{Tor}_1^R(F, R/A) = 0$ for every finitely generated left ideal A of R. [Hint: Proposition 5.3.7.]

2. Show that Tor_n^R is additive in each variable for each $n \geq 0$.

3. Let M and N be R-modules and suppose that $f : M \to N$ is an R-linear mapping. If X is a left R-module, show that

$$\text{Tor}_n^R(f, X) : \text{Tor}_n^R(M, X) \to \text{Tor}_n^R(N, X)$$

is independent of the chain map generated by f. [Hint: Exercise 2 and Proposition 11.3.1.]

4. Prove that we can let $\text{Tor}_0^R(-, N) = - \otimes_R N$ for any left R-module N and that we can set $\text{Tor}_0^R(M, -) = M \otimes_R -$ for each R-module M. [Hint: Exercise 2 and Proposition 11.3.4.]

5. Choose a projective resolution $\mathbf{P}$ of each R-module M and develop the functors $\text{Tor}_n^R(-, X)$ for a fixed left R-module X. Now choose a projective resolution $\mathbf{P}'$ for each R-module M and develop the functors $\overline{\text{Tor}}_n^R(-, X)$. Show that the functors $\text{Tor}_n^R(-, X)$ and $\overline{\text{Tor}}_n^R(-, X)$ are naturally equivalent for each $n \geq 0$. [Hint: Exercise 2 and Proposition 11.3.2.]

6. Use the projective resolution $\cdots \to 0 \to 0 \to \mathbb{Z} \xrightarrow{\alpha_1} \mathbb{Z} \xrightarrow{\alpha_0} \mathbb{Z}_k \to 0$ of $\mathbb{Z}_k$ of Example 2 in Section 11.2 to compute $\text{Tor}_0^{\mathbb{Z}}(M, \mathbb{Z}_k)$ and $\text{Tor}_1^{\mathbb{Z}}(M, \mathbb{Z}_k)$, where M is a $\mathbb{Z}$-module and k is a positive integer.

7. Let M be an R-module and suppose that N is a left R-module. Prove each of the following.

   (a) $M \otimes_R N \cong N \otimes_{R^{\text{op}}} M$.

   (b) If $\cdots \to P_n \to \cdots \to P_1 \to P_0 \to M \to 0$, is a projective resolution of M, then $H_n(\mathbf{P}_M \otimes_R N) \cong H_n(N \otimes_{R^{\text{op}}} \mathbf{P}_M)$ for all $n \geq 0$. [(a) and (b), Hint: Show that there are group isomorphisms $P_n \otimes_R N \to N \otimes_{R^{\text{op}}} P_n$ defined by $x_n \otimes y \mapsto y \otimes x_n$ for each $n \geq 0$ and that when $n \geq 1$ these maps give a chain map $\mathbf{P}_M \otimes_R N \to N \otimes_{R^{\text{op}}} \mathbf{P}_M$.]

   (c) $\text{Tor}_n^R(M, N) \cong \text{Tor}_n^{R^{\text{op}}}(N, M)$ for all $n \geq 0$. Conclude that if R is commutative, then $\text{Tor}_n^R(M, N) \cong \text{Tor}_n^R(N, M)$ for all $n \geq 0$ and all R-modules M and N.

8. Show that for any family $\{N_\alpha\}_\Delta$ of left R-modules and any family $\{M_\alpha\}_\Delta$ of R-modules that

   $$(1) \qquad \text{Tor}_n^R \left(M, \bigoplus_\Delta N_\alpha \right) \cong \bigoplus_\Delta \text{Tor}_n^R(M, N_\alpha) \qquad \text{for all } n \geq 0 \quad \text{and}$$

   $$(2) \qquad \text{Tor}_n^R \left(\bigoplus_\Delta M_\alpha, N \right) \cong \bigoplus_\Delta \text{Tor}_n^R(M_\alpha, N) \qquad \text{for all } n \geq 0.$$

Chapter 12

Homological Methods

In the previous chapter it was pointed out that the extension and torsion functors can be used to define a projective dimension, an injective dimension and a flat dimension of modules that will, in some sense, measure "how far" a module is from being projective, injective or flat. The purpose of this chapter is to establish these dimensions and to show how homological methods can be used to gain information about various rings.

12.1 Projective and Injective Dimension

Definition 12.1.1. The *projective dimension* of an R-module M, denoted by pd-M, is the smallest integer $n \geq 0$ such that $\operatorname{Ext}_R^{n+1}(M, N) = 0$ for every R-module N. If no such integer exists, then pd-$M = \infty$. Likewise, the *injective dimension* of an R-module N, denoted by id-N, is the smallest integer $n \geq 0$ such that $\operatorname{Ext}_R^{n+1}(M, N) = 0$ for every R-module M, and we set id-$N = \infty$ if no such integer n exists. The *right global projective dimension* of R is defined as

$$\text{r.gl.pd-}R = \sup\{\text{pd-}M \mid M \text{ an } R\text{-module}\}$$

and the *right global injective dimension* of R is given by

$$\text{r.gl.id-}R = \sup\{\text{id-}N \mid N \text{ an } R\text{-module}\}.$$

If an R-module M has a projective resolution of the form

$$\mathbf{P}: \ 0 \to P_n \to P_{n-1} \to \cdots \to P_0 \to M \to 0,$$

then we say that M has a *projective resolution of length n*. (Additional zeroes to the left in $\mathbf{P}$ have been suppressed.) If no shorter projective resolution of M exists, then $\mathbf{P}$ is said to be of *minimal length*. Similarly, if an R-module N has an injective resolution of the form

$$\mathbf{E}: \ 0 \to N \to E^0 \to \cdots \to E^{n-1} \to E^n \to 0,$$

then N is said to have an *injective resolution of length n*. If no shorter injective resolution of M exists, then $\mathbf{E}$ is of *minimal length*. Definitions analogous to those above, but for left R-modules, can be given so that l.gl.pd-R and l.gl.id-R will have the obvious meanings. Of course, if R is commutative, the prefixes r. and l. can be omitted.

Part (2) of the following lemma relates the projective dimension of an R-module to the kernel of a particular boundary map of a projective resolution of the module. The lemma will be used to show that any projective resolution can be used to compute the projective dimension of a module.

Lemma 12.1.2 (Dimension Shifting). *Suppose that*

$$\mathbf{P}: \cdots \to P_n \xrightarrow{\alpha_n} P_{n-1} \to \cdots \to P_0 \xrightarrow{\alpha_0} M \to 0 \quad and$$

$$\mathbf{Q}: \cdots \to Q_n \xrightarrow{\beta_n} Q_{n-1} \to \cdots \to Q_0 \xrightarrow{\beta_0} M \to 0$$

are projective resolutions of an R-module M and let $\operatorname{Ker} \alpha_n = K_n^{\mathbf{P}}$ *and* $\operatorname{Ker} \beta_n = K_n^{\mathbf{Q}}$ *for $n = 0, 1, 2, \ldots$. Then for any R-module N*

(1) $\operatorname{Ext}_R^n(K_n^{\mathbf{P}}, N) \cong \operatorname{Ext}_R^n(K_n^{\mathbf{Q}}, N)$ *and*

(2) $\operatorname{Ext}_R^1(K_{n-1}^{\mathbf{P}}, N) \cong \operatorname{Ext}_R^{n+1}(M, N)$, *where* $K_{-1}^{\mathbf{P}} = M$.

Proof. (1) Using the long form of Schanuel's lemma for projectives, we have

$$K_n^{\mathbf{P}} \oplus Q_n \oplus P_{n-1} \oplus Q_{n-2} \oplus \cdots \cong K_n^{\mathbf{Q}} \oplus P_n \oplus Q_{n-1} \oplus P_{n-2} \cdots .$$

But $\operatorname{Ext}_R^n(-, N)$ commutes with direct sums and, by Proposition 11.4.5, $\operatorname{Ext}_R^n(X, N) = 0$ for all $n \geq 1$ whenever X is a projective R-module. Hence, we have $\operatorname{Ext}_R^n(K_n^{\mathbf{P}}, N) \cong \operatorname{Ext}_R^n(K_n^{\mathbf{Q}}, N)$.

(2) Since $\operatorname{Im} \alpha_{n+1} = \operatorname{Ker} \alpha_n$ for $n = 0, 1, 2, \ldots$, $\mathbf{P}$ can be "decomposed" into short exact sequences

$$0 \to K_0^{\mathbf{P}} \to P_0 \to M \to 0$$

$$0 \to K_1^{\mathbf{P}} \to P_1 \to K_0^{\mathbf{P}} \to 0$$

$$\vdots$$

$$0 \to K_n^{\mathbf{P}} \to P_n \to K_{n-1}^{\mathbf{P}} \to 0$$

$$\vdots$$

If $\operatorname{Ext}_R^k(-, N)$ is applied to $0 \to K_j^{\mathbf{P}} \to P_j \to K_{j-1}^{\mathbf{P}} \to 0$, for $j = 0, 1, 2, \ldots$, then by selecting the appropriate sections of the resulting long exact Ext-sequence in the first variable, we see that

$$\operatorname{Ext}_R^k(P_j, N) \to \operatorname{Ext}_R^k(K_j^{\mathbf{P}}, N) \xrightarrow{\Phi^k} \operatorname{Ext}_R^{k+1}(K_{j-1}^{\mathbf{P}}, N) \to \operatorname{Ext}_R^{k+1}(P_j, N)$$

is exact for $k \geq 0$ and $j = 0, 1, 2, \ldots$, where each Φ^k is a connecting homomorphism. Since P_j is projective, $\operatorname{Ext}_R^k(P_j, N) = \operatorname{Ext}_R^{k+1}(P_j, N) = 0$, so the connecting homomorphism Φ^k is an isomorphism for each $k \geq 0$ and $j = 0, 1, 2 \dots$. Thus, for the pairs $(k, j) = (n, 0), (n-1, 1), \dots, (2, n-2), (1, n-1)$, we have isomorphisms

$$\operatorname{Ext}_R^n(K_0^{\mathbf{P}}, N) \overset{\Phi^n}{\cong} \operatorname{Ext}_R^{n+1}(M, N),$$

$$\operatorname{Ext}_R^{n-1}(K_1^{\mathbf{P}}, N) \overset{\Phi^{n-1}}{\cong} \operatorname{Ext}_R^n(K_0^{\mathbf{P}}, N),$$

$$\vdots$$

$$\operatorname{Ext}_R^2(K_{n-2}^{\mathbf{P}}, N) \overset{\Phi^2}{\cong} \operatorname{Ext}_R^3(K_{n-3}^{\mathbf{P}}, N), \quad \text{and}$$

$$\operatorname{Ext}_R^1(K_{n-1}^{\mathbf{P}}, N) \overset{\Phi^1}{\cong} \operatorname{Ext}_R^2(K_{n-2}^{\mathbf{P}}, N).$$

Hence,

$$\Phi^n \Phi^{n-1} \cdots \Phi^2 \Phi^1 : \operatorname{Ext}_R^1(K_{n-1}^{\mathbf{P}}, N) \to \operatorname{Ext}_R^{n+1}(M, N)$$

is an isomorphism. $\qquad \square$

We also have a dimension shifting lemma for injective resolutions whose proof is similar to that of the preceding dimension shifting lemma. The proof of (1) of this lemma can be effected by using the long form of Schanuel's lemma for injective modules.

Lemma 12.1.3 (Dimension Shifting). *Suppose that*

$$\mathbf{D} : 0 \to N \overset{\alpha^{-1}}{\longrightarrow} D^0 \overset{\alpha^0}{\longrightarrow} D^1 \to \cdots \to D^n \overset{\alpha^n}{\longrightarrow} D^{n+1} \to \cdots \quad \text{and}$$

$$\mathbf{E} : 0 \to N \overset{\beta^{-1}}{\longrightarrow} E^0 \overset{\beta^0}{\longrightarrow} E^1 \to \cdots \to E^n \overset{\beta^n}{\longrightarrow} E^{n+1} \to \cdots$$

are injective resolutions of an R-module N and let $\operatorname{Im} \alpha^n = C_{\mathbf{D}}^n$ *and* $\operatorname{Im} \beta^n = C_{\mathbf{E}}^n$ *for* $n = 0, 1, 2, \dots$. *Then for any R-module M*

(1) $\operatorname{Ext}_R^n(M, C_{\mathbf{D}}^n) \cong \operatorname{Ext}_R^n(M, C_{\mathbf{E}}^n)$ *and*

(2) $\operatorname{Ext}_R^1(M, C_{\mathbf{D}}^{n-1}) \cong \operatorname{Ext}_R^{n+1}(M, N)$, *where* $C_{\mathbf{D}}^{-1} = N$.

Using the shifting lemmas, we can now provide a sharpening of the tools that can be used to compute the projective dimension and the injective dimension of a module. A proof is offered for the first of the following two propositions while the proof of the second is left as an exercise.

Proposition 12.1.4. *The following are equivalent for an R-module M.*

(1) pd-$M = n$.

(2) *M has a projective resolution $\mathbf{P}$ of minimal length n.*

(3) *If* $\mathbf{Q}: \cdots \to Q_n \xrightarrow{\beta_n} Q_{n-1} \to \cdots \to Q_0 \to M \to 0$ *is a projective resolution of* M, *then* $K_{n-1} = \operatorname{Ker} \beta_{n-1}$ *is projective and*

$$\mathbf{Q}': \ 0 \to K_{n-1} \to Q_{n-1} \to \cdots \to Q_0 \to M \to 0$$

is a projective resolution of M *of minimal length* n.

(4) $\operatorname{Ext}_R^k(M, N) = 0$ *for every* R-*module* N *and all integers* $k > n$, *but* $\operatorname{Ext}_R^n(M, N) \neq 0$ *for some* R-*module* N.

Proof. (1) $\Rightarrow$ (2). Let

$$\mathbf{P}: \ \cdots \to P_n \xrightarrow{\alpha_n} P_{n-1} \to \cdots \to P_0 \to M \to 0$$

be a projective resolution of M. Part (2) of Lemma 12.1.2 gives $\operatorname{Ext}_R^1(K_{n-1}^{\mathbf{P}}, N) \cong \operatorname{Ext}_R^{n+1}(M, N) = 0$, so we see that $K_{n-1}^{\mathbf{P}} = \operatorname{Ker} \alpha_{n-1}$ is projective. Thus,

$$\mathbf{P}': \ 0 \to K_{n-1}^{\mathbf{P}} \to P_{n-1} \to \cdots \to P_0 \to M \to 0$$

is a projective resolution of M of length n. If M has a projective resolution of shorter length, then there is an integer $k < n$ such that $\operatorname{Ext}_R^k(M, N) = 0$, a clear contradiction since pd-$M = n$.

(2) $\Rightarrow$ (3). Let

$$\mathbf{P}: \ 0 \to P_n \xrightarrow{\alpha_n} P_{n-1} \to \cdots \to P_0 \to M \to 0$$

be the projective resolution of M of minimal length n. If

$$\mathbf{Q}: \ \cdots \to Q_n \xrightarrow{\beta_n} Q_{n-1} \to \cdots \to Q_0 \to M \to 0$$

is also a projective resolution of M, let $K_j^{\mathbf{P}} = \operatorname{Ker} \alpha_j$ and $K_j^{\mathbf{Q}} = \operatorname{Ker} \beta_j$. Then (1) of Lemma 12.1.2 shows that $K_j^{\mathbf{P}}$ is projective if and only if $K_j^{\mathbf{Q}}$ is projective for $j = 0, 1, 2, \ldots$. But $K_{n-1}^{\mathbf{P}} = P_n$ is projective, so

$$\mathbf{Q}': \ 0 \to K_{n-1}^{\mathbf{Q}} \to Q_{n-1} \to \cdots \to Q_0 \to M \to 0$$

is a projective resolution of M which clearly must be of minimal length n.

(3) $\Rightarrow$ (4). If (3) holds, then M has a projective resolution $\mathbf{P}$ of minimal length n. Using this projective resolution, it is obvious that $\operatorname{Ext}_R^k(M, N) = 0$ for all $k > n$ and all R-modules N. Now suppose that $\operatorname{Ext}_R^n(M, N) = 0$ for every R-module N. Then (2) of Lemma 12.1.2 shows that $\operatorname{Ext}_R^1(K_{n-2}^{\mathbf{P}}, N) \cong \operatorname{Ext}_R^n(M, N) = 0$, so $K_{n-2}^{\mathbf{P}} = \operatorname{Ker} \alpha_{n-2}$ is projective. But this implies that M has a projective resolution of length $n - 1$, a contradiction. Hence, there must exist an R-module N such that $\operatorname{Ext}_R^n(M, N) \neq 0$.

(4) $\Rightarrow$ (1). Clear. $\square$

Proposition 12.1.5. *The following are equivalent for an R-module M.*

(1) id-$N = n$.

(2) *N has an injective resolution D of minimal length n.*

(3) *If* $\mathbf{E} : \; 0 \to N \to E^0 \to \cdots \to E^{n-1} \xrightarrow{\;\beta^{n-1}\;} E^n \to \cdots$ *is an injective resolution of N, then $C^{n-1} = \operatorname{Im} \beta^{n-1}$ is injective and n is the smallest integer for which this is so.*

(4) $\operatorname{Ext}_R^k(M, N) = 0$ *for every R-module M and all integers $k > n$, but* $\operatorname{Ext}_R^n(M, N) \neq 0$ *for some R-module M.*

Proposition 12.1.4 shows that if pd-$M = n$, then any projective resolution of M can in effect be terminated at n. If pd-$M = n$ and

$$\mathbf{P} : \cdots \to P_n \xrightarrow{\;\alpha_n\;} P_{n-1} \to \cdots \to P_0 \to M \to 0$$

is a projective resolution of M, then $\operatorname{Ker} \alpha_{n-1}$ is projective, so

$$\mathbf{P'} : \; 0 \to \operatorname{Ker} \alpha_{n-1} \to P_{n-1} \to \cdots \to P_0 \to M \to 0$$

is a projective resolution of M. In view of Proposition 12.1.5, we see that similar observations hold for injective dimension and injective resolutions. So pd-M (id-M) is just the length of the shortest projective (injective) resolution of M. Thus, we can think of pd-M (id-M) as a measure of how far a module is from being projective (injective).

There is also a connection between the projective dimensions of the modules L, M and N in a short exact sequence $0 \to L \to M \to N \to 0$. Since it will be required later, we prove (2) of the following proposition and leave the proofs of the remaining parts of the proposition as exercises.

Proposition 12.1.6. *If $0 \to L \to M \to N \to 0$ is a short exact sequence of R-modules and R-module homomorphisms, then*

(1) *If pd-$L <$ pd-M, then pd-$N =$ pd-M,*

(2) *If pd-$L =$ pd-M, then pd-$N \leq 1 +$ pd-L, and*

(3) *If pd-$L >$ pd-M, then pd-$N = 1 +$ pd-L.*

Proof. (2) If pd-$L =$ pd-$M = \infty$, then (2) clearly holds, so suppose that pd-$L =$ pd-$M = n$. The short exact sequence $0 \to L \to M \to N \to 0$ and the long exact Ext-sequence give

$$0 = \operatorname{Ext}_R^{n+1}(L, X) \to \operatorname{Ext}_R^{n+2}(N, X) \to \operatorname{Ext}_R^{n+2}(M, X) = 0,$$

so $\operatorname{Ext}_R^{n+2}(N, X) = 0$ for every R-module X. Hence, pd-N is at most $n + 1$, so pd-$N \leq 1 +$ pd-L.   □

Example

1. Since $\mathbb{Z}$ is a free $\mathbb{Z}$-module, $\mathbb{Z}$ is projective, so pd-$\mathbb{Z} = 0$. However, $0 \to \mathbb{Z} \to \mathbb{Q} \to \mathbb{Q}/\mathbb{Z} \to 0 \to \cdots$ is an injective resolution of $\mathbb{Z}$ of minimal length, so id-$\mathbb{Z} = 1$. Thus, pd-$\mathbb{Z} \neq$ id-$\mathbb{Z}$.

The example above shows that there are modules M such that pd-$M \neq$ id-M. Since Ext^n_R is balanced, the value of $\text{Ext}^n_R(M, N)$ can be computed by using a projective resolution of M or by using an injective resolution of N. This leads us to suspect that r.gl.pd-$R =$ r.gl.id-R and it turns out that this is the case even though there may be modules such that pd-$M \neq$ id-M.

Proposition 12.1.7. *For any ring* R, *r.gl.pd-*$R =$ *r.gl.id-*R.

Proof. If r.gl.pd-$R = n$, then $\text{Ext}^{n+1}_R(M, N) = 0$ for every R-module M. If $\mathbf{E}$ is an injective resolution of N and $C^{n-1} = \text{Im}\,\beta^{n-1}$, (2) of Lemma 12.1.3 shows that $\text{Ext}^1_R(M, C^{n-1}) \cong \text{Ext}^{n+1}_R(M, N) = 0$. Since this holds for every R-module M, by using Proposition 11.4.6, we see that C^{n-1} is injective. Thus, id-N is at most n, so we have id-$N \leq$ r.gl.pd-R for every R-module N. Hence, r.gl.id-$R \leq$ r.gl.pd-R. The reverse inequality follows by a similar argument. $\square$

Definition 12.1.8. Since r.gl.pd-$R =$ r.gl.id-R for any ring R, this common value will now be denoted by r.gl.hd-R and called the *right global homological dimension* of R. The *left global homological dimension* of R has a similar definition.

We can now relate the right global homological dimension to specific rings.

Proposition 12.1.9. *A ring* R *is semisimple if and only if* r.gl.hd-$R = 0$.

Proof. Let N be any R-module. Then r.gl.pd-$R = 0$ if and only if $\text{Ext}^1_R(M, N) = 0$ for every R-module M which holds if and only if every R-module is projective. Proposition 6.4.7 shows that this last condition is equivalent to R being semisimple. $\square$

Because of the left-right symmetry of semisimple rings, we see that l.gl.hd-$R = 0$ if and only if r.gl.hd-$R = 0$. However, there are rings for which l.gl.hd-$R \neq$ r.gl.hd-R. For example, the matrix ring $R = \left(\begin{smallmatrix} \mathbb{Z} & \mathbb{Q} \\ 0 & \mathbb{Q} \end{smallmatrix}\right)$ is such that r.gl.hd-$R = 1$ while l.gl.hd-$R = 2$. The values of l.gl.hd-R and r.gl.hd-R measure how far R is from being semisimple with both measures being zero when R is semisimple.

Recall that a ring R is right hereditary if every right ideal of R is projective. In Proposition 5.2.15 it was established that a ring R is right hereditary if and only if factor modules of injective modules are injective which in turn is true if and only if submodules of projective modules are projective. Because of this, we have the following proposition.

Proposition 12.1.10. *A ring R is right hereditary if and only if* r.gl.hd-$R \leq 1$.

Proof. Let R be right hereditary and suppose that M is any R-module. If M is embedded in the injective module E^0, then

$$0 \to M \to E^0 \to E^0/M \to 0 \to \cdots$$

is an injective resolution of M when M is not injective and

$$0 \to M \to M \to 0 \to \cdots$$

is an injective resolution of M when M is injective. Thus, id-$M \leq 1$ for every R-module M, so we have r.gl.hd-$R \leq 1$.

Conversely, suppose that r.gl.hd-$R \leq 1$, let M be a projective R-module and let N be a submodule of M. The assumption that r.gl.hd-$R \leq 1$ implies that pd-$M/N = 0$ or pd-$M/N = 1$. If pd-$M/N = 0$, then M/N is a projective R-module, so the short exact sequence

$$0 \to N \to M \to M/N \to 0$$

splits. Hence, N is a direct summand of M and is therefore projective. Now suppose that pd-$M/N = 1$. Since M is projective, M/N has a projective resolution of the form

$$\cdots \to P_2 \to P_1 \to M \xrightarrow{\eta} M/N \to 0,$$

where η is the canonical surjection. But the assumption that pd-$M = 1$ implies that $N = \operatorname{Ker} \eta$ is projective. In either case, submodules of projective modules are projective, so R is right hereditary. □

Since a right hereditary ring is not necessarily semisimple, the inequality in Proposition 12.1.10 may be strict. If R is right hereditary but not semisimple, then r.gl.hd-$R = 1$.

Actually, we do not need to know the projective dimension of all R-modules in order to determine r.gl.hd-R. In fact, it suffices to know the projective dimension of the cyclic modules. Recall that an R-module M is cyclic if $M = xR$ for some $x \in M$. This gives an epimorphism $f : R \to xR$ such that $f(a) = xa$ for all $a \in R$ which in turn leads to an isomorphism $R/A \cong xR$, where $A = \operatorname{ann}_r(x)$. Hence, we will know the projective dimension of all cyclic R-modules if we know the projective dimension of R/A for each right ideal A of R. The following proposition is due to Auslander [49].

Proposition 12.1.11 (Auslander). *For any ring R,*

$$r.gl.hd\text{-}R = \sup\{pd\text{-}(R/A) \mid A \text{ a right ideal of } R\}.$$

Proof. If

$$\sup\{\text{pd-}(R/A) \mid A \text{ a right ideal of } R\} = \infty,$$

then it is immediate that r.gl.hd-$R = \infty$ and we are done. So suppose that

$$\sup\{\text{pd-}(R/A) \mid A \text{ a right ideal of } R\} = n.$$

Then pd-$(R/A) \leq n$ for every right ideal A of R. Hence, we see that $\text{Ext}_R^{n+1}(R/A, N) = 0$ for every right R-module N. If $\mathbf{E}$ is an injective resolution of N and $C^{n-1} = \text{Im}\,\alpha^{n-1}$, then Lemma 12.1.3 gives

$$\text{Ext}_R^1(R/A, C^{n-1}) \cong \text{Ext}_n^{n+1}(R/A, N) = 0.$$

Thus, we see by Propositions 11.4.10 and 11.4.12 that C^{n-1} is injective, so we have that id-$N \leq n$ for every right R-module N. But

$$\text{r.gl.hd-}R = \sup\{\text{id-}N \mid N \text{ an } R\text{-module}\},$$

so

$$\text{r.gl.hd-}R \leq n = \sup\{\text{pd-}(R/A) \mid A \text{ a right ideal of } R\}.$$

The reverse inequality follows from

$$\{\text{pd-}(R/A) \mid A \text{ a right ideal of } R\} \subseteq \{\text{pd-}M \mid M \text{ an } R\text{-module}\}. \qquad \square$$

Problem Set 12.1

1. Let M be an R-module and suppose that pd-$M = n$. Show that the kth kernel, $k \leq n$, in any projective resolution of M has projective dimension $n - k$.

2. (a) r.gl.hd-$R = \sup\{\text{pd-}M \mid M \text{ a cyclic } R\text{-module}\}$. [Hint: If $\sup\{\text{pd-}M \mid M$ a cyclic R-module$\} = n$, then $\text{Ext}_R^{n+1}(M, N) = 0$ for every R-module N and all cyclic M. This gives $\text{Ext}_R^{n+1}(M, N) = 0$ for all R-modules M and N.]
   (b) If r.gl.hd-$R > 0$, show that

$$\text{r.gl.hd-}R = 1 + \sup\{\text{pd-}A \mid A \text{ a right ideal of } R\}.$$

   [Hint: Since r.gl.hd-$R \geq 1$, there is a cyclic R-module N such that pd-$N \neq 0$. Consider $0 \to A \to R \to N \to 0$ and show that $1 + \text{pd-}A = \text{pd-}N$.]

3. If R is a principal ideal domain, deduce that gl.hd-$R \leq 1$.

4. Prove Proposition 12.1.5. [Hint: Dualize the proof of Proposition 12.1.4.]

5. Complete the proof of Proposition 12.1.7.

6. Let $0 \to L \to M \to N \to 0$ be a short exact sequence of R-modules and R-module homomorphisms. Show that pd-$M \leq \max(\text{pd-}L, \text{pd-}N)$ with equality holding unless pd-$N = 1 + \text{pd-}L$. [Hint: Use Proposition 12.1.6 and consider the three cases pd-$L \leq \text{pd-}M$, pd-$L = \text{pd-}M$ and pd-$L \geq \text{pd-}M$.]

7. Let $\{M_\alpha\}_\Delta$ be a family of R-modules. Prove each of the following.

   (a) pd-$(\bigoplus_\Delta M_\alpha) = \sup\{\text{pd-}M_\alpha \mid \alpha \in \Delta\}$. Use this to show that $\bigoplus_\Delta M_\alpha$ is projective if and only if each M_α is projective. [Hint: For each $\alpha \in \Delta$, construct an exact sequence

   $$0 \to K_{\alpha,n} \to P_{\alpha,n-1} \to \cdots \to P_{\alpha,1} \to P_{\alpha,0} \to M_\alpha \to 0$$

   where $P_{\alpha,k}$ is projective for $k = 0, 1, 2, \ldots, n-1$. Consider the exact sequence

   $$0 \to \bigoplus_\Delta K_{\alpha,n} \to \bigoplus_\Delta P_{\alpha,n-1} \to \cdots \to \bigoplus_\Delta P_{\alpha,1} \to \bigoplus_\Delta P_{\alpha,0} \to \bigoplus_\Delta M_\alpha \to 0$$

   constructed in the obvious way.]

   (b) id-$(\prod_\Delta M_\alpha) = \sup\{\text{id-}M_\alpha \mid \alpha \in \Delta\}$. Use this result to prove that $\prod_\Delta M_\alpha$ is injective if and only if each M_α is injective.

   (c) If R is right noetherian, then id-$(\bigoplus_\Delta M_\alpha) = \sup\{\text{id-}M_\alpha \mid \alpha \in \Delta\}$.

8. Show that r.gl.hd-$R = \infty$ if and only if there is an R-module M such that pd-$M = \infty$. [Hint: If r.gl.hd-$R = \infty$, then for every $n \geq 0$, there is an R-module M such that pd-$M \geq n$. If one of these modules has infinite projective dimension, then we are done. If not, then for each $n \geq 0$, there is an R-module M_n such that pd-$M_n = k$ with $k \geq n$.]

9. Recall that a Dedekind domain is an integral domain that is hereditary. Prove that if R is a Dedekind domain with quotient field Q, then an R-module N is injective if and only if $\text{Ext}_R^1(Q/R, N) = 0$.

12.2 Flat Dimension

We will now develop the flat dimension of an R-module M and the global flat dimension of R. These concepts will be applied to specific rings and modules and compared to the dimensions given in the preceding section.

Definition 12.2.1. The *flat dimension* of an R-module M, denoted by fd-M, is the smallest integer n such that $\text{Tor}_{n+1}^R(M, N) = 0$ for every left R-module N. The *right global flat dimension* of R, denoted by r.gl.fd-R, is given by

$$\text{r.gl.fd-}R = \sup\{\text{fd-}M \mid M \text{ an } R\text{-module}\}.$$

The flat dimension of M is also called the *weak dimension* of M and the right global flat dimension of R is often referred to as the *right weak global dimension* of R. The flat dimension of a left R-module and the *left global flat dimension* of R are defined and denoted in the obvious way.

Since Proposition 11.5.2 shows that $\mathrm{Tor}_n^R(M, N) = 0$ for every left R-module N and every $n \geq 1$ if and only if M is a flat R-module, fd-M can be viewed as a measure of how far M is from being flat. Another result that follows immediately from Proposition 11.5.4, is that a ring R is regular if and only if r.gl.fd-$R = 0$. Thus, the right global flat dimension of R can be thought of as a measure of how far R is from being a regular ring.

Proposition 12.2.2. *For any ring R, l.gl.fd-R = r.gl.fd-R.*

Proof. If r.gl.fd-$R = \infty$, then we clearly have l.gl.fd-$R \leq$ r.gl.fd-R. So suppose that r.gl.fd-$R = n$. Then $\mathrm{Tor}_{n+1}^R(M, N) = 0$ for every R-module M and every left R-module N. But for a given left R-module N, $\mathrm{Tor}_{n+1}^R(M, N) = 0$ for every R-module M implies that fd-N is at most n. Since this is true for every left R-module N, we see that l.gl.fd-$R \leq$ r.gl.fd-R. The reverse inequality follows by symmetry.   □

Because of Proposition 12.2.2 the prefixes $l.$ and $r.$ can be omitted from l.gl.fd-R and r.gl.fd-R, respectively, and the common value can be denoted simply by gl.fd-R. We will now refer to gl.fd-R as the *global flat dimension* of R.

The bifunctor $\mathrm{Tor}_n^R : \mathbf{Mod}_R \times_R \mathbf{Mod} \to \mathbf{Ab}$ was developed in Chapter 11 using projective resolutions. We now show how Tor_n^R is related to flat resolutions of modules.

Definition 12.2.3. An exact sequence of R-modules and R-module homomorphisms

$$\mathbf{F} : \cdots \to F_n \xrightarrow{\alpha_n} F_{n-1} \to \cdots \to F_0 \xrightarrow{\alpha_0} M \to 0$$

is said to be a *flat resolution* of M if F_n is a flat R-module for $n = 0, 1, 2, \ldots$.

Since every projective module is flat and since every module has a projective resolution, we see that flat resolutions of modules do exist. However, there are flat resolutions that are not projective resolutions. For example, $\cdots \to 0 \to \mathbb{Z} \to \mathbb{Q} \to \mathbb{Q}/\mathbb{Z} \to 0$, where the maps are the obvious ones, is a flat resolution of the $\mathbb{Z}$-module $\mathbb{Q}/\mathbb{Z}$ that is not a projective resolution of $\mathbb{Q}/\mathbb{Z}$. This follows since $\mathbb{Z}$ is free, hence projective and therefore flat. Now $\mathbb{Q}$ is a flat $\mathbb{Z}$-module, but Proposition 5.2.16 shows that $\mathbb{Q}$ is not a projective $\mathbb{Z}$-module since it is not free.

Proposition 12.2.4. *The following are equivalent for an R-module M.*

(1) fd-$M = n$.

(2) *M has a flat resolution F of minimal length n.*

(3) *If* $\mathbf{G} : \cdots \to G_n \xrightarrow{\beta_n} G_{n-1} \to \cdots \to G_0 \to M \to 0$ *is a flat resolution of* M, *then* $K_{n-1} = \mathrm{Ker}\,\beta_{n-1}$ *is flat and*

$$\mathbf{G}' : 0 \to K_{n-1} \to G_{n-1} \to \cdots \to G_0 \to M \to 0$$

is a flat resolution of M *of minimal length* n.

(4) $\mathrm{Tor}_k^R(M, N) = 0$ *for every* R-*module* N *and all integers* $k > n$, *but* $\mathrm{Tor}_n^R(M, N) \neq 0$ *for some* R-*module* N.

Proof. (1) $\Rightarrow$ (2). Let

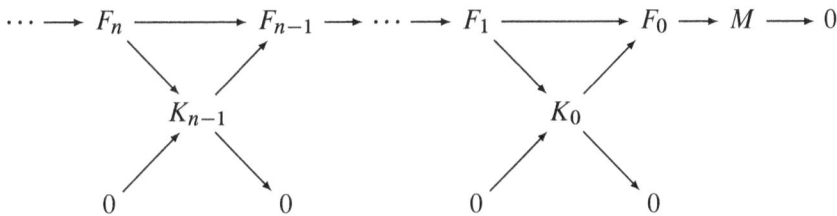

be a flat resolution of M, where the K are the kernels of the boundary maps with $K_{-1} = M$, and suppose that N is a left R-module. Then the long exact Tor-sequence in the first variable applied to $0 \to K_j \to F_j \to K_{j-1} \to 0$ produces a sequence

$$\mathrm{Tor}_{k+1}^R(F_j, N) \to \mathrm{Tor}_{k+1}^R(K_{j-1}, N) \xrightarrow{\Phi_{k+1}} \mathrm{Tor}_k^R(K_j, N) \to \mathrm{Tor}_k^R(F_j, N)$$

for $k \geq 0$ and $n = 0, 1, 2, \ldots$, where each Φ_{k+1} is a connecting homomorphism. Since F_j is flat, $\mathrm{Tor}_{k+1}^R(F_j, N) = \mathrm{Tor}_k^R(F_j, N) = 0$, so Φ_{k+1} is an isomorphism for each $k \geq 0$ and $j = 0, 1, 2 \ldots$. Thus, for the pairs $(k, j) = (n, 0)$, $(n-1, 1), \ldots, (2, n-2), (1, n-1)$, we have isomorphisms

$$\mathrm{Tor}_{n+1}^R(M, N) \stackrel{\Phi_{n+1}}{\cong} \mathrm{Tor}_n^R(K_0, N),$$

$$\mathrm{Tor}_n^R(K_0, N) \stackrel{\Phi_n}{\cong} \mathrm{Tor}_{n-1}^R(K_1, N),$$

$$\vdots$$

$$\mathrm{Tor}_3^R(K_{n-3}, N) \stackrel{\Phi_3}{\cong} \mathrm{Tor}_2^R(K_{n-2}, N), \quad \text{and}$$

$$\mathrm{Tor}_2^R(K_{n-2}, N) \stackrel{\Phi_2}{\cong} \mathrm{Tor}_1^R(K_{n-1}, N).$$

Hence,

$$\Phi_{n+1}^{-1} \cdots \Phi_3^{-1} \Phi_2^{-1} : \mathrm{Tor}_1^R(K_{n-1}, N) \to \mathrm{Tor}_{n+1}^R(M, N)$$

is an isomorphism. If fd-$M = n$, then $\text{Tor}_{n+1}^R(M, N) = 0$, so $\text{Tor}_1^R(K_{n-1}, N) = 0$ for every left R-module N. Thus, Proposition 11.5.2 shows that K_{n-1} is flat. Therefore, M has a flat resolution of length n. If M has a flat resolution

$$0 \to F_k \to F_{k-1} \to \cdots \to F_0 \to M \to 0$$

of length k, $k < n$, it follows that $\text{Tor}_{k+1}^R(M, N) \cong \text{Tor}_1^R(F_k, N) = 0$ for every left R-module N, a contradiction since n is the smallest integer such that $\text{Tor}_{n+1}^R(M, N) = 0$ for every left R-module N. So M has a flat resolution of minimal length n.

(2) $\Rightarrow$ (3). Left as an exercise.

The implications (3) $\Rightarrow$ (4) and (4) $\Rightarrow$ (1) follow easily.   □

If R is right perfect, then it follows from Proposition 7.2.29 that an R-module is projective if and only if it is flat. Over these rings fd-$M =$ pd-M for every R-module M, so r.gl.hd-$R =$ gl.fd-R. This leads to the more general question of how do gl.fd-R and r.gl.hd-R compare in general?

Proposition 12.2.5. *For any ring R,*

$$\text{gl.fd-}R \leq \min\{\text{l.gl.hd-}R, \text{r.gl.hd-}R\}.$$

Proof. If l.gl.hd-$R =$ r.gl.hd-$R = \infty$, there is nothing to prove, so suppose that $\min\{\text{l.gl.hd-}R, \text{r.gl.hd-}R\} = n$. If r.gl.hd-$R = n$ and M is any R-module, then every projective resolution of M has length at most n. If $\mathbf{P}$ is a projective resolution of M of minimal length $k \leq n$, then $\mathbf{P}$ is a flat resolution of M, but as a flat resolution $\mathbf{P}$ may not be of minimal length. Hence, fd-$M \leq$ pd-M, so fd-$M \leq n$ for every R-module M. Thus, gl.fd-$R \leq$ r.gl.hd-R. A similar argument works if l.g.hd-$R = n$, so we also have gl.fd-$R \leq$ l.gl.hd-R. Hence, gl.fd-$R \leq \min\{\text{l.gl.hd-}R, \text{r.gl.hd-}R\}$.   □

We saw in the previous section that r.gl.hd-R can be computed using the cyclic R-modules. The same is true for gl.fd-R.

Proposition 12.2.6. *For any ring R*

$$\text{gl.fd-}R = \sup\{\text{fd-}(R/A) \mid A \text{ a left ideal of } R\}$$
$$= \sup\{\text{fd-}(R/A) \mid A \text{ a right ideal of } R\}.$$

Proof. Since l.gl.fd-$R =$ r.gl.fd-$R =$ gl.fd-R, it suffices to show that

$$\text{gl.fd-}R = \sup\{\text{fd-}(R/A) \mid A \text{ a right ideal of } R\}.$$

Using Exercise 3, the argument is similar to the proof of Proposition 12.1.11.   □

Proposition 12.2.7. *If R is a right noetherian ring, then gl.fd-$R =$ r.gl.hd-R.*

Proof. Propositions 12.1.11 and 12.2.6 show that gl.fd-R and r.gl.hd-R can be computed using R/A as A varies through the right ideals of R. So it suffices to show that if R is right noetherian, then fd-$(R/A) = $ pd-(R/A). If R is right noetherian and A is a right ideal of R, then the finitely generated R-module R/A has a projective resolution

$$\mathbf{P}: \cdots \to P_n \xrightarrow{\alpha_n} P_{n-1} \to \cdots \to P_0 \to R/A \to 0$$

where all the projective modules and all the kernels of the boundary maps are finitely generated. So if fd-$(R/A) = n$, then since $\mathbf{P}$ is also a flat resolution of R/A, we have that $\operatorname{Ker}\alpha_{n-1}$ is finitely generated and flat. Also since R is right noetherian, $\operatorname{Ker}\alpha_{n-1}$ is easily shown to be finitely presented. So if $0 \to K \to F \to \operatorname{Ker}\alpha_{n-1} \to 0$ is a finite presentation of $\operatorname{Ker}\alpha_{n-1}$, then we can, without loss of generality, assume that K is a submodule of F. Proposition 5.3.11 now shows that there is an R-linear map $f : F \to \operatorname{Ker}\alpha_{n-1}$ such that $f(x_i) = x_i$ for $i = 1, 2, \ldots, n$, where $x_1, x_2, \ldots, x_n$ is a set of generators of $\operatorname{Ker}\alpha_{n-1}$. Thus, f is a splitting map for the canonical injection $\operatorname{Ker}\alpha_{n-1} \to F$ and so the sequence $0 \to K \to F \to \operatorname{Ker}\alpha_{n-1} \to 0$ splits. Hence, $\operatorname{Ker}\alpha_{n-1}$ is projective, so pd-(R/A) is at most n. Consequently, pd-$(R/A) \leq$ fd-(R/A) and since it is always the case that fd-$(R/A) \leq$ pd-(R/A), we have fd-$(R/A) = $ pd-(R/A). $\qquad\square$

Clearly, if we switch sides in the proof of Proposition 12.2.7, we have that if R is left noetherian, then gl.fd-$R = $ l.gl.hd-R. Consequently, we have the following proposition.

Proposition 12.2.8. *If R is a noetherian ring, then* gl.fd-$R = $ l.gl.hd-$R = $ r.gl.hd-R.

Problem Set 12.2

1. If

$$\mathbf{F}: \cdots \to F_n \xrightarrow{\alpha_n} F_{n-1} \to \cdots \to F_0 \xrightarrow{\alpha_0} M \to 0 \quad \text{and}$$

$$\mathbf{G}: \cdots \to G_n \xrightarrow{\beta_n} G_{n-1} \to \cdots \to G_0 \xrightarrow{\beta_0} M \to 0$$

   are flat resolutions of M, prove that $\operatorname{Ker}\alpha_n$ is flat if and only if $\operatorname{Ker}\beta_n$ is flat. [Hint: Consider the functor $(-)^+ : \mathbf{Mod}_R \to {}_R\mathbf{Mod}$ and recall that M is a flat R-module if and only if M^+ is an injective left R-module.]

2. (a) Show that an analogue of Schanuel's lemma does not hold for flat resolutions, that is, find two flat resolutions

$$0 \to K_1 \to F_1 \to M \to 0 \quad \text{and} \quad 0 \to K_2 \to F_2 \to M \to 0$$

of M such that $K_1 \oplus F_2 \ncong K_2 \oplus F_1$. [Hint: Consider

$$0 \to \mathbb{Z} \to \mathbb{Q} \to \mathbb{Q}/\mathbb{Z} \to 0 \quad \text{and} \quad 0 \to K \to F \to \mathbb{Q}/\mathbb{Z} \to 0,$$

where F is a free $\mathbb{Z}$-module.]

(b) Prove (2) $\Rightarrow$ (3) of Proposition 12.2.4.

3. Prove that the following are equivalent for an R-module M.

   (a) M is a flat R-module.

   (b) $\mathrm{Tor}_n^R(M, R/A) = 0$ for every left ideal A of R and all $n \geq 1$.

   (c) $\mathrm{Tor}_1^R(M, R/A) = 0$ for every left ideal A of R.

   (d) $\mathrm{Tor}_n^R(M, R/A) = 0$ for every finitely generated left ideal A of R and all $n \geq 1$.

   (e) $\mathrm{Tor}_1^R(M, R/A) = 0$ for every finitely generated left ideal A of R.

   [Hint: Proposition 5.3.7.]

4. Complete the proof of Proposition 12.2.6. [Hint: Exercise 3 and Proposition 12.1.11.]

5. Show that for any ring R,

$$\mathrm{gl.fd}\text{-}R = \sup\{\mathrm{fd}\text{-}(R/A) \mid A \text{ a finitely generated left ideal of } R\}$$
$$= \sup\{\mathrm{fd}\text{-}(R/A) \mid A \text{ a finitely generated right ideal of } R\}.$$

   [Hint: Exercise 3 and Proposition 12.2.6.]

6. If $\mathrm{gl.fd}\text{-}R > 0$, prove that

$$\mathrm{gl.fd}\text{-}R = 1 + \sup\{\mathrm{fd}\text{-}A \mid A \text{ a finitely generated left ideal of } R\}$$
$$= 1 + \sup\{\mathrm{fd}\text{-}A \mid A \text{ a finitely generated right ideal of } R\}.$$

   [Hint: Since $\mathrm{r.gl.hd}\text{-}R \geq 1$, Exercise 6 indicates that there is a finitely generated right ideal A of R such that $\mathrm{fd}\text{-}(R/A) \neq 0$. Consider $0 \to A \to R \to R/A \to 0$ and show that $1 + \mathrm{fd}\text{-}A = \mathrm{fd}\text{-}(R/A)$.]

7. Show that $\mathrm{gl.fd}\text{-}R \leq 1$ if and only if every finitely generated right (left) ideal of R is flat.

8. Show that each of the following hold for a family of R-modules $\{M_\alpha\}_\Delta$.

   (a) $\mathrm{fd}\text{-}\bigoplus_\Delta M_\alpha = \sup\{\mathrm{fd}\text{-}M_\alpha \mid \alpha \in \Delta\}$. [Hint: For each $\alpha \in \Delta$, construct an exact sequence

$$0 \to K_{\alpha,n} \to P_{\alpha,n-1} \to \cdots \to P_{\alpha,1} \to P_{\alpha,0} \to M_\alpha \to 0$$

where $P_{\alpha,k}$ is projective for $k = 0, 1, 2, \ldots, n-1$. Consider the exact sequence

$$0 \to \bigoplus_{\Delta} K_{\alpha,n} \to \bigoplus_{\Delta} P_{\alpha,n-1} \to \cdots \to \bigoplus_{\Delta} P_{\alpha,1} \to \bigoplus_{\Delta} P_{\alpha,0} \to \bigoplus_{\Delta} M_{\alpha} \to 0$$

constructed in the obvious way.]

(b) If R is right noetherian, then fd-$\prod_{\Delta} M_{\alpha} = \sup\{\text{fd-}M_{\alpha} \mid \alpha \in \Delta\}$.

9. In general, l.gl.hd-$R \neq$ r.gl.hd-R. If R is a perfect ring, decide whether or not l.gl.hd-$R =$ r.gl.hd-R.

12.3 Dimension of Polynomial Rings

We can now investigate the relation between the right global dimension of R and the right global dimension of the polynomial ring $R[X_1, X_2, \ldots, X_n]$ in n commuting indeterminates. In particular, we will show that

$$\text{r.gl.hd-}R[X_1, X_2, \ldots, X_n] = n + \text{r.gl.hd-}R.$$

The technique used will be to show that r.gl.hd-$R[X] = 1+$r.gl.hd-R with the general case following by induction. The following lemma will be useful in establishing this result.

Lemma 12.3.1.

(1) *For any family* $\{M_{\alpha}\}_{\Delta}$ *of R-modules,* pd-$(\bigoplus_{\Delta} M_{\alpha}) = \sup\{\text{pd-}M_{\alpha} \mid \alpha \in \Delta\}$.

(2) *Let* $f : R \to S$ *be a ring homomorphism and suppose that M is an S-module, If M is made into an R-module by pullback along f, then* pd-$M_R \leq$ pd-$M_S +$ pd-S_R.

Proof. (1) For each $\alpha \in \Delta$, construct an exact sequence

$$0 \to K_{\alpha,n} \to P_{\alpha,n-1} \to \cdots \to P_{\alpha,1} \to P_{\alpha,0} \to M_{\alpha} \to 0,$$

where each $P_{\alpha,i}$ is projective. This gives rise to an exact sequence

$$0 \to \bigoplus_{\Delta} K_{\alpha,n} \to \bigoplus_{\Delta} P_{\alpha,n-1} \to \cdots \to \bigoplus_{\Delta} P_{\alpha,1} \to \bigoplus_{\Delta} P_{\alpha,0} \to \bigoplus_{\Delta} M_{\alpha} \to 0$$

and $\bigoplus_{\Delta} P_{\alpha,i}$ is projective if and only if $P_{\alpha,i}$ is projective for each $\alpha \in \Delta$. Hence, pd-$(\bigoplus_{\Delta} M_{\alpha}) = n$ if and only if $\bigoplus_{\Delta} K_{\alpha,n}$ is projective if and only if $K_{\alpha,n}$ is projective for each $\alpha \in \Delta$. But if $K_{\alpha,n}$ is projective for each $\alpha \in \Delta$, then pd-$M_{\alpha} \leq n$ for each $\alpha \in \Delta$, so $\sup\{\text{pd-}M_{\alpha} \mid \alpha \in \Delta\} \leq n$. Thus, $\sup\{\text{pd-}M_{\alpha} \mid \alpha \in \Delta\} \leq$ pd-$(\bigoplus_{\Delta} M_{\alpha})$. A similar argument gives pd-$(\bigoplus_{\Delta} M_{\alpha}) \leq \sup\{\text{pd-}M_{\alpha} \mid \alpha \in \Delta\}$, so pd-$(\bigoplus_{\Delta} M_{\alpha}) = \sup\{\text{pd-}M_{\alpha} \mid \alpha \in \Delta\}$.

(2) If pd-$M_S = \infty$ or pd-$S_R = \infty$, then there is nothing to prove, so suppose that both are finite and let pd-$S_R = m$. We proceed by induction on pd-M_S. If pd-$M_S = 0$, then M_S is projective and a direct summand of a free S-module $S^{(\Delta)}$. It follows that M_R is also a direct summand of $S^{(\Delta)}$ when $S^{(\Delta)}$ is viewed as an R-module. If $S_R^{(\Delta)} = M_R \oplus N_R$, then using (1), we have

$$\text{pd-}M_R \leq \text{pd-}(M_R \oplus N_R) = \text{pd-}S_R^{(\Delta)} = \text{pd-}S_R = \text{pd-}M_S + \text{pd-}S_R.$$

Finally, suppose that pd-$M_R \leq$ pd-$M_S +$ pd-S_R holds for all S-modules such that pd-$M_S \leq n, n \geq 0$. If M is an S-module such that pd-$M_S = n + 1$, then we can construct an exact sequence $0 \to K_S \xrightarrow{f} F_S \to M_S \to 0$, where F is a free S-module. If $\cdots \to P_n \xrightarrow{\alpha_n} P_{n-1} \to \cdots \to P_0 \xrightarrow{\alpha_0} K_S \to 0$ is an S-projective resolution of K_S, then $\cdots \to P_n \xrightarrow{\alpha_n} P_{n-1} \to \cdots \to P_0 \xrightarrow{f\alpha_0} F_S \to M_S \to 0$ is an S-projective resolution of M_S. Since pd-$M_S = n + 1$, $\text{Ker}\,\alpha_n$ is a projective S-module and so pd-$K_S = n$. Thus, by our induction hypothesis, pd-$K_R \leq$ pd-$K_S +$ pd-$S_R = n + m$. Now $0 \to K_R \to F_R \to M_R \to 0$ is exact and pd-$F_R =$ pd-$S_R^{(\Delta)} =$ pd-$S_R = m$, so pd-$F_R <$ pd-S_R. Thus, we have an exact sequence $0 \to K_R \to F_R \to M_R \to 0$ such that pd-$F_R <$ pd-S_R, so in view of Proposition 12.1.6 we have pd-$M_R = 1 +$ pd-K_R. Hence, pd-$M_R \leq 1 + n + m =$ pd-$M_S +$ pd-S_R and we are done.   □

As in [40], we call $a \in R$ *normal* if $aR = Ra$. Of course if a is in the center of R, then a is normal. Unless stated otherwise, a will henceforth be a fixed normal element of R that is neither a zero divisor nor a unit in R. The assumption that a is normal in R means that aR is an ideal of R and that Ma is a submodule of M for any R-module M. We also have that aR is a projective R-module, since assuming that a is not a zero divisor gives $aR \cong R$.

The ring R/aR will now be denoted by S

and since a is a nonunit, we have $S \neq 0$. Finally, if M is an S-module, then we can view M as an R-module by pullback along the canonical map $\eta : R \to S$.

Lemma 12.3.2. *The following hold for the rings R and S.*

(1) *If M is a nonzero S-module, then M is not a projective R-module.*

(2) *If F is a nonzero free S-module, then pd-$F_R = 1$.*

Proof. (1) Let M be a nonzero S-module. If M is projective as an R-module, then there is an R-linear embedding $f : M \to R^{(\Delta)}$ for some set Δ. Suppose that $x \in M$, $x \neq 0$, and let $f(x) = (a_\alpha)$. Since $Ma = 0$, we have $0 = f(xa) = f(x)a = (a_\alpha a)$, so $a_\alpha a = 0$ for all $\alpha \in \Delta$. Hence, $a_\alpha = 0$ for all $\alpha \in \Delta$ since a is not a zero divisor in R. Thus, $f(x) = 0$, so $x = 0$, a contradiction. Therefore, M cannot be a projective R-module.

(2) Since R/aR is a nonzero S-module, (1) shows that R/aR is not a projective R-module. Now $aR \cong R$ and $0 \to aR \to R \to S \to 0$ is a projective resolution of S_R, so it follows that pd-$S_R = 1$. Thus, we see from Lemma 12.3.1 that if F is any free S-module, then pd-$F_R \leq 1$. But from (1) we have pd-$F_R = 1$. $\square$

To prove the next proposition we make use of the fact that if P is a projective R-module, then P/Pa is a projective S-module. Indeed, if P is a projective R-module, then there is a free R-module $F \cong R^{(\Delta)}$ and an R-module Q such that $F = P \oplus Q$. Since $Fa \cong (aR)^{(\Delta)}$, it follows that $F/Fa \cong (R/aR)^{(\Delta)} = S^{(\Delta)}$, so F/Fa is a free S-module. Now $Fa = Pa \oplus Qa$, so we see that $F/Fa = P/Pa \oplus Q/Qa$. Hence, P/Pa is a projective S-module, since direct summands of projective modules are projective.

Proposition 12.3.3 (Change of Rings). *If M is a nonzero right S-module such that* pd-$M_S = n$, *then* pd-$M_R = 1 + n$.

Proof. The proof is by induction on n. If $n = 0$, then (2) of Lemma 12.3.1 indicates that pd-$M_R \leq$ pd-S_R. Since S_S is a free S-module, (2) of Lemma 12.3.2 shows that pd-$S_R = 1$. Hence, pd-$M_R \leq 1$. Moreover, by using (1) of the same lemma, we see that pd-$M_R \neq 0$, so pd-$M_R = 1$.

Next, make the induction hypothesis that if M is any nonzero S-module, such that pd-$M_S = k$, then pd-$M_R = k + 1$ for all $0 \leq k < n$. If pd-$M_S = n$, then M_S has an S-projective resolution

$$0 \to P_n \to \cdots \to P_1 \to F_S \overset{\alpha_0}{\to} M_S \to 0$$

of minimal length n, where F_S is a free S-module. If $K_S = \operatorname{Ker} \alpha_0$, then pd-$K_S = n - 1$ since

$$0 \to P_n \to \cdots \to P_1 \to K_S \to 0$$

is an S-projective resolution of K_S which must be of minimal length. Thus, the induction hypothesis gives pd-$K_R = n$. If $n > 1$, we see that pd-$K_R > 1 =$ pd-F_R, where $1 =$ pd-F_R is (2) of Lemma 12.3.2. Hence, (3) of Proposition 12.1.6 gives pd-$M_R = n + 1$. Thus, it only remains to treat the case $n = 1$, so suppose that pd-$M_S = 1$. Then M_S has a projective resolution of the form

$$0 \to K_S \to F_S \to M_S \to 0$$

in $\mathbf{Mod}_S$, where F is a free S-module. Since pd-$K_S = 0$, the induction hypothesis gives pd-$K_R = 1$ and, as before, pd-$F_R = 1$. Hence, pd-$K_R =$ pd-F_R, so we see from (2) of Proposition 12.1.6 that pd-$M_R \leq 2$. We claim that equality holds. If not, then pd-$M_R \leq 1$ and we have already seen in (1) of Lemma 12.3.2 that pd-$M_R \neq 0$, so pd-$M_R = 1$. Hence, M_R has a projective resolution of the form

$$0 \to K'_R \to F'_R \to M_R \to 0,$$

where F'_R a free R-module and K'_R is a projective R-module. Since $Ma = 0$, $F'a \subseteq K'$, so we have an exact sequence

$$0 \to K'/F'a \to F'/F'a \to M \to 0$$

in $\mathbf{Mod}_S$. The observation given in the paragraph immediately preceding this proposition shows that $F'/F'a$ is a free S-module and the assumption that pd-$M_S = 1$ means that that $K'/F'a$ is a projective S-module. Therefore, the exact sequence

$$0 \to F'a/K'a \to K'/K'a \to K'/F'a \to 0$$

splits in $\mathbf{Mod}_S$. Since a is not a zero divisor in R, $F'/K' \cong F'a/K'a$ in $\mathbf{Mod}_R$, so the composition map $M \to F'/K' \to (F'a)/(K'a)$ produces an R-linear isomorphism that is also S-linear. Hence, $K'/K'a \cong M \oplus K'/F'a$. Using the observation given in the paragraph preceding the proposition again, we see that $K'/K'a$ is a projective S-module which indicates that M is a projective S-module. But this cannot be the case since we are assuming that pd-$M_S = 1$. Therefore, pd-$M_R \neq 1$, so pd-$M_R = 2$ and we are finished. $\square$

Corollary 12.3.4. *If* r.gl.hd-$S = n$, *then* r.gl.hd-$R \geq 1 + n$.

Proof. If r.gl.hd-$S = n$, then pd-$M_S \leq n$ for each right S-module M. Hence, by the proposition, pd-$M_R \leq 1 + n$ for each S-module M and so

$$\begin{aligned}
\text{r.gl.hd-}R &= \sup\{\text{pd-}M_R \mid M \text{ an } R\text{-module}\} \\
&\geq \sup\{\text{pd-}M_R \mid M \text{ an } S\text{-module}\} \\
&= 1 + n.
\end{aligned}$$
                                                                         $\square$

Corollary 12.3.5. *If* r.gl.hd-$R = n$, *then*

$$r.gl.hd\text{-}R[X] \geq 1 + n.$$

Proof. Note that X is a nonzero normal element of $R[X]$ that is clearly neither a unit nor a zero divisor in $R[X]$. Moreover, $R[X]/XR[X] \cong R$, so the result follows from the preceding corollary. $\square$

The assumption that pd-$M_S < \infty$ in Proposition 12.3.3 cannot be eliminated. For example, let $R = \mathbb{Z}$ and $a = 4$, so that $S = R/aR = \mathbb{Z}_4$. Now

$$\cdots \to \mathbb{Z}_4 \xrightarrow{\alpha_n} \mathbb{Z}_4 \xrightarrow{\alpha_{n-1}} \mathbb{Z}_4 \to \cdots \to \mathbb{Z}_4 \xrightarrow{\alpha_0} \mathbb{Z}_2 \to 0$$

is a $\mathbb{Z}_4$-projective resolution of $\mathbb{Z}_2$, where $\alpha_n([a]) = [2a]$ for $n \geq 1$ and $\alpha_0([a]) = [a]$. Since $\operatorname{Ker} \alpha_{n-1} = \{[0], [2]\}$ is not a projective $\mathbb{Z}_4$-module for $n \geq 1$, it follows from Proposition 12.1.4 that pd-$(\mathbb{Z}_2)_{\mathbb{Z}_4} = \infty$. As a $\mathbb{Z}$-module, $\mathbb{Z}_2$ has

$$0 \to \mathbb{Z} \xrightarrow{\alpha_1} \mathbb{Z} \xrightarrow{\alpha_0} \mathbb{Z}_2 \to 0$$

as a $\mathbb{Z}$-projective resolution of minimal length, where $\alpha_1(a) = 2a$ and $\alpha_0(a) = [a]$ for each $a \in \mathbb{Z}$. Thus, pd-$(\mathbb{Z}_2)_{\mathbb{Z}} = 1$ and so we have pd-$(\mathbb{Z}_2)_{\mathbb{Z}} \neq 1 +$ pd-$(\mathbb{Z}_2)_{\mathbb{Z}_4}$.

Definition 12.3.6. Suppose that X is an indeterminate that commutes with elements of M and with elements of R. If M is an R-module, let $M[X]$ denote the set of formal polynomials of the form $p_M(X) = \sum_{\mathbb{N}_0} x_i X^i$, where $x_i = 0$ for almost all $i \in \mathbb{N}_0$. Since $x_i = 0$ for almost all $i \in \mathbb{N}_0$, $p_M(X) = \sum_{\mathbb{N}_0} x_i X^i$ may also be written as $p_M(X) = \sum_{i=0}^n x_i X^i$ if additional clarity is required. If these polynomials are added in the obvious way, then $M[X]$ is an additive abelian group. Moreover, if $p_R(X) = \sum_{\mathbb{N}_0} a_i X^i$ is a polynomial in $R[X]$ and if we set $p_M(X) p_R(X) = \sum_{\mathbb{N}_0} x_i^* X^i$, where $x_i^* = x_i a_0 + x_{i-1} a_1 + \cdots + x_0 a_i$ for each $i \in \mathbb{N}_0$, then $M[X]$ becomes an $R[X]$-module. Note also that if $f : M \to N$ is an R-linear mapping and $p_M(X) = \sum_{\mathbb{N}_0} x_i X^i \in M[X]$, then $f(p_M(X)) = \sum_{\mathbb{N}_0} f(x_i) X^i$ is a polynomial in $N[X]$. Thus, if $\mathscr{F} : \mathbf{Mod}_R \to \mathbf{Mod}_{R[X]}$ is such that $\mathscr{F}(M) = M[X]$ and $\mathscr{F}(f) : M[X] \to N[X]$, where $\mathscr{F}(f)(p_M(X)) = f(p_M(X))$ for each $p_M(X) \in M[X]$, then $\mathscr{F}$ is a functor called the *polynomial functor* from $\mathbf{Mod}_R$ to $\mathbf{Mod}_{R[X]}$.

Lemma 12.3.7.

(1) *The polynomial functor* $\mathscr{F} : \mathbf{Mod}_R \to \mathbf{Mod}_{R[X]}$ *is exact.*

(2) $M[X] \cong M \otimes_R R[X]$ *as $R[X]$-modules, for any R-module M.*

(3) *If F is a free R-module, then $F[X]$ is a free $R[X]$-module.*

(4) *Every free $R[X]$-module is a free R-module.*

(5) *An R-module P is R-projective if and only if $P[X]$ is $R[X]$-projective.*

Proof. (1) Left as an exercise.

(2) The map $\varphi : M[X] \to M \otimes_R R[X]$ defined by $\varphi(\sum_{\mathbb{N}_0} x_i X^i) = \sum_{\mathbb{N}_0} (x_i \otimes X^i)$ is an $R[X]$-linear mapping. Since the map $M \times R[X] \to M[X]$ given by $(x, X^i) \mapsto x X^i$ is R-balanced, the definition of a tensor product produces a group homomorphism $\psi : M \otimes_R R[X] \to M[X]$ such that $\psi(\sum_{\mathbb{N}_0} (x_i \otimes X^i)) = \sum_{\mathbb{N}_0} x_i X^i$. A routine calculation shows that φ and ψ are $R[X]$-linear, that $\psi\varphi = id_{M[X]}$ and that $\varphi\psi = id_{M \otimes_R R[X]}$. Hence, φ is an $R[X]$-isomorphism.

(3) Suppose that F is a free R-module with basis $\{x_\alpha\}_\Delta$, then each x_α can be viewed as a constant polynomial $\bar{x}_\alpha$ in $F[X]$. It follows that $\{\bar{x}_\alpha\}_\Delta$ is a basis for $F[X]$ as an $R[X]$-module and so $F[X]$ is a free $R[X]$-module.

(4) Since R is (isomorphic to) a subring of $R[X]$, $M[X]$ can be viewed as an R-module. Now suppose that F is a free $R[X]$-module. Then $F_{R[X]} \cong (R[X])^{(\Delta)}$ for some set Δ. But $R[X]$ is a free R-module with basis $\{X^i\}_{\mathbb{N}_0}$, so $F_R \cong (R^{(\mathbb{N}_0)})^{(\Delta)} = R^{(\mathbb{N}_0 \times \Delta)}$. Hence, F is also a free R-module.

(5) If P is a projective R-module, then there is a free R-module F such that $F = P \oplus Q$ for some R-module Q. Since the polynomial functor is exact, we see that $F[X] = P[X] \oplus Q[X]$ and (3) indicates that $F[X]$ is a free $R[X]$-module. Hence, $P[X]$ is a projective $R[X]$-module.

Conversely, suppose that $P[X]$ is a projective $R[X]$-module. Then there is a free $R[X]$-module F such that $F = P[X] \oplus Q$ for some $R[X]$-module Q. This gives $F_R = P[X]_R \oplus Q_R$ and (4) shows that F_R is a free R-module. Hence, $P[X]$ is a projective R-module. Now the map $P[X] \to P^{(\mathbb{N}_0)}$ defined by $\sum_{\mathbb{N}_0} x_i X^i \mapsto (x_i)$ is an isomorphism and so $P^{(\mathbb{N}_0)}$ is a projective R-module. But a direct summand of a projective R-module is projective, so P is R-projective.   □

Lemma 12.3.8. *For any R-module M, pd-M_R = pd-$M[X]_{R[X]}$.*

Proof. Suppose first that pd-$M_R = n$. Then M has an R-projective resolution of the form

$$0 \to P_n \to P_{n-1} \to \cdots \to P_0 \to M \to 0.$$

Since the polynomial functor is exact, the sequence

$$0 \to P_n[X] \to P_{n-1}[X] \to \cdots \to P_0[X] \to M[X] \to 0$$

is exact in $\mathbf{Mod}_{R[X]}$. But each P_k is a projective R-module and (4) of the previous lemma shows that $P_k[X]$ is a projective $R[X]$-module for $k = 0, 1, 2, \ldots, n$. But this may not be the shortest $R[X]$-projective resolution of $M[X]$ and so pd-$M[X]_{R[X]} \le$ pd-M_R.

Conversely, if pd-$M[X]_{R[X]} = n$, then there is an $R[X]$-projective resolution of $M[X]$ of the form

$$0 \to Q_n \to Q_{n-1} \to \cdots \to Q_0 \to M[X] \to 0.$$

Moreover, since $R \subseteq R[X]$, each Q_k can be viewed as an R-module. As indicated earlier, $R[X]$ is a free left R-module, so $R[X] \cong \bigoplus_{i=0}^{\infty} R_i$, where $R_i = R$ for $i = 0, 1, 2, \ldots$. But tensor products preserve isomorphisms and tensor products commute with direct sums, so $Q_j \otimes_R R[X] \cong Q_j \otimes_R (\bigoplus_{i=0}^{\infty} R_i) \cong \bigoplus_{i=0}^{\infty} (Q_j \otimes_R R_i) \cong \bigoplus_{i=0}^{\infty} Q_{j,i}$ with $Q_{j,i} = Q_j$ for each j. But each Q_j is a projective $R[X]$-module, so $\bigoplus_{i=0}^{\infty} Q_{j,i}$ and hence $Q_j \otimes_R R[X]$ is a projective $R[X]$-module. Part (5) of the previous lemma gives $Q_j[X] \cong Q_j \otimes_R R[X]$, so $Q_j[X]$ is a projective $R[X]$-module for each j. Hence, (2) of the same lemma indicates that each Q_j is a projective

R-module. Now $M[X] \cong M^{(\mathbb{N}_0)}$, so we have an exact sequence

$$0 \to Q_n \to Q_{n-1} \to \cdots \to Q_0 \to M^{(\mathbb{N}_0)} \to 0$$

which is an R-projective resolution of $M^{(\mathbb{N}_0)}$. Since this may not be the shortest R-projective resolution of $M^{(\mathbb{N}_0)}$, pd-$(M^{(\mathbb{N}_0)})_R \le$ pd-$M[X]_{R[X]}$. But (1) of Lemma 12.3.1 gives pd-$(M^{(\mathbb{N}_0)})_R =$ pd-M_R, so pd-$M_R \le$ pd-$M[X]_{R[X]}$. Hence, pd-$M_R =$ pd-$M[X]_{R[X]}$.

If either dimension is infinite, then it clearly must be the case that the other dimension is infinite as well. Indeed, if pd-$M_R = \infty$ and pd-$M[X]_{R[X]}$ is finite, then, as above, we can show that pd-$M_R \le$ pd-$M[X]_{R[X]}$, a contradiction. Hence if pd-$M_R = \infty$, then pd-$M[X]_{R[X]} = \infty$. Similarly, if pd-$M[X]_{R[X]} = \infty$, then pd-$M_R = \infty$ and this completes the proof. $\qquad\square$

The following proposition is a generalization of a well-known result of Hilbert. Hilbert was the first to prove the proposition when the ring is a field.

Lemma 12.3.9. *If M is an $R[X]$-module, then there is an exact sequence $0 \to M[X] \to M[X] \to M \to 0$ in $\mathbf{Mod}_{R[X]}$.*

Proof. Due to the $R[X]$-isomorphism $M[X] \cong M \otimes_R R[X]$ of (2) in Lemma 12.3.7, it suffices to show that if M is an $R[X]$-module, then there is an exact sequence

$$0 \to M \otimes_R R[X] \to M \otimes_R R[X] \to M \to 0$$

in $\mathbf{Mod}_{R[X]}$.

If M is an $R[X]$-module and $\rho : M \times R[X] \to M$ is defined by $\rho(x, X^i) = xX^i$, then ρ is R-balanced and it follows that we have an $R[X]$-epimorphism $f : M \otimes_R R[X] \to M$ given by $f(\sum_{i=0}^{n}(x_i \otimes X^i)) = \sum_{i=1}^{n} x_i X^i$. Next, consider the $R[X]$-homomorphism $g : M \otimes_R R[X] \to M \otimes_R R[X]$ defined by

$$g\left(\sum_{i=0}^{n}(x_i \otimes X^i)\right) = x_0 X \otimes 1 + \sum_{i=1}^{n}(x_i X - x_{i-1}) \otimes X^i - x_n \otimes X^{n+1}$$

and note that

$$f\left(x_0 X \otimes 1 + \sum_{i=1}^{n}(x_i X - x_{i-1}) \otimes X^i - x_n \otimes X^{n+1}\right)$$

$$= x_0 X + \sum_{i=1}^{n}(x_i X - x_{i-1})X^i - x_n X^{n+1}$$

$$= 0.$$

Hence Im $g \subseteq$ Ker f.

We claim that $\operatorname{Ker} f \subseteq \operatorname{Im} g$. If $\sum_{i=0}^{n}(y_i \otimes X^i) \in \operatorname{Ker} f$, then we need to find $\sum_{i=0}^{n}(x_i \otimes X^i) \in M \otimes_R R[X]$ such that $g(\sum_{i=0}^{n}(x_i \otimes X^i)) = \sum_{i=0}^{n}(y_i \otimes X^i)$. If $\sum_{i=0}^{n}(x_i \otimes X^i)$ is such an element of $M \otimes_R R[X]$, then we must have

$$x_0 X \otimes 1 + \sum_{i=1}^{n}(x_i X - x_{i-1}) \otimes X^i - x_n \otimes X^{n+1} = \sum_{i=0}^{n}(y_i \otimes X^i).$$

If we set

$$y_0 = x_0 X, \quad y_1 = x_1 X - x_0, \quad y_2 = x_2 X - x_1, \ldots,$$
$$y_n = x_n X - x_{n-1} \quad \text{and} \quad y_n = -x_n,$$

then these equations can be solved recursively to find the x_i for which $\sum_{i=0}^{n}(x_i \otimes X^i) \in M \otimes_R R[X]$ is such that $g(\sum_{i=0}^{n-1}(x_i \otimes X^i)) = \sum_{i=0}^{n}(y_i \otimes X^i)$. Thus, $\operatorname{Ker} f \subseteq \operatorname{Im} g$ and so we have $\operatorname{Im} g = \operatorname{Ker} f$. Therefore, the sequence

$$M \otimes_R R[X] \xrightarrow{g} M \otimes_R R[X] \xrightarrow{f} M \to 0$$

is exact in $\mathbf{Mod}_{R[X]}$.

Finally, we claim that g is injective. If $\sum_{i=0}^{n}(x_i \otimes X^i) \in \operatorname{Ker} g$, then

$$g\left(\sum_{i=0}^{n}(x_i \otimes X^i)\right)$$
$$= x_0 X \otimes 1 + \sum_{i=1}^{n}(x_i X - x_{i-1}) \otimes X^i - x_n \otimes X^{n+1}$$
$$= 0, \quad \text{so}$$
$$f\left(x_0 X \otimes 1 + \sum_{i=1}^{n}(x_i X - x_{i-1}) \otimes X^i - x_n \otimes X^{n+1}\right)$$
$$= x_0 X + \sum_{i=1}^{n}(x_i X - x_{i-1})X^i - x_n X^{n+1}$$
$$= 0.$$

But this gives

$$x_n = x_n X - x_{n-1} = x_{n-1} X - x_{n-2} = \cdots = x_1 X - x_0 = 0$$

and so it follows that $x_i = 0$ for $i = 0, 1, 2, \ldots, n$. Therefore, g is injective, so the sequence

$$0 \to M \otimes_R R[X] \xrightarrow{g} M \otimes_R R[X] \xrightarrow{f} M \to 0$$

is exact in $\mathbf{Mod}_{R[X]}$.   □

Proposition 12.3.10. *If R is any ring, then*

$$\text{r.gl.hd-}R[X] = 1 + \text{r.gl.hd-}R.$$

Proof. Due to Lemma 12.3.8 we can assume that r.gl.hd-R and r.gl.hd-$R[X]$ are finite. Also note that Corollary 12.3.5 indicates that r.gl.hd-$R[X] \geq 1 + $r.gl.hd-$R$, so we are only required to show that r.gl.hd-$R[X] \leq 1 + $r.gl.hd-$R$.

If M is an $R[X]$-module, then Lemma 12.3.9 indicates that there is an exact sequence

$$0 \to M[X] \to M[X] \to M \to 0$$

in **Mod**$_{R[X]}$. Furthermore, (2) of Proposition 12.1.6 shows that pd-$M_{R[X]} \leq 1 + $pd-$M[x]_{R[X]}$. But Lemma 12.3.8 indicates that pd-M_R = pd-$M_{R[X]}$ and so pd-$M_{R[X]} \leq 1 + $pd-$M_R \leq 1 + $r.gl.hd-$R$. Therefore, r.gl.hd-$R[X] \leq 1 + $r.gl.hd-$R$. □

Corollary 12.3.11. *If R is any ring, then*

$$\text{r.gl.hd-}R[X_1, X_2, \ldots, X_n] = n + \text{r.gl.hd-}R$$

for any n ≥ 1

Corollary 12.3.12 (Hilbert). *If K is a field, then*

$$\text{r.gl.hd-}K[X_1, X_2, \ldots, X_n] = n.$$

Problem Set 12.3

1. Show that the polynomial functor $\mathcal{F} : \mathbf{Mod}_R \to \mathbf{Mod}_{R[X]}$ is exact. [Hint: If $0 \to L \to M \to N \to 0$ is exact in **Mod**$_R$, it suffices to show that $0 \to \mathcal{F}(L) \to \mathcal{F}(M) \to \mathcal{F}(N) \to 0$ is exact in **Mod**$_{R[X]}$.]

2. If F is a free R-module with basis $\{x_\alpha\}_\Delta$, then each x_α can be viewed as a constant polynomial $\bar{x}_\alpha$ in $F[X]$. Show that $\{\bar{x}_\alpha\}_\Delta$ is a basis for $F[X]$ as an $R[X]$-module.

3. Prove Corollaries 12.3.11 and 12.3.12.

12.4 Dimension of Matrix Rings

We now turn our attention to how the right global dimension of a ring R relates to the right global dimension of the $n \times n$ matrix ring $\mathbb{M}_n(R)$.

Let $f : R \to \mathbb{M}_n(R)$ be defined by $f(a) = I_a$, where I_a is the $n \times n$ matrix with a on the main diagonal and zeroes elsewhere. Since $I_a + I_b = I_{a+b}$, $I_a I_b = I_{ab}$ and I_1 is the identity matrix of $\mathbb{M}_n(R)$, f is an embedding, so we can consider R

to be a subring of $\mathbb{M}_n(R)$. Thus, any $\mathbb{M}_n(R)$-module is also an R-module and any $\mathbb{M}_n(R)$-module homomorphism is an R-linear mapping. Observe also that if M is an R-module, then $M^{(n)}$ is an $\mathbb{M}_n(R)$-module with the $\mathbb{M}_n(R)$-action on $M^{(n)}$ given by

$$(x_1, x_2, \ldots, x_n)(a_{ij}) = \left(\sum_{i=1}^{n} x_i a_{i1}, \sum_{i=1}^{n} x_i a_{i2}, \ldots, \sum_{i=1}^{n} x_i a_{in} \right).$$

Finally, we see that as R-modules $\mathbb{M}_n(R) \cong R^{(n^2)}$ via the map

$$(a_{ij}) \mapsto (r_1, r_2, \ldots, r_n),$$

where $r_i = (a_{i1}, a_{i2}, a_{i3}, \ldots, a_{in})$ for $i = 1, 2, 3, \ldots, n$.

With these observations in mind, we have the following.

Lemma 12.4.1. *The following hold for any R-module M.*

(1) *If*

$$0 \to L \xrightarrow{f} M \xrightarrow{g} N \to 0$$

is exact in $\mathbf{Mod}_R$, then

$$0 \to L^{(n)} \xrightarrow{\oplus f} M^{(n)} \xrightarrow{\oplus g} N^{(n)} \to 0$$

is exact in $\mathbf{Mod}_{\mathbb{M}_n(R)}$, where the map $\oplus f : M^{(n)} \to N^{(n)}$ is given by $\oplus f((x_1, x_2, \ldots, x_n)) = (f(x_1), f(x_2), \ldots, f(x_n))$. The map $\oplus g$ is defined similarly and $(\oplus g)(\oplus f) = \oplus gf$.

(2) *M is a projective R-module if and only if $M^{(n)}$ is a projective $\mathbb{M}_n(R)$-module.*

(3) $\text{pd-}M^{(n)}_{\mathbb{M}_n(R)} = \text{pd-}M_R$.

Proof. (1) Straightforward.

(2) First, note that if F is a free R-module, then there is a set Δ such that $F \cong R^{(\Delta)}$. But then $F^{(n^2)} \cong (R^{(\Delta)})^{(n^2)} \cong (R^{(n^2)})^{(\Delta)} \cong \mathbb{M}_n(R)^{(\Delta)}$, so $F^{(n^2)}$ is a free $\mathbb{M}_n(R)$-module. Now suppose that M is a projective R-module. Then there is a free R-module F such that the sequence $0 \to K \to F \to M \to 0$ is split exact in $\mathbf{Mod}_R$. It follows that $0 \to K^{(n^2)} \to F^{(n^2)} \to M^{(n^2)} \to 0$ is split exact in $\mathbf{Mod}_{\mathbb{M}_n(R)}$. But $F^{(n^2)}$ is a free $\mathbb{M}_n(R)$-module, so $M^{(n^2)}$ is a projective $\mathbb{M}_n(R)$-module. Since $M^{(n)}$ is isomorphic to a direct summand of $M^{(n^2)}$, it follows that $M^{(n)}$ is a projective $\mathbb{M}_n(R)$-module.

Conversely, suppose that $M^{(n)}$ is a projective $\mathbb{M}_n(R)$-module. To show that M is a projective R-module, it suffices to show that any row exact diagram

$$
\begin{array}{ccc}
 & & M \\
 & \overset{h}{\nearrow} & \downarrow f \\
L & \xrightarrow{\ \ g\ \ } & N \longrightarrow 0
\end{array}
$$

of R-modules and R-module homomorphisms can be completed commutatively by an R-linear mapping $h : M \to L$. Such a diagram gives rise to a row exact commutative diagram

$$
\begin{array}{ccc}
 & & M^{(n)} \\
 & \overset{h^*}{\nearrow} & \downarrow \oplus f \\
L^{(n)} & \xrightarrow{\ \oplus g\ } & N^{(n)} \longrightarrow 0
\end{array}
$$

of $\mathbb{M}_n(R)$-modules and $\mathbb{M}_n(R)$-module homomorphisms with h^* given by the $\mathbb{M}_n(R)$-projectivity of $M^{(n)}$. If $h : M \to L$ is such that $h = \pi_1 h^* i_1$, where $i_1 : M \to M^{(n)}$ is the first canonical injection and $\pi_1 : L^{(n)} \to L$ is the first canonical projection, respectively, then h is the required map. Thus, M is a projective R-module whenever $M^{(n)}$ is a projective $\mathbb{M}_n(R)$-module.

(3) If pd-$M_R = \infty$, then it is obvious that pd-$M^{(n)}_{\mathbb{M}_n(R)} \le$ pd-M_R, so suppose that pd-$M_R = m$. If

$$0 \to P_m \to P_{m-1} \to \cdots \to P_0 \to M \to 0$$

is a projective resolution of M, then because of (1) and (2)

$$0 \to (P_m)^{(n)} \to (P_{m-1})^{(n)} \to \cdots \to (P_0)^{(n)} \to M^{(n)} \to 0$$

is an $\mathbb{M}_n(R)$-projective resolution of $M^{(n)}$. Thus, pd-$M^{(n)}_{\mathbb{M}_n(R)}$ is at most m, so pd-$M^{(n)}_{\mathbb{M}_n(R)} \le$ pd-M_R. Next, we show that pd-$(M^{(n)})_R \le$ pd-$M^{(n)}_{\mathbb{M}_n(R)}$. Suppose that pd-$M^{(n)}_{\mathbb{M}_n(R)} \le m$ and let

$$0 \to P_m \to P_{m-1} \to \cdots \to P_0 \to M^{(n)} \to 0$$

be an $\mathbb{M}_n(R)$-projective resolution of $M^{(n)}$. Then it follows from (2) that P_k is a projective R-module for $k = 0, 1, 2, \ldots, m$, so pd-$(M^{(n)})_R$ is at most m. Hence, pd-$(M^{(n)})_R \le$ pd-$M^{(n)}_{\mathbb{M}_n(R)}$. Part (1) of Lemma 12.3.1 gives pd-$(M^{(n)})_R =$ pd-M_R and so we have pd-$M_R \le$ pd-$M^{(n)}_{\mathbb{M}_n(R)}$. Thus, pd-$M_R =$ pd-$M^{(n)}_{\mathbb{M}_n(R)}$.

$\square$

Proposition 12.4.2. *For any ring R, r.gl.hd-R = r.gl.hd-$\mathbb{M}_n(R)$.*

Proof. For each R-module M we have, by (3) of Lemma 12.4.1, that

$$\text{pd-}M_R = \text{pd-}M^{(n)}{}_{\mathbb{M}_n(R)} \le \text{r.gl.hd-}\mathbb{M}_n(R),$$

so r.gl.hd-$R \le$ r.gl.hd-$\mathbb{M}_n(R)$. For the reverse inequality, note that

$$\text{pd-}M^{(n)}{}_{\mathbb{M}_n(R)} = \text{pd-}M_R \le \text{r.gl.hd-}R.$$

If M is an $\mathbb{M}_n(R)$-module such that pd-$M_{\mathbb{M}_n(R)} >$ r.gl.hd-R, then viewing M as an R-module, we have

$$\text{pd-}M^{(n)}{}_{\mathbb{M}_n(R)} = \text{pd-}M_R \le \text{r.gl.hd-}R.$$

But pd-$M_{\mathbb{M}_n(R)}$ = pd-$M^{(n)}{}_{\mathbb{M}_n(R)}$, so pd-$M_{\mathbb{M}_n(R)} \le$ r.gl.hd-R, a contradiction. Thus, no such $\mathbb{M}_n(R)$-module can exist, so r.gl.hd-$\mathbb{M}_n(R) \le$ r.gl.hd-R. $\square$

Corollary 12.4.3. *The following hold for any ring R.*

(1) *R is a semisimple ring if and only if $\mathbb{M}_n(R)$ is a semisimple ring.*

(2) *R is a right hereditary ring if and only if $\mathbb{M}_n(R)$ is right hereditary.*

Remark. The results of this section are part of a theory known as *Morita Theory* [26], [31] that describes how the equivalence of module categories can arise. If R and S are rings, recall that a functor $\mathscr{F} : \mathbf{Mod}_R \to \mathbf{Mod}_S$ is a category equivalence, denoted by $\mathbf{Mod}_R \approx \mathbf{Mod}_S$, if there is a functor $\mathscr{G} : \mathbf{Mod}_S \to \mathbf{Mod}_R$ such that $\mathscr{G}\mathscr{F} \approx \mathbf{Id}_{\mathbf{Mod}_R}$ and $\mathscr{F}\mathscr{G} \approx \mathbf{Id}_{\mathbf{Mod}_S}$. If M is an R-module, then we have seen that $M^{(n)}$ is an $M_n(R)$-module and, moreover, if $f : M \to N$ is an R-linear mapping, then $\oplus f : M^{(n)} \to N^{(n)}$ defined by $\oplus f((x_1, x_2, \ldots, x_n)) = (f(x_1), f(x_2), \ldots, f(x_n))$ is an $M_n(R)$-linear map. Hence, we have a functor $\mathscr{F} : \mathbf{Mod}_R \to \mathbf{Mod}_{M_n(R)}$. We have also seen that if M is an $M_n(R)$-module, then M is an R-module since R embeds in $M_n(R)$. So consider $\mathscr{G} : \mathbf{Mod}_{M_n(R)} \to \mathbf{Mod}_R$ such that if M is a module in $\mathbf{Mod}_{M_n(R)}$, then $\mathscr{G}(M) = ME_{11}$, where E_{11} is the matrix unit with 1_R is the first row and first column an zeroes elsewhere. Then for any $a \in R$ it follows that $ME_{11}a = MaE_{11}$ for any $a \in R$, so ME_{11} is also an R-module. If $f : M \to N$ is an $M_n(R)$-linear map, then $f(ME_{11}) = f(M)E_{11} \subseteq NE_{11}$ and so f induces an R-homomorphism $\mathscr{G}(f) : \mathscr{G}(M) \to \mathscr{G}(N)$. Thus, $\mathscr{G} : \mathbf{Mod}_{M_n(R)} \to \mathbf{Mod}_R$ is a functor and it follows that $\mathscr{G}\mathscr{F} \approx \mathbf{Id}_{\mathbf{Mod}_R}$ and $\mathscr{F}\mathscr{G} \approx \mathbf{Id}_{\mathbf{Mod}_{M_n(R)}}$. Hence, $\mathbf{Mod}_R \approx \mathbf{Mod}_{M_n(R)}$. If $\mathscr{F} : \mathbf{Mod}_R \to \mathbf{Mod}_S$ is an equivalence of categories, then a property $\mathcal{P}$ of R-modules M and R-module homomorphisms f in $\mathbf{Mod}_R$ is said to be a *Morita invariant for modules* if $\mathcal{P}$ is also a property of the modules $\mathscr{F}(M)$ and S-module homomorphisms $\mathscr{F}(f)$ in $\mathbf{Mod}_S$. The following are some of the Morita invariants for modules: M is artinian, M is noetherian, M is finitely generated, M is projective and M has projective dimension n. If $\mathbf{Mod}_R \approx \mathbf{Mod}_S$, then R and S are said to be *Morita equivalent rings*, denoted by $R \overset{M}{\approx} S$. If $\mathcal{P}$ is a property of a ring R, then $\mathcal{P}$ is said to be a *Morita invariant for rings*, if whenever $R \overset{M}{\approx} S$, $\mathcal{P}$ is also a property of S. Artinian, noetherian, prime, semiprime, semisimple, right hereditary and r.gl.hd $= n$ are Morita invariants for rings. Since $R \overset{M}{\approx} M_n(R)$, Lemma 12.4.1, Proposition 12.4.2 and Corollary 12.4.3 are just special case of the Morita Theory for the equivalence of module categories.

Problem Set 12.4

1. Prove (1) of Lemma 12.4.1.

2. Prove that if $0 \to L \xrightarrow{f} M \xrightarrow{g} N \to 0$ is a split short exact sequence in $\mathbf{Mod}_R$, then $0 \to L^{(n)} \xrightarrow{\oplus f} M^{(n)} \xrightarrow{\oplus g} N^{(n)} \to 0$ splits in $\mathbf{Mod}_{M_n(R)}$ for any integer $n \geq 1$.

3. In the proof of (2) of Lemma 12.4.1, it was indicated that $M^{(n)}$ is isomorphic to a direct summand of $M^{(n^2)}$. Deduce that this is the case.

4. Show that the map $h = \pi_1 h^* i_1$ given in the proof of (2) of Lemma 12.4.1 makes the diagram

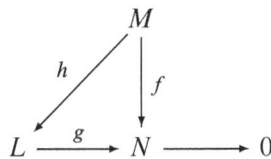

commute.

5. Verify the isomorphisms $F^{(n^2)} \cong (R^{(\Delta)})^{(n^2)} \cong (R^{(n^2)})^{(\Delta)} \cong M_n(R)^{(\Delta)}$ given in the proof of (2) of Lemma 12.4.1.

12.5 Quasi-Frobenius Rings Revisited

We saw in Chapter 10 that if a ring R is quasi-Frobenius, then finitely generated left and right R-modules are reflexive. Now that homological methods are at hand, we will use these methods to show that the converse holds for rings that are left and right noetherian. We continue with the notation and terminology established in Chapter 10.

More on Reflexive Modules

Before we can consider left and right noetherian rings over which finitely generated left and right modules are reflexive, we need several results. We begin with the following proposition.

Proposition 12.5.1. *Suppose that M is an (a left) R-module and that $\varphi_M : M \rightarrow M^{**}$ is the canonical map. Then $\varphi_M^* : M^{***} \rightarrow M^*$ and $\varphi_{M^*} : M^* \rightarrow M^{***}$ are such that $\varphi_M^* \varphi_{M^*} = id_{M^*}$.*

Proof. If $g \in M^*$, then $\varphi_M^* \varphi_{M^*}(g) = \varphi_M^*(f_g) = f_g \varphi_M$. Hence, if $x \in M$, then $[\varphi_M^* \varphi_{M^*}(g)](x) = [f_g \varphi_M](x) = f_g(f_x) = f_x(g) = g(x)$. Thus, $\varphi_M^* \varphi_{M^*}(g) = g$, so $\varphi_M^* \varphi_{M^*} = id_{M^*}$.   □

The proof of the following corollary is left as an exercise.

Corollary 12.5.2. *The following hold for each (left) R-module M.*

(1) M^* *is torsionless.*

(2) *If M is reflexive, then so is M^*.*

Definition 12.5.3. A submodule N of an R-module M is said to be a *closed submodule* of M if $N = \operatorname{ann}_r^M(\operatorname{ann}_\ell^{M^*}(N))$.

We will need the following lemmas.

Lemma 12.5.4. *The following are equivalent for an R-module M.*

(1) M *is torsionless.*

(2) M *is cogenerated by R.*

Proof. (1) $\Rightarrow$ (2). If M is torsionless, then $\varphi_M : M \rightarrow M^{**}$ is an injection. Now $0 = \operatorname{Ker} \varphi_M = \bigcap_{M^*} \operatorname{Ker} g$, so for each nonzero $x \in M$ there is a $g \in M^*$ such that $g(x) \neq 0$. If $\phi : M \rightarrow \prod_{M^*} R_g$, where $R_g = R$ for each $g \in M^*$, is such that $\phi(x) = (g(x))$, then ϕ is a monomorphism.

(2) $\Rightarrow$ (1). If M is cogenerated by R, then there is a monomorphism $\phi : M \rightarrow \prod_\Delta R_\alpha$, where $R_\alpha = R$ for each $\alpha \in \Delta$. Let $\pi_\alpha : \prod_\Delta R_\alpha \rightarrow R_\alpha$ be the canonical projection for each $\alpha \in \Delta$. Then $\pi_\alpha \phi(x) \neq 0$ for at least one $\alpha \in \Delta$ and $\pi_\alpha \phi \in M^*$. Hence, if $x \in M$, $x \neq 0$, then $\varphi_M : M \rightarrow M^{**}$ is such that $\varphi_M(x) = f_x$ and $f_x(\pi_\alpha \phi) = \pi_\alpha \phi(x) \neq 0$ for at least one $\alpha \in \Delta$. Hence, $x \neq 0$ gives $f_x \neq 0$, so φ_M is an injection.   □

Corollary 12.5.5. R_R *is a cogenerator for* $\mathbf{Mod}_R$ *if and only if every R-module is torsionless.*

Clearly, Lemma 12.5.4 and its corollary holds for left R-modules.

Lemma 12.5.6. *Let $M \xrightarrow{h} N$ be an epimorphism with kernel K and consider the short exact sequence $0 \rightarrow K \xrightarrow{i} M \xrightarrow{h} N \rightarrow 0$, where i is the canonical injection. Then $h^*(N^*) = \operatorname{ann}_\ell^{M^*}(K)$, where $0 \rightarrow N^* \xrightarrow{h^*} M^* \xrightarrow{i^*} K^*$ is the exact sequence* *obtained by applying the duality functor* $(-)^* = \text{Hom}_R(-, R)$. *Furthermore, N is torsionless if and only if K is closed in M.*

Proof. If $h^*(f) \in h^*(N^*)$, then $h^*(f)(K) = fh(K) = 0$, so $h^*(N^*) \subseteq \text{ann}_\ell^{M^*}(K)$. Conversely, suppose that $f \in \text{ann}_\ell^{M^*}(K)$. Then $f(K) = 0$, so let $\bar{f} : M/K \to R$ be the map induced by f. If $\bar{h}$ is the isomorphism induced by h and $h(x) \in N$, then $\bar{h}^{-1} : N \to M/K$ is such that $\bar{h}^{-1}(h(x)) = x + K$. Thus, $g = \bar{f}\bar{h}^{-1} \in N^*$ and for $x \in N$ we have

$$h^*(g)(x) = \bar{f}\bar{h}^{-1}h(x) = \bar{f}(x + K) = f(x).$$

Hence, $h^*(g) = f$ which shows that $f \in h^*(N^*)$. Therefore, $\text{ann}_\ell^{M^*}(K) \subseteq h^*(N^*)$, so $h^*(N^*) = \text{ann}_\ell^{M^*}(K)$, as asserted.

Suppose that N is torsionless and let $x \in \text{ann}_r^M(\text{ann}_\ell^{M^*}(K))$. Since $K \subseteq \text{ann}_r^M(\text{ann}_\ell^{M^*}(K))$, we need only show that $x \in K$. If $x \notin K$, then $h(x) \neq 0$, so since N is torsionless, Lemma 12.5.4 shows that there is an embedding $\phi : N \to \prod_\Delta R_\alpha$, where $R_\alpha = R$ for each $\alpha \in \Delta$. Moreover, there is an $f \in N^*$ such that $h^*(f)(x) = fh(x) \neq 0$. But $h^*(f) \in h^*(N^*) = \text{ann}_\ell^{M^*}(K)$ and $x \in \text{ann}_r^M(\text{ann}_\ell^{M^*}(K))$, so $h^*(f)(x) = 0$. This contradiction shows that $x \in K$ and so $K = \text{ann}_r^M(\text{ann}_\ell^{M^*}(K))$ when N is torsionless.

Finally, suppose that $K = \text{ann}_r^M(\text{ann}_\ell^{M^*}(K))$. If $y \in N$, $y \neq 0$, let $x \in M$ be such that $h(x) = y$. Then $x \notin K$, so $x \notin \text{ann}_r^M(\text{ann}_\ell^{M^*}(K))$. Thus, there is an $f \in \text{ann}_\ell^{M^*}(K)$ such that $f(x) \neq 0$. But $h^*(N^*) = \text{ann}_\ell^{M^*}(K)$, so there is a $g \in N^*$ such that $h^*(g) = f$. Hence, $gh(x) \neq 0$ and so $g(y) \neq 0$. It follows that $\phi : N \to \prod_{N^*} R_g$ defined by $\phi(y) = (g(y))$ is an embedding, so Lemma 12.5.4 shows that N is torsionless. $\qquad\square$

Lemma 12.5.7. *If $M \xrightarrow{h} N \to 0$ is an exact sequence of R-modules, then in the dual sequence $0 \to N^* \xrightarrow{h^*} M^*$, $h^*(N^*)$ is a closed submodule of M^*.*

Proof. If $K = \text{Ker } h$, then $0 \to K \xrightarrow{i} M \xrightarrow{h} N \to 0$ is exact, where i is the canonical injection, so $0 \to N^* \xrightarrow{h^*} M^* \xrightarrow{i^*} \text{Im } i^* \to 0$ is exact. The result will follow from the preceding lemma if we can show that $\text{Im } i^* \subseteq K^*$ is torsionless. Corollary 12.5.2 indicates that K^* is torsionless and submodules of torsionless modules are torsionless, so we are done. $\qquad\square$

Remark. If M and N are R-modules, then $\text{Hom}_R(M, N)$ is a left R-module via $(af)(x) = f(xa)$ for each $f \in \text{Hom}_R(M, N)$ and all $x \in M$ and $a \in R$. It follows that if $0 \to L \to M \to N \to 0$ is a short exact sequence of R-modules and

R-module homomorphisms, then the long exact Ext-sequence

$$0 \to \mathrm{Hom}_R(N, X) \to \mathrm{Hom}_R(M, X) \to \mathrm{Hom}_R(L, X)$$
$$\to \mathrm{Ext}^1_R(N, X) \to \mathrm{Ext}^1_R(M, X) \to \mathrm{Ext}^1_R(L, X) \to \cdots$$

is a sequence of left R-modules and left R-module homomorphisms. Similarly, a short exact sequence $0 \to L \to M \to N \to 0$ of left R-modules and left R-module homomorphisms yields a long exact Ext-sequence of R-modules.

Proposition 12.5.8. *If M is a finitely generated torsionless R-module, then there is a finitely generated torsionless left R-module N such that the following sequences are exact.*

(1) $0 \to M \xrightarrow{\varphi_M} M^{**} \to \mathrm{Ext}^1_R(N, R) \to 0$

(2) $0 \to N \xrightarrow{\varphi_N} N^{**} \to \mathrm{Ext}^1_R(M, R) \to 0$

Proof. (1) Since M is a finitely generated R-module, there is a finitely generated projective R-module P and an epimorphism $f : P \to M$. If $K = \mathrm{Ker}\, f$, then we have an exact sequence

$$0 \to K \to P \to M \to 0.$$

Taking duals, we see that $0 \to M^* \xrightarrow{f^*} P^*$ is exact and P^* is, by Proposition 10.2.2, a finitely generated projective left R-module. Hence, if $N = \mathrm{Coker}\, f^*$, then we have a finitely generated left R-module N and an exact sequence

$$0 \to M^* \xrightarrow{f^*} P^* \xrightarrow{\eta} N \to 0, \tag{12.1}$$

where η is the natural surjection. Lemma 12.5.7 shows that $f^*(M^*)$ is a closed submodule of P^*, so it follows from Lemma 12.5.6 that N is torsionless. Using the short exact sequence (12.1), we get an exact sequence

$$0 \to N^* \xrightarrow{\eta^*} P^{**} \xrightarrow{f^{**}} M^{**} \to \mathrm{Ext}^1_R(N, R) \to \mathrm{Ext}^1_R(P^*, R) \to \cdots.$$

But P^* is projective, so $\mathrm{Ext}^1_R(P^*, R) = 0$. Thus, we have the exact sequence

$$0 \to N^* \xrightarrow{\eta^*} P^{**} \xrightarrow{f^{**}} M^{**} \to \mathrm{Ext}^1_R(N, R) \to 0. \tag{12.2}$$

Next, consider the commutative diagram

$$
\begin{array}{ccc}
P & \xrightarrow{\;f\;} & M \\
\downarrow{\scriptstyle \varphi_P} & & \downarrow{\scriptstyle \varphi_M} \\
P^{**} & \xrightarrow{f^{**}} & M^{**}
\end{array}
$$

Since P is finitely generated and projective, Proposition 10.2.2 indicates that P is reflexive, so φ_P is an isomorphism. Now M is torsionless, so φ_M is a monomorphism and M is isomorphic to $\varphi_M(M) = \operatorname{Im} f^{**}$. But (12.2) gives $\operatorname{Coker} f^{**} \cong \operatorname{Ext}^1_R(N, R)$, so we have the exact sequence

$$0 \to M \xrightarrow{\varphi_M} M^{**} \to \operatorname{Ext}^1_R(N, R) \to 0 \tag{12.3}$$

which proves (1).

For the proof of (2), use (12.2) to form the exact sequence

$$0 \to N^* \xrightarrow{\eta^*} P^{**} \xrightarrow{f^{**}} \operatorname{Im} f^{**} \to 0$$

which can be replaced by

$$0 \to N^* \xrightarrow{\eta^*} P^{**} \xrightarrow{f^{**}} M \to 0 \tag{12.4}$$

since, as we have seen, $M \cong \operatorname{Im} f^{**}$. Next, let $X = \operatorname{Im} f^*$. Then (12.1) gives the exact sequence

$$0 \to X \to P^* \xrightarrow{\eta} N \to 0, \tag{12.5}$$

so we can now repeat the proof of (1) with (12.5) in place of the original sequence $0 \to K \to P \to M \to 0$. With this done, (12.4) shows that M can be used to play the role previously played by N. Under these changes (12.3) becomes

$$0 \to N \xrightarrow{\varphi_N} N^{**} \to \operatorname{Ext}^1_R(M, R) \to 0$$

and we have (2). $\qquad\qquad\qquad\qquad\qquad\qquad\qquad\qquad\qquad\qquad\qquad\qquad\qquad\qquad\qquad\square$

We are now in a position to prove the main result of this section.

Proposition 12.5.9. *If R is a left and right noetherian ring, then the following are equivalent.*

(1) *R is left and right self-injective.*

(2) *Every finitely generated left and right R-module is reflexive.*

Proof. Since R is left and right noetherian, if R is also left and right self-injective, then, in view of Proposition 10.2.14, R is a QF-ring. Hence, (1) $\Rightarrow$ (2) is Proposition 10.2.16, so we are only required to show (2) $\Rightarrow$ (1). If M is a finitely generated R-module, then M is, by hypothesis, reflexive, so M is finitely generated and torsionless. Therefore, by Proposition 12.5.8, there is a finitely generated torsionless left R-module N and an exact sequence

$$0 \to N \xrightarrow{\varphi_N} N^{**} \to \operatorname{Ext}^1_R(M, R) \to 0.$$

But we are assuming that every finitely generated left and right R-module is reflexive. Hence, φ_N is an isomorphism and so $\operatorname{Ext}_R^1(M, R) = 0$. Since this holds for every finitely generated R-module, it holds for every cyclic R-module. Thus, Proposition 11.4.6 shows that R is right self-injective. A similar proof shows that R is left self-injective, so (2) $\Rightarrow$ (1).   □

Corollary 12.5.10. *The following are equivalent.*

(1) *R is a QF-ring.*

(2) *R is left and right noetherian and finitely generated left and right R-modules are reflexive.*

(3) *R is left and right noetherian and left and right self-injective.*

Proof. (1) $\Rightarrow$ (2) follows from the definition of a QF-ring and Proposition 10.2.16.
(2) $\Rightarrow$ (3) is Proposition 12.5.9.
(3) $\Rightarrow$ (1) follows from Proposition 10.2.14.   □

We now know conditions under which finitely generated left and right R-modules are reflexive. The following proposition gives information concerning when finitely generated torsionless R-modules are reflexive. To prove the proposition we need the following lemma.

Lemma 12.5.11. *If $0 \to M_1 \to P \to M \to 0$ and $0 \to N \to E \to N_1 \to 0$ are short exact sequences where P and E are projective and injective R-modules, respectively, then $\operatorname{Ext}_R^1(M_1, N)$ and $\operatorname{Ext}_R^1(M, N_1)$ are isomorphic.*

Proof. The long exact Ext-sequences give

$$\cdots \to \operatorname{Ext}_R^1(P, N) \to \operatorname{Ext}_R^1(M_1, N) \to \operatorname{Ext}_R^2(M, N) \to \operatorname{Ext}_R^2(P, N) \to \cdots$$

and

$$\cdots \to \operatorname{Ext}_R^1(M, E) \to \operatorname{Ext}_R^1(M, N_1) \to \operatorname{Ext}_R^2(M, N) \to \operatorname{Ext}_R^2(M, E) \to \cdots .$$

But

$$\operatorname{Ext}_R^1(P, N) = \operatorname{Ext}_R^2(P, N) = \operatorname{Ext}_R^1(M, E) = \operatorname{Ext}_R^2(M, E) = 0, \quad \text{so}$$

$$\operatorname{Ext}_R^1(M_1, N) \cong \operatorname{Ext}_R^2(M, N) \cong \operatorname{Ext}_R^1(M, N_1).$$ □

Proposition 12.5.12. *If R is a left and right noetherian ring, then the following are equivalent.*

(1) *id-$_R R \leq 1$.*

(2) *Every finitely generated torsionless R-module is reflexive.*

Proof. (1) $\Rightarrow$ (2). Let M be a finitely generated torsionless R-module. Then by Proposition 12.5.8 there is a finitely generated torsionless left R-module N such that the sequence

$$0 \to M \xrightarrow{\varphi_M} M^{**} \to \text{Ext}_R^1(N, R) \to 0$$

is exact. Since id-$_R R \le 1$, R as a left R-module has an injective resolution of the form $0 \to R \to E^0 \to E^1 \to 0$. We also see that since N is a finitely generated and torsionless left R-module, N can be viewed as a submodule of a finitely generated free left R-module F. Hence, we also have an exact sequence $0 \to N \to F \to F/N \to 0$. Now Lemma 12.5.11 shows that $\text{Ext}_R^1(N, R)$ and $\text{Ext}_R^1(F/N, E^1)$ are isomorphic and Proposition 11.4.6 shows that $\text{Ext}_R^1(F/N, E^1) = 0$. Thus, $\text{Ext}_R^1(N, R) = 0$, so φ_M is an isomorphism and, consequently, M is reflexive.

(2) $\Rightarrow$ (1). Let N be a finitely generated left R-module. Then we have an exact sequence $0 \to K \to P \to N \to 0$, where P is a finitely generated projective left R-module. Note that K is also finitely generated since R is left noetherian and since P is reflexive, P is torsionless, so K is torsionless. Hence, by (2) of the left-hand version of Proposition 12.5.8, we have an exact sequence

$$0 \to M \xrightarrow{\varphi_M} M^{**} \to \text{Ext}_R^1(K, R) \to 0,$$

where M is a finitely generated torsionless R-module. But we are assuming that all such R-modules are reflexive and so φ_M is an isomorphism. Therefore, $\text{Ext}_R^1(K, R) = 0$. Next, construct a short exact sequence of left R-modules $0 \to R \to E \to X \to 0$, where E is injective. We will have id-$_R R \le 1$, if we can show that X is an injective left R-module. By considering the short exact sequences

$$0 \to K \to P \to N \to 0 \quad \text{and} \quad 0 \to R \to E \to X \to 0$$

and invoking the left-hand version of Proposition 12.5.11, we have

$$\text{Ext}_R^1(N, X) \cong \text{Ext}_R^1(K, R).$$

Hence, $\text{Ext}_R^1(N, X) = 0$ for every finitely generated left R-module N. But then $\text{Ext}_R^1(R/A, X) = 0$ for all left ideals A of R and so, by the left-hand versions of Propositions 11.4.10 and 11.4.12, X is an injective left R-module. Hence, id-$_R R \le 1$. $\quad\square$

In conclusion, we offer, without proof, the following characterization of QF-rings. A proof can be found in [14].

Proposition 12.5.13. *The following are equivalent.*

(1) *R is a QF-ring.*

(2) *Every projective R-module is injective.*

(3) *Every injective R-module is projective.*

Problem Set 12.5

1. Prove Corollary 12.5.2.

2. Prove each of the following.

   (a) If $\varphi_M : M \to M^{**}$ is the canonical map, then $\operatorname{Ker} \varphi_M = \bigcap_{g \in M^*} \operatorname{Ker} g$.

   (b) Submodules of torsionless modules are torsionless.

   (c) A direct product of torsionless modules is torsionless. Conclude from (b) and (c) that a direct sum of torsionless modules is torsionless.

3. In the proof of Proposition 12.5.12, we used the fact that if N is finitely generated and torsionless, then N can be viewed as a submodule of a finitely generated free R-module F. Given the conditions Proposition 12.5.12, prove that this is the case. [Hint: Lemma 12.5.4.]

4. Is a direct summand of a reflexive R-module reflexive?

5. Prove that the exact sequence $0 \to M^* \xrightarrow{\varphi_{M^*}} M^{***} \to \operatorname{Coker} \varphi_{M^*} \to 0$ splits.

6. Prove that the following are equivalent.

   (a) Torsionless R-modules are reflexive.

   (b) $\operatorname{Ext}^1_R(M, R) = 0$ for all torsionless left R-modules M.

   (c) $\operatorname{Ext}^2_R(M, R) = 0$ for all left R-modules M.

   (d) $\operatorname{id-}_R R \leq 1$.

Appendix A

Ordinal and Cardinal Numbers

Ordinal Numbers

If X is a well-ordered set with "enough elements", then X has a first element, a second element, a third element and so on. Ordinal numbers can be viewed as numbers that represent the position of an element in a well-ordered set. When defining ordinal numbers, one goal is to capture this sense of position.

Let X be a well-ordered set. A subset $S \subseteq X$ is said to be a *segment* of X if x, $x' \in X$ and $x' \leq x$, then $x \in S$ implies that $x' \in S$. If $S \subsetneq X$ and x is the first element of $X - S$, then $S_x = \{x' \in X \mid x' < x\}$ is a segment of X referred to as an *initial segment*. If X and Y are well-ordered sets, then a bijective function $f : X \to Y$ is said to be an *order isomorphism* if $x \leq x'$ in X implies that $f(x) \leq f(x')$ in Y. Two well-ordered sets X and Y are called *order isomorphic* if there is an order isomorphism $f : X \to Y$. If X and Y are order isomorphic, then we write $X \approx Y$. The notation $X < Y$ will indicate that X is order isomorphic to an initial segment of Y and $X \leq Y$ will mean that $X < Y$ or $X \approx Y$.

If the integers are defined by the sets

$$0 = \varnothing$$
$$1 = \{0\}$$
$$2 = \{0, 1\}$$
$$3 = \{0, 1, 2\}$$
$$\vdots$$
$$n = \{0, 1, 2, \ldots, n - 1\}$$
$$\vdots$$

then the ordering

$$0 \subsetneq \{0\} \subsetneq \{0, 1\} \subsetneq \{0, 1, 2\} \subsetneq \cdots \subsetneq \{0, 1, 2, \ldots, n - 1\} \subsetneq \cdots$$

can be used to define the usual order

$$0 < 1 < 2 < \cdots < n < \cdots$$

on the set $\mathbb{N}_0$. Under this ordering, the set n is well ordered for $n = 0, 1, 2, \ldots$. Next, consider the set n and let

$$S_0 = \{x \in n \mid x < 0\} = \varnothing = 0$$
$$S_1 = \{x \in n \mid x < 1\} = \{0\} = 1$$
$$S_2 = \{x \in n \mid x < 2\} = \{0, 1\} = 2$$

$$\vdots$$

$$S_{n-1} = \{x \in n \mid x < n-1\} = \{0, 1, 2, \ldots, n-2\} = n-1.$$

Then n is a well-ordered set such that $k = S_k$ for each $k \in n$. This observation motivates the following definition.

Definition. A well-ordered set α is said to be an *ordinal number* if $x = S_x$ for each $x \in \alpha$. Furthermore, if X is a well-ordered set such that $X \approx \alpha$, then we say that X has ordinal number α and write $\mathrm{ord}(X) = \alpha$. If $\mathrm{ord}(X) = \alpha$ and X is a finite set, then α is a *finite ordinal number*. Otherwise, α is an *infinite ordinal number*.

Each of the sets n is an ordinal number and if $\mathbb{N}_0 = \{0, 1, 2, \ldots\}$ is given the usual order, then

$$\omega \text{ will denote the ordinal number of } \mathbb{N}_0.$$

Also, if X is a well-ordered set and $x \in X$ is such that $S_x \approx n$, then x occupies the $(n+1)th$ *position* in X, for $n = 0, 1, 2, \ldots$. For example, if $X = \{a, b, c, d, \ldots\}$ is a well-ordered set, where $a < b < c < d < \cdots$, then S_a is order isomorphic to 0, so a occupies the first position in X, S_b is order isomorphic to 1, so b occupies the second position in X and so on. It follows that if X and Y are well-ordered sets that are order isomorphic, then $x \in X$ occupies the same position in X as $y \in Y$ occupies in Y if $S_x \approx S_y$. Furthermore, if X is a well-ordered set, then there is a unique ordinal number α such that $\mathrm{ord}(X) = \alpha$ or, more briefly,

Every well-ordered set has a unique ordinal number.

With this in mind, addition, multiplication and exponentiation of ordinal numbers can be defined as follows.

1. **Ordinal Number Addition.** If α and β are ordinal numbers, then the well orderings on α and β can be used to establish a well ordering of the set $(\alpha \times \{1\}) \cup (\beta \times \{2\})$. If $(x, 1)$ and $(x', 1)$ are in $\alpha \times \{1\}$, let $(x, 1) \leq (x', 1)$ if $x \leq x'$ in α. Similarly, for $(y, 2)$ and $(y', 2)$ in $\beta \times \{2\}$. Finally, if we set $(x, 1) \leq (y, 2)$ for all $(x, 1) \in \alpha \times \{1\}$ and $(y, 2) \in \beta \times \{2\}$, then $\leq$ is a well ordering of $(\alpha \times \{1\}) \cup (\beta \times \{2\})$. With this ordering of $(\alpha \times \{1\}) \cup (\beta \times \{2\})$, $\alpha + \beta$ is defined as $\alpha + \beta = \mathrm{ord}((X \times \{1\}) \cup (Y \times \{2\}))$.

2. **Ordinal Number Multiplication.** If α and β are ordinal numbers, define the order $\leq$ on $\alpha \times \beta$ as follows: $(x, y) \leq (x', y')$ if and only if $(x, y) = (x', y')$ or $(x, y) < (x', y')$. The order $(x, y) < (x', y')$ means that $(x, y) < (x', y')$ when $y < y'$ and if $y = y'$, then $(x, y) < (x', y')$ if $x < x'$. It follows that $\alpha \times \beta$ is well ordered under this ordering and $\alpha\beta$ is defined as $\alpha\beta = \text{ord}(\alpha \times \beta)$.

3. **Ordinal Number Exponentiation.** For this definition, we need the concept of a limit ordinal. If β is an ordinal number that does not have a last element, β is said to be a *limit ordinal*. For example, $\omega = \text{ord}(\{0, 1, 2, \dots\})$ is a limit ordinal. Thus, if β is a limit ordinal, then $\beta \neq \alpha + 1$ for any ordinal number α. If α and β are ordinal numbers and β is not a limit ordinal, then α^β is defined by $\alpha^0 = 1, \alpha^1 = \alpha$ and $\alpha^{\beta+1} = \alpha^\beta \alpha$ for all ordinal numbers $\beta \geq 1$. If β is a limit ordinal, then $\alpha^\beta = \sup\{\alpha^\delta \mid \delta \text{ is an ordinal number such that } \delta < \beta\}$.

Proposition. *The trichotomy property holds for the class of ordinal numbers. That is, if α and β are ordinal numbers, then exactly one of $\alpha < \beta$, $\alpha = \beta$ and $\beta > \alpha$ holds.*

Ordinal addition shows that the ordinal numbers of the following well-ordered sets (each ordered in the obvious manner) are distinct

$$\omega = \text{ord}(\{0, 1, 2, \dots\})$$
$$\omega + 1 = \text{ord}(\{0, 1, 2, \cdots ; 0'\})$$
$$\omega + 2 = \text{ord}(\{0, 1, 2, \dots ; 0', 1'\})$$
$$\vdots$$
$$\omega + n = \text{ord}(\{0, 1, 2, \dots ; 0', 1', \dots, n - 1'\})$$
$$\vdots$$
$$\omega + \omega = \omega 2 = \text{ord}\{0, 1, 2, \dots ; 0', 1', 2', \dots\}$$
$$\omega 2 + 1 = \text{ord}\{0, 1, 2, \dots ; 0', 1', 2', \dots ; 0''\}$$
$$\vdots$$

Hence, the class of ordinal numbers forms a chain with a "front end" that looks like

$$0 < 1 < 2 < \cdots < \omega < \omega + 1 < \cdots < \omega 2 < \omega 2 + 1 < \cdots .$$

The first limit ordinal in the chain is ω, $\omega 2$ is the second and so on. Moreover, the ordering $\leq$ of $\mathbb{N}_0$ as a set of ordinal numbers agrees with the usual ordering $\leq$ of $\mathbb{N}_0$.

If **Ord** denotes the class of ordinal numbers, then since any set can be well ordered, the assumption that **Ord** is a set will lead to a contradiction. So **Ord** is a proper class that is linearly ordered by $\leq$. Also, a class can be linearly ordered by a given order and yet not be well ordered by this ordering as is pointed out by the usual ordering on $\mathbb{R}$. However, **Ord** is well ordered by $\leq$.

Proposition. *The proper class* **Ord** *is well ordered by* $\leq$*, that is, every nonempty class of ordinal numbers has a first element.*

Proof. Suppose that $\mathcal{O}$ is a nonempty class of ordinal numbers. If $\alpha \in \mathcal{O}$ and α is the first element of $\mathcal{O}$, then there is nothing to prove. If α is not the first element of $\mathcal{O}$, let $\mathcal{B} = \{\beta \in \mathcal{O} \mid \beta < \alpha\}$ and suppose that $\phi : \mathcal{B} \to \alpha$ is such that $\phi(\beta) = x$, where $x \in \alpha$ is such that $\beta \approx S_x = S_{\phi(\beta)}$. Then $\{\phi(\beta) \mid \beta \in \mathcal{B}\}$ is a nonempty subset of α and α is well ordered by an ordering $\leq$. Hence, $\{\phi(\beta) \mid \beta \in \mathcal{B}\}$ has a first element, say $\phi(\delta)$. If $\beta \in \mathcal{B}$, then $\phi(\delta) \leq \phi(\beta)$ and $S_{\phi(\delta)}$ and $S_{\phi(\beta)}$ are initial segments of α such that $S_{\phi(\delta)} \subseteq S_{\phi(\beta)}$. If $S_{\phi(\delta)} = S_{\phi(\beta)}$, then $\delta = \beta$ and if $S_{\phi(\delta)} \subsetneq S_{\phi(\beta)}$, then it follows that $S_{\phi(\delta)}$ is an initial segment of $S_{\phi(\beta)}$ which means, of course, that $\delta < \beta$. Thus, $\delta \leq \beta$ for each $\beta \in \mathcal{B}$. But $\mathcal{O}$ is linearly ordered, so we have $\delta \leq \gamma$ for each $\gamma \in \mathcal{O}$. Therefore, δ is the first element of $\mathcal{O}$. $\qquad\square$

Corollary. *The set* $\mathbb{N}_0$ *is well ordered under the usual ordering on* $\mathbb{N}_0$.

Neither addition nor multiplication of ordinal numbers is commutative. For example, $2\omega \neq \omega 2$ and $\omega \neq \omega + 1$ but $1 + \omega = \omega$. Moreover, the familiar laws of exponentiation may not hold. If α, β and γ are ordinal numbers, then $(\alpha\beta)^{\gamma}$ may be different from $\alpha^{\gamma}\beta^{\gamma}$. For instance, $(2 \cdot 2)^{\omega} = 4^{\omega} = \omega$ and $2^{\omega}2^{\omega} = \omega\omega = \omega^2$.

Finally, if X is any nonempty set, then X can be well ordered, so X can be viewed as a set of ordinal numbers $\{0, 1, 2, \ldots, \omega, \omega + 1, \ldots\}$ such that $\alpha \in \{0, 1, 2, \ldots, \omega, \omega + 1, \ldots\}$ if and only if $\alpha < \mathrm{ord}(X)$. That is, there is an order preserving bijection $\{0, 1, 2, \ldots, \omega, \omega + 1, \ldots\} \to X$. Indeed, if α_0 is the first element of X, map 0 to α_0 and if α_1 is the first element of $X - \{\alpha_0\}$, map 1 to α_1 and so on.

One important aspect of ordinal numbers is that these numbers allow us to extend induction as practiced with the integers to the ordinal numbers, a process known as *transfinite induction*.

Proposition (The Principle of Transfinite Induction). *Suppose that X is a well-ordered class and let $S(x)$ be a statement that is either true or false for each $x \in X$. If $S(y)$ true for each $y < x$ implies that $S(x)$ is true, then $S(x)$ is true for each $x \in X$.*

Proof. Let X and $S(x)$ be as in the statement of the proposition and suppose that $S(y)$ true for each $y < x$ implies that $S(x)$ is true. We claim that this means that $S(x)$ is true for each $x \in X$. Indeed, if there is an element $x \in X$ such that $S(x)$ is false, let $F = \{x \in X \mid S(x) \text{ is false}\}$. Then F is nonempty and so has a first element, say x_0. Consequently, $S(y)$ is true for each $y < x_0$. But our assumption now implies that $S(x_0)$ is true and we have a contradiction. Therefore, an $x \in X$ cannot exist such that $S(x)$ is false, so $S(x)$ is true for all $x \in X$. $\qquad\square$

Since **Ord** is well ordered by $\leq$, the Principle of Transfinite Induction holds over **Ord**.

(1) **Principle of Transfinite Induction over Ord.** *Let $S(\beta)$ be a statement that is either true or false for each $\beta \in$ **Ord**. If $S(\alpha)$ true for each $\alpha \in$ **Ord** such that $\alpha < \beta$ implies that $S(\beta)$ is true, then $S(\beta)$ is true for each $\beta \in$ **Ord**.*

In particular, the Principle of Transfinite Induction holds for the well-ordered set of ordinal numbers $\mathbb{N}_0$ and the result is the familiar induction as practiced with the integers. Since $\mathbb{N}_0 \subseteq$ **Ord** and since the ordering $\leq$ on **Ord** induces the usual order $\leq$ on $\mathbb{N}_0$, transfinite induction can be viewed as extending induction as practiced with the integers to the class of ordinal numbers.

It is well known that the following are equivalent:

(2) *Let $S(n)$ be a statement that is either true or false for each $n \in \mathbb{N}_0$. If $S(0)$ is true and if $S(k)$ true implies that $S(k + 1)$ is true for each $k \in \mathbb{N}_0$, then $S(n)$ is true for each $n \in \mathbb{N}_0$.*

(3) *Let $S(n)$ be a statement that is either true or false for each $n \in \mathbb{N}_0$. If $S(j)$ true for all $0 \leq j < k$ implies that $S(k)$ is true, then $S(n)$ is true for each $n \in \mathbb{N}_0$.*

In view of (2), one is tempted to formulate a Principle of Transfinite Induction for the ordinal numbers as follows:

*Let $S(\beta)$ be a statement that is either true or false for each $\beta \in$ **Ord**. If $S(0)$ is true and if $S(\alpha)$ true implies that $S(\alpha + 1)$ is true, then $S(\beta)$ is true for each $\beta \in$ **Ord**.*

The statement, "$S(0)$ true and $S(\alpha)$ true implies that $S(\alpha + 1)$ true" does not imply that $S(\beta)$ is true for each $\beta \in$ **Ord** The difficulty is with the limit ordinals. For example, ω is a limit ordinal and there is no ordinal α such that $\alpha + 1 = \omega$. So the condition does not provide a way to "reach" ω from any ordinal $\alpha < \omega$ to show that $S(\omega)$ is true. However, an additional step can be added, as shown in (4) below, that will give an alternate form of the Principle of Transfinite Induction over **Ord**.

(4) **Principle of Transfinite Induction over Ord** (Alternate Form). *Let $S(\beta)$ be a statement that is either true or false for each $\beta \in$ **Ord**. If the following two conditions are satisfied, then $S(\beta)$ is true for all ordinal numbers β.*

(a) *$S(0)$ is true and if $S(\alpha)$ true, then $S(\alpha + 1)$ is true for each non-limit ordinal α.*

(b) *If β is a limit ordinal and $S(\alpha)$ is true for each $\alpha < \beta$, then $S(\beta)$ is true.*

Cardinal Numbers

Our development of ordinal numbers captured the notion of the position of an element in a set. The following development of cardinal numbers corresponds to the intuitive notion that two sets X and Y have the same "number" of elements if there is a one-to-one correspondence among their elements without regard to an order on X and Y.

If the relation $\sim$ is defined on the proper class of all sets by $X \sim Y$ if there is bijection $f : X \to Y$, then $\sim$ is an equivalence relation. If card(X) is the equivalence class determined by a set X, then we say that a set Y has the same *cardinal number* as X if $Y \in \text{card}(X)$. For example, if we set

$$0 = \text{card}(\varnothing), \quad 1 = \text{card}(\{0\}), \quad 2 = \text{card}(\{0, 1\}), \ldots$$
$$n = \text{card}(\{0, 1, 2, \ldots, n - 1\}), \quad \ldots,$$

then $\text{card}(X) = n$ for each set $X \in \text{card}(\{0, 1, 2, \ldots, n - 1\})$. (Additional details regarding this method of establishing cardinal numbers can be found in [20].) If $\text{card}(X) = n$, then we say that X has *finite cardinal number n*. There are also *infinite cardinal numbers*. For example,

$\aleph_0 = \text{card}(\mathbb{N}_0)$ is the first infinite cardinal,

$\aleph_1 = \text{card}(\wp(\mathbb{N}_0))$ is the second,

$\aleph_2 = \text{card}(\wp(\wp(\mathbb{N}_0)))$ is the third,

$$\vdots$$

$\aleph_k = \text{card}(\wp^k(\mathbb{N}_0))$ is the kth, where $\wp^k$ is $\wp$ composed with itself k times,

$$\vdots$$

Hence, if $X \in \text{card}(\wp^k(\mathbb{N}_0))$, then we write $\text{card}(X) = \aleph_k$ and say that X has cardinal number $\aleph_k$.

Let $a = \text{card}(X)$ and $b = \text{card}(Y)$ be cardinal numbers. If there is an injective function $f : X \to Y$, then we write $a \leq b$ with $a < b$ holding when such an injective function exists, but there is no bijective function from X to Y. If $a = \text{card}(X)$ and $b = \text{card}(\wp(X))$, then the fact that there is no surjective function from X to $\wp(X)$ together with the observation that the function $X \to \wp(X)$ given by $x \mapsto \{x\}$ is injective shows that $a < b$. Hence, there is no largest cardinal number and it follows that the class **Card** of cardinal numbers is a proper class.

The *trichotomy property* also holds for cardinal numbers, so if a and b are cardinal numbers, then one and only one of the following holds.

$$(1)\ a < b, \quad (2)\ a = b, \quad (3)\ a > b.$$

Moreover, the class of cardinal numbers is a chain

$$0 < 1 < 2 < 3 < \cdots < \aleph_0 < \aleph_1 < \aleph_2 < \cdots$$

and

Every set has a unique cardinal number.

Addition, multiplication and exponentiation can be defined on the class of cardinal numbers as follows:

1. **Cardinal Number Addition.** If $a = \text{card}(X)$ and $b = \text{card}(Y)$, then $a + b = c$, where $c = \text{card}((X \times \{1\}) \cup (Y \times \{2\}))$.

2. **Cardinal Number Multiplication.** If $a = \text{card}(X)$ and $b = \text{card}(Y)$, then $ab = c$, where $c = \text{card}(X \times Y)$.

3. **Cardinal Number Exponentiation.** If X and Y are sets and if $a = \text{card}(X)$ and $b = \text{card}(Y)$, then $a^b = c$, where $c = \text{card}(X^Y)$.

Proposition. *Each of the following holds in* **Card.**

1. *If a and b are cardinal numbers and a is infinite, then $a + b = \max\{a, b\}$.*

2. *If a and b are cardinal numbers, a is infinite and $b \neq 0$, then $ab = \max\{a, b\}$.*

3. *If a, b and c are cardinal numbers, then $(a^b)^c = a^{bc}$.*

Problem Set

1. If X and Y are finite sets and if there is a bijection $f : X \to Y$, prove that $\text{ord}(X) = \text{ord}(Y)$ regardless of how X and Y are well ordered.

2. Prove that the class **Ord** of all ordinal numbers is a proper class.

3. Prove that $n + \omega \neq \omega + n$, where n is a positive integer, and that $2\omega = (1 + 1)\omega \neq \omega(1 + 1) = \omega 2$. Conclude that neither addition nor multiplication of ordinal numbers is commutative.

4. Let α, β and γ be ordinal numbers. Show that
   (a) $\alpha + (\beta + \gamma) = (\alpha + \beta) + \gamma$
   (b) $\alpha(\beta\gamma) = (\alpha\beta)\gamma$
   (c) $\alpha(\beta + \gamma) = \alpha\beta + \alpha\gamma$
   (d) $(\beta + \gamma)\alpha \neq \beta\alpha + \gamma\alpha$.
   (e) $\alpha\beta = 0$ if and only if $\alpha = 0$ or $\beta = 0$.
   (f) If $\alpha + \beta = \alpha + \gamma$, then $\beta = \gamma$. Give an example where $\beta + \alpha = \gamma + \alpha$ fails to imply that $\beta = \gamma$.

5. (a) If α and β are ordinal numbers, show that $\alpha \leq \alpha + \beta$ and that equality holds if and only if $\beta = 0$.
   (b) Suppose that α and β are ordinal numbers such that $\alpha \leq \beta$. If β is an infinite ordinal number, prove that $\alpha + \beta = \beta$.

6. If X is a nonempty set, show that X can be viewed as a set of ordinal numbers $\{0, 1, 2, \ldots, \omega, \omega + 1, \ldots\}$ such that $\alpha \in \{0, 1, 2, \ldots, \omega, \omega + 1, \ldots\}$ if and only if $\alpha < \text{ord}(X)$. That is, show that there is an order preserving bijection $X \to \{0, 1, 2, \ldots, \omega, \omega + 1, \ldots\}$.

7. Prove each of the following.

   (a) Addition (multiplication) of cardinal numbers is commutative.

   (b) If a, b and c are cardinal numbers, then $(a + b)c = ac + bc$ and $a(bc) = (ab)c$.

   (c) For all cardinal numbers a, $a + 0 = a$, $a0 = 0$ and $a1 = a$.

   (d) If a and b are cardinal numbers, then $ab = 1$ if and only $a = 1$ and $b = 1$.

8. If a is an infinite cardinal number, show that there is no cardinal number b such that $a + b = 0$.

9. (a) If a and b are cardinal numbers and $a \leq b$, show that $a + b = b$ whenever b is an infinite cardinal.

   (b) If a and b are cardinal numbers, $a \neq 0$, $a \leq b$, and b is an infinite cardinal, prove that $ab = b$.

   Conclude from (a) and (b) that if a is an infinite cardinal, then $a + a = a$ and $aa = a$.

   (c) If a, b, and c are cardinal numbers, prove that

   (i) $a^{b+c} = a^b a^c$

   (ii) If $a \leq b$, then $a^c \leq b^c$.

   (iii) $(ab)^c = a^c b^c$

   (iv) $(a^b)^c = a^{bc}$

10. Show that if $\text{card}(X) = m$ and $\text{card}(Y) = n$, where $m, n \in \mathbb{N}$, then $\text{card}(X^Y) = m^n$. Conclude that exponentiation as defined for cardinal arithmetic produces the usual exponentiation of positive integers.

11. If X and Y are sets and $f : X \to Y$ is a surjective function, show that $\text{card}(Y) \leq \text{card}(X)$.

12. Let X be any set. If $f : X \to \wp(X)$ is a function, show that the set $\{x \in X \mid x \notin f(x)\}$ has no preimage in X. Deduce that $\text{card}(X) < \text{card}(\wp(X))$ and conclude that there is no largest cardinal number and that **Card** is a proper class.

Bibliography

Books

[1] Adkins, W. and Weintraub, S., *Algebra*, Springer-Verlag, New York, Berlin, 1992.

[2] Anderson, F. and Fuller, K., *Rings and Categories of Modules*, Springer-Verlag, New York, Berlin, 1974.

[3] Blyth, T., *Categories*, Longman Group Limited, New York, London, 1986.

[4] Bourbaki, N., *Algebra I, Chapters 1–3*, Springer-Verlag, New York, Berlin, 1989.

[5] Bourbaki, N., *Algebra II, Chapters 4–7*, Springer-Verlag, New York, Berlin, 1989.

[6] Bourbaki, N., *Commutative Algebra II, Chapters 1–7*, Springer-Verlag, New York, Berlin, 1989.

[7] Cameron, P., *Sets, Logic and Categories*, Springer-Verlag, New York, Berlin, 1999.

[8] Cartan, H. and Eilenberg, S., *Homological Algebra*, Princeton University Press, Princeton, 1956.

[9] Divinsky, N., *Rings and Radicals*, University of Toronto Press, Toronto, Mathematical Expositions, No. 14, 1965.

[10] Dummit, D. and Foote, R., *Abstract Algebra*, 2nd edition, Prentice Hall, Upper Saddle River, N.J., 1999.

[11] Dung, N., Huynh, D., Smith, P. and Wisbauer, R., *Extending Modules*, Pitman Research Notes in Mathematics Series, No. 313, Longman Scientific & Technical, Essex, 1994.

[12] Enochs, E. and Overtoun, M., *Relative Homological Algebra*, Walter de Gruyter, New York, 2000.

[13] Faith, C., *Algebra: Rings, Modules and Categories I*, Springer-Verlag, New York, Berlin, 1973.

[14] Faith, C., *Algebra: Rings, Modules and Categories II, Springer-Verlag, New York, Berlin, 1976.*

[15] Goodearl, K., *Singular Torsion and Splitting Properties*, Memoirs of the American Mathematical Society, No. 124, Providence, 1972.

[16] Goodearl, K., *Von Neumann Regular Rings, 2nd Edition*, Krieger Publishing Company, Malabar, FL, 1991.

[17] Gray, M., *A Radical Approach to Algebra*, Addison Wesley, Reading, MA, 1970.

[18] Hilton, P. and Stammbach, U., *A Course in Homological Algebra*, Springer-Verlag, New York, Berlin, 1971.

[19] Hu, S., *Introduction to Homological Algebra*, Holden-Day, Inc., San Francisco, London, 1968.

[20] Hungerford, T., *Algebra*, Springer-Verlag, New York, Berlin, 1974.

[21] Jacobson, N., *Basic Algebra*, W. H. Freeman, San Francisco, 1974

[22] Jacobson, N., *Structure of Rings*, American Mathematical Society, Providence, 1964.

[23] Jans, J., *Rings and Homology*, Holt, RineHart and Winston, Inc., New York, London, 1964.

[24] Kasch, F., Modules and Rings, Academic Press, New York, London, 1983.

[25] Lam, T., *A First Course in Noncommutative Rings*, Springer-Verlag, New York, Berlin, 1991.

[26] Lam, T., *Lectures on Modules and Rings*, Springer-Verlag, New York, Berlin, 1998.

[27] Lambek, J., *Lectures on Rings and Modules*, Blaisdell Publishing Company, Toronto, London, 1966.

[28] Lang, S., *Algebra*, Addison-Wesley Publishing Company, Inc., Reading Massachusetts, 1965.

[29] Larsen, M. and McCarthy, P., *Multiplicative Theory of Ideals*, Academic Press, New York, London, 1971,

[30] MacLane, S., *Homology*, Springer-Verlag, New York, Berlin, 1975.

[31] McConnell, J. C. and Robson, J. C., *Noncommutative Noetherian Rings*, Graduate Studies in Mathematics, Volume 30, American Mathematical Society, Providence, 2001.

[32] Mitchell, B., *Theory of Categories*, Academic Press, New York, London, 1965.

[33] Mohamed, S., and Müller, B., *Continuous and Discrete Modules*, Cambridge University Press, Cambridge, 1990.

[34] Nagata, M., *Local Rings*, Wiley Interscience Publishers, New York-London, 1962.

[35] Năstăsescu, C. and Van Oystaeyen, F., *Graded and Filtered Rings and Modules*, Springer-Verlag, Berlin, Heidelberg, New York, 1979.

[36] Năstăsescu, C. and Van Oystaeyen, F., *Methods of Graded Rings*, Springer-Verlag, New York, Berlin, 2004.

[37] Northcott, D., *A First Course on Homological Algebra*, Cambridge University Press, London, 1973.

[38] Osborne, M., *Basic Homological Algebra*, Springer-Verlag, New York, Berlin, 2000.

[39] Osofsky, B., Rutgers University Ph.D. Thesis, Rutgers University, New Brunswick, 1964.

[40] Passman, D., *A Course in Ring Theory*, AMS Chelsea Publishing, American Mathematical Society, Providence, 2004.

[41] Popescu, N., *Abelian Categories with Applications to Rings and Modules*, Academic Press, New York, London, 1973.

[42] Rotman, T., *An Introduction to Homological Algebra 2nd Edition*, Springer-Verlag, New York, Berlin, 2009.

[43] Stenström, B., *Rings of Quotients*, Springer-Verlag, New York, Berlin, 1975.

[44] Szasz, F., *Radicals of Rings*, John Wiley and Sons, Inc., Hoboken, NJ, 1981.

[45] Wisbauer, R., *Foundations of Modules and Rings*, Gordon and Breach Science Publishers, New York, 1991.

[46] Xu, Jinzhong, *Flat Covers of Modules*, Springer-Verlag, Lecture Notes in Mathematics, New York, Berlin, 1996.

Journal Articles

[47] Artin, E., *Zur Theorie der hyperkomplexen Zahlen*, Abh. Math. Sem. Univ., Hamburg, **5** (1927), 251–260.

[48] Artin, E., *Zur Arithmetik hyperkomplexer Zahlen*, Abh. Math. Sem. Univ., Hamburg, **5** (1927), 261–289.

[49] Auslander, M., *On the dimension of modules and algebras III*, Nogoya Math. J., **9** (1955), 67–77.

[50] Baer, R., *Abelian groups which are direct summands of every containing abelian group*, Bull. Amer. Math. Soc., **46** (1940), 800–806.

[51] Bass, H., *Finitistic dimension and a homological generalization of semiprimary rings*, Trans. Amer. Math. Soc., **95** (1960), 466–488.

[52] Bergman, G., *A ring primitive on the right but not on the left*, Proc. Amer. Math. Soc., **15** (1964), 473–475.

[53] Björk, J., *Rings satisfying a minimum condition on principal ideals*, J. Reine Angew, Math., **236** (1969), 112–119.

[54] Chase, S., *Direct products of modules*, Proc. Amer. Math. Soc., **97** (1960), 457–473.

[55] Eckmann, B. and Schopf, A., *Über injektive moduln*, Archiv. der Math., **4** (1953), 75–78.

[56] Faith, C., and Huynh, D., *When self-injective rings are QF: A report on a problem*, J. Algebra & Applications, **1** (2002), 75–105.

[57] Goldie, A., *Semiprime rings with minimum condition*, Proc. Amer. Math. Soc., **10** (1960), 201–220.

[58] Goldie, A., *The structure of prime rings under ascending chain conditions*, Proc. London Math. Soc. **8** (1958), 589–608.

[59] Hopkins, C., *Rings with minimum condition on left ideals*, Ann. of Math., **40** (1939), 712–730.

[60] Jacobson, N., *The radical and semisimplicity for arbitrary rings*, Amer. J. Math., **67** (1945), 300–320.

[61] Johnson, R. and Wong, E., *Quasi-injective modules and irreducible rings*, J. London. Math. Soc., **36** (1961), 260–268.

[62] Jonah, D., *Rings with minimum condition for principal right ideals have maximum condition for principal left ideals*, Math. Zeit., **113** (1970), 106–112.

[63] Kaplansky, I., *Projective modules*, Math. Ann., (1958), 372–377.

[64] Kertesz, A., *Noethersche Ringe, die artinsch sind*, Acta Sci. Math., **31** (1970), 219–221.

[65] Koehler, A., *Quasi-projective covers and direct sums*, Proc. Amer. Math. Soc., **24** (1970), 655–658.

[66] Krull, W., *Die Idealtheorie in Ringen ohne Endlicheitsbedingungen, Mathematische Annalen*, **101** (1929), no.1, 729–744.

[67] Lesieur, L. and Croisot, R., *Sur les anneaux premiers noetheriens à gauche*, Ann. Sci. École Norm. Sup., **76** (1959), 161–183.

[68] Matlis, E., *Injective modules over Noetherian rings*, Pacific J. Math., **8** (1958), 511–528.

[69] Ore, O., *Linear equations in noncommutative fields*, Ann. of Math., **32** (1931), 463–477.

[70] Papp, Z., *On algebraically closed modules*, Publ. Math. Debrecen, **6** (1959), 311–327.

[71] Wedderburn, J., *On hypercomplex numbers*, Proc. Lond. Math. Soc., **6** (1908), 77–117.

[72] Wu, L. and Jans, J., *On quasi-projectives*, Illinois J. Math., **11** (1967), 439–448.

List of Symbols

Index

www.ingramcontent.com/pod-product-compliance
Lightning Source LLC
Chambersburg PA
CBHW081224220326
41598CB00037B/6867